# Contemporary Synthetic Glycoscience: A Theme Issue Dedicated to the Memory of Hans Paulsen

# Contemporary Synthetic Glycoscience: A Theme Issue Dedicated to the Memory of Hans Paulsen

Guest Editor

**Joachim Erich Thiem**

MDPI

Basel • Beijing • Wuhan • Barcelona • Belgrade • Novi Sad • Cluj • Manchester

*Guest Editor*
Joachim Erich Thiem
Department of Chemistry
University of Hamburg
Hamburg
Germany

*Editorial Office*
MDPI AG
Grosspeteranlage 5
4052 Basel, Switzerland

This is a reprint of the Special Issue, published open access by the journal *Molecules* (ISSN 1420-3049), freely accessible at: https://www.mdpi.com/journal/molecules/special_issues/BSM30Y2G7M.

For citation purposes, cite each article independently as indicated on the article page online and as indicated below:

| Lastname, A.A.; Lastname, B.B. Article Title. *Journal Name* **Year**, *Volume Number*, Page Range. |
|---|

**ISBN 978-3-7258-6518-5 (Hbk)**
**ISBN 978-3-7258-6519-2 (PDF)**
**https://doi.org/10.3390/books978-3-7258-6519-2**

# Contents

# About the Editor

**Joachim Erich Thiem**

Joachim Erich Thiem is a Professor at the Department of Chemistry, Faculty of Science, University of Hamburg, Germany. His research interests are in synthetic carbohydrate chemistry, preparative chemoenzymatic synthesis, and the use of carbohydrates as chemical feedstock. As Emeritus Professor, these subjects are still followed but at a slightly lower intensity. As a director of the DFG-funded Collaborative Research Grant 470 "Glycostructures in Biosystems" (1997–2009), he coordinated many groups studying chemistry, biology, physics, and medicinal aspects of glycostructures in living systems. He continues to shape the field as a Guest Editor for MDPI Special Issues and as an Editorial Board Member for the journal *Molecules*, section "Bioorganic Chemistry"; his extensive publication record of more than 550 papers documents pioneering contributions to stereoselective glycosylation, enzymatic carbohydrate chemistry, and the deployment of sugar-derived building blocks in applied and medicinal chemistry.

# Preface

This Special Issue of *Molecules* is dedicated to Professor Dr. Hans Paulsen for his contributions to the field of glycoscience.

His scientific work at the heart of carbohydrate chemistry contributed seminal findings to the synthesis of many and various carbohydrate derivatives, to conformations of monosaccharide and later oligosaccharide derivatives, as well as to the understanding of the exo-anomeric effect. His group studied acyloxonium rearrangements in carbohydrate chemistry, phosphorus-containing saccharides, carbohydrates with branched functional chains, aminoglycoside antibiotics, and oligosaccharide, glycolipid, and O- and N-glycoprotein syntheses.

Hans Paulsen's scientific work, with more than 500 publications, has received many of the highest national and international awards. He passed away on 3 October 2024, and will be remembered as a great scientist and highly valued colleague.

Former colleagues and friends are asked for contributions to this commemorative Special Issue dedicated to honoring his achievements in glycoscience.

**Joachim Erich Thiem**
*Guest Editor*

*molecules*

MDPI

*Editorial*

# Contemporary Synthetic Glycoscience: A Theme Issue Dedicated to the Memory of Hans Paulsen

**Joachim Thiem**

Institute of Organic Chemistry, Department of Chemistry, Faculty of Science, University of Hamburg, Martin-Luther-King-Platz 6, D-20146 Hamburg, Germany; joachim.thiem@uni-hamburg.de

Annually, the biomass of the Earth, estimated to be about 200 billion tons, is nourished by photosynthesis and degraded via various pathways. The primary product, glucose, is transformed into a huge number of distinct and structurally complex mono-, di-, tri-, and oligosaccharides, which provide glyco-based molecules for nourishment, as well as structural units. Together, these constitute about 95% of the total biomass, and the remaining 5% is transformed into all the other organic components involved in Earth's life-cycles.

In addition, nature has provided an overwhelmingly large number of complex glycosylated components, which underpin the life and health of all animate beings. Thus, fascinating questions involving the operating sequences and mechanisms underpinning all life on Earth have encouraged chemists to synthesize and study increasingly complex glucose derivatives.

One of the dominant figures in this area was Hans Paulsen, based in Hamburg, Germany, who, over his decades-long his career, dedicated himself and his research group to syntheses in carbohydrate chemistry.

This Special Issue compiles the work of former colleagues of Hans Paulsen from around the world, who exchanged ideas with him extensively and contributed seminal studies covering a variety of exciting developments in contemporary glycoscience. In this introduction, I will briefly comment on the studies in synthetic and analytical glycoscience compiled herein.

As has been known for decades, the expression of glycosyltransferases in cancer cells generates tumor-associated carbohydrate antigens (TACAs) [1]. For example, aberrant sialylation on cell surfaces is known to be associated with cancer progression. The over-sialylation of glycoproteins on cancer cells increases the number of interactions with specific sialic acid-binding immunoglobulin-type lectins (Siglecs) on immune cells. This results in the suppression of the immune response against cancer cells, known as immune checkpoint inhibition [2,3]. Wong et al., who have, for a number of decades, been among the leading research groups employing chemo-enzymatic syntheses for highly complex hetero-oligosaccharides [4], contribute a number of convincing scalable chemo-enzymatic approaches to the synthesis of DSGb5 and complex sialylated hetero-oligosaccharides. These sialylated glycans from cancer-associated glycolipids were used by Wong et al. to compose glycan microarrays [5].

Cardiac glycosides [6] have been used for centuries in clinics for the treatment of heart failure and cardiac rhythm disorders [7,8]. Yu et al., who are among the leading research groups devoted to the synthesis of complex steroid glycosides [9], demonstrate the total syntheses of the structurally challenging cardiac glycosides acospectoside A and acovenoside B [10].

Glycosphingolipids (GSLs) in cell membranes function as important regulators of essential biological processes such as cellular recognition and signaling, cell differentiation, and proliferation [11–14]. Thus, the contribution of Guo et al. focuses on syntheses of β-lactosylceramide (β-LacCer) analogs as key intermediates in the study of complex GSL biosynthetic processes [15].

Brucellosis constitutes a serious zoonotic bacterial—and highly contagious—disease in many animals, including valuable cattle [16,17]. The binding of murine monoclonal antibodies (mAbs) to hexasaccharide fragments of epitopes of the Brucella *O*-antigen is soon to be explored in computational study [18]. Bundle, Wood et al. were able identify stable binding modes, which further our understanding of the recognition mechanism [19]. In the future, designing synthetic glycomimetics and engineering epitope-specific antibodies *en route* to vaccine design might be possible.

Since the original Koenigs–Knorr glycosylation in 1901 (124 years ago [20]), a compelling number of improvements on, and alternatives to, this seminal reaction have published. Almost every scientist working in the synthesis on glycosylated components is required to employ this reaction, and each will be likely to affirm the following statement from Hans Paulsen, made in 1982: *"Although we have now learned to synthesize oligosaccharides, it should be emphasized that each oligosaccharide synthesis remains an independent problem, whose resolution requires considerable systematic research and a good deal of know-how. There are no universal reaction conditions for oligosaccharide syntheses"* [21]. One of the more recent modifications made in chemical glycosylation is the "4K reaction" [22]. The studies of Demchenko et al. demonstrate further improvements to this glycosylation modification, employing iron(III) triflate as an efficient activator of thioglycosides in this reaction pathway [23].

An important issue in glycobiology concerns bacterial adhesion as the initial step of infectious diseases. A closer understanding of the mechanisms involved could be of interest in preventing the adhesion of bacteria, for example, by employing carbohydrate-based inhibitors to facilitate further breakthroughs in diagnostics and therapy [24]. Of particular interest are functional glycomimetics, which are suited to studying the details of carbohydrate recognition, such as the varying distances between multiple glycoligands [25]. In their contribution, Lindhorst et al. employed xylopyranosides as a scaffold for the presentation of two mannoside ligands of the bacterial lectin FimH. The chair conformation of xylosides could be switched between alternative conformations, thus flipping the ligands from a diequatorial into a diaxial position. Further, such conformational switches may function as advanced smart molecular tools with which to study structural binding conditions in carbohydrate recognition [26].

Essential for membrane functions, cell recognition, and the maintenance of the nervous system are gangliosides such as GM2 (GalNAcβ1-4Galβ1-4Glcβ-Cer). Some rare genetic lysosomal storage diseases—e.g., Tay-Sachs disease (TSD) and Sandhoff disease (SD)—occur due to deficient β-*N*-acetyl-hexosaminidase activity, leading to decreased catabolism of β-*N*-acetyl-hexosamine-containing ganglioside GM2 in the lysosomes and, consequently, causing damage to cells and tissues, as well as severe neurological symptoms [27,28]. A number of contemporary studies have aimed to express and purify human B4GALNT1, the enzyme that transfers the GalNAc residue in the β1-4 linkage to the Galβ1-4 residue of the gangliosides GM3, GD3, and other GSLs. A number of studies have focused on characterizing its activity and exploring its structural features via protein modeling and substrate docking. Brockhausen et al., for a complex of B4GALNT1 docked with its donor substrate, UDP-GalNAc, were able determine the relevant amino acids near the docking site that were likely involved in UDP-GalNAc binding and catalysis [29]. Thus, further studies *en route* to drug development, including the kinetics of inhibition, toxicity, and drug delivery to neurons accumulating GM2, are to follow.

Cell membranes are embedded with a glycocalix consisting of a complex mixture of glycoproteins, glycolipids, complex oligosaccharides, glycoconjugates, and proteoglycans. Thus, cellular processes, such as bacterial and viral infection, cancer metastasis, the modulation and activation of the immune system, tissue differentiation and development, and further intercellular recognition events, are controlled by glycoconjugates [30–34]. For mimicking the glycocalyx, the self-assembled monolayer (SAM) formation of carbohydrate derivatives on gold nanoparticles was selected, allowing for the study of carbohydrate–protein and carbohydrate–carbohydrate interactions. The novel modular approach reported here includes facile and rapid syntheses for linking spacers and carbohydrate derivatives, enhancing binding events. Immobilization was performed on biorepulsive aminoxy-substituted gold nanoparticles via the oxime formation of aldehyde-functionalized mono-, di-, and complex trisaccharides. Uniform gold nanoparticles could be obtained, and effective immobilization and binding studies were presented for concanavalin A. Through this novel approach, Thiem et al. revealed a number of advantageous perspectives to be taken on various biomimetic studies of carbohydrates and carbohydrate-based array development for diagnostics and screening [35].

The innate immune system is sensitive to lipopolysaccharides (LPS), which play a pivotal role in immune response [36]. Lipid A, composed of a mono- or di-phosphorylated di-glucosamine core with multiple fatty acid chains, anchors LPS into the bacterial membrane and is largely responsible for the toxicity of Gram-negative bacteria [37]. In their contribution, D'Orazio, Lay et al. present syntheses and a computational evaluation of a library of glycolipid analogues to B. fragilis lipid A, revealing a rationally designed approach to developing novel anti-inflammatory agents [38].

Pathogenic bacteria often utilize cell surface sialylation to evade the host immune system via molecular mimicry of the host sialo-glycoconjugates [39]. Human pathogen Neisseria meningitidis serotype B (NmB) expresses both the sialylated capsule and surface lipooligosaccharides—pivotal virulence factors [40]. In NmB's sialylation pathway, CMP-sialic acid synthetase (CSS) is essential for sialic acid transfer. Along various routes (compare also e.g., [41]), functionalized derivatives of neuraminic acid β-methyl glycoside (Neuβ2Me) were synthesized as candidates for the inhibition of N. meningitidis CSS. Direct interaction with the enzyme was confirmed by saturation transfer difference (STD) NMR [42,43]. In their contribution, Jennings, Münster-Kühnel, and von Itzstein et al. provide data encouraging the further development of potential inhibitors for the treatment of bacterial meningitis [44].

Nature makes sucrose annually in virtually limitless quantities, the majority of which is consumed via the food industry. Further uses include the production of bioethanol, biodegradable surfactants, and polymers [45]. Application of this cheap raw material with 100% optical purity and eight genuine stereogenic centers for the synthesis of various sophisticated products is of particular interest [46]. In their review, Jaraosz et al. focused on approaches to the use of sucrose in the design of fine chemicals, the macrocyclic scaffold, and further-modified derivatives for various applications [47].

For the colonization of human organs, pathogenic bacteria use a multilayer cell envelope largely composed of glycans and their conjugates [48,49]. Most Gram-negative and some Gram-positive bacteria comprise a capsule largely composed of glycans and their conjugates [50]. Thus, the synthesis of spacer-armed phosphooligosaccharides, structurally related to the capsular phosphoglycans of pathogenic bacteria, is of particular interest (e.g., [51]).

In their extended review, Nifantiev et al. summarize our current understanding of the preparation and immunogenicity of neoglycoconjugates based on synthetic phospho-oligosaccharides to foster prospects for the development of conjugate vaccines on the basis of synthetic phosphooligosaccharide antigens [52].

Acknowledging Hans Paulsen and his collaborators for their seminal study of glyco-protein biosynthesis, the final review by Brockhausen sheds light on their many individual and decisive contributions. The stepwise development is demonstrated via the continual interaction of synthetic and biochemical glycoscientists, resulting in cleverly designed components [53].

This Special Issue contains ten contributions discussing challenging topics pertaining to both syntheses [5,10,15,23,26,38,44] and analyses [19,29]. Additionally, the review articles included discuss the relevant aspects of sucrose-based components [47] and the immunogenicity of neoglycoconjugates [52]. The final feature review describes the numerous synthetic contributions of the Paulsen group to solving questions pertaining to glycoprotein biosynthesis [53]. In this Special Issue, a broad overview of contemporary glycoscience research is compiled, indicating the significant influence of this subject in natural products and heterocyclic chemistry, as well as its value in addressing broader issues in biology and medicine.

**Funding:** This research received no external funding.

**Conflicts of Interest:** The author declares no conflicts of interest.

## References

1. Hakomori, S. Tumor-associated carbohydrate antigens. *Annu. Rev. Immunol.* **1984**, *2*, 103–126. [CrossRef]
2. Duan, S.; Paulson, J.C. Siglecs as Immune Cell Checkpoints in Disease. *Annu. Rev. Immunol.* **2020**, *38*, 365–395. [CrossRef]
3. Smith, B.A.H.; Bertozzi, C.R. The clinical impact of glycobiology: Targeting selectins, Siglecs and mammalian glycans. *Nat. Rev. Drug Discov.* **2021**, *20*, 217–243. [CrossRef] [PubMed]
4. Tsai, T.I.; Lee, H.Y.; Chang, S.H.; Wang, C.H.; Tu, Y.C.; Lin, Y.C.; Hwang, D.R.; Wu, C.Y.; Wong, C.H. Effective sugar nucleotide regeneration for the large-scale enzymatic synthesis of Globo H and SSEA4. *J. Am. Chem. Soc.* **2013**, *135*, 14831–14839. [CrossRef]
5. Liao, K.-S.; Zhou, Y.; Chung, C.; Kung, C.-C.; Ren, C.-T.; Wu, C.-Y.; Lou, Y.-W.; Chuang, P.-K.; Imre, B.; Hsieh, Y.S.Y.; et al. Chemical and Enzymatic Synthesis of DisialylGb5 and Other Sialosides for Glycan Array Assembly and Evaluation of Siglec-Mediated Immune Checkpoint Inhibition. *Molecules* **2025**, *30*, 2264. [CrossRef] [PubMed]
6. Albrecht, H.P. Cardiac glycosides. In *Naturally Occurring Glycosides*; Ikan, R., Ed.; John Wiley & Sons Ltd.: New York, NY, USA, 1999; p. 83.
7. Botelho, A.F.M.; Pierezan, F.; Soto-Blanco, B.; Melo, M.M. A review of cardiac glycosides: Structure, toxicokinetics, clinical signs, diagnosis and antineoplastic potential. *Toxicon* **2019**, *158*, 63–68. [CrossRef] [PubMed]
8. Hou, Y.; Shang, C.; Meng, T.; Lou, W. Anticancer potential of cardiac glycosides and steroid-azole hybrids. *Steroids* **2021**, *171*, 108852. [CrossRef]
9. Liu, B.; Bi, S.; Wang, J.; Xu, P.; Yu, B. Synthesis of acovenosides: Cardiac glycosides with potent antitumor activities. *Org. Lett.* **2024**, *26*, 8725–8729. [CrossRef]
10. Liu, B.; Xu, P.; Yu, B. Total Synthesis of Cardenolides Acospectoside A and Acovenoside B. *Molecules* **2025**, *30*, 2297. [CrossRef]
11. Hakomori, S.I. Structure and function of glycosphingolipids and sphingolipids: Recollections and future trends. *Biochim. Biophys. Acta* **2008**, *1780*, 325–346. [CrossRef]
12. Guo, Z. The structural diversity of natural glycosphingolipids (GSLs). *J. Carbohydr. Chem.* **2022**, *41*, 63–154. [CrossRef] [PubMed]
13. Ando, H.; Komura, N. Recent progress in the synthesis of glycosphingolipids. *Curr. Opin. Chem. Biol.* **2024**, *78*, 102423. [CrossRef]
14. Wennekes, T.; van den Berg, R.J.; Boot, R.G.; van der Marel, G.A.; Overkleeft, H.S.; Aerts, J.M. Glycosphingolipids—Nature, function, and pharmacological modulation. *Angew. Chem. Int. Ed. Engl.* **2009**, *48*, 8848–8869. [CrossRef] [PubMed]

15. Mohamed, B.; Rohokale, R.; Yan, X.; Ghanim, A.M.; Osman, N.A.; Abdel-Fattah, H.A.; Guo, Z. Synthesis of Azide-Labeled β-Lactosylceramide Analogs Containing Different Lipid Chains as Useful Glycosphingolipid Probes. *Molecules* **2025**, *30*, 2667. [CrossRef] [PubMed]
16. Maudlin, I.; Weber, S. *The Control of Neglected Zoonotic Diseases: A Route to Poverty Alleviation*; WHO/SDE/FOS/2006.1; World Health Organization: Geneva, Switzerland, 2006; p. 83.
17. Corbel, M.J. *Brucellosis in Humans and Animals*; World Health Organization: Geneva, Switzerland, 2006.
18. Ganesh, N.V.; Sadowska, J.M.; Sarkar, S.; Howells, L.; McGiven, J.; Bundle, D.R. Molecular recognition of Brucella A and M antigens dissected by synthetic oligosaccharide glycoconjugates leads to a disaccharide diagnostic for brucellosis. *J. Am. Chem. Soc.* **2014**, *136*, 16260–16269. [CrossRef]
19. Sood, A.; Bundle, D.R.; Woods, R.J. Towards Understanding the Basis of Brucella Antigen–Antibody Specificity. *Molecules* **2025**, *30*, 2906. [CrossRef]
20. Koenigs, W.; Knorr, E. Ueber einige Derivate des Traubenzuckers und der Galactose. *Ber. Deutsch. Chem. Ges.* **1901**, *34*, 957–981. [CrossRef]
21. Paulsen, H. Advances in selective chemical syntheses of complex oligosaccharides. *Angew. Chem. Int. Ed. Engl.* **1982**, *21*, 155–173. [CrossRef]
22. Demchenko, A.V.; De Meo, C. The 4K reaction. *Carbohydr. Res.* **2024**, *538*, 109102. [CrossRef]
23. Dent, A.R.; DeSpain, A.M.; Demchenko, A.V. Cooperatively Catalyzed Activation of Thioglycosides with Iodine and Iron(III) Trifluoromethane-sulfonate. *Molecules* **2025**, *30*, 3058. [CrossRef]
24. Fares, M.; Imberty, A.; Titz, A. Bacterial Lectins: Multifunctional Tools in Pathogenesis and Possible Drug Targets. *Trends Microbiol.* **2025**, *33*, 839–852. [CrossRef] [PubMed]
25. Müller, C.; Despras, G.; Lindhorst, T.K. Organizing Multivalency in Carbohydrate Recognition. *Chem. Soc. Rev.* **2016**, *45*, 3275–3302. [CrossRef]
26. Jaeschke, S.O.; vom Sondern, I.; Lindhorst, T.K. Bivalent Inhibitors of Mannose-Specific Bacterial Adhesion: A Xylose-Based Conformational Switch to Control Glycoligand Distance. *Molecules* **2025**, *30*, 3074. [CrossRef]
27. Ledeen, R.; Wu, G. Gangliosides of the nervous system. *Methods Mol. Biol.* **2018**, *1804*, 19–55. [PubMed]
28. Furukawa, K.; Takamiya, K.; Furukawa, K. Beta 1,4-N-acetylgalactosaminyltransferase-GM2/GD2 synthase: A key enzyme to control the synthesis of brain-enriched complex gangliosides. *Biochim. Biophys. Acta* **2002**, *1573*, 356–362. [CrossRef]
29. Abidi, I.; Kocev, A.N.; Babulic, J.L.; Capicciotti, C.J.; Walia, J.; Brockhausen, I. Characterization of Human Recombinant β1,4-GalNAc-Transferase B4GALNT1 and Inhibition by Selected Compounds. *Molecules* **2025**, *30*, 3615. [CrossRef]
30. Ratner, D.M.; Adams, E.W.; Disney, M.D.; Seeberger, P.H. Tools for glycomics: Mapping interactions of carbohydrates in biological systems. *ChemBioChem* **2004**, *15*, 1375–1383. [CrossRef]
31. Varki, A. Biological roles of oligosaccharides: All of the theories are correct. *Glycobiology* **1993**, *3*, 97–130. [CrossRef]
32. Haltiwanger, R.S.; Lowe, J.B. Role of Glycosylation in Development. *Annu. Rev. Biochem.* **2004**, *73*, 491–537. [CrossRef]
33. Ohtsubo, K.; Marth, J.D. Glycosylation in cellular mechanisms of health and disease. *Cell* **2006**, *126*, 855–867. [CrossRef] [PubMed]
34. Simanek, E.E.; McGarvey, G.J.; Jablonowski, J.A.; Wong, C.-H. Selectin minus sign Carbohydrate Interactions: From Natural Ligands to Designed Mimics. *Chem. Rev.* **1998**, *98*, 833–862. [CrossRef]
35. Kopitzki, S.; Thiem, J. Synthesis and Characterization of Glyco-SAMs on Gold Nanoparticles: A Modular Approach Towards Glycan-Based Recognition Studies. *Molecules* **2025**, *30*, 3765. [CrossRef]
36. Di Lorenzo, F.; Duda, K.A.; Lanzetta, R.; Silipo, A.; De Castro, C.; Molinaro, A. A Journey from Structure to Function of Bacterial Lipopolysaccharides. *Chem. Rev.* **2022**, *122*, 15767–15821. [CrossRef]
37. Raetz, C.R.H.; Whitfield, C. Lipopolysaccharide Endotoxins. *Annu. Rev. Biochem.* **2002**, *71*, 635–700. [CrossRef]
38. Kenneth, D.; Santi, C.M.; Tanda, F.; Izzo, A.; Civera, M.; D'Orazio, G.; Lay, L. Synthesis and Docking Studies of Glycolipids Inspired by Bacteroides fragilis Lipid A. *Molecules* **2025**, *30*, 3927. [CrossRef]
39. Schauer, R.; Kamerling, J.P. Exploration of the sialic acid world. *Adv. Carbohydr. Chem. Biochem.* **2018**, *75*, 1–213. [PubMed]
40. Harrison, L.H.; Trotter, C.L.; Ramsay, M.E. Global epidemiology of meningococcal disease. *Vaccine* **2009**, *27* (Suppl. 2), B51–B63. [CrossRef] [PubMed]
41. Zbiral, E.; Schreiner, E.; Christian, R. Synthesis of the 4-acetamido-4-deoxy analogue of N-acetylneuraminic acid and its behaviour towards CMP-sialate synthase. *Carbohydr. Res.* **1989**, *194*, c15–c18. [CrossRef]
42. Mayer, M.; Meyer, B. Characterization of ligand binding by saturation transfer difference NMR spectroscopy. *Angew. Chem. Int. Ed. Engl.* **1999**, *38*, 1784–1788. [CrossRef]
43. Meyer, B.; Peters, T. NMR spectroscopy techniques for screening and identifying ligand binding to protein receptors. *Angew. Chem. Int. Ed. Engl.* **2003**, *42*, 864–890. [CrossRef] [PubMed]

44. Chopra, P.; Führing, J.; Ng, P.; Haselhorst, T.; Dyason, J.C.; Rose, F.J.; Thomson, R.J.; Gerardy-Schahn, R.; Grice, I.D.; Jennings, M.P.; et al. Exploring the Scope of Functionalized N-Acylneuraminic Acid β-Methyl Glycosides as Inhibitors of Neisseria meningitidis CMP-Sialic Acid Synthetase. *Molecules* **2025**, *30*, 4329. [CrossRef]
45. Queneau, Y.; Jarosz, S.; Lewandowski, B.; Fitremann, J. Sucrose chemistry and applications of sucrochemicals. *Adv. Carbohydr. Chem. Biochem.* **2008**, *61*, 217–292.
46. Jarosz, S.; Pakulski, Z. The roadmap for sucrose—A very inexpensive raw product. *Adv. Carbohydr. Chem. Biochem.* **2024**, *86*, 1–13. [PubMed]
47. Jarosz, S.; Pakulski, Z. Sucrose-Based Macrocycles: An Update. *Molecules* **2025**, *30*, 2721. [CrossRef]
48. Fivenson, E.M.; Dubois, L.; Bernhardt, T.G. Co-Ordinated Assembly of the Multilayered Cell Envelope of Gram-Negative Bacteria. *Curr. Opin. Microbiol.* **2024**, *79*, 102479. [CrossRef] [PubMed]
49. Ribet, D.; Cossart, P. How Bacterial Pathogens Colonize Their Hosts and Invade Deeper Tissues. *Microbes Infect.* **2015**, *17*, 173–183. [CrossRef] [PubMed]
50. Cescutti, P. Bacterial Capsular Polysaccharides and Exopolysaccharides. In *Microbial Glycobiology*; Holst, O., Brennan, P.J., von Itzstein, M., Moran, A.P., Eds.; Academic Press: San Diego, CA, USA, 2010; pp. 93–108.
51. Khatuntseva, E.A.; Nifantiev, N.E. Glycoconjugate Vaccines for Prevention of Haemophilus Influenzae Type b Diseases. *Russ. J. Bioorg. Chem.* **2021**, *47*, 26–52. [CrossRef]
52. Khatuntseva, E.A.; Kamneva, A.A.; Yashunsky, D.V.; Nifantiev, N.E. Synthesis Immunogenicity of Pseudo-Oligosaccharides Structurally Related to Repeating Units of Capsular Phosphoglycans of Human Pathogens. *Molecules* **2025**, *30*, 3068. [CrossRef]
53. Brockhausen, I. Hans Paulsen: Contributions to the Investigations of Glycoprotein Biosynthesis. *Molecules* **2025**, *30*, 3735. [CrossRef]

molecules

MDPI

*Review*

# Sucrose-Based Macrocycles: An Update

**Sławomir Jarosz * and Zbigniew Pakulski**

Institute of Organic Chemistry, Polish Academy of Sciences, Kasprzaka 44/52, 01-224 Warsaw, Poland; zbigniew.pakulski@icho.edu.pl
* Correspondence: slawomir.jarosz@icho.edu.pl

**Abstract**

Sucrose is by far the most abundant disaccharide found in nature, consisting of two simple hexose units: D-glucose and D-fructose. This exceptionally inexpensive and widely accessible raw material is produced in virtually limitless quantities. The vast majority is consumed in the food industry either in its native form—as commercial table sugar—or, to a lesser extent, as the basis for artificial sweeteners such as palatinose and sucralose. Beyond its dietary use, sucrose serves as a feedstock for the production of bioethanol, liquid crystals, biodegradable surfactants, and polymers. However, the application of this valuable and extremely cheap raw material (100% optical purity and eight stereogenic centers with precisely defined stereochemistry) in the synthesis of more sophisticated products remains surprisingly limited. In this short review, we focus on the strategic use of the sucrose scaffold in the design and synthesis of fine chemicals. Special attention will be paid to macrocyclic derivatives incorporating the sucrose backbone. These water-soluble structures show promise as molecular receptors within biological environments, offering unique advantages in terms of solubility, biocompatibility, and stereochemical precision.

**Keywords:** sucrose in fine organic synthesis; macrocyclic derivatives with sucrose scaffold; molecular cages based on sucrose and cycloveratrylene; complexing properties of macrocycles with sucrose scaffold; sucrose-based molecular switches

## 1. Introduction

Sucrose (**1**; α-D-glucopyranosyl-β-D-fructofuranoside), a very cheap raw material produced in plants (sugar cane and sugar beets) from the primary products of photosynthesis, is available in practically unlimited quantities. The majority of commercially available sucrose is consumed in its native form as table sugar, primarily in the food industry. To a lesser extent, it also serves as a precursor for the industrial-scale production of sucrose-derived sweeteners such as palatinose and sucralose. Beyond its applications in food technology, sucrose plays an increasingly important role in industrial chemistry. One of its major non-food uses is in the production of bioethanol, where it serves as a fermentable sugar source. Moreover, sucrose has emerged as a versatile starting material in the synthesis of high-value compounds, including liquid crystals, biodegradable surfactants, polymers, and, potentially, cytotoxic antitumor agents (Figure 1) [1].

Despite its exceptional advantages—such as 100% optical purity and eight stereogenic centers with precisely defined stereochemistry—the utilization of sucrose in the synthesis of complex fine chemicals remains surprisingly underexplored. The potential of this structurally rich and sustainable molecule for advanced chemical synthesis has only been modestly realized. Some of these synthetic applications have been discussed in our recent reviews [2,3].

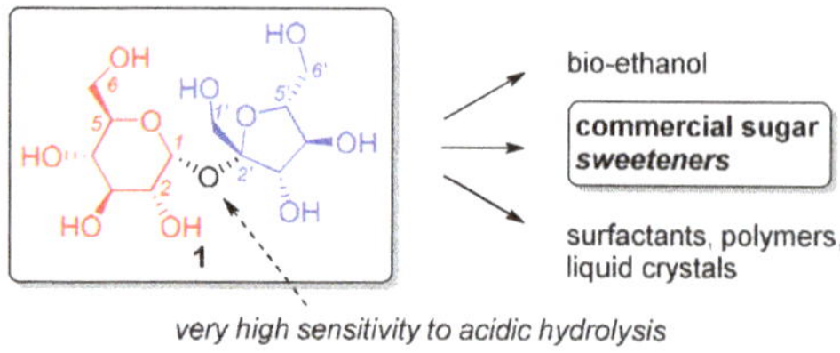


**Figure 1.** Sucrose [**1**; α-D-glucopyranosyl-β-D-fructofuranoside]: its presence in food chemistry and as a starting material for industrial biodegradable products.

The conformation of sucrose in both the solid state and in solution indicated that the primary hydroxyl groups at the C6 position of the glucose moiety and the C6′ position of the fructose moiety are in close spatial proximity due to strong intramolecular hydrogen bonding, which suggests the feasibility of covalently linking these two positions via a suitable linker (Figure 2) [4].

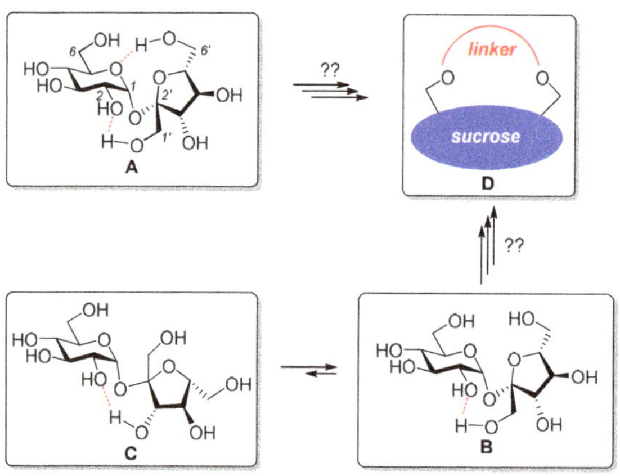


**Figure 2.** Conformation of sucrose in solid state (**A**) and in solution (**B**,**C**). Possible macrocyclic target with sucrose scaffold (**D**).

Such a possibility may open a route to highly functionalized macrocyclic derivatives that are soluble in water, which would allow these compounds to be used as receptors in biological environments. The attempt to utilize sucrose-based macrocycles requires several steps (Figure 3):

- The protection of all secondary hydroxyl groups with temporary blocks (preferably benzyl) that could be easily removed after the desired transformations without the destruction of a highly sensitive glycosidic bond (Figure 3; compounds **2** and **3**). This strategy requires temporary protection of the primary hydroxyl groups (C6, C6′ and eventually C1′), benzylation of all remaining OH groups, and the removal of temporary blocks from C6, C6′ (and eventually C1′) positions. The selective protection of the primary OH groups is well known and can be achieved using very expensive bulky silyl chlorides or much cheaper trityl chlorides. However, the removal of the trityl blocks from per-benzylated intermediates was not trivial since this process can ***only*** be realized by careful acidic hydrolysis.

- The connection of the terminal positions (C6, C6′ and eventually C1′) should be connected via a proper link (**4** and **6**).
- Total deprotection, i.e., the removal of all temporary blocks, which should provide macrocyclic derivatives (potential receptors) soluble in water (**5** and **7**).

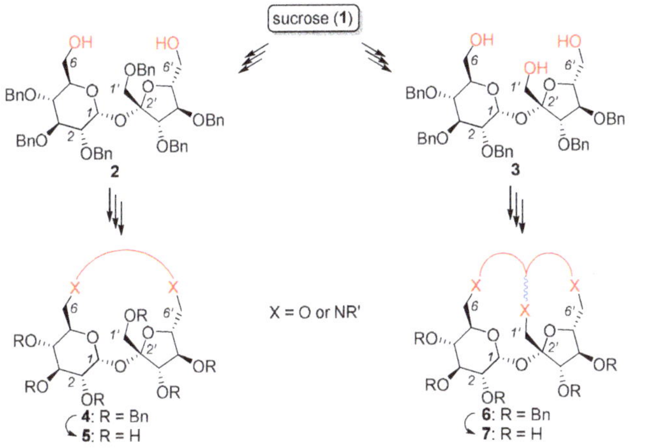

**Figure 3.** Synthetic plan for the preparation of soluble sucrose-based receptors.

During our work on the application of sucrose as a valuable synthon for macrocyclic, water-soluble derivatives, we describe a convenient methodology for the synthesis of so-called "sucrose diol" (**2**) and "sucrose triol" (**3**) and other useful partially protected derivatives of this disaccharide; examples are shown in Scheme 1.

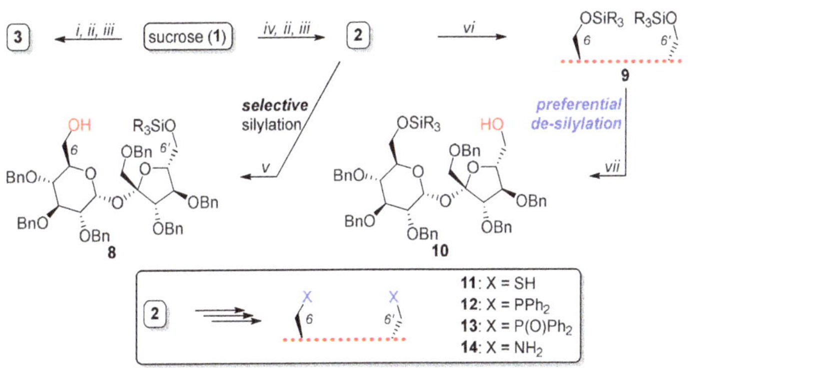

**Scheme 1.** Synthesis of sucrose diol **2**, sucrose triol **3**, and their partially protected derivatives **11–14**. Reagents and conditions: *i.* 3 equiv. TrCl; *ii.* BnBr, NaH; *iii.* careful detritylation; *iv.* 6 equiv. TrCl; *v.* 1 equiv. $R_3SiCl$; *vi.* excess of $R_3SiCl$; *vii.* $F^{(-)}$.

Selective silylation of diol **2** afforded product **8** with the free C6-OH group with a very small amount of double-protected compound **9**. This high selectivity was rather expected in view of the early observation by Khan, who reported that the reaction of free sucrose with *tert*-BDPS-Cl afforded 49% of the mono-silylated product in which the silyl group was located at the C6′ (fructose part) position [5].

The double silylated compound **9** could be obtained in high yield by the exhaustive silylation of diol **2**. The most important (not expected) observation in our study was the ***selective de-silylation*** of **9**, which opened a convenient route to regioisomeric mono-alcohol **10** with the free C6′-OH (fructose part) group (Scheme 1). The sulfur, amino-, and phosphorus derivatives were also available, as shown in Scheme 1 [2,3].

## 2. Sugar Derived Macrocyclic Derivatives with Sucrose Scaffold

The best-known macrocyclic derivatives with sugar platform are—undoubtedly—cyclodextrins, which were obtained from the enzymatic degradation of starch by the end of the XIXth century by French scientist Antoine Villiers. In the second half of the XXth century, the method for their industrial preparation was elaborated by Hungarian chemists. These cyclic oligosaccharides have found widespread applications in chemistry, medicine, and daily life [6,7]. Other macrocyclic sugar-based derivatives are also known [8]. We decided to use sucrose—the most common disaccharide—as a platform to build macrocyclic derivatives, which, being soluble in water, could act as a receptor in a biological environment.

### *2.1. Synthesis of Macrocyclic Derivatives from "Sucrose Diol"* 2

Diol **2** served as a starting material for the preparation of the analogs of crown and aza-crown ethers. Several such targets built on a sucrose platform were prepared by our group; model examples are shown in Scheme 2 [9,10].

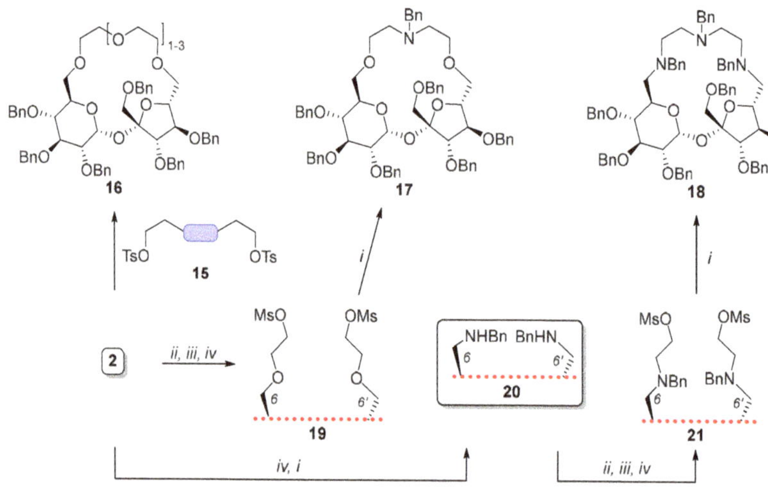


**Scheme 2.** Synthesis of aza-crown ether analogs with sucrose scaffold from diol **2**. Reagents and conditions: *i.* $BnNH_2$; *ii.* $BrCH_2CO_2R$; *iii.* $LiAlH_4$; *iv.* MsCl.

For example, diol **2** was treated with polyethylene glycol di-tosylates (**15**), affording crown ether analog **16**. The preparation of aza-crowns required slightly different options. Thus, the elongation of **2** by two carbon atoms at each side (→ **19**) followed by a reaction with benzylamine provided macrocycle **17** with one nitrogen atom in the ring. The synthesis of macrocycles with more nitrogen atoms was also possible and was exemplified by compound **18**. It was prepared in a few steps from diol **2** by its conversion into di-amine **20**, which was elongated to **21** (similarly to **19**) and finally treated with benzylamine, affording **18**.

In aza-macrocycles **17** and **18**, the nitrogen atom in the ring is protected with the benzyl group; it would have been much more convenient to have this atom free, which would have allowed us to introduce various substituents modifying the properties of the target and such a possibility is shown in Scheme 3 [11]. Starting from alcohol **8**, it was possible to introduce different units at the position C6 (via a reaction with chloroacetonitrile) and C6′ (via a reaction with $^{t}$butyl bromoacetate after the deprotection of 6′-OH); this intermediate was subsequently reduced to aminoalcohol **22**. This compound was converted, under Garegg's conditions (imidazole, $I_2$, $Ph_3P$), to iodoamine **23,** which spontaneously cyclized to aza-crown **24**. The alkylation of the nitrogen atom afforded a series of crown ether analogs with a sucrose scaffold [12].

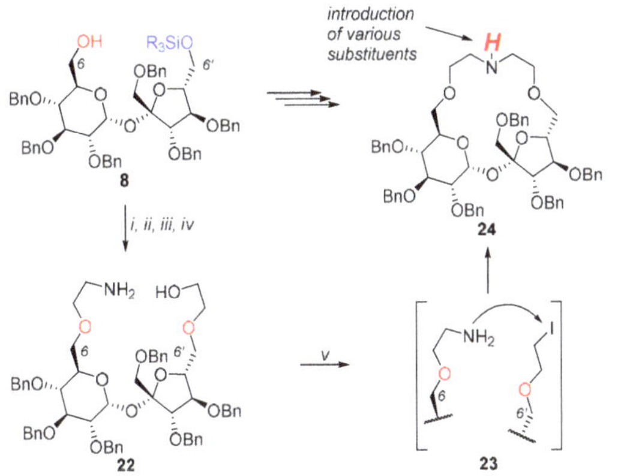


**Scheme 3.** Efficient synthesis of macrocyclic sucrose template. Reagents and conditions: *i.* $ClCH_2CN$; *ii.* $F^{(-)}$; *iii.* $BrCH_2CO_2{}^{t}Bu$; *iv.* $LiAlH_4$; *v.* imidazole, $PPh_3$, $I_2$.

The macrocyclic derivatives presented in Schemes 2 and 3 are "symmetrical", i.e., they have the same atoms (oxygen or nitrogen) at both "ends" of sucrose (positions C6 and C6′). The synthesis of "unsymmetrical" aza-crown analogs was, however, possible, as shown in Scheme 4. Alcohol **8** was converted into **25** and further into aminoalcohol **26**. The cyclization of this intermediate, induced by a Garegg's reagent, provided the first "unsymmetrical" azacrown analog **27**. The second macrocyclic analog **28** was prepared analogously from regioisomeric alcohol **10** [12].

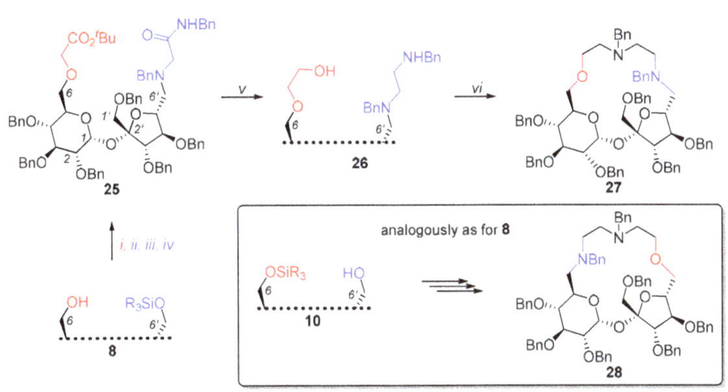


**Scheme 4.** Synthesis of "unsymmetrical" analogs of aza-crowns with sucrose scaffold. Reagents and conditions: *i.* $BrCH_2CO_2{}^{t}Bu$; *ii.* $F^{(-)}$; *iii.* Swern oxidation; *iv.* $BnNHCH_2C(O)NHBn$, $NaBH_3CN$; *v.* $LiAlH_4$; *vi.* imidazole, $PPh_3$, $I_2$.

We also elaborated on the convenient method for the synthesis of macrocyclic sucrose-based derivatives containing sulfur, which is shown in Scheme 5. Sucrose (**1**) was selectively chlorinated according to the Whistler procedure [13]; then, it was perbenzylated to **29** [14]. The reaction of this intermediate with 2-mercaptoethanol provided diol **30** (X = OH), which was chlorinated (to **31**; X = Cl) under Appel conditions. The treatment of this dichloride with either amines (e.g., $BnNH_2$) or di-amines ($BnNHCH_2CH_2NHBn$) afforded macrocyclic derivatives **32** or **33**, respectively, with a sucrose scaffold [15].

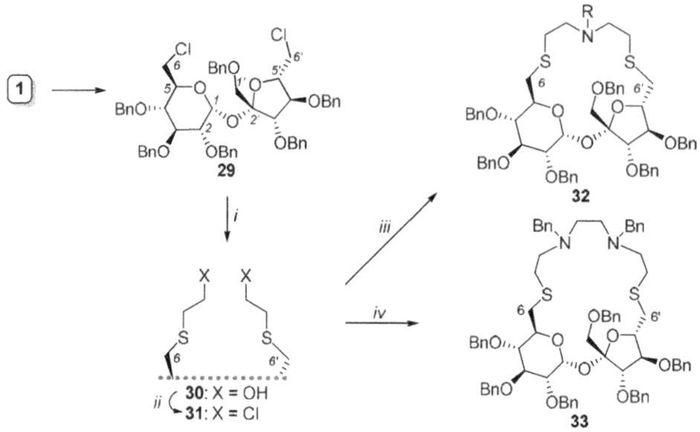

**Scheme 5.** Efficient preparation of sucrose macrocycles with two sulfur atoms. Reagents and conditions: *i.* NaI, 2-mercaptoethanol, DMF; *ii.* $CCl_4$, $PPh_3$; *iii.* $RNH_2$; *iv.* $BnNHCH_2CH_2NHBn$.

### 2.2. Synthesis of Other Complex Macrocyclic Sucrose-Based Derivatives

One of the first sucrose-based macrocyclic derivatives—that was not the crown analog—was prepared by us in 2010, as shown in Scheme 6 [16]. Sucrose diol **2** was converted using a few well-defined steps into azido-alcohol **34**, which—upon reaction with di-acetylene **35**—provided precursor **36** (R = H) at an 80% yield. The activation of both hydroxyl groups as mesylates (**37**; R = Ms), followed by their reaction with ethylenediamine, afforded macrocycle **38** at a 5% yield. However, when this process was carried out in the presence of a template (L-phenylglycine methyl ester hydrochloride), this yield increased to 25%.

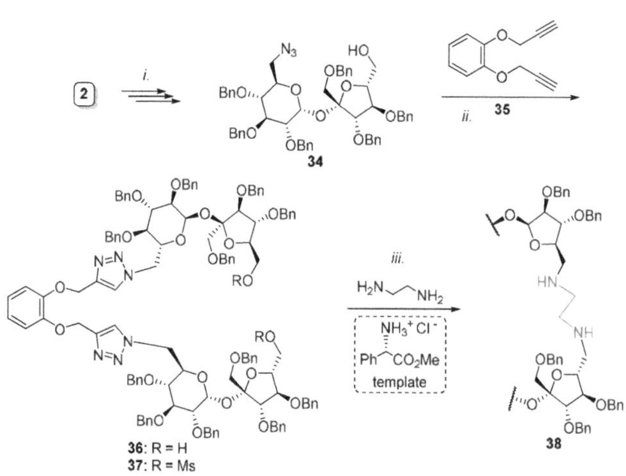

**Scheme 6.** The synthesis of complex macrocyclic derivative **38**. *i.* Ref. [9]; *ii.* $CuSO_4$, Na-ascorbate, *tert*-BuOH/$H_2O$, 80% and then, MsCl, $Et_3N$, DMAP, 95%; *iii.* $Na_2CO_3$, MeCN, reflux, 48h: 5% without a template and 25% with a template.

Soon after this pioneering work, we introduced a method for the synthesis of other dimers with a sucrose scaffold, as shown in Scheme 7. Diol **39** (permethylated analog of **2**) was converted after mesylation into di-amino derivative **40,** which—upon reaction either with di-acid chloride **41a** or **41b**—gave dimer **42** (head-to-tail dimer a or b) together with its head-to-head analog **43** [17].

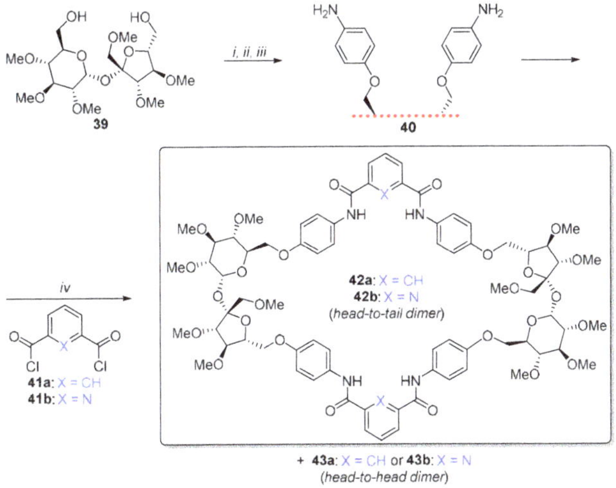


**Scheme 7.** The synthesis of sucrose dimers from **39**. Reagents and conditions: *i*. MsCl, $Et_3N$, 90%; *ii*. *p*-nitrophenol, $K_2CO_3$; *iii*. $H_2$, 10%Pd/C; *iv*. $Et_3N$, **42a** + **43a** (62%) or **42b** + **43b** (54%); both dimers HT and HH were formed at a ratio of 1:1.

The synthesis of urea- and thiourea sucrose dimers is shown in Scheme 8 [18,19]. The treatment of diamine **14** with triphosgene or thiophosgene afforded intermediates **44a** or **44b**, which were further reacted with diamine **14** to afford urea and thiourea derivatives **45** (head-to-head) and **46** (head-to-tail) dimers. These macrocycles show high affinity to chloride and acetate anions.

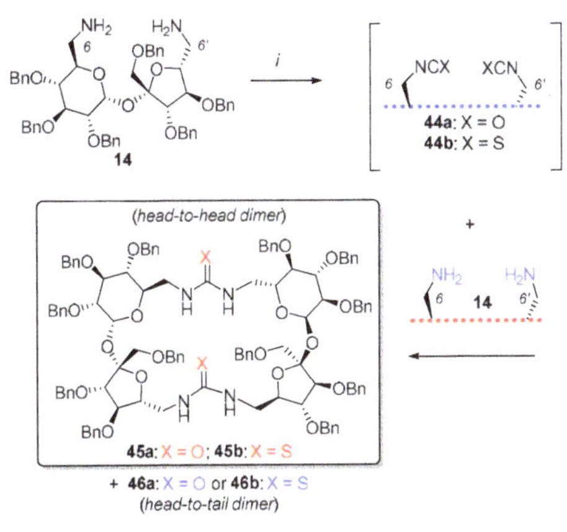


**Scheme 8.** Synthesis of urea and thiourea sucrose dimers. Reagents and conditions: *i*. triphosgene → **44a** or thiophosgene → **44b;** then, **14**, THF, $Et_3N$, −10 °C high dilution (90% for **45a** + **46a**; ref. [18]) or $Na_2CO_3$, AcOEt/$H_2O$, rt (84% for **45b** + **46b**; ref. [19]).

### 2.3. Synthesis of Macrocyclic Derivatives from "Sucrose Triol" 3

After the successful preparation of the relatively simple analogs of crown and aza-crown ethers (some of which showed interesting complexing properties; see the next chapter), we turned our attention to more sterically demanding molecules in which the molecular cavity responsible for complexation would be much more rigid (see models **6** and/or **7** in Figure 3). Such a possibility requires the connection of all three "ends" of the sucrose molecule, i.e., positions C1′, C6, and C6′. The first example of such a complex derivative, shown in Scheme 9, was prepared by us in 2019 [20]. Triol **3** was used as the starting material for the preparation of sucrose-based cryptands; this was the first (and up-to-date the only one) example of such derivatives. The elongation of all three terminal positions of triol **3** by a –$CH_2CH_2$-O-$CH_2CH_2$-I unit (→ **47**) followed by a reaction with tripodal amine **48** afforded cryptand **49** at a very high yield (45.5%).

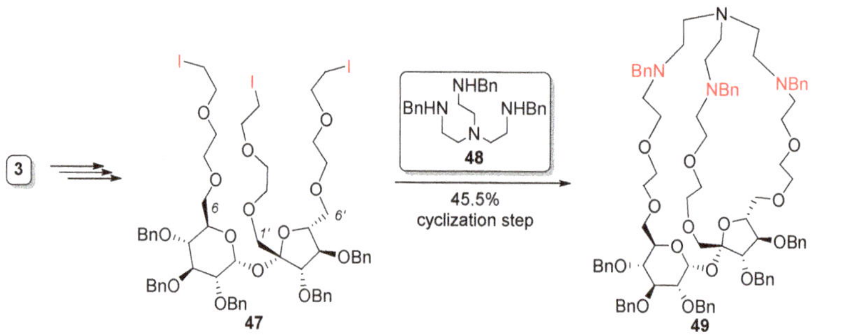


**Scheme 9.** First synthesis of sucrose-based cryptand.

A special case is represented, however, by the objects in which the sucrose platform is connected to a tripodal *chiral* unit. Thus, the reaction of a *racemic* $C_3$-symmetrical cyclotriveratrylene unit (CTV, **50**) with elongated (by two carbon atoms at all terminal positions) sucrose triiodide **51** provided two (out of four possible) diastereoisomeric cages, ***M*-52** and ***P*-52,** as shown in Scheme 10. The benzyl groups protecting the sucrose backbone were easily removed by simple hydrogenation-affording capsules that were soluble in water (***M*-53** and ***P*-53**) [21]. The formation of only two diastereoisomeric cages (**52**) from **51** and ***rac*-50** resulted (most likely) from the fact that the ethylene linker connecting both units was too short.

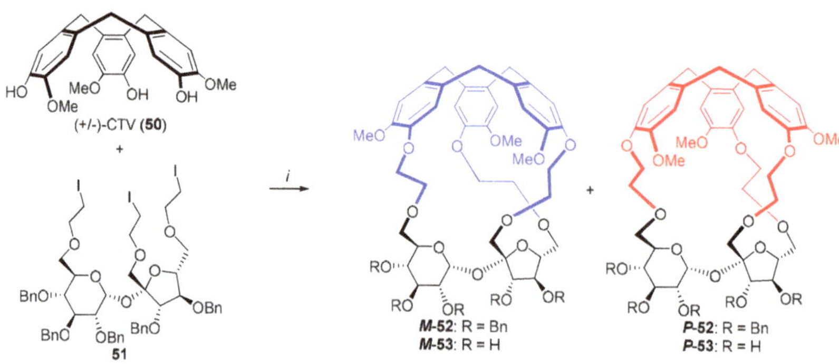


**Scheme 10.** Synthesis of molecular cages **52** and **53** from elongated sucrose triiodide **51** and racemic CTV (**50**). Reagents and conditions: *i*. $Cs_2CO_3$, acetonitrile, and reflux.

For longer than the ethylene linker, the possibility of the formation of all four possible stereoisomers increased significantly. Indeed, the reaction of compound **54** (another analog of sucrose triol **3**) with racemic CTV (**50**) afforded four diastereoisomeric cages (***P*-55**, ***M*-55**, ***P*-56**, and ***M*-56**), as shown in Scheme 11 [22,23].

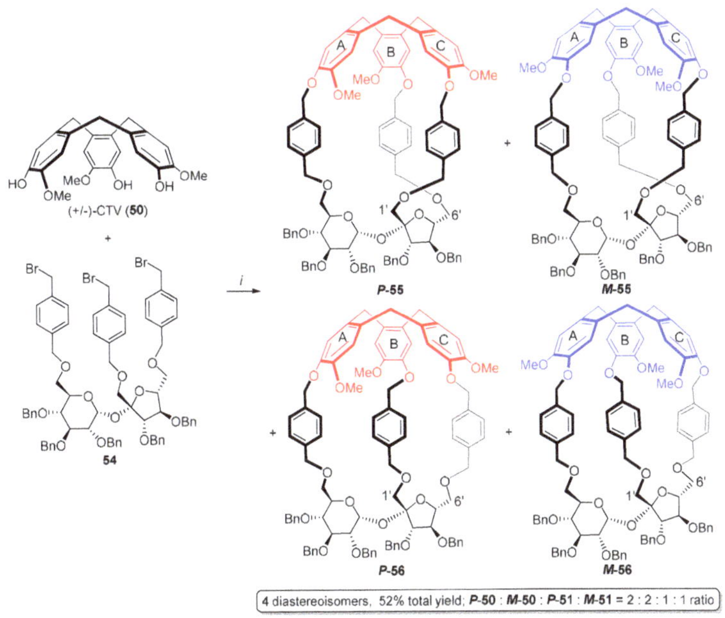


**Scheme 11.** The formation of all four possible distereoisomeric cages in the reaction of *rac*-CTV (**50**) with a "longer" analog **54**. Reagents and conditions: *i.* $Cs_2CO_3$, acetonitrile, and reflux.

Another example of such molecular cages containing CTV and sucrose moieties connected via the naphthalene linkers is shown in Figure 4 (only two out of four isomers formed are shown). These cages were found to be selective and efficient fluorogenic sensors for the detection of acetylcholine or choline. Compound ***P*-57** has a better affinity for choline over acetylcholine, while cage ***M*-57** exhibits a higher constant association for acetylcholine over choline [23].

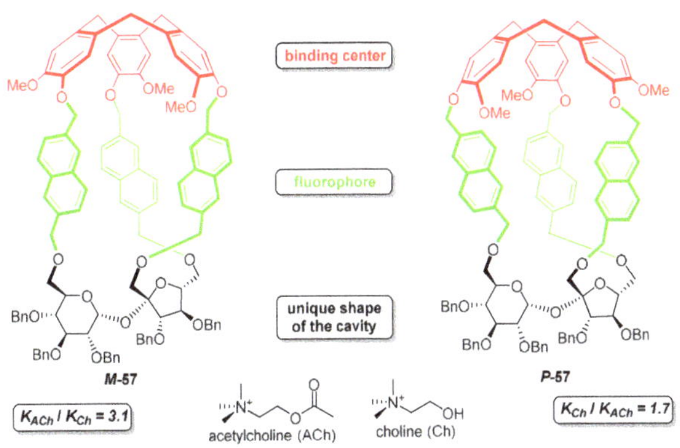


**Figure 4.** Complexing ability of two (out of four) molecular cages, **57**, against acetylcholine and choline.

Molecular switches built on the sucrose platform were also available, as shown in Scheme 12. Diol **2** was converted in a few simple steps into di-iodide **58,** which—upon its reaction with azobenzene **59**—gave macrocycle **60** at a high yield with the trans isomer strongly predominated over the cis one. After irradiation with green light, isomer trans-**60** underwent conversion into the cis isomer, while the irradiation of cis-**60** with blue light (or just heating) induced its conversion into the more thermodynamically stable trans isomer [24]. The complexing properties of both isomers towards achiral cations were rather low but slightly differed from each other, as shown in Scheme 12.

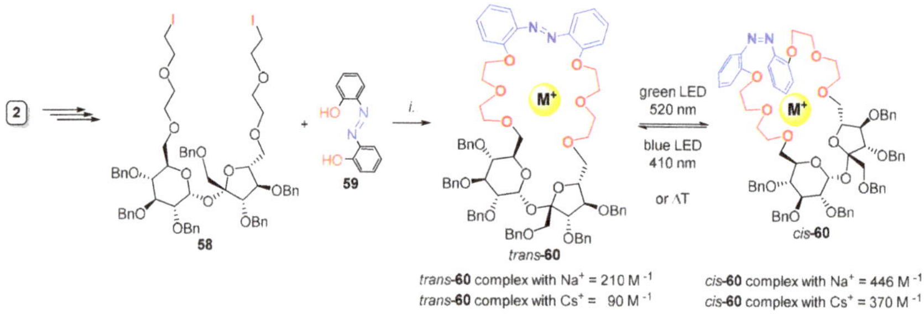


**Scheme 12.** Synthesis of molecular switches with sucrose scaffold. *i.* $Cs_2CO_3$, MeCN, 82 °C, 6h, 85%.

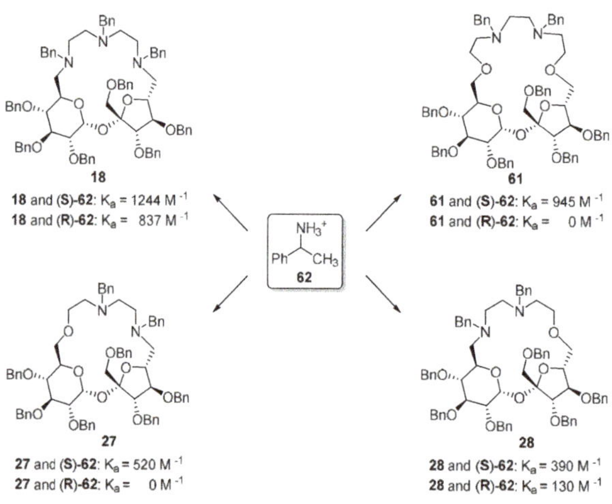


**Figure 5.** Association constants of aza-crowns **18** and **61** and unsymmetrical aza-crowns (**27**, **28**) with both enantiomers of α-phenylethylamine (**62**).

## 3. Complexing Properties

Macrocycles with a sucrose scaffold presented above (crown, aza-crown analogs, and cryptand **49**) show complexing properties towards achiral cations (mostly inorganic) or anions (ureas **45a** and **45b**). However, the application of such highly functionalized, enantiomerically pure receptors to complex achiral guest is not interesting; we performed these studies only to check if such compounds have the complexing abilities. The application of these sucrose-based hosts for the complexation of chiral guests was, however, quite promising. We proved that they are able to distinguish both enantiomers of the phenylethylamine cation; thus, the ***enantioselective*** complexation of chiral guests is possible, at least for simple model guests. The first interesting results were reported in our earlier study (from 2008) on the synthesis of sucrose-base receptors. The preferred complexation of (S)-phenylethylamine cation (**62**) was noted by receptor **18,** while receptor **61** formed a strong complex with the (S)-enantiomer only (Figure 5) [9]. This selectivity towards the (S)-**62** over the (R)-**62** was also noted for "unsymmetrical" receptors **27** and **28** (Figure 5) [12].

## 4. Conclusions

We have proved that sucrose, a very cheap raw material available in almost unlimited quantities, can be used (besides in "commercial" applications) as a building block for the synthesis of fine products with high added value. Several types of macrocyclic derivatives with sucrose scaffold are available. Some of them show interesting complexing properties and are able to differentiate between the two enantiomers of simple chiral guests.

The sucrose platform in our syntheses was fully protected with benzyl groups, which could be removed (as we proved in a few cases) under mild conditions, not affecting the very sensitive glycosidic bond. Such "free" (not protected) receptors, which are soluble in water, might find an application in studies of the complexation of chiral guests in biological environments.

Especially interesting is the field of molecular cages incorporating sucrose and the CTV unit. These two partners can be connected by appropriate linkers of different lengths, which opens a convenient way for cages with various cavities to be able to differentiate between important biologically active compounds.

Also, the molecular switches built from sucrose and azobenzene open a promising field of study. Depending on the light (green or blue), such derivatives change their configuration (from trans to cis and vice versa), thus modifying their cavity and, hence, the complexing properties towards various cations.

As shown in this short note, sucrose can also find an application in the synthesis of fine chemicals, which can be useful in many fields. Thus, this disaccharide is available in practically unlimited quantities, which is ignored in most typical organic laboratories, and can be used as a starting material for the synthesis of various types of optically pure fine chemicals.

**Author Contributions:** Conceptualization, S.J.; writing—original draft preparation, S.J.; writing—review and editing, S.J. and Z.P. All authors have read and agreed to the published version of the manuscript.

**Funding:** This research received no external funding.

**Institutional Review Board Statement:** Not applicable.

**Informed Consent Statement:** Not applicable.

**Data Availability Statement:** Data sharing is not applicable.

**Conflicts of Interest:** The authors declare no conflicts of interest.

## Abbreviations

The following abbreviations are used in this manuscript:

| | |
|---|---|
| CTV | cycloveratrylene |
| *tert*-BDPS-Cl | *tert*-butyldiphenyl silyl chloride |
| TrCl | triphenylmethyl chloride (trityl chloride) |

## References


1. Queneau, Y.; Jarosz, S.; Lewandowski, B.; Fitremann, J. Sucrose chemistry and applications of sucrochemicals. *Adv. Carbohydr. Chem. Biochem.* **2008**, *61*, 217–292. [CrossRef]
2. Jarosz, S.; Pakulski, Z. The roadmap for sucrose—A very inexpensive raw product. *Adv. Carbohydr. Chem. Biochem.* **2024**, *86*, 1–13. [CrossRef] [PubMed]
3. Jarosz, S.; Sokołowska, P.; Szyszka, Ł. Synthesis of fine chemicals with high added value from sucrose: Towards sucrose-based macrocycles. *Tetrahedron Lett.* **2020**, *61*, 151888. [CrossRef]
4. Lichtenthaler, F.; Immel, W.; Kreis, S.U. Old roots-new branches: Evolution of the structural representation of sucrose. In *Carbohydrates as Organic Raw Materials*; VCH Verlagsgesellschaft: Weinheim, Germany, 1991; pp. 1–32. ISBN 3-527-28280-7.
5. Karl, H.; Lee, C.K.; Khan, R. Synthesis and reactions of tert-butyldiphenylsilyl ethers of sucrose. *Carbohydr. Res.* **1982**, *101*, 31–38. [CrossRef]
6. Szejtli, J. Introduction and general overview of cyclodextrin chemistry. *Chem. Rev.* **1998**, *98*, 1743–1758. [CrossRef]
7. Crini, G. Review: A history of cyclodextrins. *Chem. Rev.* **2014**, *114*, 10940–10975. [CrossRef]
8. Xie, J.; Bogliotti, N. Synthesis and applications of carbohydrate-derived macrocyclic compounds. *Chem. Rev.* **2014**, *114*, 7678–7739. [CrossRef]
9. Lewandowski, B.; Jarosz, S. Chiral recognition of α-phenylethylamine by sucrose-based macrocyclic receptors. *Chem. Comm.* **2008**, *47*, 6399–6401. [CrossRef]
10. Jarosz, S.; Lewandowski, B. Synthesis and complexation properties towards the ammonium cation of aza-coronand analogues containing sucrose. *Carbohydr. Res.* **2008**, *343*, 965–969. [CrossRef]
11. Potopnyk, M.A.; Lewandowski, B.; Jarosz, S. Novel sucrose-based macrocyclic receptors for enantioselective recognition of chiral ammonium cations. *Tetrahedron Asymmetry* **2012**, *23*, 1474–1479. [CrossRef]
12. Potopnyk, M.A.; Jarosz, S. Synthesis and complexing properties of "unsymmetrical" sucrose-based receptors. *Eur. J. Org. Chem.* **2013**, *2013*, 5117–5126. [CrossRef]
13. Whistler, R.L.; Anisuzzaman, A.K.M. Preferential halogenation of the primary hydroxyl group. *Methods Carbohydr. Chem.* **1980**, *8*, 227–231.
14. Jarosz, S.; Listkowski, A. Synthesis of macrocyclic derivatives containing sucrose unit. *J. Carbohydr. Chem.* **2003**, *22*, 753–763. [CrossRef]
15. Gajewska, A.; Osuch-Kwiatkowska, A.; Jarosz, S. Synthesis of sulfur containing macrocycles with sucrose scaffold. *Carbohydr. Res.* **2019**, *486*, 107825. [CrossRef]
16. Lewandowski, B.; Jarosz, S. Amino-acid templated assembly of sucrose-derived macrocycles. *Org. Lett.* **2010**, *12*, 2532–2535. [CrossRef]
17. Potopnyk, M.A.; Cmoch, P.; Jarosz, S. Short synthesis of diamide-linked sucrose macrocycles. *Org. Lett.* **2012**, *14*, 4258–4261. [CrossRef]
18. Łęczycka-Wilk, K.; Dąbrowa, K.; Cmoch, P.; Jarosz, S. Chloride-templated macrocyclization and anion-binding properties of $C_2$-symmetric macrocyclic ureas from sucrose. *Org. Lett.* **2017**, *19*, 4596–4599. [CrossRef]
19. Łęczycka-Wilk, K.; Ulatowski, F.; Cmoch, P.; Jarosz, S. "Choose-a-size" control in the synthesis of sucrose based urea and thiourea macrocycles. *Org. Biomol. Chem.* **2018**, *16*, 6063–6069. [CrossRef]
20. Sokołowska, P.; Kowalski, M.; Jarosz, S. First synthesis of cryptands with sucrose scaffold. *Beilstein J. Org. Chem.* **2019**, *15*, 210–217. [CrossRef]
21. Szyszka, Ł.; Cmoch, P.; Butkiewicz, A.; Potopnyk, M.A.; Jarosz, S. Synthesis of cyclotriveratrylene-sucrose-based capsules. *Org. Lett.* **2019**, *21*, 6523–6528. [CrossRef]
22. Szyszka, Ł.; Cmoch, P.; Górecki, M.; Ceborska, M.; Potopnyk, M.A.; Jarosz, S. Chiral molecular cages based on cyclotriveratrylene and sucrose units connected with *p*-phenylene linkers. *Eur. J. Org. Chem.* **2021**, *2021*, 897–906. [CrossRef]
23. Szyszka, Ł.; Górecki, M.; Cmoch, P.; Jarosz, S. Fluorescent molecular cages with sucrose and cyclotriveratrylene units for the selective recognition of choline and acetylcholine. *J. Org. Chem.* **2021**, *86*, 5129–5141. [CrossRef] [PubMed]
24. Sokołowska, P.; Dąbrowa, K.; Jarosz, S. Visible-light responsive sucrose-containing macrocyclic host for cations. *Org. Lett.* **2021**, *23*, 2687–2692. [CrossRef] [PubMed]

molecules

MDPI

Review

# Synthesis and Immunogenicity of Pseudo-Oligosaccharides Structurally Related to Repeating Units of Capsular Phosphoglycans of Human Pathogens†

**Elena A. Khatuntseva, Anastasia A. Kamneva, Dmitry V. Yashunsky and Nikolay E. Nifantiev ***

N.D. Zelinsky Institute of Organic Chemistry, Russian Academy of Sciences, Leninsky Prospect 47, Moscow 119991, Russia; glyco@ioc.ac.ru (E.A.K.); nastyakamneva@icloud.com (A.A.K.); yashunsky1959@yandex.ru (D.V.Y.)

* Correspondence: nen@ioc.ac.ru

† This review is dedicated to the memory of Professor Hans Paulsen.

**Abstract**

This review focuses on the synthesis of spacer-armed phosphooligosaccharides structurally related to the capsular phosphoglycans of pathogenic bacteria, including the *Haemophilus influenzae* serotypes a, b, c, and f, *Neisseria meningitidis* serogroups a and x, the *Streptococcus pneumoniae* serotypes 6a, 6b, 6c, 6f, 19a, and 19f, and the *Campylobacter jejuni* serotype HS:53, strain RM1221, in which the phosphodiester linkage is a structural component of a phosphoglycan backbone. Also, in this review, we summarize the current knowledge on the preparation and immunogenicity of neoglycoconjugates based on synthetic phospho-oligosaccharides. The discussed data helps evaluate the prospects for the development of conjugate vaccines on the basis of synthetic phosphooligosaccharide antigens.

**Keywords:** phosphoglycans; phosphodiester linkage; capsular polysaccharides; synthesis; neoglycoconjugates; immunogenicity; conjugate glycovaccines; human pathogens

## 1. Introduction

Among many smart tools for the colonization of human organs, pathogenic bacteria use a multilayer cell envelope [1,2], which is largely composed of glycans and their conjugates. The outermost thick layer found in the majority of Gram-negative and some Gram-positive bacteria is called a glycan capsule, the composition and structure of which depend on the particular bacterial species [3]. Most capsular glycans are long-chain linear negatively charged CPSs or phosphoglycans. In the course of infection, this part of the bacterial outer shell first comes into contact with the components of innate and adaptive immunity. The anionic glycans of pathogenic bacteria shield them from the action of the components of the complement system, phagocytes and cationic antimicrobial peptides, which are secreted by immune and epithelial cells and destabilize the membrane of pathogenic bacteria [4,5]. Thus, it was established [6] that the phosphoglycan capsule of *Hib* prevents phagocyte attachment and the subsequent uptake of these bacteria. Moreover, *Hib* can survive and multiply after having been engulfed by a phagocyte in the acidic medium of phagolysosomes [7]. On the one hand, these biopolymers act as both protective shields and adhesive agents, enabling the bacterial evasion of host immunity [8], and on the other hand, they represent a target for the host immune system.

It is known that a key step in the adaptive immune response is the production of antigen-specific antibodies, and the presence of IgG indicates the development of immunological memory to surface biopolymers [9]. Immunological memory enables the body to quickly recognize the antigen on the surface of the pathogen during the second and subsequent contact and more effectively activate the body's defenses. This phenomenon encouraged the development of a whole range of antibacterial prophylactic vaccines. To date, the most effective type of vaccines for the prevention of bacterial infections are conjugate vaccines [10–16], in which a capsular glycan of the targeted pathogen is covalently linked to a protein carrier. The use of glycoconjugates helps circumvent the problem of the low immunogenicity of CPS and directs immunity along the T-dependent pathway [17–20].

The majority of commercial conjugate vaccines are currently manufactured on the basis of CPS produced in bacterial cell cultures. This approach has significant disadvantages, which are the laborious and operationally challenging steps of manufacturing, sophisticated quality control, and the high cost of producing pathogenic microbiological material. At present, an advanced approach to glycan conjugate vaccines is being developed, which employs synthetic OS antigens as an alternative to bacterial CPS [21–26].

In the first step of conjugate OS vaccine development, it is important to specify the particular bacterial glycoantigen to be mimicked. The probability factor speaks in favor of regular glycans, which have a repeating unit. These include the CPS of both Gram-positive and Gram-negative bacteria, lipopolysaccharide *O*-antigens of Gram-negative bacteria, and the cell-wall teichoic acids and lipoteichoic acids of Gram-positive bacteria. Today, commercial glycoconjugate vaccines incorporate bacterial glycan antigens exclusively.

In the second step, the optimum length of the OS, which is sufficient for the induction of protective immunity, has to be specified. On the one hand, the use of shorter OSs can significantly reduce production costs, and on the other hand, the OS chain has to be long enough to mimic the polymer and minimize the possible influence of the terminal monosaccharide residue. Identification of the minimum length of the OS antigen is performed in laboratory animals, which are immunized with the conjugates of synthetic oligosaccharides of different lengths, and the interaction of induced antibodies with bacterial antigens or directly with bacteria is investigated. It is commonly accepted [27] that the minimum length of an effective glycoantigen is 3–4 repeating units. A reliable algorithm that can predict the optimum length of OS antigens has not yet been developed. Today, the identification of protective glycotopes for conjugate vaccine candidates can be defined by the screening of glycan arrays, which encompass a series of synthetic OSs related to capsular glycans [28,29] in combination with conformational studies [29,30]. Alternatively, the development of the anti-Hib vaccine Quimi-Hib® (Centro de Ingeniería Genética y Biotecnología (CIGB), Republic of Cuba), which is a unique commercial synthetic OS-based conjugate vaccine, circumvented the problem of the choice of the optimum antigen length by the production of a protein-conjugated homologous phosphooligosaccharide obtained by oligomerization, with 7–8 repeating units as the average length of antigens [31]. The choice of the optimum structure of an OS antigen for a glycovaccine is complicated by the possibility of structural changes in a synthetic antigen after the injection of the vaccine into the recipient's body. Partial lysis or migration of acetyl groups can occur in lymphs (pH 7.4–9.0) or endolysosomes of follicular B-lymphocytes [32–34], which are known to have an acidic environment.

In a number of human pathogens, a phosphodiester bond is involved in the formation of the main polymer chain. Phosphoglycans of this type were found, for example, in the cell wall of the human pathogens *Hia*, *Hib*, *Hic*, and *Hif*, the *S. pneumonia* serotypes 6A, 6B, 11A, 17F, 19A, and 19F, the *N. meningitidis* serogroups *A* and *X*, the *Campylobacter jejuni* serotypes HS53 (strain RM1221) and HS1 (strain ATCC 43429), and *Escherichia coli* K100 and K2 (Figure 1) [35–38]. Today, eleven bacterial pathogens of this type (*Hib*, *Men A*, and the *S. pneumonia* serotypes 6A, 6B, 10A, 11A, 15B, 18C, 19A, 19F, and 23F) are targeted by preventive vaccination, with commercial conjugate vaccines based on the corresponding capsular phosphooligosaccharides [21].

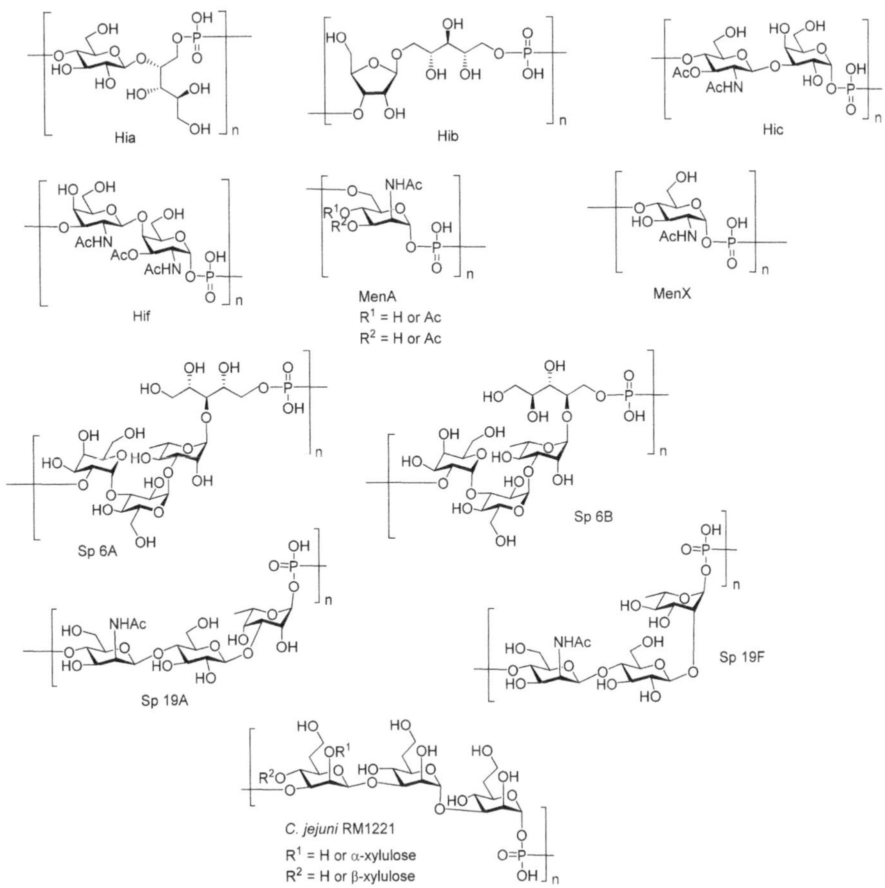


**Figure 1.** Examples of bacterial capsular phosphoglycans.

In this review, we focus on the synthesis of spacer-armed mono-, di-, and oligomeric antigens structurally related to bacterial capsular phosphoglycans (Figure 1), in which the repeating units are connected via a phosphodiester bond, and consider the preparation and immunogenicity of neoglycoconjugates on the basis of these antigens.

Phosphoglycans are the common glycocalyx components of pathogenic Gram-negative and Gram-positive bacteria [35,36,39], yeast [39], and protozoan parasites [39,40]. In a living cell, phosphodiester-linked carbohydrates are arranged via the transfer of a hexose 1-phosphate to a glycan acceptor. In the presence of enzymes, which belong to the Stealth enzyme family, a hydroxyl group of a glycan acceptor attacks the phosphoester group in the nucleotide donor, and finally, a phosphodiester interglycosidic bridge unit is formed. It was established [41] (Figure 2) that in *MenX* bacteria, the hydroxyl group of *N*-acetyl glucosamine acceptor attacks the P-atom of uridine-5′-diphosphate-*N*-acetyl glucosamine to form an α-glycosyl phosphodiester, and the uridine monophosphate moiety serves as a leaving group.

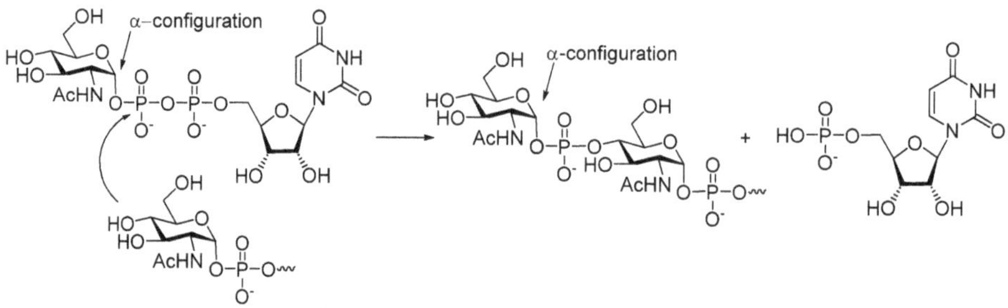


**Figure 2.** Biosynthesis of *MenX* phosphoglycan involves nucleotide glycosyl donor uridine-5′-diphosphate-*N*-acetyl glucosamine and is catalyzed by hexose phosphotransferase [41].

A similar method of the establishment of a phosphodiester intersaccharide bridge is used in laboratory practice for the preparation of phosphooligosaccharides related to bacterial phosphoglycans. For the efficient and stereoselective formation of a phosphodiester linkage, a phosphodiester synthon is first introduced into one of the saccharide blocks, and the resulting product is reacted with a free hydroxyl group of another saccharide block.

As a rule, phosphodiesters, along with their diverse precursors (mono- and diphosphates, H-phosphonates, and phosphamidites; Figure 3), decompose under the conditions of a glycosylation reaction. Therefore, the retrosynthetic analysis of phosphoglycans suggests the formation of phosphodiester bridges within the latest steps of the synthetic route after the glycosidic linkages are already established. Usually, the preparation of phospho-oligosaccharides related to natural biopolymers is a multi-step synthesis, which can be performed following a linear, convergent, or oligomerization pathway via the formation of an O-P-O tether between selectively protected and activated saccharide blocks.

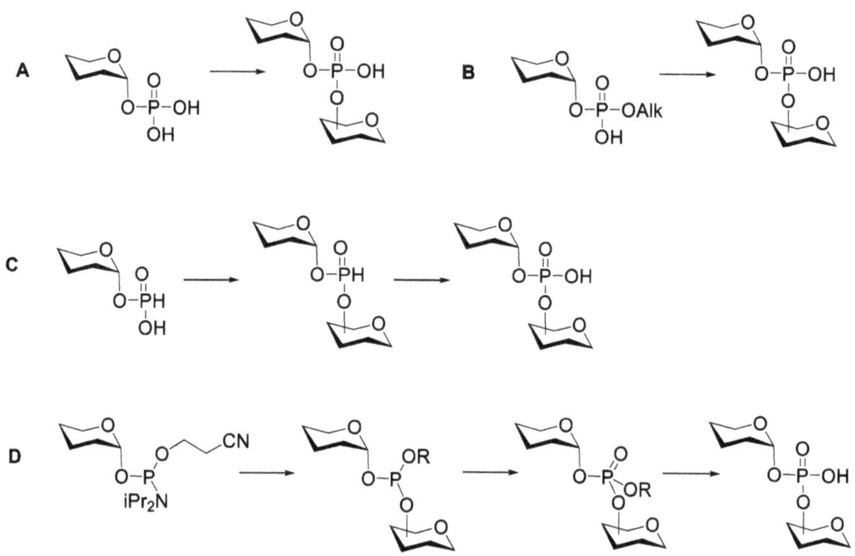


**Figure 3.** Three main types of synthetic approaches used in the preparation of glycosyl phosphodiesters.

The conventional synthetic blocks for the preparation of glycosylphosphodiesters are monophosphates (Figure 3A), diphosphates (Figure 3B), H-phosphonates (Figure 3C), and phosphoramidites (Figure 3D). Previously developed methods of condensation involved phosphorus (V) chemistry (Figure 3A,B) and were promoted by *N,N′*-dicyclohexylcarbodiimide and 1-(2,4,6-triisopropylbenzenesulfonyl)-3-nitro-1H-1,2,4-triazole. In the late 1980s, methods A and B gave way to fast, efficient, and convenient techniques that employed H-phosphonates (Figure 3C) and phosphoramidites (Figure 3D).

The H-phosphonate condensation general procedure [42,43] was first proposed in the 1950s for oligonucleotide synthesis by Todd et al. [44], and it was further developed [45,46] and adapted for solid-phase synthesis [47] and customized to the needs of phosphoglycan chemistry [39,48]. The phosphoramidite method was first proposed by van Boom [49] and was later successfully applied in the solid-phase preparation of long-chain oligophosphodiesters [50] and the P-modified analogs of glycosylphosphate oligomers [51]. In addition to current methods, novel approaches are being actively developed, which are aimed at the preparation of glycosylphosphates with a predetermined anomeric configuration and the synthesis of the stabilized mimetics of phosphoglycans [52–58].

One of the key features of bacterial phosphoglycans and synthetic phosphooligosaccharides is their susceptibility to degradation via hydrolytic cleavage, transesterification, and rearrangement. Thus, PRP (Figure 1) was found to degrade spontaneously [59,60] in aqueous media. This molecule is destabilized by a hydroxyl group at C-2 (D-ribose), which is located in close proximity to the phosphodiester moiety and promotes the depolymerization and formation of cyclophosphate and phosphate monoester terminal groups [61]. In vaccine production, the inherent tendency of PRP to autolyze results in the loss of manufactured phosphoglycan and conjugate preparations [60]. Additionally, the low stability of PRP imposes significant limitations on the use of liquid Hib vaccines, especially in view of the acceleration of the degradation process in the presence of an alum adjuvant [62]. Stability studies conducted by Cintra et al. [60] showed that the rate of PRP depolymerization accelerates substantially with an increase in pH in the range 5.41–7.55 and with a rise in temperature in the interval of 28–40 °C. As a result, it may be assumed that in a host organism, PRP intensely degrades into fragments, which neutralize anti-PRP protective antibodies, thus hampering the immune response. In a similar way, partial PRP destruction after immunization with a conjugate *Hib* vaccine may result in the loss of Hib epitopes and reduce the level of protective anti-Hib antibodies. Two more factors that affect the stability of PRP are the presence of $Na^+$ [60] and $Ca^{2+}$ [59] cations. Similar to PRP, the capsular phosphoglycan of *Hif* (Figure 1) was shown to decompose under mild conditions [59].

MenA capsular phosphoglycan is especially susceptible to hydrolysis. It is assumed that the hydrolytic destruction of α-glycosylphosphodiester linkage occurs by two pathways. One of these includes the formation of an oxocarbenium ion, and within another pathway, the phosphodiester bond is cleaved with the assistance of the axial NAc group located at C-2 of the ManNAc, and a thermodynamically stable oxazoline is formed [63].

## 2. Synthesis of Pseudo-Oligosaccharides Structurally Related to PRP

Highly effective infant immunization with conjugate *Hib* vaccines, which are composed of a partially depolymerized *Hib* capsular phosphoglycan and a protein carrier (see review [15]), inspired researchers to develop conjugate vaccines using oligomer synthetic PRP fragments (Figure 4). Several series of PRP-related oligomers were synthesized, which were equipped with different types and locations of amino-spacers for convenient attachment to a protein carrier (compounds **1–15**, Figure 4).

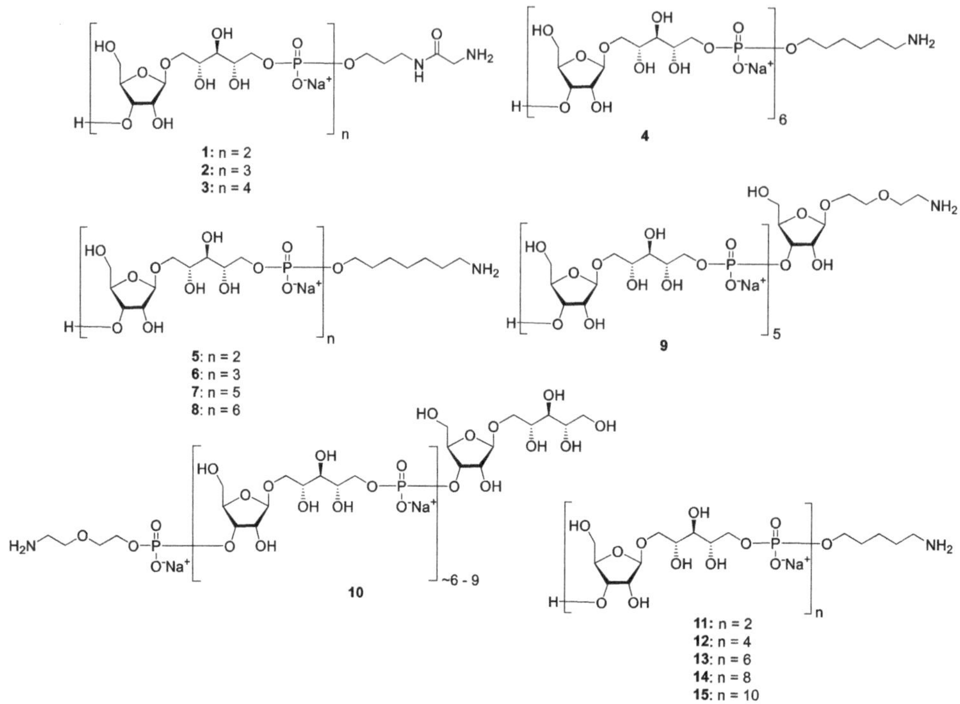

**Figure 4.** Synthetic pseudo-oligosaccharides structurally related to PRP.

In pseudo-oligosaccharide **13** with a *N*-glycyl-aminopropyl spacer, which comprises two, three, or four residues of PRP repeating units [64], as well as in hexamer **4** [65] and oligomers **5–8** [66], the ω-aminospacer is linked to the D-ribose fragment via a phosphodiester bond. In pentamer **9** [67], the spacer is attached immediately to C-1 of a D-ribose residue as an aglycone. In the oligomeric mixture **10** [68] and in most of the representative oligomer series, which comprises the individual compounds **11–15** [69], the spacer is connected to a D-ribose via a phosphodiester bridge. In this review, basic synthetic blocks and strategies used in the preparation of compounds **1–15** are briefly considered, with a focus on the arrangement of a phosphodiester linkage and the introduction of a ω-aminospacer. A more detailed consideration of the syntheses of pseudo-oligosaccharides **1–15** was presented in our earlier review [15].

Between 1988 and 1992, van Boom and his research group at Leiden University synthesized spacer-armed PRP-related fragments for the first time. Dimer **1** [64,70], trimer **2** [64,70], and tetramer **3** [71] (Figure 4) were prepared by a sequential elongation of the oligomer chain from the non-reducing end, and the final equipment with an aminospacer. In brief, the phosphorylation of a free 3-OH group in the selectively protected riboside **16** with bis(benzotriazol-1-yl) (2-chlorophenyl) phosphate **17** produced the key activated phosphotriester building block **18** (Scheme 1). The condensation of compound **18** with the selectively protected disaccharide **19** in the presence of *N*-methylimidazole in Py, followed by the removal of the 1-*O*-propenyl group, was carried out in a sequential manner, one, two, or three times, followed by the final capping with phosphotriester **20**. Total deprotection afforded the series of conjugation-ready compounds **1–3**.

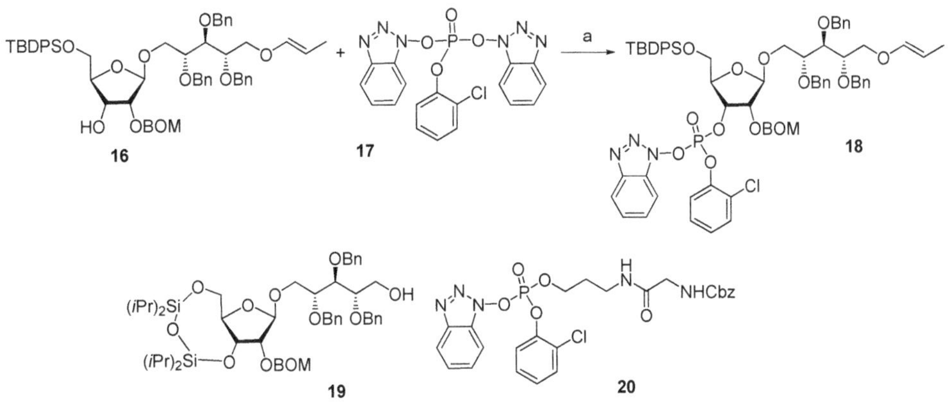

**Scheme 1.** Key building blocks for the synthesis of spacer-armed PRP-related oligomers (**1–3**). Reagents and conditions: (**a**) Py/dioxane, r.t., 30 min [64,70,71].

The spacer-armed trimer **2** and tetramer **3** were conjugated to TT (Figure 5) to create the corresponding neoglycoconjugates **21** and **22**. The antigenic properties of oligomers **2** and **3** as components of neoglycoconjugates **21** and **22** were examined [71] in competitive inhibition ELISA experiments. A conjugate of bacterial PRP and tyramine was used as a coating antigen, and normal human serum and bacterial PRP were used as positive controls.

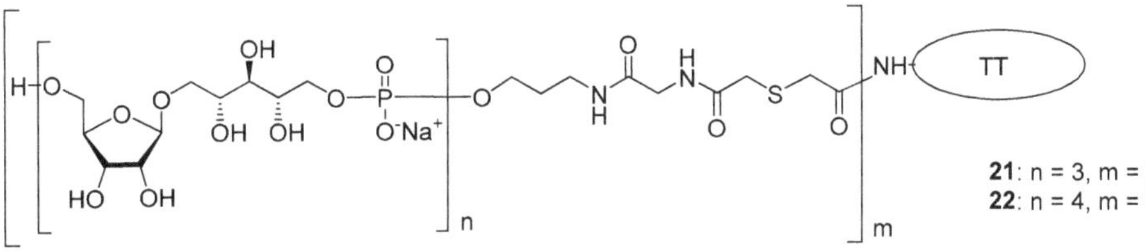

**Figure 5.** Conjugates **21** and **22** obtained from PRP oligomers **2** and **3** with TT [71].

Total inhibition was observed when bacterial PRP and conjugate **22** (tetramer + TT, Figure 5) at a concentration of 25 μg/mL were used as inhibitors, and conjugate **21** with the trimeric PRP–antigen was less effective. Compared to conjugates, the inhibitory capacity of the unconjugated oligomers **2** and **3** was much lower, and at a concentration of 25 μg/mL, it did not reach 40%.

The immunogenicity of glycoconjugates **21** and **22** was studied [71] in mice at a dose of 1 μg of phosphoglycan per mouse. IgG antibodies in immune sera were analyzed in ELISA using a bacterial PRP–tyramine conjugate as a coating antigen and normal human serum as a positive control. Conjugate **21** with a trimeric antigen was found to be low immunogenic in a mouse model. In IgG ELISA, the mean optical density value for the sera of immunized mice was less than two times that of the corresponding value obtained for the sera of mice that received PBS.

For the immune sera from mice vaccinated with conjugate **22**, which comprised tetrameric residues, the mean optical density value was three times that of the negative control, thus unambiguously evidencing the immunogenicity of conjugate **22**. Immunization experiments were conducted using preparations formulated with $AlPO_4$ or without an adjuvant. It was shown that the presence of $AlPO_4$ had no effect on the result of immunization [71].

To date, there is no animal model that reliably predicts the immunogenicity of glycoconjugate preparations in humans, and the choice of experimental animals for the examination of the immune activity of *Hib* preparations is usually at the discretion of the researcher. In the blood sera of immunized mice and guinea pigs, the concentration of antibodies directed against *Hib* antigens does not reach 1 mg/mL, and these animals are not recommended for laboratory studies of conjugate vaccines [66].

As shown in Scheme 1, the key reaction in the assembly of the synthetic *Hib* phosphooligosaccharides **1–3** is the arrangement of the O-P-O linkage, which can be applied iteratively. As no new chiral centers or other types of isomerism are created, this reaction sequence can be fulfilled using a solid-phase approach. This approach significantly decreases the number of laborious steps of the chromatographic separation of the product. In 1989, van Boom synthesized the spacer-armed hexamer **4** using controlled-pore glass as a solid support [65]. First, the selectively protected pseudo-disaccharide **23** was phosphitylated with chlorophosphoramidite **24** (Scheme 2), and the resulting phosphoramidite **25** was used for a stepwise extension of the oligomer chain starting from ribosylribitol **26**, which was immobilized on a glass support. After each step of chain elongation, the phosphorus atom was oxidized with iodine in an acetonitrile/water/collidine mixture to obtain the corresponding phosphotriester, and the terminal DMTr protective group was removed to provide a free hydroxyl group for the next phosphitylation step. After the sequential attachment of five repeating units and spacer **27** and the removal of the protective groups, the target spacer-armed hexamer **4** was obtained.

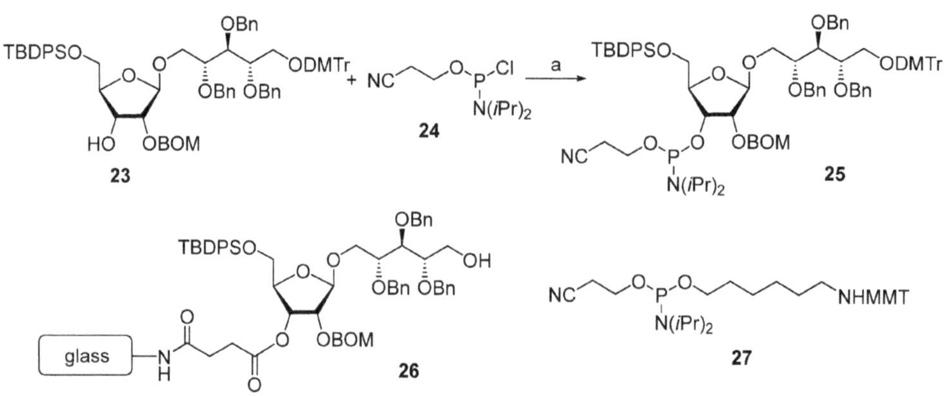

**Scheme 2.** Key building blocks for the synthesis of hexamer **4**. Reagents and conditions: (**a**) diisopropylethylamide/dichloromethane, 85% yield [65].

In 1992, Kandil et al. [66] synthesized the series of spacer-armed Hib oligomers **5–8** on a PEG support using a strategy similar to the synthesis of hexamer **4** (Scheme 3). In this synthesis, the tert-butyldiphenylsilyl and benzyloxymethyl protective groups in the starting pseudo-disaccharide **28** were replaced with easily removable benzyl groups. Thus, the sequential condensation of phosphoramidite **29** with PEG-immobilized pseudo-disaccharide **30** and detritylation was repeated five times. Next, the pre-spacer **31** was attached, and finally, total deprotection resulted in the formation of hexamer **6**. The latter was conjugated to synthetic peptides structurally related to the *Hib* outer membrane protein, and one of these preparations showed immunogenicity comparable to that shown by the commercial *Hib* vaccine [72].

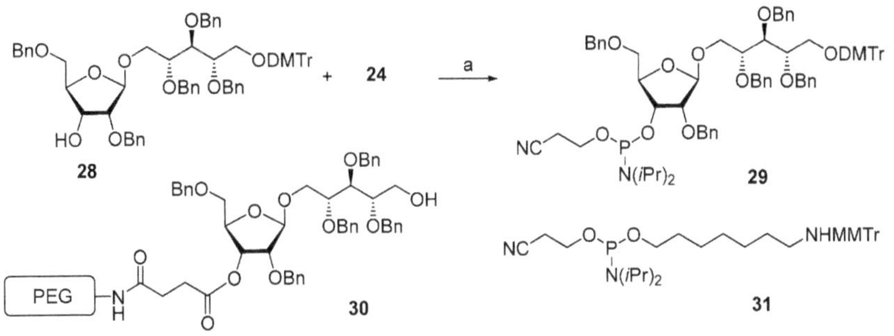

**Scheme 3.** Key building blocks used in the synthesis of the spacer-armed oligomers **5**–**8**. Reagents and conditions: (**a**) 1H-tetrazole/acetonitrile [66].

The same year, a group of Swedish researchers under the leadership of Norberg published [67] the synthesis of the *Hib* pentamer **9** on a polystyrene support (Scheme 4). In this work, the H-phosphonate method was used for the establishment of a phosphodiester bond, which subsequently became the most popular in the preparation of oligophosphoglycans related to the glycocalyx components of bacteria and parasites [39,56]. The selectively protected disaccharide **32** was reacted with 5,5-dimethyl-2-chloro-1,3,2-dioxaphosphorinane 2-oxide (compound **33**) in Py and the presence of phosphonic acid to obtain H-phosphonate **34** with an 84% yield.

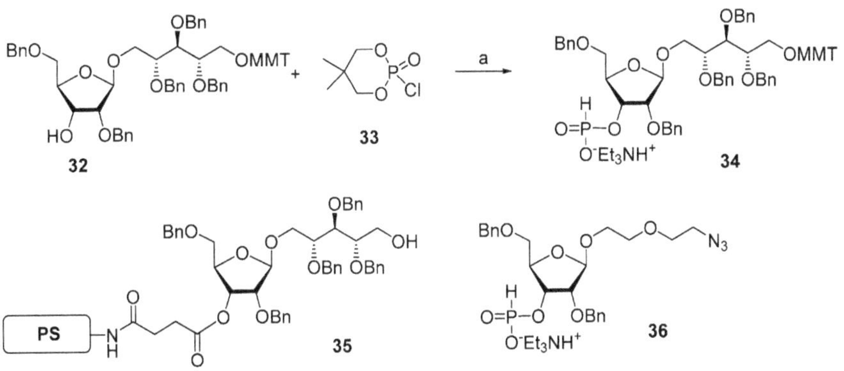

**Scheme 4.** Key building blocks for the synthesis of pentamer **9** [67]. Reagents and conditions: (**a**) $H_3PO_3$, Py, yield of 84%.

The activation of H-phosphonates for subsequent attachment to alcohols can be achieved in the presence of the acyl chlorides of sterically hindered acids, e.g., PivCl or 1-adamantanecarbonyl chloride. In this reaction, the formation of by-products depends significantly on the order of the addition of reagents. For example, the activation of H-phosphonate by PivCl in Py in the absence of alcohol led to the formation of unwanted bis-acylated phosphites, and the use of a large excess of PivCl in solid-phase synthesis resulted in the pivaloylation of the support [67]. The search for the optimum conditions for the attachment of H-phosphonate **34** to be selectively protected and immobilized on the solid support monomer **35** resulted in the discovery of the most efficient proportion—5 eq. of PivCl and 5 eq. of H-phosphonate **34**—that provided the adduct with a 95% yield. The authors noted [67] that the yield decreased with each chain extension step and dropped to 86% in the fourth cycle. The optimum sequence of reagent addition was found as follows:

in the first step, H-phosphonate was activated by the addition of PivCl in Py, and then, the mixture was added to the immobilized alcohol. As a result, the yield of the pentamer increased to 96%. In the final step, the chain was capped with 2-(2-azidoethoxy) ethyl riboside **36**, H-phosphonate was oxidized with iodine in 2% aqueous Py, and the protective groups were removed to obtain oligomer **9**.

The multistep synthetic pathways discussed above are based on the linear sequential elongation of an oligomeric chain. As a rule, laboratory processes of this type cannot be scaled up to a commercial scale because of the high cost. For profitable manufacturing, a convergent synthetic scheme is needed. Under the leadership of Verez-Bencomo and Roy, a convergent synthesis of a mixture of homologous PRP oligomers (Figure 4, compound **10**) was developed and then scaled up to the commercial production of the anti-*Hib* vaccine Quimi-Hib® (Heber Biotec, S.A., Republic of Cuba). In this synthesis, polycondensation of a bifunctional monomer is employed for the establishment of O-P-O linkages as a key step of oligomer synthesis in the place of stepwise elongation (Scheme 5).

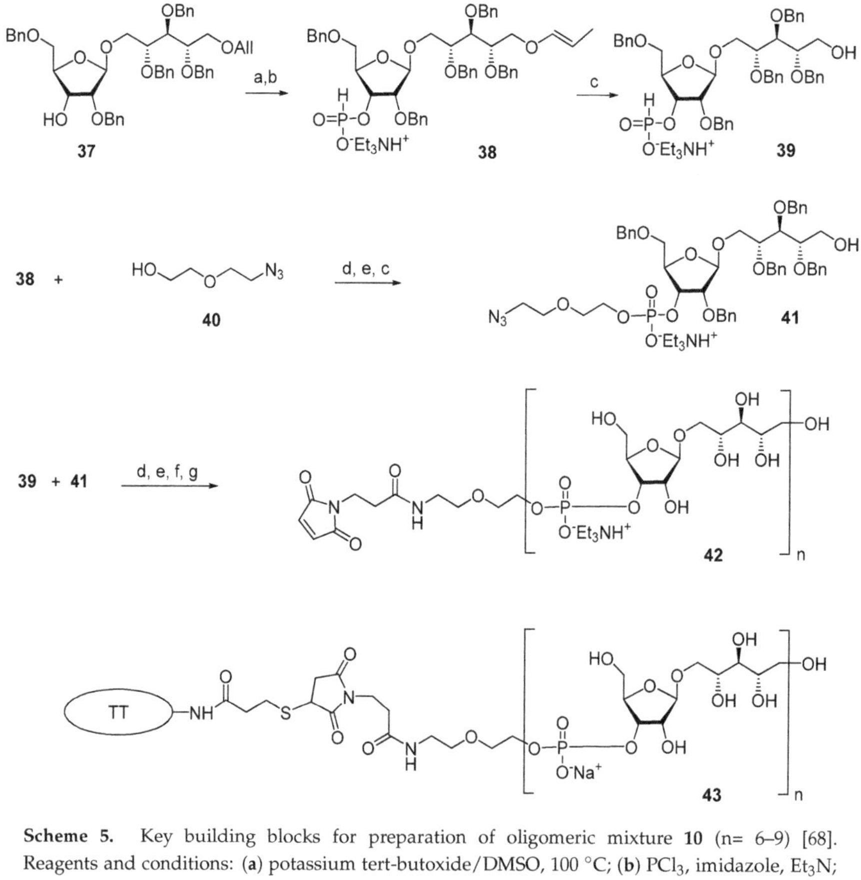


**Scheme 5.** Key building blocks for preparation of oligomeric mixture **10** (n= 6–9) [68]. Reagents and conditions: (**a**) potassium tert-butoxide/DMSO, 100 °C; (**b**) $PCl_3$, imidazole, $Et_3N$; (**c**) $CH_3COOH/H_2O$, 80 °C; (**d**) PivCl/Py; (**e**) $I_2$, $Py/H_2O$; (**f**) $H_2$, Pd/C, EtOH-$H_2O$-EtOAc-AcOH at 1.5 atm; (**g**) SMP.

In brief, the allyl group of the selectively protected pseudo-disaccharide **37** is isomerized into the 1-*O*-propenyl group by the action of potassium tert-butoxide in DMSO. The resulting isomer is converted to H-phosphonate **38** by the action of $PCl_3$ and imidazole, and finally, the 1-*O*-propenyl group is removed under acidic conditions to obtain the key heterobifunctional monomer **39**. The interaction of **38** with the diethylene glycol spacer **40** in the presence of PivCl/Py, followed by oxidation with iodine in aqueous Py, and the further removal of the 1-*O*-propenyl group, creates alcohol **41**, which serves as a terminal unit in the polycondensation. The polycondensation of **39** and **41** is carried out in the presence of PivCl in Py, followed by oxidation of the mixture of H-phosphonates into the corresponding phosphodiester mixture, reduction of the azido groups, *N*-acetylation, and total deprotection. The obtained mixture of the spacer-equipped oligomer **10** is activated by the action of SMP to obtain a conjugation-ready mixture of maleimide **42**. The conjugation of the activated oligophosphodiesters with thiolated TT results in the production of a set of conjugate **43** with the average weight ratio PRP:TT of 1:2.6 [31,68].

On the basis of the mixture of conjugate **43**, an anti-*Hib* vaccine, Quimi-Hib®, was developed, which successfully passed all the required preclinical [73] and clinical trials [31,74]. The antigenicity of synthetic PRP oligomer **10** was compared to the antigenicity of bacterial PRP in ELISA experiments [31]. A conjugate of a mixture of the spacer-equipped oligomer **10** with HSA (**10**-HSA), prepared similarly to conjugate **43**, and a conjugate of partially depolymerized bacterial PRP (PRPDp30) with HSA were used as coating antigens. The PRPDp30-HSA conjugate was obtained by the reductive amination of the mixture of bacterial PRP depolymerized via periodate oxidation to a length of ~30 monomeric units and conjugation with HSA [68]. Standard rabbit anti-PRP antibodies were obtained by the immunization of animals with two licensed commercial conjugate vaccines based on bacterial PRP. These were Hiberix® (PRP-TT), which contains PRP activated by cyanogen bromide and conjugated to TT via an adipic acid dihydrazide linker, and Vaxem-Hib® (PRP-CRM197), which is a product of the conjugation of partially depolymerized bacterial PRP with an avDP10 repeating unit and the protein carrier CRM197 via an adipic acid dihydrazide spacer. The correlation coefficient for the titers obtained in the ELISA experiments with standard rabbit sera and the coating antigens 10-HSA and PRPDp30-HSA was in the high-value range (0.972–0.978), indicating the similarity of the antigenic properties of the synthetic oligosaccharide **10** and *Hib* CPS [75].

For the preliminary evaluation of the immunological activity of the conjugate **43** in vivo, rats, mice, and rabbits were chosen as experimental animals. Rats were immunized subcutaneously twice at an interval of 4 weeks with a dose of conjugate **43** containing 2 μg of the ligand **10** [75]. Mice were immunized intraperitoneally three times at an interval of 2 weeks with a dose of conjugate **43** containing 2.5 μg of ligand **10**. Rabbits were immunized with conjugate **43** using both two-step and three-step regimens at a dose of 5 μg of ligand **10** per animal. The efficiency of immunization was assessed by the evaluation of IgG titers, which were calculated as the logarithm of the highest dilution at which the light absorbance of the diluted serum sample is twice as high as that of the negative control. For the negative control, the animals were injected with PBS. A PRP-HSA conjugate was used as a coating antigen in the ELISA experiments. The study of the PRP-specific antibodies showed that the immune response to conjugate **43** in rodents was weak and inconsistent. In contrast, the immunization of rabbits with conjugate **43** efficiently evoked PRP-specific antibodies, as shown in the inhibition ELISA experiments with bacterial PRP as the inhibitor [75].

In a phase 1 clinical trial [31], more than 100 children aged 4 to 5 years were immunized with a single dose of a vaccine preparation formulated on the basis of conjugate **43** without an adjuvant. Comparative studies of the immunogenicity of synthetic antigens in conjugate **43** and partly depolymerized bacterial PRP in Vaxem-Hib® showed that the average anti-PRP IgG titers were similar. It was found that PRP-specific IgG antibodies induced in children by immunization with conjugate **43** had bactericidal activity comparable to that stimulated by the action of Vaxem-Hib®. In the course of phase 2 clinical trials [31], 1141 infants were immunized three times at 2, 4, and 6 months of age with or without $AlPO_4$, and the positive control group was immunized with Vaxem-Hib®. Antibody tests showed that 99.7% of infants had serum anti-PRP IgG concentrations >1 μg/mL, which is known to provide prolonged protection against Hib infection [76], and the mean concentration of anti-PRP IgG was as high as 27.4 μg/mL.

It is noteworthy that in Quimi-Hib®, different-sized oligomeric antigens were present, thus circumventing the problem of the elucidation of the optimal length of an oligomeric PRP antigen. The success of the development and application of the cost-effective vaccine Quimi-Hib® with a synthetic *Hib*-antigen inspired scientists to search for the shortest protective *Hib*-antigen, which is likely to be in the range of chain lengths that are obtained during the polycondensation of **39**.

One of the novel and efficient approaches to the preparation of PRP-related oligomeric phosphoglycans suggested by Seeberger et al. [69] involved the elongation of the chain by two PRP repeating units at once (Scheme 6). The applied strategy effectively shortened the task of the oligomer assembly and raised the degree of convergence.

For the preparation of a selectively protected and activated bifunctional dimeric key unit, the universal precursor **44** was deallylated, and the resulting diol was regioselectively 4,4′-dimethoxytritylated at the primary hydroxyl group to obtain compound **45**, which was then converted to H-phosphonate **46** by the action of $PCl_3$, $Et_3N$, and imidazole. For the preparation of alcohol **47**, the universal precursor **44** was levulinoylated and deallylated. Alcohol **47** was reacted with H-phosphonate **46** in PivCl/Py to obtain the key dimeric precursor **48** with readily removable orthogonal levulinoyl and dimethoxytrityl protective groups. The de-4,4′-dimethoxytritylation of pseudo-tetrasaccharide **48** in the presence of trichloroacetic acid afforded alcohol **49** with a free terminal hydroxyl group in the ribitol residue. The delevulinoylation of universal precursor **48** by the action of hydrazinium acetate resulted in the formation of alcohol **50**, which was converted into H-phosphonate **51**. The combination of H-phosphonate **51** and primary alcohol **49**, followed by oxidation, yielded a tetramer, which was, in turn, de-4,4′-dimethoxytritylated and reacted with H-phosphonate **46** for the elongation of the chain or with the 5-azidopentyl derivative **52** for chain termination. Using this elegant strategy, PRP-related tetramer (85%), hexamer (85%), octamer (83%), and decamer (80%) were prepared, which were subjected to sequential detritylation, the attachment of a 5-azidopentyl spacer under the action of H-phosphonate **52**, the transformation of the azido group into an amino group, and total deprotection to obtain the spacer-armed oligomers **12–15** (Figure 4). After activation, oligomers **12–15** were reacted with TT to obtain conjugates **53–56** [69].

The antigenicity of oligomers **11–15** (Figure 4) was examined using a glycan microarray. Compounds **11–15** were immobilized on a *N*-hydroxysuccinimide hydrogel surface on glass slides and reacted with polyclonal anti-*Hib* hyperimmune rabbit sera or standard human sera, which were mixed with different concentrations of the WHO PRP standard as an inhibitor. The subsequent addition of a fluorescent secondary antibody revealed the dependence of the fluorescence intensity on the PRP concentration, thus proving the presence of antibodies specific to the synthetic PRP-oligomers **11–15**. Hexamer **13**, octamer **14**, and decamer **15** interacted with standard human sera in a similar way, binding with tetramer **12** was weaker, and in the case of dimer **11**, the adsorption of the antibodies was even less effective. In contrast, rabbit hyperimmune sera interacted with all five oligomers equally well [69].

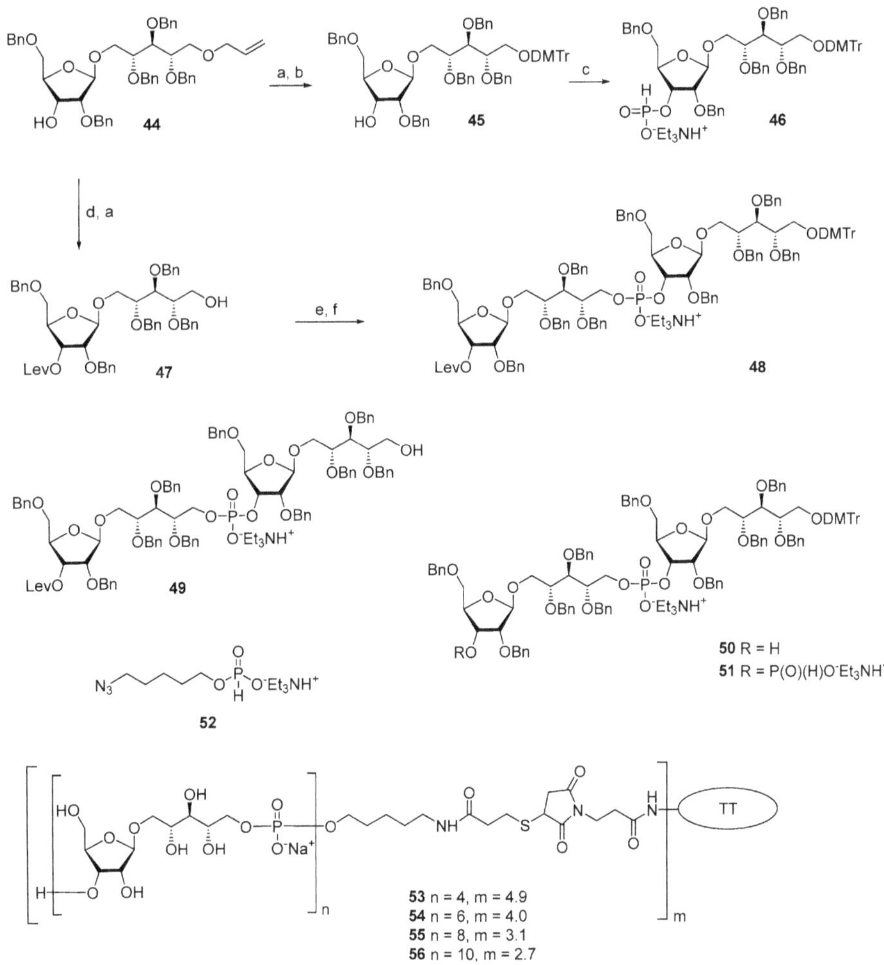

**Scheme 6.** Key building blocks for preparation of oligomers **12–15** [69]. Reagents and conditions: (**a**) $Pd(PPh_3)_4$, 1,3-dimethylbarbituric acid, methanol, 75%; (**b**) DMTCl, DMAP, $CH_2Cl_2$, 93%; (**c**) $PCl_3$, $Et_3N$, imidazole, $CH_2Cl_2$, 0 °C, 85%; (**d**) levulinic acid, DMAP, $CH_2Cl_2$, 93%; (**e**) **46**, PivCl, Py, 0 °C; (**f**) $I_2$, Py/$H_2O$, 85%.

The immunogenicity of conjugates **53–56** was studied using an animal model. Rabbits were immunized with these conjugates at a dose of 5 μg of an oligomer without an adjuvant, and the positive control group received ActHib® (PRP-TT). Bacterial PRP was used as a coating antigen. Immunization with conjugates **53** and **55**, based on tetramer **12** and octamer **14** ligands, respectively, resulted in a high level of PRP-specific antibodies. Conjugates **54** and **56**, which comprised hexamer **13** and decamer **15**, induced substantially lower levels of PRP-specific antibodies [69]. Therefore, in a series of conjugates equipped with antigens with an even number of repeating units, the tetramer was the most promising synthetic antigen candidate for the design of an anti-*Hib* vaccine.

In a continuation of this work, Seeberger et al. synthesized [77] a series of mimetics of synthetic PRP-related dimer, tetramer, hexamer, and octamer compounds (**11–14**), which comprised 2-deoxy-, 2-deoxy-2-fluoro-, 2-deoxy-2-(*N*,*N*-dimethyl)-carbamoyloxy-, and

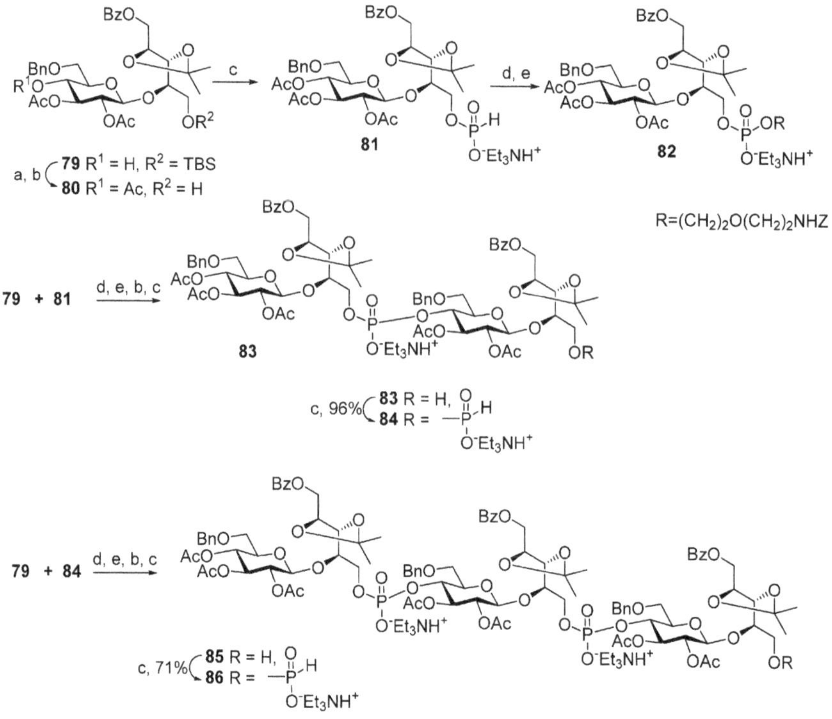

Scheme 8. Key blocks for synthesis of oligomers structurally related to *Hia* phosphoglycans [96]. Reagents and conditions: (**a**) acetic anhydride, DMAP, Py; (**b**) HF (aq.), Py, 60 °C, 97% for two stages; (**c**) $H_3PO_3$, PivCl, Py, 74%; (**d**) $ZNH(CH_2)_2O(CH_2)_2OH$, PivCl, Py; (**e**) $I_2$, Py/$H_2O$.

For the choice of the optimum strategy for the assembly of oligomers **57–61** (Scheme 7) [95], two alternative ways were considered. The first strategy involved pseudo-disaccharide **69** with a free hydroxyl group in the ribitol residue, which was reacted with 2-cyanoethyl *N,N,N′,N′*-tetraisopropylphosphordiamidite (compound **70**) in the presence of diispropylammonium tetrazolide to obtain the selectively protected phosphoramidite **71** as a universal precursor with a 95% yield. The following condensation of phosphoramidite **71** and glycoside **72** with a free hydroxyl group at C-4 (Glc), oxidation of a phosphite into a phosphodiester, and detritylation created the conjugation-ready dimer **73** with a 73% yield over three steps.

Another way of preparing phosphotriester **73** (Scheme 7) employed phosphoramidite **74** with a pro-phosphodiester group located at C-4 (Glc). The sequential combination of phosphoramidite **74** with alcohol **69**, oxidation, and detritylation resulted in the formation of the phosphodiester block **73** with a 79% yield over three steps [94,95]. Therefore, both assembly strategies were equally effective. Starting with phosphoramidite **74**, protected pseudo-oligosaccharides **75–77** were obtained and then converted to ligands **58–61** by interaction with the pre-spacer phosphoramidite **78** and total deprotection. The authors pointed out that in comparison to the H-phosphonate approach, the phosphoramidite protocol was more efficient for the construction of longer phosphodiester-linked chains. Oligomers **57–61** were *N*-acylated with a di(*N*-hydroxysuccinimidyl) adipate linker, and the activated esters were conjugated to CRM197. The resulting conjugates contained 13–25 copies of the oligomeric antigen per CRM197 unit. Rats were immunized with three doses, which contained 2 μg of the glycan antigen, and the IgG antibodies were analyzed in the sera using ELISA. Conjugates of trimer **59** and pentamer **61** with HSA were used as coating antigens. All the conjugates were immunogenic and induced similar levels of IgG antibodies, which recognized immobilized synthetic antigens. The researchers noted that the immunogenicity data for the CRM197-based conjugates of oligomers **57–61** was independent of the length of the oligomeric antigen [95].

Compounds **62–64** [96] were prepared (Scheme 8) starting from the universal precursor pseudo-disaccharide **79**. The acetylation of compound **79** resulted in acetate **80**, which was phosphitylated to obtain key monomeric H-phosphonate **81** with a 97% yield. The condensation of H-phosphonate **81** with a pre-spacer *N*-carbamoyl aminopropanol, as shown in Scheme 8, created phosphodiester **82** (yield: 34%), which, after the removal of the protective groups, was transformed into the spacer-armed target monomer **62** (Figure 6). To obtain dimer **83**, alcohol **79** was condensed with H-phosphonate **81** with a 77% yield and then oxidized. Phosphodiester **83** was, in turn, converted into H-phosphonate **84**. The interaction of H-phosphonate **84** with alcohol **79** produced phosphodiester **85**, which was converted into H-phosphonate **86**. Spacer-armed derivatives were obtained by the interaction of H-phosphonates **84** and **86** with *N*-carbamoyl aminopropanol to obtain, after total deprotection, the spacer-armed dimer **63** and trimer **64**, respectively (Figure 6). It is interesting to note that the efficiency of condensation with N-protected aminopropanol increased in the series **81** > **84** > **86** with the rise in the number of phosphodiester fragments present in these H-phosphonates, whereas numerous experimental data suggest the opposite [39].

In 2001, Oscarson et al. synthesized the amino spacer-armed oligomeric fragments of *Hic* (compounds **65** and **66**) and *Hif* (compounds **67** and **68**) phosphoglycans [97] (Figure 6), in which the repeating disaccharide units are linked via a phosphodiester bridge. Similar to the block syntheses of *Hib* and *Hia* fragments described above, the establishment of a phosphodiester linkage between selectively protected oligosaccharide blocks was used for the chain elongation. However, unlike *Hib* and *Hia*, in *Hic* and *Hif* capsular phosphoglycans, the anomeric carbons in hexoses are involved in the formation of a phosphodiester bridge. Therefore, the development of a synthetic strategy for the preparation of *Hic* and *Hif* fragments has to be developed with respect to the possibility of the formation of an unwanted anomer. At the step of chain assembly, it can be particularly difficult to provide stereocontrol in the formation of a C-1-O-P bond. Instability of the anomeric phosphodiester linkage also has to be considered, which makes it possible to use anomeric phosphodiesters as glycosyl donors in the glycosylation reaction. A convenient strategy for the preparation of the desired anomer includes the initial stereocontrolled glycosylation of a phosphodiester synthon, which provides an intermediate product with the target anomeric configuration, and the subsequent coupling of this compound in mild conditions for the prevention of anomerization. These conditions are met in the H-phosphonate protocol, as hexose hemiacetals retain the anomeric configuration, while they are transformed into H-phosphonates upon the action of triimidazolyl phosphine prepared in situ from $PCl_3$ and imidazole in the presence of $Et_3N$ [39].

For the preparation of the spacer-armed oligomers **65** and **66** (Figure 6) [97], which are structurally related to *Hic* phosphoglycans, hemiacetal **87** with an axial hydroxyl group at C-1 was converted into the key α-H-phosphonate 88 (yield: 89%), which was condensed with the selectively protected alcohol **89** and oxidized with $I_2$ (Scheme 9A) to provide phosphodiester **90** (yield: 71%). The desilylation of phosphodiester **90** produced alcohol **91**, which was readily condensed with α-H-phosphonate **88**, and the phosphite group was oxidized to produce phosphate **92** in a moderate yield (36%). After total deprotection, the reduction of the azido group, and the N-acetylation of compounds **91** and **92**, the target spacer-armed phosphooligosaccharides **65** and **66** were obtained.

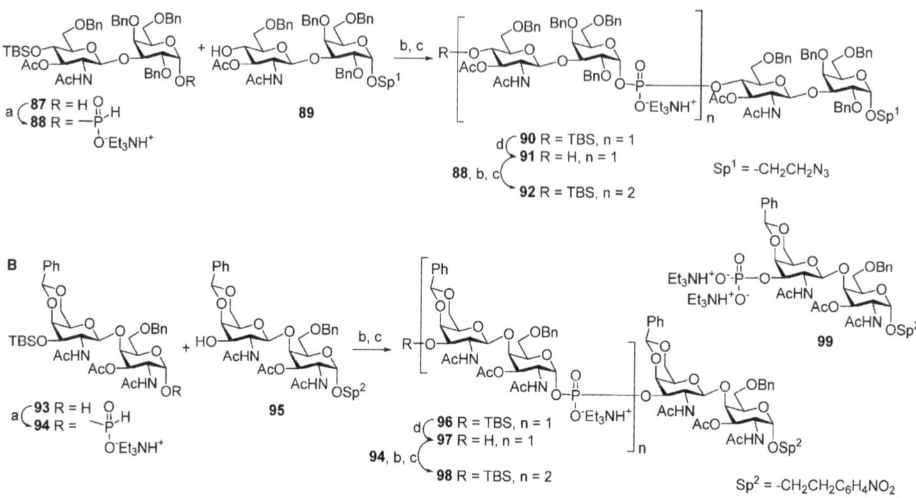


**Scheme 9.** (**A**) Key blocks **87–92** for the synthesis of oligomers [97] **65** and **66** related to *Hic* phosphoglycans; (**B**) key blocks **93–98** for the synthesis of oligomers **67** and **68** related to *Hif* phosphoglycans. Reagents and conditions: (**a**) PivCl, Py; (**b**) $PCl_3$, imidazole, $Et_3N$, 85%; (**c**) $I_2$, $Py/H_2O$, -40 °C; (**d**) $Et_3N(HF)_3$.

Analogously, the favorable α-configuration of the GalNAc hemiacetal residue in disaccharide **93** (Scheme 9B) paved the way to the preparation of the spacer-armed phosphooligosaccharides **67** and **68** related to *Hif* phosphoglycans [97]. Upon action with $PCl_3$ and imidazole, hemiacetal **93** was transformed into α-H-phosphonate **94** with a retention of configuration (yield: 87%). The condensation of α-H-phosphonate **94** with disaccharide **95** and the subsequent oxidation resulted in the efficient formation of phosphodiester **96** (yield: 81%), which was desilylated to obtain alcohol **97**. Similar to the condensation of compounds **91** and **88**, the interaction of **97** and **94** was less efficient, and phosphodiester **98** was obtained with a 37% yield. Most likely, low yield in this reaction is associated with the high lability of the phosphodiester group in acceptors **91** and **97** in the conditions of oxidation with $I_2$ [39]. The authors [97] observed the formation of a significant amount (up to 40%) of 3′-O-phosphate **99**, which indicates the lability of a glycosyl-O-phosphodiester tether under oxidation conditions. The low yields of attachment of the second glycosyl-phosphodiester residue for variants A and B, shown in Scheme 9, suggest that the stability of a phosphodiester bond is largely determined by the condensation and oxidation conditions and is less dependent on the structure of disaccharide residues. The target spacer-armed phosphooligosaccharides **67** and **68** were obtained after complete deblocking of the compounds and the reduction of the nitrophenyl residue into an aniline residue.

In view of the development of an anti-*Hia* vaccine, two aspects have to be addressed. Similar immunogenic properties of synthetic *Hia* antigens from monomer to pentamer can be connected with the low stability of phosphodiester linkages in aqueous solutions. In this case, the design of a vaccine may require a mimetic structure for the antigen. Also, it is advisable to use a phosphoramidite-based strategy for the preparation of phosphooligosaccharide antigens for industrial production, as the considered examples demonstrate the advantage of the phosphoramidite-based method over the H-phosphonate procedure. Meanwhile, anti-*Hic and anti-Hif* conjugate vaccines are not considered to be of high importance in contemporary glycoscience, as researchers have not referred to this topic for more than 20 years.

## 4. Synthesis of Phosphooligomers Related to Capsular Phosphoglycan of MenA

In 2010, a conjugate polysaccharide meningococcal monovalent vaccine, MenAfriVac® (MenA-TT), was licensed [98,99]. In contrast to meningococcal polysaccharide vaccines, the conjugate preparation was found to be highly effective. For example, the results of mass vaccination campaigns in African meningitis belt countries that were carried out in 2010–2015 showed more than a 99% decrease in the incidence of *MenA*-associated meningitis [100]. Today, *MenA* phosphoglycan is a component of a number of polyvalent conjugate meningococcal vaccines, including Menactra® (A-meningococcal component MenA-diphtheria toxoid), Menveo® (A-meningococcal component MenA-CRM197), and Nimerix® (A-meningococcal component MenA-TT) [14].

Polymer chains of *MenA* phosphoglycan are highly labile in aqueous media [63]. This intrinsic property of *MenA* phosphoglycan creates the need for the cold-chain transport of conjugate anti-*MenA* vaccines [101], which significantly increases the cost per dose and, in some cases, is an insurmountable obstacle to the use of this type of preparation [102]. Also, this feature imposes additional requirements on the production of both monovalent vaccines and polyvalent vaccines in a convenient liquid form. In this regard, considerable research efforts have been directed towards the synthesis of oligomeric antigens structurally related to *MenA* CPS fragments or corresponding mimetics with a view to preparing commercial conjugate vaccines. The use of anti-*MenA* conjugate preparations with a synthetic oligoside antigen, the structure of which fully corresponds to *MenA* CPS fragments, could allow the monitoring of the structural integrity of this preparation during storage and transportation, and the design of an antigen using ManNAc phosphate mimetics obtained by replacing the oxygen atom with the methylene group in the pyranose ring with (carba-analogs) or by replacing the anomeric oxygen atom with a methylene group (C-phosphonates) will make the antigen structure resistant to hydrolysis.

It is important to emphasize that in *MenA* bacterial phosphoglycans, ~80% of 3-OH groups and ~10% of 4-OH groups in ManNAc residues are acetylated [103]. A study of phosphoglycan structures present on the surface of a living *MenA* bacterial cell [104], which was conducted using high-resolution magic-angle spinning, showed the presence of acetyl substituents at 50–60% of 3-OH groups and 25–30% of 4-OH groups. The acetylation of *MenA* phosphoglycans is known to be an important antigenicity factor [105,106].

Spacer-armed synthetic antigens structurally related to the *MenA* capsular phosphoglycan and the corresponding phosphono- and carba-analogs are attractive synthetic compounds considering their potential use as ligands in marketed meningococcal vaccines. Since the first time a phosphodiester bridge was arranged between two αManNAc residues (Figure 7, compounds **100** and **101**) in 1993 by the Shibaev group [107], a vast number of *MenA*-related mono-, di-, and oligosaccharides **102–109** (Figure 7) have been synthesized [108–112]. However, the preparation of αManNAc anomeric phosphodiesters remains a challenge, as these compounds comprise a highly labile linkage between C-1 and O-1 of αManNAc, which is, in addition, destabilized by the presence of the NHAc group at C-2 of αManNAc. As a result, researchers have focused significant efforts on the design of hydrolytically stable phosphono-mimetics (compounds **111–115**) [113–115] and carba-mimetics (compounds **116–127**) [116–118]. However, new challenges are emerging, which are associated with the establishment of C-C bonds and the stereoselective formation of new chiral centers. The majority of publications that describe the preparation of oligomeric antigens related to *MenA* phosphoglycans concern nonacetylated molecules, whereas bacterial *MenA* phosphoglycans comprise the acetyl groups at O-3/O-4 (Figure 6), which are important for the antigenic properties of the polymer [105,106]. Additionally, a number of synthetic *MenA*-related antigens were reported (compounds **107–109** and **127**), which are totally or partially acetylated at O-3 of αManNAc, with a view to bringing their antigenic properties closer to those of the bacterial antigen. Synthetic fragments of *MenA* phosphoglycans with 3,4-di-O-acetylated αGManNAc residues are not considered in this review.

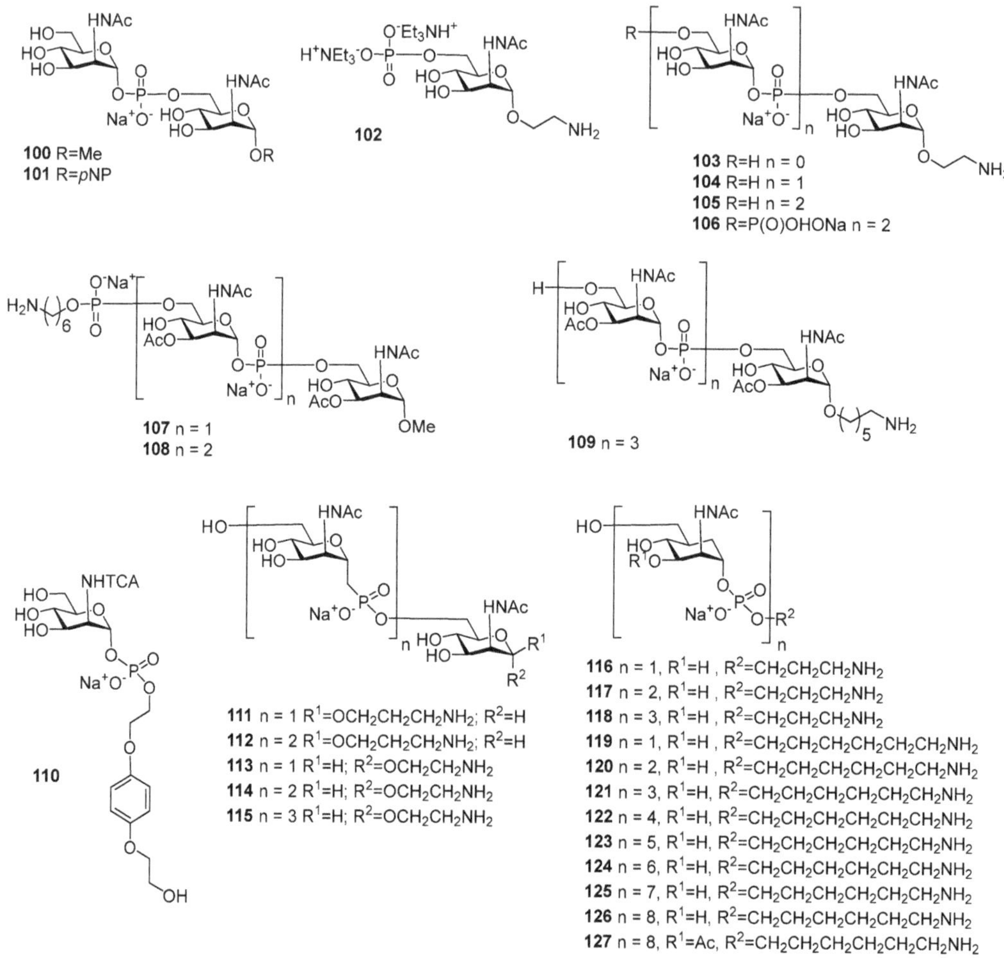


**Figure 7.** Synthetic oligosaccharides structurally related to *MenA* phosphoglycans.

As the capsular MenA glycan is an αManNAc(1→(-$PO_3$)→6) polymer, the H-phosphonate method is applicable, provided that 1-OH has the axial orientation in the selectively protected H-phosphonate ManNAc or 2-deoxy-2-azido mannose precursor, as described above for oligomers related to *Hic* and *Hif* phosphoglycans (Scheme 9). A number of convenient and efficient procedures have been developed for the preparation of this type of compound. The common protocol suggests phosphitylation with tri(1-imidazolyl) phosphine (Figure 8, compound **128**), which is obtained in situ by the interaction of $PCl_3$ and imidazole in the presence of $Et_3N$ and salicylchlorophosphite **129** (2-chloro-4H-1,3,2-benzodioxaphosphin-4-one; Figure 8). Alternatively, within a phosphoramidite protocol, chloroanhydrides **130** and **131** (Figure 8) were especially designed for automated synthesis to produce stable phosphoramidites as synthons of α-mannosamine phosphodiesters [108].

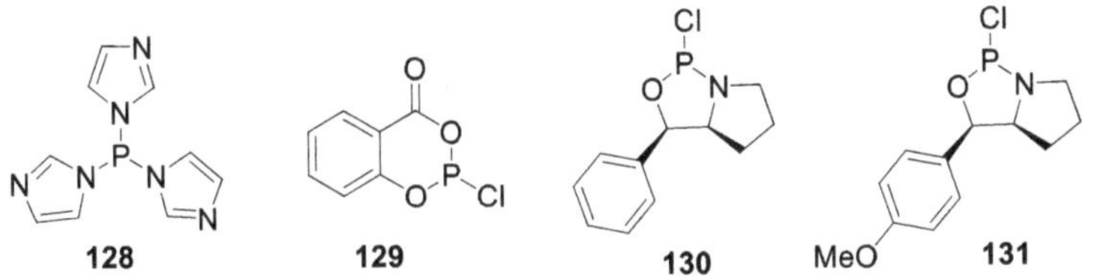


**Figure 8.** Reagents used for synthesis of H-phosphonates.

For the $PCl_3$/imidazole protocol, the key prerequisite for the preparation of ManNAc or 2-deoxy-2-azido mannose α-H-phosphonates is the α-configuration of the 1-OH group in the corresponding selectively protected hemiacetal. This feature imposes certain restrictions on the use of the H-phosphonate protocol. Hemiacetal **132** (Table 1) with a 5:1 ratio of α/β anomers was quantitatively converted into the 5:1 mixture of α- and β-isomeric H-phosphonate **133** upon the action of $PCl_3$ and imidazole in acetonitrile at 0 °C, followed by hydrolysis with an aqueous solution of $Et_3NH{\cdot}HCO_3$ at 20 °C (Table 1, entry 1) [107]. In a similar way, phosphitylation of the mixture of hemiacetal **134** with a 3.5:1 ratio of α/β anomers in analogous conditions proceeded with the retention of the configuration of the anomeric center and resulted in the formation of the mixture of H-phosphonate **135**, with the α:β ratio 3.5:1 and an 88% yield (Table 1, entry 2) [110]. In similar conditions, the interaction of the α-anomer of hemiacetal **136** (Table 1, entry 3) [111] afforded α-H-phosphonate **137** with a 97% yield, and hemiacetal **138** with a dibenzyl phosphate group at C-6 was converted into α-H-phosphonate **139** (Table 1, entry 4) [111]. The phosphitylation of hemiacetal **140** (Table 1) with an α-configuration of the anomeric center upon the action of $(PhO)_2P(O)H$ in Py, followed by hydrolysis, resulted in a mixture of α- and β-isomer **141** in a ratio of 93:3 (Table 1, entry 5) [112]. Similarly, the phosphitylation of α-hemiacetal **142** in these conditions formed α-H-phosphonate **143** with a 78% yield (Table 1, entry 6) [109]. Anomeric mixtures of hemiacetals, which contain considerable quantities of β-isomers, can be efficiently converted into α-H-phosphonates with chlorophosphite **129** (Figure 8) [110,119]. For example, the phosphitylation of the anomeric mixture **144**, with a ratio of α- and β-anomers of 4.5:1, was transformed into α-H-phosphonate **145** by the action of chlorophosphite **129** in a mixture of dioxane and triethylamine with an 88% yield (Table 1, entry 8) [110]. Similarly, in these conditions, hemiacetals **142** and **146** were converted into the corresponding α-H-phosphonates **143** and **147** at 91 and 90% yields, respectively (Table 1, entries 7 and 9) [110,119]. Unlike H-phosphonates, phosphoramidites are rarely used as synthetic blocks for the establishment of phosphodiester bonds between α-ManNAc residues because of their utmost lability. Two successful examples of ManNAc phosphoramidites are compounds **149** and **150**, which are stabilized by rigid oxazaphospholidine substituents. These compounds were obtained from the anomeric mixture of hemiacetal **148** by the action of tricyclic chlorophosphoroamidites **130** and **131** at yields of 36 and 39% (Table 1, entries 10 and 11), respectively [108]. The generation of α-H-phosphonates and the anomerization of the α/β mixtures of hemiacetals or the stereoselective cleavage of the unwanted β-isomer in the presence of $H_3PO_3$ [107] or AgOTf [110] were ineffective.

**Table 1.** Synthesis of H-phosphonates and phosphoramidites from selectively protected 2-deoxy-2-azido-D-mannose and ManNAc hemiacetals.

| Entry | Starting Hemiacetal (Even Numbers) and Target H-Phosphonate (Odd Numbers) | α-:β Ratio of Starting Hemiacetal | Reagents and Conditions | α-:β Ratio and Yield of the Target *H*-Phosphonate | References |
|---|---|---|---|---|---|
| 1 | 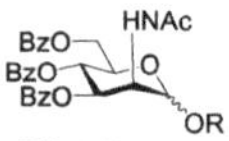 **132** R=H **133** R=P(H)(O)O⁻Et$_3$NH$^+$ | 5:1 | a, b * | 5:1, 100% | [107] |
| 2 | 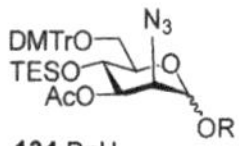 **134** R=H **135** R=P(H)(O)O⁻Et$_3$NH$^+$ | 3.5:1 | a, b * | 3.5:1, 88% | [110] |
| 3 | 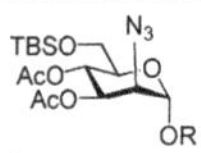 **136** R=H **137** R=P(H)(O)O⁻Et$_3$NH$^+$ | Only α | a, b * | Only α, 97% | [111] |
| 4 | 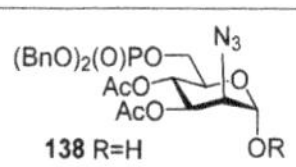 **138** R=H **139** R=P(H)(O)O⁻Et$_3$NH$^+$ | Only α | a, b * | Only α, the yield is not available | [111] |
| 5 | 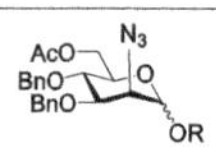 **140** R=H **141** R=P(H)(O)O⁻Et$_3$NH$^+$ | Only α | c, d * | 93:3, 84% | [112] |
| 6 | 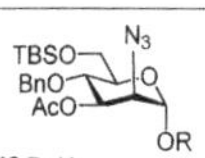 **142** R=H **143** R=P(H)(O)O⁻Et$_3$NH$^+$ | Only α | c, d * | Only α, 78% | [109] |
| 7 | 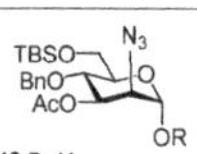 **142** R=H **143** R=P(H)(O)O⁻Et$_3$NH$^+$ | Only α | e, c * | Only α, 88–91% | [110,119] |
| 8 | 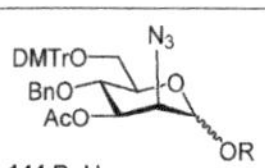 **144** R=H **145** R=P(H)(O)O⁻Et$_3$NH$^+$ | 4.5:1 | e, c * | Only α, 88% | [110] |
| 9 | 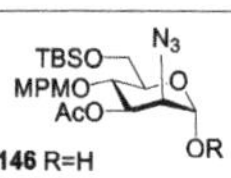 **146** R=H **147** R=P(H)(O)O⁻Et$_3$NH$^+$ | Only α | e, c * | Only α, 90% | [119] |
| 10 | 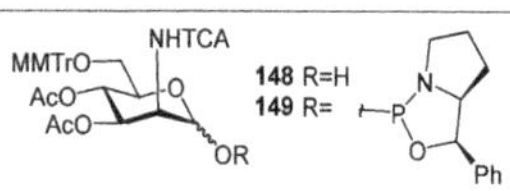 **148** R=H **149** R= | 84:16 | f * | Only α, 39% | [108] |
| 11 | 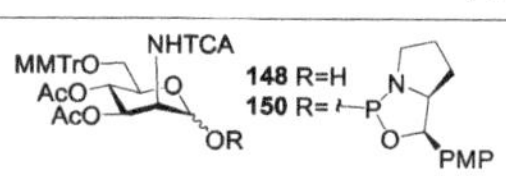 **148** R=H **150** R= | 84:16 | g * | Only α, 36% | [108] |

* Reagents and conditions: (**a**) PCl$_3$/imidazole, Et$_3$N, MeCN, 0 °C; (**b**) 0.5 M aq. Et$_3$NH·HCO$_3$, pH 7–8; (**c**) 7 eq. (PhO)$_2$P(O)H, Py, 84%; (**d**) 50% (vol.) Et$_3$N aq. solution; (**e**) **129** (Figure 8), Et$_3$N, dioxane, 0 °C; (**f**) **130** (Figure 8), 3 eq, Et$_3$N, 7 eq, THF, −78 °C→r.t., 3 h; (**g**) **131** (Figure 8), 3 eq, Et$_3$N, 7 eq, THF, −78 °C→r.t., 3 h.

The majority of syntheses of the spacer-armed oligomers related to *MenA* phosphoglycans (Figure 7) were performed using a straightforward and efficient protocol suggested by Van Boom [120], which constitutes the condensation of H-phosphonates with alcohols, with the subsequent oxidation of phosphites into phosphodiesters. In the first step, the dissolved alcohol is activated by the addition of sterically hindered chloroanhydride, e.g., PivCl, followed by the addition of H-phosphonate in Py and stirring for 5–30 min. In the second step, the intermediate products are subjected to oxidation with a 0.5 M solution of $I_2$ in a Py:$H_2O$ mixture within the temperature interval from −40 °C to 0 °C for 30 min. The first syntheses of the non-acetylated fragments of *MenA* phosphoglycans as methyl glycoside **100** and nitrophenyl ether **101** were accomplished by the Shibaev group [107] using the H-phosphonate protocol (Figure 7). Then, monomers **102** and **103**, dimer **104**, and trimer **105** (Figure 7), related to *MenA* phosphoglycans and equipped with a spacer carrying a primary amino group for conjugation with protein carriers, were synthesized by Pozsgay et al. [112]. The condensation of H-phosphonate **141** with alcohol **151**, with subsequent oxidation, resulted in the formation of phosphodiester **152** with a yield of 95% over two steps (Scheme 10). The successive reduction of an azido group into the amino group and N-acetylation yielded mannosaminyl phosphodiester **153**, which was converted into alcohol **154** by chemoselective 6-O′-deacetylation. The condensation of H-phosphonate **141** and alcohol **154** was less effective (a yield of 74% over two steps) in connection with the destruction of the phosphodiester bond in the reaction conditions. The authors noted that attempts at further elongation of the chain were unsuccessful. Phosphoester **102** and phosphodiesters **103**, **104**, and **105** were converted into conjugates with HSA. The antigenic properties of the conjugates were confirmed by a double immunodiffusion assay with the *MenA* phosphoglycan as a positive reference and chemically modified HSA as a negative reference, and anti-*MenA* horse serum [112].

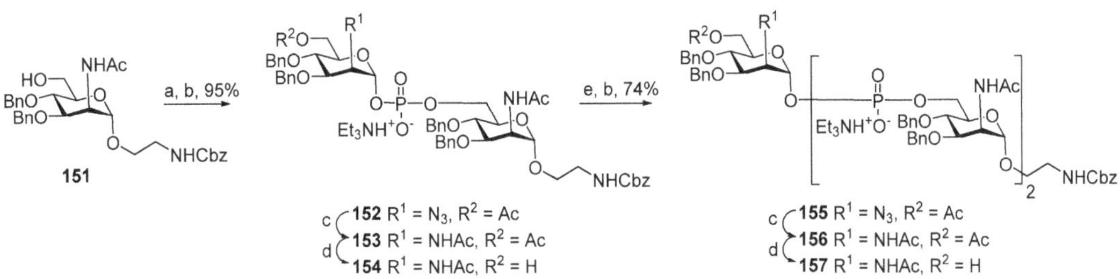


**Scheme 10.** Key building blocks for the synthesis of oligomers **103**, **104**, and **105** related to *MenA* phosphoglycans [112]. Reagents and conditions: (**a**) **141** (0.8 eq.), PivCl (2.2 eq.), 23 °C, 30 min. Py; (**b**) $I_2$ (2.2 eq.), Py:$H_2O$ 95:1, 0 °C, 30 min, 95% yield over 2 steps; (**c**) $NiCl_2{\cdot}6H_2O$ (4.8 eq.), $NaBH_4$ (8.0 eq.), methanol, 0 °C → 10 °C, 30 min; (**d**) NaOMe/MeOH; (**e**) **141** (1 eq.), PivCl (1.5 eq.), Py, 23 °C, 5 min.

In 2005, the non-acetylated spacer-armed *MenA* phosphoglycan-related oligomer **105** (Figure 7), which comprises three ManNAc residues connected by phosphodiester bridges and trimer **106** (Figure 7), composed of three non-acetylated *MenA* repeating units, was synthesized by the Oscarson group [111]. In contrast to the abovementioned synthetic strategy designed by the Pozsgay group, which suggested the reduction of the azido group and N-acetylation after each chain elongation step (Scheme 10), 2-azido glycoside **158** was used as a starting unit for further chain elongation, and 2-azido H-phosphonate **137** (Table 1, entry 3) was applied as a repeating unit synthon without intermediate $N_3$→NHAc transformation (Scheme 11). The condensation of H-phosphonate **137** and acceptor **158** with a free hydroxyl group at C-6, followed by oxidation, resulted in the efficient formation of phosphodiester **159** (yield: 96%), which was then 6′-O-desilylated to create a new active site in acceptor **160**. However, the attachment of the next monomer unit **137** to phosphodiester **160** was less effective and, after oxidation, formed the corresponding phosphodiester **161** with a 62% yield. With a view to synthesizing trimer **106** with a phosphoester substituent at C-6, acceptor **160** was reacted with phosphodiester **139** (Table 1, entry 4) and a dibenzylphosphate group at C-6 to obtain the trimeric precursor **162** with a 59% yield. A reduction in all azido groups, total N-acetylation, and deprotection in compounds **161** and **162** afforded the spacer-armed conjugation-ready antigens **105** and **106** [111].

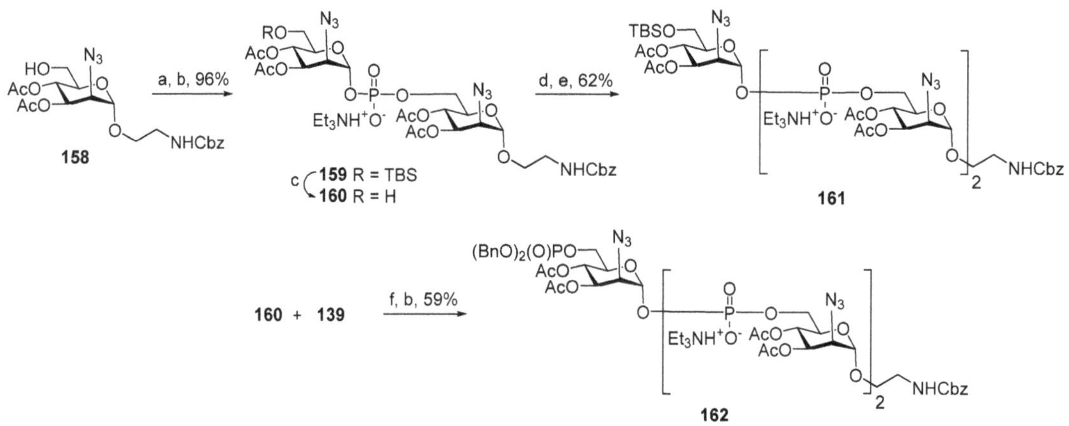

**Scheme 11.** Key building blocks for synthesis of spacer-armed antigens **105** and **106** related to MenA phosphoglycans [111]. Reagents and conditions: (**a**) **137** (1.3 eq.), PivCl (3.8 eq.), Py, r.t., 60 min; (**b**) $I_2$ (1.2 eq.), Py:$H_2O$ 49:1, −40 °C → 0 °C, 30 min, 96% yield over 2 steps; (**c**) $Et_3N{\cdot}3HF$ (5 eq.), THF, 91% yield; (**d**) **137** (1.3 eq.), PivCl (2.5 eq.), r.t., 120 min. Py; (**e**) $I_2$ (1.2 eq.), Py:$H_2O$ 49:1, −40 °C → −10 °C., 62% yield over 2 steps; (**f**) PivCl, Py.

3-O-Acetylated dimer **107** and trimer **108** related to *MenA* capsular phosphoglycans were synthesized as methyl glycosides (Figure 7) [110]. The phosphitylation of acceptors **163**, **164**, and **165**, which comprised one, two, or three 2-deoxy-2-azido mannose residues with H-phosphonate **145**, followed by oxidation and detritylation, afforded phosphodiesters **164**, **165**, and **166** with 88%, 74%, and 68% yields, respectively. The alcohols obtained were subjected to phosphitylation with H-phosphonate **167** and subsequent oxidation (Scheme 12). The connection of H-phosphonate **167** to the 6-OH of the terminal monosaccharide was effective only for the shorter alcohols **164** and **165** (88% over two steps) and resulted in the preparation of protected di- and trimers **168** and **169** (Scheme 12). The transformation of azido groups into amino groups and the total N-acetylation and debenzylation of compounds **168** and **169** formed the spacer-armed dimer **107** and trimer **108**, which carry acetyl groups at O-3 of each ManNAc residue, as they do in bacterial *MenA* phosphoglycans. In the homologous series **164**–**166**, attempts to phosphitylate the longest alcohol **166** with H-phosphonate **145** in order to obtain a tetramer or with the pre-spacer H-phosphonate **167** were not successful, in contrast to compounds **164** and **165**.

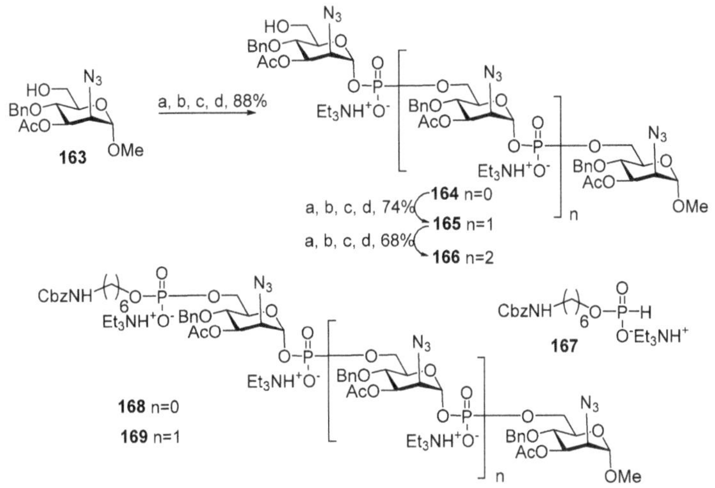


**Scheme 12.** Key building blocks for synthesis of spacer-armed oligomers **107** and **108** [110] related to *MenA* capsular phosphoglycans. Reagents and conditions: (**a**) **145** (1.2 eq.), PivCl (1.8 eq.), Py, 20 °C, 30 min; (**b**) $Et_3N$ (4.8 eq.), −40 °C, Py:$H_2O$ 95:5; (**c**) −40 °C, $I_2$ (2.4 eq.), 30 min; (**d**) 10 °C, 1% TFA in dichloromethane.

In general, the examples of the synthesis of oligomers related to *MenA* capsular phosphoglycans discussed above are in line with the tendency outlined in an earlier review [39], which discussed the decrease in the efficiency of phosphitylation with H-phosphonates with the increase in the number of already formed phosphodiester bonds and suggested the lability of phosphodiester fragments in the conditions of the H-phosphonate protocol. However, in 2017, a group of Indian researchers synthesized an aminohexyl glycoside of oligomer **109** related to capsular phosphoglycans, which contained four 3-O-acetylated ManNAc residues, using a linear, synthetic strategy (Scheme 13) [109]. In a sequence of phosphitylation steps, H-phosphonate **143** was used as a universal monomer (Table 1, entry 6). The establishment of the first phosphodiester linkage between alcohol **170** and H-phosphonate **143** afforded compound **171** (yield: 92%), which was, in turn, desilylated to form acceptor **172**. The phosphitylation of alcohol **172** with H-phosphonate **143**, followed by oxidation, resulted in the formation of compound **173** with two phosphodiester bridges (yield: 89%). The desilylation of compound **173** formed acceptor **174**, which was phosphitylated with H-phosphonate **143** and, after oxidation, compound **175** with three phosphodiester bridges was obtained with a 76% yield.

After the transformation of azido groups into NHAc groups and the desilylation, debenzylation, and deprotection of the spacer amino group, the spacer-armed ligand **109** was conjugated to TT. The antigenic properties of oligomer **109** were evaluated in a competitive ELISA experiment. Both oligomer **109** and its conjugate with TT in the concentration range 12.5–400 μg glycan/mL were found to neutralize anti-MenA rabbit antiserum and inhibit the binding of antibodies to the bacterial *MenA* phosphoglycan used as a coating antigen. In comparison to the conjugate, oligomer **109** showed lower inhibition [109]. It can be concluded that the improvement of synthetic protocols made it possible to obtain oligomeric antigens related to *MenA* capsular phosphoglycans, which contain up to four ManNAc residues. However, for the efficient application of antigens of this type in immunodiagnostic tests and vaccine production, the hydrolytic lability issue has to be addressed.

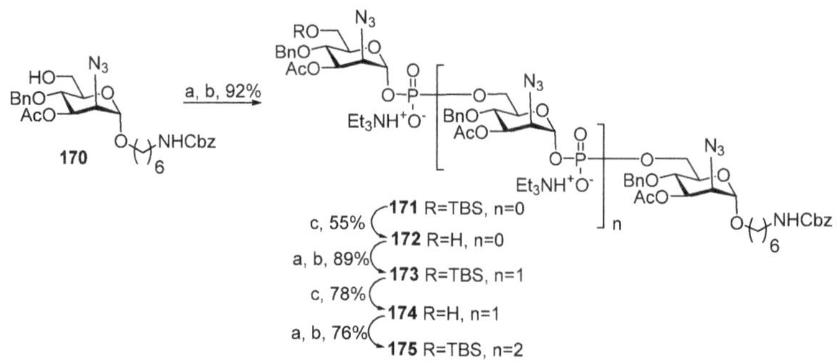

**Scheme 13.** Key building blocks for the synthesis of oligomer **109** structurally related to *MenA* phosphoglycans [109]. Reagents and conditions: (**a**) **143** (2 eq.), PivCl (3 eq.), Py, 20 °C, 90 min; (**b**) −40 °C, solution of $I_2$ (5 eq.) in Py/$H_2O$ 20:1, 150 min, −40 °C → −10 °C, 30 min; (**c**) $Et_3N{\cdot}3HF$ (5 eq.), THF.

## 5. Synthesis of Glycomimetics of MenA Capsular Phosphoglycans

Today, all anti-*MenA* conjugate vaccines include a lyophilized bacterial *MenA* component except for the fully liquid commercial vaccine preparation Menactra®. The presence of the *MenA* antigen in the dissolved form substantially shortens the shelf life for Menactra® to 18 months at 2–8 °C compared to the 4-year shelf life of MenQuadfi® and the 3-year shelf life of Menjugate® and MenAfriVac® in these conditions. With a view to preparing hydrolytically stable MenA antigens, a number of oligomeric analogs were designed, in which NHAc groups were replaced with trichloroacetamide groups as they are not likely to undergo transformation into oxazolines. Another type of mimetics is compounds with the isosteric replacement of one of the hemiacetal oxygens with a methylene group (phosphono- and carba-analogs).

A comparative conformational analysis of a hexapyranose ring in 2-deoxy-2-acetamido mannohexapyranosyl phosphate **176** and its phosphono-analog **177** and carba-analog **178** (Figure 9) in a study of conformations in combination with NMR experiments showed that for compound **176**, 4C1 was almost the only populated conformer, whereas for phosphonate **177**, the proportion of pyranose ring conformers other than 4C1 was 4%, and for the carba-analog **178**, this proportion rose to 7% [121]. The most populated 4C1 conformer for compound **178** was confirmed by quantum mechanics and molecular dynamics calculations [122]. The molecular dynamics calculations performed for a decamer of a *MenA* capsular phosphoglycan repeating unit and its carba-analog showed that, despite a number of conformational and dynamic variations, the carba-mimetics of the fragments of *MenA* capsular phosphoglycans can be considered candidate antigens for the construction of anti-*MenA* vaccine preparations [123].

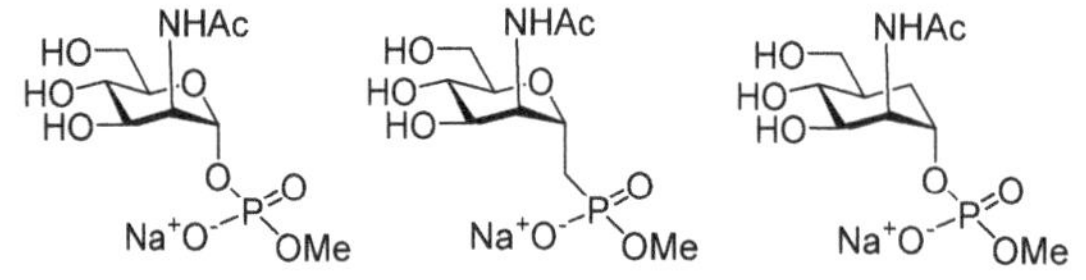

**Figure 9.** Sodium salt phosphodiesters of α-ManNAc (**176**) and its phosphono-analog (**177**) and carba-analog (**178**) as model compounds for comparative conformational analysis.

As mentioned previously, the axial NHAc group at C-2 of the ManNAc residue contributes largely to the lability of *MenA* capsular phosphoglycans and their fragments via neighboring group participation. One of the possible ways to circumvent this obstacle is the replacement [55] of the acetamide group at C-2 with the trichloroacetamide group, which is not likely to form oxazolines. To study the possibility of the preparation of the N-trichloroacetamide mimetics of ManNHAc, oxazaphospholidines **149** and **150** (Table 1) were obtained in a solid-phase synthesis. Bis(2-hydroxyethyl) hydroquinone **179** (Scheme 14) [108] was immobilized on a glass support and phosphitylated with phosphoramidites **149** and **150**. The phosphitylation of hydroquinone **179** with phosphoramidite **149** in the presence of 1-(cyanomethyl) pyrrolidinium trifluoromethanesulfonate (compound **180**) resulted in the formation of phosphite **181**, which carried a 2-phenylpyrrolydine residue as a result of the cycle opening. Phosphotriester **181** was subjected to mild oxidation with DCSO (compound **182**) and desilylated with TFA/TES to produce phosphotriester 183, with a free 6-OH group for the following elongation of the oligomeric chain. Finally, the cleavage of the 2-phenylpyrrolydine ether was fulfilled in the presence of a base, followed by deacetylation and removal of the solid support by the action of MeONa/MeOH.

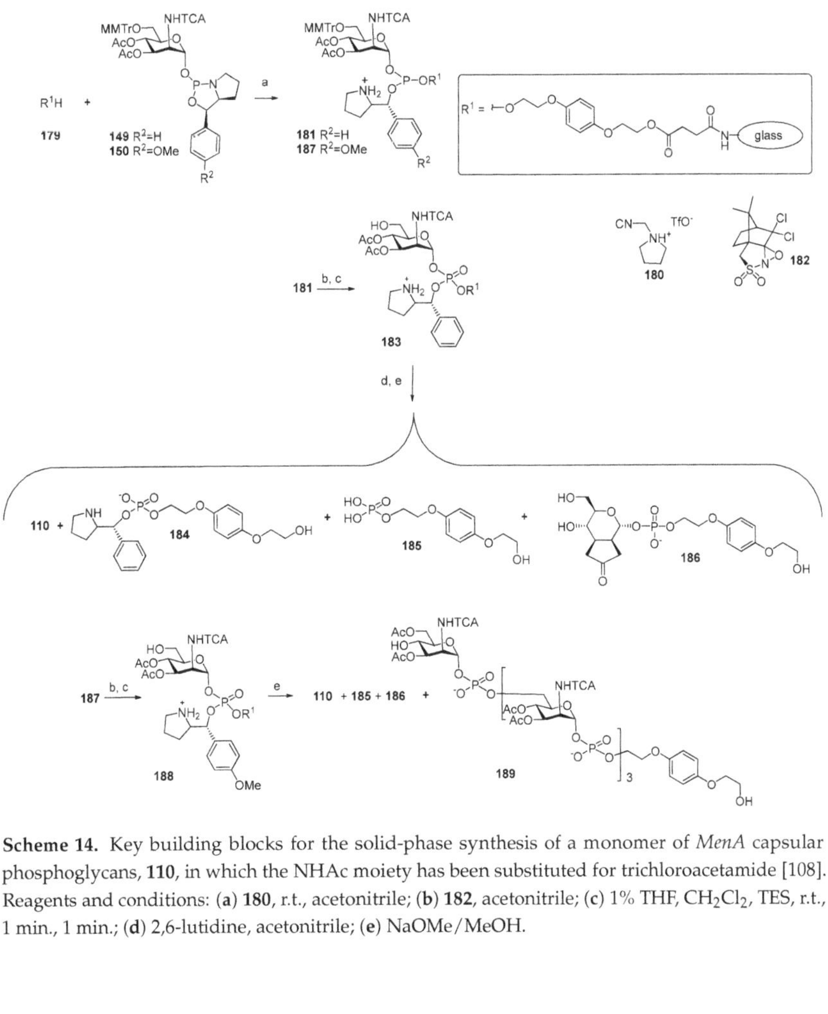

**Scheme 14.** Key building blocks for the solid-phase synthesis of a monomer of *MenA* capsular phosphoglycans, **110**, in which the NHAc moiety has been substituted for trichloroacetamide [108]. Reagents and conditions: (**a**) **180**, r.t., acetonitrile; (**b**) **182**, acetonitrile; (**c**) 1% THF, $CH_2Cl_2$, TES, r.t., 1 min., 1 min.; (**d**) 2,6-lutidine, acetonitrile; (**e**) NaOMe/MeOH.

However, researchers faced considerable difficulties at the step of the cleavage of the 2-phenylpyrrolydine ether. After the support was removed, analysis of the products showed that the use of DBU for the cleavage of the ether did not provide target phosphodiester **110**, and the identified products **184–186** did not contain a ManNAc residue. When the weaker bases of $Et_3N$ or 2,6-lutidine were used as basic catalysts, the target product **110** was obtained with a moderate yield (17%) and low conversion.

In order to replace the step of the basic hydrolysis of the O(P)-protective group for acidic hydrolysis, phosphoramidite **150** with a p-MeO-Ph residue in the place of the Ph residue, as in compound **149**, was used. Phosphoramidite **150** was attached to a solid support, and phosphite **187** was oxidized with DCSO (compound **182**) to yield phosphotriester **188**, followed by the simultaneous acid hydrolysis of the trityl group and removal of the 2-(p-methoxybenzyl) pyrrolidine moiety. However, after deacetylation and removal of the solid support with 50 mM NaOMe/MeOH, the unwanted products **186** and **187** were detected. Without taking into account inefficient deprotection, this method provided the preparation of tetramer **189**, which was O-acetylated and N-trichloroacetylated. The authors noted that the developed method in the current state is not suitable for the synthesis of oligomers [108].

In 2005, the preparation of the first isosteric hydrolysis-resistant phosphono-analog **111** structurally related to *MenA* phosphoglycans was published by Lay et al. [115]. The interaction of iodide **190**, in which the iodomethylene group is arranged axially, with trimethylphosphite (Scheme 15) afforded phosphonodiester **191**, which was further converted to ester **192** by the action of triethylamine and thiophenol.

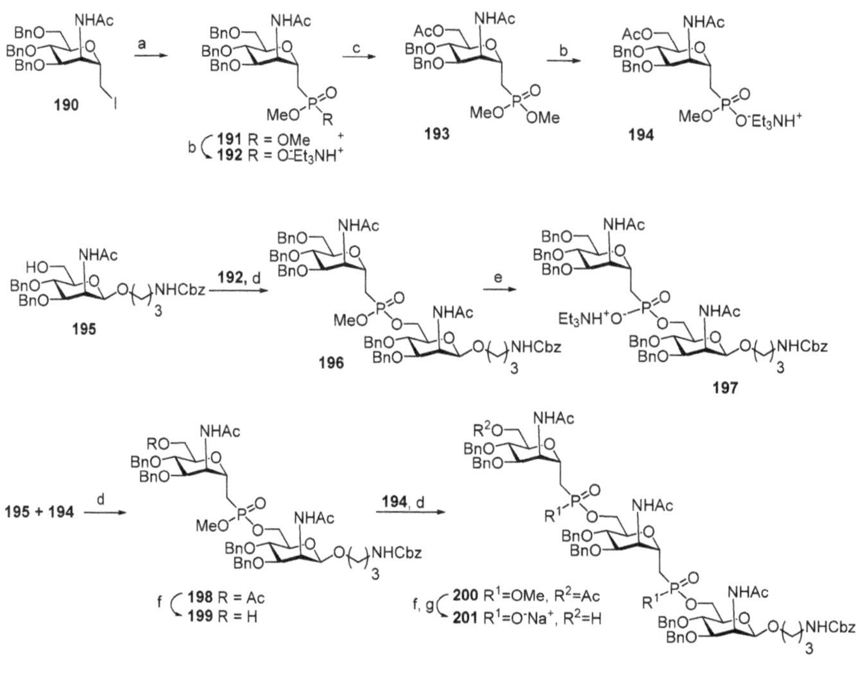

**Scheme 15.** Key blocks for the synthesis of spacer-armed phosphono-analogs of *MenA* phosphoglycans, **111** and **112** (Figure 6) [113,115]. Reagents and conditions: (**a**) $P(OMe)_3$, 100 °C, vacuum; (**b**) $Et_3N$, PhSH, THF, r.t., 62%; (**c**) $ZnCl_2$, $Ac_2O$:AcOH 2:1, r.t., 16 h, 92%; (**d**) $Ph_3P$, DIAD, THF, 0 °C, 24 h, yield 97% for **197**, 90% for **198**, 83% for 200; (**e**) $Et_3N$, PhSH, toluene, 110 °C, 36 h, Amberlite IR120 ($Na^+$) 95%; (**f**) NaOMe/MeOH; (**g**) DBU, PhSH, THF, r.t., 8 h, Amberlite IR120 ($Na^+$), 95%.

Similarly, 6-O-acetylated phosphonodiester **193** was converted to the corresponding ester **194** by partial hydrolysis. Mitsunobu reaction conditions were used for the condensation of ester **192** with the selectively protected ManNAc **195**, which formed phosphonodiester **196** (yield: 97%) [113]. Partial hydrolysis of phosphonodiester **196** into phosphonoester **197** was also efficient. 6′-O-Acetylated phosphonodiester **198** was synthesized by the condensation of **195** with phosphonoester **194**, with a 90% yield. By deacetylation, phosphonodiester 198 was converted into alcohol **199** and condensed with phosphonoester **194** to form compound **200** with two phosphonodiester bridges and an 83% yield. Chemoselective hydrolysis of the methyl phosphonate moieties in compound **200** formed phosphonoester **201**. Total deprotection of phosphonoester **197** with two ManNAc residues and **201** with three ManNAc residues resulted in phosphono-analogs **111** and **112** (Figure 6) [113].

A similar strategy in combination with Mitsunobu reaction modification, where $Ph_3P$ is replaced with tris(4-chlorophenyl) phosphine, was used for the synthesis of the series of phosphono-analogs **113–115** of *MenA* phosphoglycans (Scheme 16) [114]. The 6″-O-deacetylation of phosphonodiester **202** resulted in alcohol **203**, which was then reacted with universal monosaccharide block **194** in modified Mitsunobu conditions to obtain the phosphono-analog **204** of *MenA* phosphoglycans with three phosphonodiester-bridged fragments (yield: 87%). Partial hydrolysis of diester moieties afforded compound **205** (yield: 75%), and after total deprotection, phosphono-analog **115** was obtained with three pseudo-ManNAc residues.

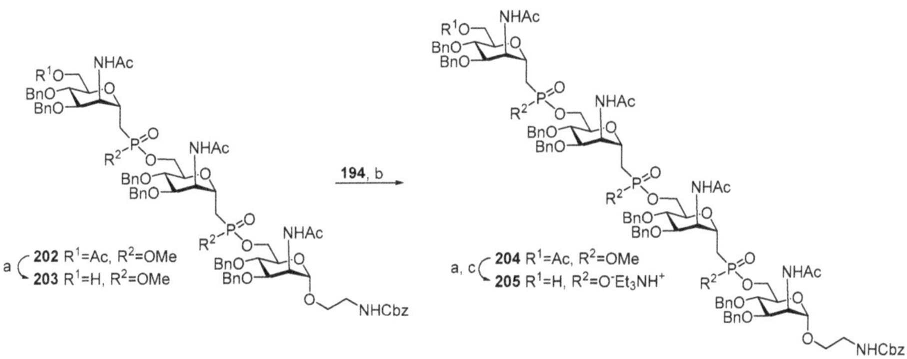


**Scheme 16.** Key blocks for the synthesis of phospho-analog **115** (Figure 6) related to *MenA* phosphoglycans [114]. Reagents and conditions: (**a**) 1 M KOH/MeOH, 84%; (**b**) (p-$ClC_6H_4)_3P$, DIAD, THF, 0 °C, 30 min, 87%; (**c**) DBU, PhSH, acetonitrile, r.t., 75%.

The antigenic properties of ligands **111** and **112** [113] were studied in competitive ELISA experiments with anti-*MenA* human antisera. Native *MenA* phosphoglycans were used as a coating antigen and positive control, and *MenY* phosphoglycans were used as a negative control. For both ligands, the EC50 was about $10^{-3}$ mg/mL, which is three orders higher than the EC50 of $6.6 \times 10^{-6}$ mg/mL for *MenA* phosphoglycans.

Aminopropyl glycosides **111** and **112** and 3-aminopropyl β-D-ManNAc were transformed into conjugates with HSA using the squarate procedure [124] (Figure 10). One series consisted of conjugates **206–208** (Figure 10) with the maximum saccharide/protein molar ratio, and another series of conjugates was composed of compounds **209–211** (Figure 10) with the saccharide/protein molar ratio being half of the value achieved for conjugates **206–208**. Competitive ELISA experiments with the mouse polyclonal anti-*MenA* antisera and *MenA* capsular phosphoglycans as a coating antigen showed that at an inhibitor concentration of 1 mg/mL, inhibition using the fully loaded conjugate **206** was 55%. For conjugates **207** and **208**, it reached 65%, whereas *MenA* capsular phosphoglycans provided 100% inhibition. Inhibition with the half-loaded conjugates **209–211** was 5–15% lower than for the fully loaded conjugates with the same antigen type [124]. Conjugates **206–211** were used for the immunization of mice at a dose of 2 μg/mouse and efficiently evoked IgG antibodies, which is a reliable marker of the induction of thymus-dependent immune responses. Quantitative elucidation of the level of anti-*MenA* IgG antibodies developed against conjugates **206–211** showed that immunization with the half-loaded conjugates **209–211** was more efficient compared to the fully loaded conjugates **206–208**. It is important to note that the level of induced anti-*MenA* antibodies was similar for the half-loaded conjugates **209–211**, regardless of the length of the pseudo-oligosaccharide antigen. The authors concluded that this result indicated the antibodies' recognition of the ManNAc epitope [124].

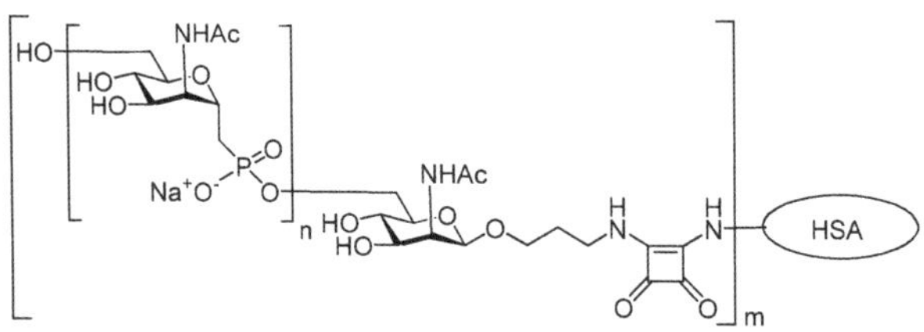


**Figure 10.** Conjugates of amino-spacered C-phosphono analogs of *MenA* capsular phosphoglycans [124].

In contrast to C-phosphono mimetics, in which the methylene group stands in the place of anomeric oxygen, in carba-analogs, the decrease in electrophilicity of the carbonyl carbon, along with the increase in the stability of the phosphodiester linkage, is achieved by the replacement of the ring hemiacetal oxygen of ManNAc with a methylene group [125].

Synthesis of the series of carba-analogs **116–118** (Figure 7) of aminopropyl glycosides of a monomer, a dimer, and a trimer of *MenA* capsular phosphoglycans, and the advanced series of carba-analogs **119–126**, from monomers to octamers, as aminohexyl glycosides, was performed by the Lay group [116–118]. Carba-analogs **116–118** [118] were obtained by the sequential elongation of a pseudo-oligosaccharide chain, starting from the spacer-equipped monomer (Scheme 17). For the preparation of monomer **116**, the universal orthogonally protected precursor **212** was transformed into alcohol **213**, which was phosphitylated with H-phosphonate **214** and oxidized to obtain the spacer-armed phosphodiester **215** with an 81% yield.

For the preparation of oligomers **117** and **118**, the universal precursor **212** was desilylated and converted into H-phosphonate **216**, which was used as a universal monomer block for chain elongation. The interaction of H-phosphonate **216** with alcohol **213** and the subsequent oxidation formed phosphodiester **217** (yield: 82%), which was deacetylated to obtain alcohol **218**. The condensation of alcohol **218** with H-phosphonate **216** resulted in pseudo-trisaccharide **219** (yield: 81%), which was deacetylated to produce alcohol **220**. Finally, pre-spacer **214** was introduced into alcohols **218** and **220** to obtain the spacer-armed, protected dimers **221** and **222** with yields of 85% and 57%, respectively. The high efficiency of the phosphitylation of alcohol **218**, which already includes a phosphodiester bridge with H-phosphonates **214** and **216**, evidences the increase in stability of the phosphodiester linkage in protected carba-analogs compared to phosphodiester-linked oligosaccharides. However, the phosphitylation of **220**, which already comprised two phosphodiester moieties, was less efficient, indicating the limitations of the application of the H-phosphonate procedure to the linear synthesis of longer oligomers. The target carba-analogs **116–118** of monomers, dimers, and trimers related to *MenA* phosphoglycans were obtained after the total deprotection of compounds **215**, **221**, and **222** [118].

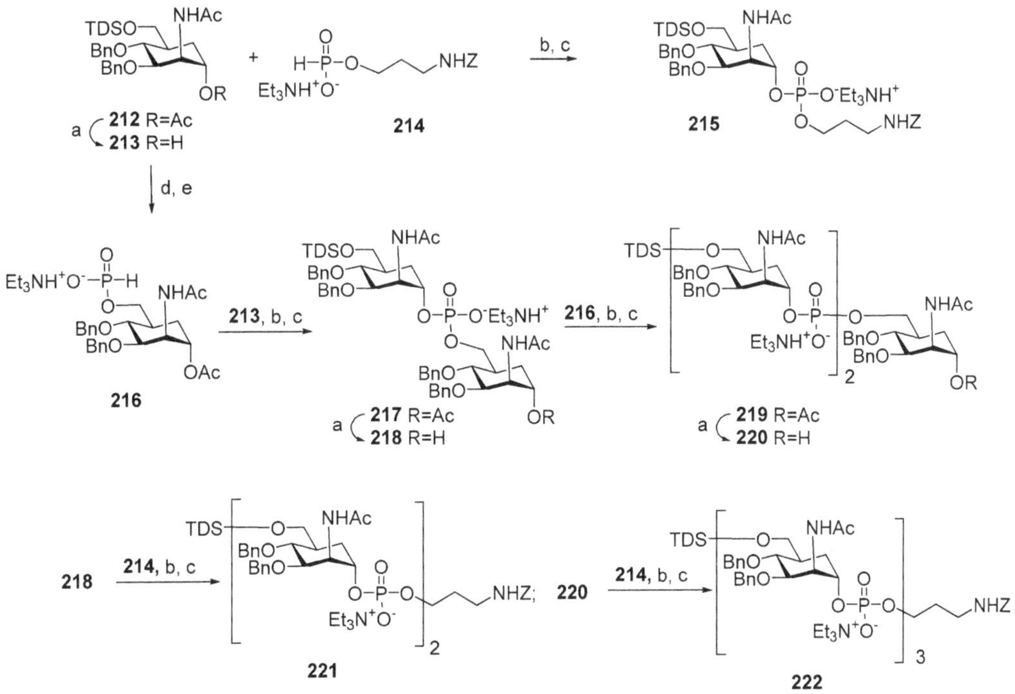

**Scheme 17.** Key blocks for preparation of spacer-armed carba-analogs of oligomers **116–118** related to *MenA* CPS [118]. Reagents and conditions: (**a**) NaOMe/MeOH, r.t., 4 h, yield of compound **213**—84%, **218**—87%, **220**—70%; (**b**) Py, PivCl, r.t., 45 min; (**c**) $I_2$, Py -$H_2O$ 19:1, r.t., 15 min, yield for compound **215**—81%, **217**—82%, **219**—81%, **221**—85%, **222**—57%; (**d**) $Bu_4NF$, THF, r.t., 3 h; (**e**)–**129**, $CH_3CN$-Py 3:1, r.t., 45 min, yield 82%.

For the synthesis of the advanced series of carba-analogs **119–126**, from monomers to octamers, the Lay group used the strategy of a step-by-step chain extension using the spacer-equipped monomer as a starting compound and phosphitylation with a selectively protected monomeric phosphoramidite as the key step (Scheme 18) [116]. Alcohol **223** was converted into the universal phosphoramidite block **224** under the action of chlorophosphoroamidite **24**. The interaction of alcohol **223** with the phosphoramidite pre-spacer **225** in the presence of DCI, followed by oxidation with DCSO and detritylation, yielded phosphotriester **226** with a free hydroxyl group for further phosphitylation, which was used as a starting compound for chain elongation. The sequential execution of phosphitylation with the universal monomer block **224**, oxidation, and detritylation afforded the protected compounds **227–233** with excellent yields. After total deblocking, the spacer-armed carba-analogs **119–126**, structurally related to MenA capsular phosphoglycans, were obtained. For better resemblance of the octamer antigen to the natural structure, octamer **126** was N-Boc-protected, subjected to random monoacetylation, and N-deblocked, and the mixture of oligoacetates **234** was obtained, which contained a certain amount of oligomer **127** related to MenA capsular phosphoglycans [116].

The stability of compound **126** was studied using an accelerated stability test and fragments of natural and deacetylated *MenA* phosphoglycans, with avDP15 as a reference compound. During four weeks of keeping these samples in buffered 5 mM sodium acetate at pH 7 and 37 °C, a drastic decrease in chain length for the deacetylated bacterial oligosaccharide was observed, and the natural sample with acetyl groups was subjected to partial depolymerization, and for compound **126**, no trace of decomposition was detected [116].

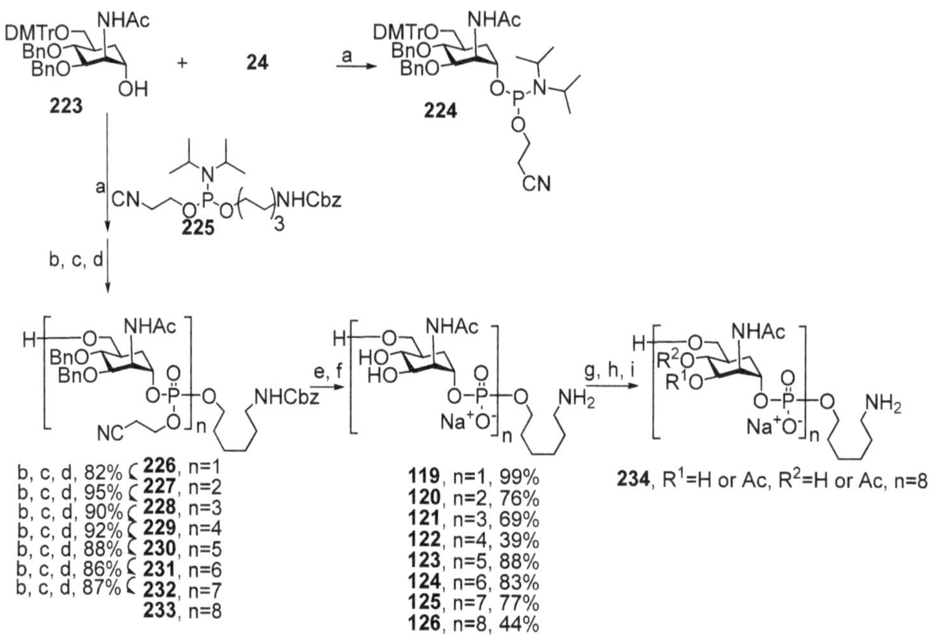


**Scheme 18.** Key blocks for synthesis of spacer-armed carba-analogs **119–126** related to *MenA* capsular phosphoglycans [116]. Reagents and conditions: (**a**) DIPEA, $CH_2Cl_2$, r.t., 94%; (**b**) DCI, MeCN; (**c**) DCSO, MeCN; (**d**) TCA, $CH_2Cl_2$, $H_2O$, 94% over 3 steps; (**e**) $NH_4OH$, $H_2O$, dioxane; (**f**) $H_2$, Pd, $H_2O$, AcOH, 44%; (**g**) $(Boc)_2O$, $NaHCO_3$, r.t. 16 h; (**h**) $Ac_2O$/imidazole, 9 d.; (**i**) TFA, r.t., 1 h.

Competitive ELISA experiments with polyclonal immune mouse anti-*MenA* sera and *MenA* capsular phosphoglycans as a coating antigen were used for the assessment of the antigenic properties of the mono-, di-, and trimeric pseudo-oligosaccharides **116–118** [118]. *MenA* capsular phosphoglycans and their derivatives *MenA* avDP15 and *MenA* avDP3 were used as positive controls. The highest inhibition rate was found for *MenA* capsular phosphoglycans and *MenA* avDP15, with an IC50 of $5.15 \times 10^{-6}$ and $4.3 \times 10^{-3}$ mM, respectively. For the carba-dimer **117**, the IC50 was about 0.16–0.091 mM, which is less than the IC50 ($4.3 \times 10^{-2}$) found for *MenA* avDP3. For the carba-monomer **116** and carba-trimer **118**, only 30% inhibition was achieved [118]. The synthetic carba-analogs **116–118** were converted into conjugates with CRM197 (compounds **235–237**, Figure 11) and HSA (compounds **238–240**, Figure 11). Mice were immunized three times with conjugates **235–237**, **241**, and **242**, which comprised *MenA* avDP5 and *MenA* avDP15 antigens, respectively, at a dose of 2 μg of antigen per mouse. The induced antibodies were analyzed in ELISA experiments using the HSA-based conjugates **238–240** as coating antigens and *MenA* capsular phosphoglycans. The immunogenicity of conjugates **235**, **236**, and **237** was evaluated using the HSA-conjugated coating antigens **238**, **239**, and **240**, respectively. In this experiment, the immunogenicity of conjugate **238**, which comprised the monomeric antigen **116**, was somewhat lower compared to conjugates **239** and **240** with dimeric and trimeric antigens, respectively. In ELISA experiments with *MenA* capsular phosphoglycans as a coating antigen, it was found that only conjugate **240** with a trimeric antigen evoked antibodies specific to the natural phosphoglycans, and conjugates **238** and **239** did not induce anti-*MenA* immunity. However, the level of IgG antibodies induced by immunization with **240** was 2–3 orders lower than that induced by conjugates **241** and **242**, which contained fragments of natural phosphoglycans with intact acetyl substituents. These data are in line with the results of an in vitro bactericidal assay, which showed the intensity of the complement-mediated lysis of *N. meningitidis* [118].

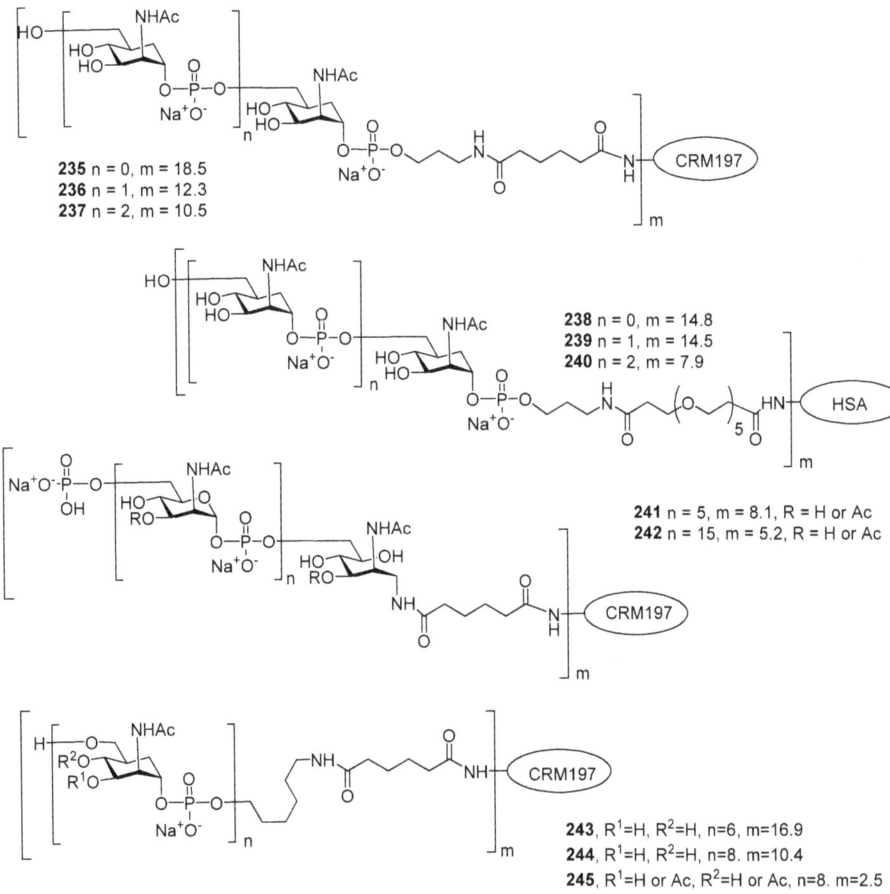


**Figure 11.** Neoglycoconjugates of mono- and oligomeric carba-analogs and partially fragmented *MenA* CPS with protein carrier CRM197 and HSA [116].

The carbocyclic hexamer **124** and the non-acetylated and acetylated octamers **126** and **234** were conjugated to CRM197 to form compounds **243**, **244**, and **245**, respectively (Figure 11). Conjugates **243–245** were used for the immunization of mice, and another group of mice was immunized with a conjugate composed of partially depolymerized

*MenA* phosphoglycans as a positive control. ELISA analysis of IgG antibodies on plates coated with *MenA* capsular phosphoglycans showed that immunization with conjugate **245**, based on the acetylated carba-octamer, efficiently induced anti-*MenA* immunity. For this conjugate, the IgG titers were comparable to the titers registered for conjugate **242** based on fragmented bacterial antigens, whereas the titers measured in animals immunized with conjugates **243** and **244**, which carried non-acetylated synthetic antigens, were two orders lower [116].

The urgency of the development of anti-*MenA* conjugate vaccines is evidenced by the considerable efforts of several research groups aimed toward antigenic and non-hydrolyzable synthetic phosphooligosaccharides. However, the antigenicity studies of phosphono- and carba-mimetics show that these oligomers do not induce anti-*MenA* immunity and, therefore, are unable to prevent MenA infection. Another obstacle in the way of the design of the semisynthetic anti-MenA vaccine is the importance of acetyl groups at the O3/O4 of ManNAc for the protective properties of the vaccines. The introduction of these substituents increases production costs.

To date, synthetic *MenA* antigens cannot compete with multiple conjugate anti-*MenA* vaccine preparations with a bacterial antigen. However, the repeating unit of *MenA* phosphoglycans is a simple monosaccharide, and it is a suitable target for production on automated synthesis platforms. We expect that the development of a stable mimetic with high antigenicity, in combination with automated synthesis technology, will help create an affordable anti-*MenA* semisynthetic conjugate vaccine.

## 6. Synthesis of Phosphooligomers Related to MenX Capsular Phosphoglycans

High anti-*MenA* vaccination coverage in the African meningitis belt in 2010–2015 resulted in a significant reduction of *MenA*-associated invasive diseases [100,126]. At the same time, outbreaks of invasive diseases caused by *MenX* were registered in Sub-Saharan Africa [127,128]. The clinical and epidemiological characteristics of diseases associated with *MenX* are similar to those of *MenA*. Over 90% of cases are registered during the dry season, the median age of patients is 9.2 years, and the mortality rate is 11.9% [129]. As a result, the development of a conjugated anti-*MenX* vaccine became an urgent issue [130].

For the elucidation of the minimal immunogenic epitope of *MenX* capsular phosphoglycans, a bactericidal mAb, *MenX*.01, was obtained, which induced bactericidal killing at a physiological concentration of 1 μg/mL in the first step. The interaction of this antibody with fragments of the natural *MenX* capsular phosphoglycans of different lengths showed that the minimal immunogenic epitope comprises 5–6 repeating units [131].

With a view to developing a conjugated anti-*MenX* glycovaccine with a fully synthetic glycan epitope, phosphodiester **246** [132] and the series of spacer-armed mono-, di-, and trimeric phosphoglycans **247–249** [129,133], were synthesized. In addition, a mixture of oligomeric phosphoglycan **250**, with an average length of 12 repeating units, was prepared using a combined chemo-enzymatic protocol, and pseudo-tetrasaccharide **251** was used as a glycoside with an aminohexyl spacer attached as an aglycon (Figure 12) [134].

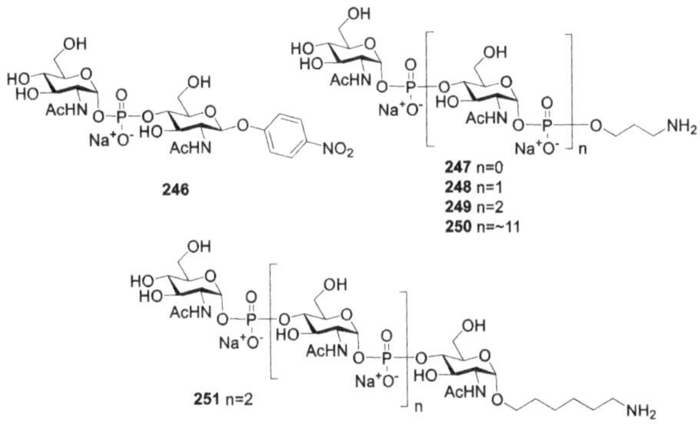

**Figure 12.** Synthetic phosphooligosaccharides **246–251** structurally related to *MenX* capsular phosphoglycans.

The preparation of the synthetic oligosaccharide fragments of *MenX* capsular phosphoglycans was most commonly performed using an H-phosphonate procedure for the establishment of the phosphodiester linkage. In *MenX* phosphoglycans, the GlcNAc residue was incorporated in the polymer chain as an α-anomer. In synthetic fragments, the anomeric configuration of GlcNAc was determined with the configuration of the H-phosphonate, which, in turn, depends on the configuration of C-1 in the starting hemiacetal.

In 1991, pseudo-disaccharide **246** was synthesized for the first time by the Shibaev group (Scheme 19) [132]. Hemiacetal **252** with the NHAc group at C-2 was transformed into α-H-phosphonate **253** with a 95% yield under the action of in situ generated tris(imidazol-1-yl) phosphine. The condensation of H-phosphonate **253** and alcohol **254**, followed by oxidation, afforded phosphodiester **255** (yield: 72%), which was deblocked to obtain pseudo-disaccharide **246** (Figure 12).

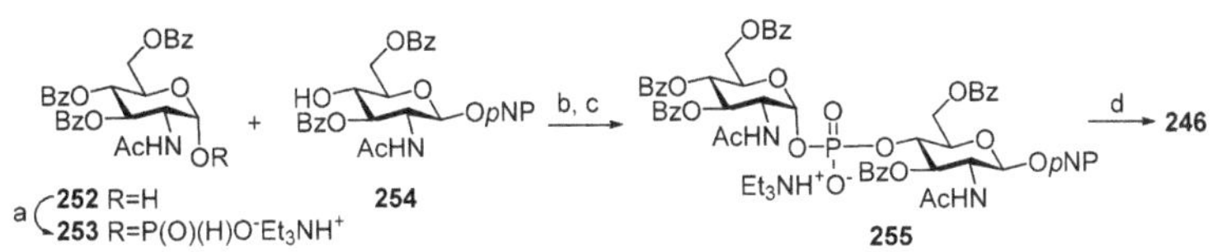

**Scheme 19.** Key blocks for the synthesis of pseudo-disaccharide fragment **246** structurally related to *MenX* capsular phosphoglycans [132]. Reagents and conditions: (**a**) $PCl_3$ (14 eq.), imidazole (4.3 eq.), $Et_3N$ (15 eq.), acetonitrile, 0 °C, 30 min, 95%; (**b**) PivCl (2.5 eq.), Py, 20 °C, 30 min. 72%; (**c**) Py:$H_2O$ 95:5; $I_2$ (2.0 eq.), 20 °C, 10 min; (**d**) MeONa/MeOH, 71%.

In 2013, the series of spacer-armed oligomers **247 –250** was synthesized by the Lay group [133]. Unsuccessfully, 2-deoxy-2 azido hemiacetals **256** and **257**, intended for the preparation of the corresponding H-phosphonates **258** and **259**, were obtained as anomeric mixtures, thus challenging the preparation of H-phosphonates **258** and **259** as α-anomers. In order to provide the α-stereoselectivity of phosphonylation, the reaction of hemiacetals **256** and **257** with chlorophosphite **129** was conducted in the presence of phosphoric acid, which is known to destroy the β-anomer [39] (Scheme 20). The reaction took more than a week and afforded α-H-phosphonates **258** and **259** with 41% and 52% yields, respectively.

The phosphitylation of benzyl N-(3-hydroxypropyl) carbamate with benzylidenated H-phosphonate 258, followed by oxidation, resulted in phosphodiester **260** (yield: 62%), and the condensation of this alcohol with 4-O-acetylated H-phosphonate **259** formed phosphodiester **262** (yield: 64%), which was deacetylated to afford acceptor **262**. The condensation of acceptor **262** with 4,6-O-benzylidenated H-phosphonate **258** and 4-O-acetylated H-phosphonate **259** was conducted with a moderate yield of phosphodiester **263** (45%) and **264** (40%). After deacetylation, pseudo-disaccharide **264** was converted into alcohol **265**, which was reacted with 4,6-O-benzylidenated H-phosphonate **258** to afford trimer **266** with a 33% yield. The authors attribute the low efficiency of the phosphodiester linkage formation in trimer **266** to the presence of the azido group at C-2 of the hexapyranose ring. Total deprotection of compounds **260**, **263**, and **266** and transformation of the azido group into NHAc produced the target spacer-armed mono-, di-, and trimeric phosphodiesters **247**, **248**, and **249**, structurally related to MenX capsular phosphooligoglycan [133].

Unlike 2-deoxy-2-azido hemiacetals **256** and **257**, hemiacetal **267** with an NHAc group at C-2 exists as an α-anomer and is readily converted into α-H-phosphonate **268** under the action of chlorophosphite **129** (Scheme 20). Further improvement was achieved when the process was conducted in a microreactor, which provided a reduction in the reaction time to 30 min. and an increase in efficiency of the formation of α-H-phosphonate 268 to 76% [129].

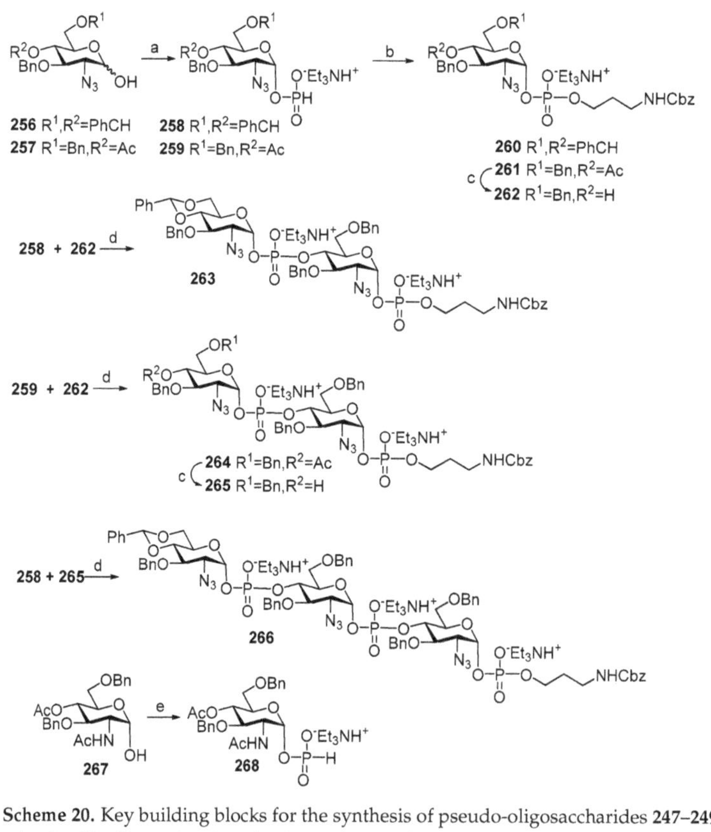

**Scheme 20.** Key building blocks for the synthesis of pseudo-oligosaccharides **247–249** structurally related to *MenX* capsular phosphoglycans [129,133]. Reagents and conditions: (**a**) **129**, 2 M $H_3PO_3$/Py, 0 °C → 40 °C, 8–9 days, yield for **258**— 41% (+45% of starting material **256**), yield for **259**—52% (+38% of starting **257**); (**b**) $HOCH_2CH_2CH_2CH_2NHCbz$, PivCl (2.5 eq), Py, 45 min, followed by 0.5 M I2 in Py:$H_2O$ 19:1 (2.5 eq.), −40 °C, 10 min, yield over 2 steps for compound **260**—62%, for compound **261**—64%; (**c**) MeONa/MeOH, (**d**) PivCl (2.5 eq.), Py, 45 min, followed by 0.5 M $I_2$ in Py:$H_2O$ 19:1 (2.5 eq.); 1 M triethylammonium bicarbonate buffer; (**e**) **129**, 1M triethylammonium bicarbonate buffer, batch 62% yield, microreactor 76% yield.

In the synthesis of a mixture of the spacer-armed oligomer **250** with avDP12, the trimeric phosphoglycan **249** was extended using an enzymatic method. The mixture of trimer **249** and a nucleotide GlcNAc-UDP was passed through a HiTrap® column with immobilized recombinant capsule polymerase CsxA [135], followed by fractionation of the resulting oligomers.

An advanced method of phosphonylation [134] of an anomeric mixture of hemiacetal **259** with chlorophosphite **129** was used by Indian researchers for the preparation of pseudo-tetrasaccharide **251** as an aminohexyl glycoside (Figure 12). Interaction of the hemiacetal mixture **259** with a phosphonylating agent $(PhO)_2POH$ in Py in the presence of $Et_3N$, followed by treatment with $H_3PO_3$ for 4 days, afforded the desired α-H-phosphonate **261** with a 65% yield [134].

In the first step of the assembly of pseudo-tetrasaccharide **251**, the selectively protected aminohexyl glycoside **269** was condensed with H-phosphonate **259**, and after oxidation, the pseudo-tetrasaccharide **270** was formed. By sequential deacetylation, condensation with H-phosphonate **259**, and oxidation, the protected pseudo-tetrasaccharide **271** was obtained and further deblocked to yield the target pseudo-tetrasaccharide **271** (Scheme 21). In the course of the synthesis, the efficiency of the formation of the phosphodiester linkage decreased in each subsequent step (yields of 86% > 60% > 53%) [134], reflecting the tendency of the phosphodiester bond to be destroyed in the conditions of the H-phosphonate procedure.

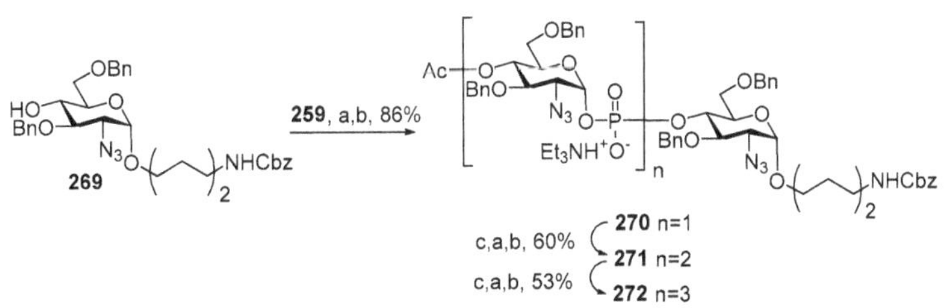


**Scheme 21.** Key building blocks for the synthesis of pseudo-tetrasaccharide **251** (Figure 12) structurally related to *MenX* capsular phosphoglycans [134]. Reagents and conditions: (**a**) PivCl (2–3 eq. relative to 259), Py, 45 min; (**b**) 0.5 M $I_2$ in Py:$H_2O$ 9.75:0.25; (**c**) MeONa/MeOH.

The spacer-armed pseudo-saccharides **247–251** (Figure 12) were conjugated to protein carriers, and the immunogenic and antigenic properties of the resulting neoglycoconjugates were studied. Oligomers **247–249** were attached to CRM197 to yield conjugates **273–275** (Figure 13) and the corresponding conjugates **277–279** with HSA (Figure 13). Conjugates **273–275** were used for the immunization of mice at a dose of 0.3 µg of glycan per mouse. In addition, conjugate **275** was used at a dose of 1 µg per mouse. For positive controls, a group of mice was immunized with a conjugate of the partially fragmented *MenX* phosphoglycan avDP15 with CRM197 (*MenX*DP15-CRM197) at a dose of 0.3 µg and 1 µg of phosphoglycans, and bacterial MenX phosphoglycans were used as a coating antigen in ELISA experiments.

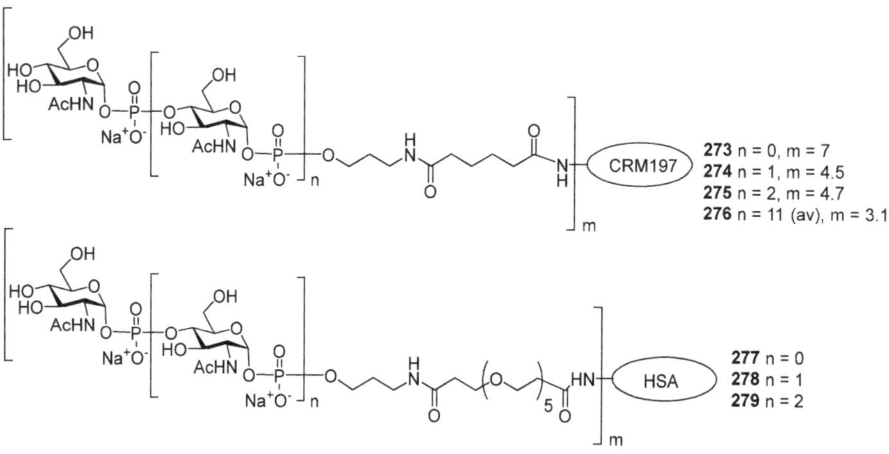


**Figure 13.** Neoglycoconjugates **273–275** prepared using synthetic *MenX*-related antigens **247–249** and CRM197 protein carrier and corresponding neoglycoconjugates **277–279** with HSA protein carrier [129].

Conjugates **273** and **274** did not induce anti-*MenX* IgG antibodies. Immunization with conjugate **275** with trimeric antigens and *MenX*DP15-CRM197 efficiently elicited anti-*MenX* IgG antibodies, yet the intensity of the specific immune response for conjugate **275** was significantly weaker. For these conjugates, the immune response did not depend on the dose [133].

In the ELISA experiments, the immunogenic capacity of conjugates **273–275** was assessed using HSA conjugates **277–279** as coating antigens. It was found that the hyperimmune antisera of mice immunized with conjugates **273** (monomeric ligand) and **274** (dimeric ligand) contained negligible amounts of IgG antibodies to synthetic antigens. In contrast, immunization with conjugate **279** elicited anti-trimer IgG antibodies. An assessment of bactericidal activity in the rSBA tests showed that the pooled mouse sera obtained from animals immunized with conjugates **273** and **274** revealed that these conjugates did not elicit bactericidal antibodies. For conjugate **276**, rSBA titers were 16 times lower compared to *MenX*DP15-CRM197. Therefore, the length of the synthetic oligomers **247–250** (Figure 12), which are incorporated into conjugates **273–275**, was too small to effectively induce an anti-*MenX* immune response [133].

The mixture of the spacer-armed oligomer **250** with avDP12 was conjugated with the protein carrier CRM197. The resulting conjugate **276** was used for the immunization of mice, and the group of mice immunized with *MenX*DP15-CRM197 was used as a positive control. The antigenicity of conjugates **276** and *MenX*DP15-CRM197 was assessed in ELISA experiments with *MenX* capsular phosphoglycans as a coating antigen. It was found that conjugates **276** and *MenX*DP15-CRM197 demonstrate similar antigenic properties. However, SBA titers evidence the slightly greater bactericidal activity of antibodies elicited by *MenX*DP15-CRM197 [135].

## 7. Synthesis of Pseudo-Oligosaccharides Structurally Related to Phosphoglycans of *S. Pneumonia*

On a global scale, one in five infants under 1 year of age has had pneumonia, and $\frac{3}{4}$ of these pneumonia cases are caused by bacteria belonging to the *S. pneumoniae* species [136]. The prevention of community-acquired diseases associated with the most virulent S. pneumoniae serotypes is provided by a number of polyvalent conjugated pneumococcal vaccines, which comprise a number of pneumococcal capsular glycans conjugated to protein carriers.

Today, the pharmaceutical industry provides a whole range of anti-pneumococcal conjugated vaccines, including PCV7 (the *S. pneumoniae* serotypes 4, 6B, 9V, 14, 19F, 18C, and 23F), PCV10 (the *S. pneumoniae* serotypes 1, 4, 5, 6B, 7F, 9V, 14, 18C, 19F, and 23F) and (the *S. pneumoniae* serotypes 1, 5, 6A, 6B, 7F, 9V, 14, 19A, 19F, and 23F), PCV13 (the *S. pneumoniae* serotypes 1, 3, 4, 5, 6A, 6B, 7F, 9V, 14, 19A, 19F, 18C, and 23F), PCV15 (the *S. pneumoniae* serotypes 1, 3, 4, 5, 6A, 6B, 7F, 9V, 14, 18C, 19A, 19F, 22F, 23F, and 33F), and PCV20 (*S. pneumoniae* 1, 3, 4, 5, 6A, 6B, 7F, 8, 9V, 10A, 11A, 12F, 14, 15B, 18C, 19A, 19F, 22F, 23F, and 33F). Among the capsular glycan antigens used in these vaccines, five biopolymers, in particular *S. pneumoniae* 6A, *S. pneumoniae* 6B, *S. pneumoniae* 10A, *S. pneumoniae* 19A, and *S. pneumoniae* 19F, comprise phosphodiester bonds in the main chain. *S. pneumoniae* 6A and *S. pneumoniae* 6B phosphoglycans are built of pseudo-tetrasaccharide repeating units connected by phosphodiester linkage. The repeating unit of *S. pneumoniae* 10A phosphoglycans is a pseudo-heptaglycosyl phosphate, and the phosphoglycans of *S. pneumoniae* 19A and *S. pneumoniae* 19F are composed of trisaccharide phosphate and hexasaccharide phosphate repeating units [137].

Although the anti-pneumococcal vaccines for the prevention of *S. pneumoniae* 6A, *S. pneumoniae* 6B, *S. pneumoniae* 19A, and *S. pneumoniae* 19F have been available for more than 10 years, these serotypes are still frequently identified as the causative agents in pneumococcal pneumonia [136] and meningitis [138] and colonize the upper respiratory tract as opportunistic pathogens [139]. One of the factors hindering the broad coverage of people with anti-pneumococcal vaccination programs is high costs [9], which can be reduced by the replacement of bacterial antigens with their synthetic analogs [22].

The *S. pneumoniae* strains of serotype 6 are the common causative agents of invasive disease. The group is subdivided into serotypes 6A and 6B, which both have a capsule composed of phosphoglycans with Rha–ribitol–Gal–Glu subunits, which differ in the type of Rha–ribitol linkage. In serotype 6A, it is Rha1→3ribitol, and in serotype 6B, it is Rha1→4ribitol. It was reported that point mutation between *S. pneumoniae* 6A and *S. pneumoniae* 6B, as well as recombination, can compromise the linkage specificity and mediate serotype change [140]. As a result, it is reasonable to include both glycan antigens in combined vaccines. The surveillance of invasive pneumococcal disease in different countries evidences the immense importance of pneumococcal conjugated vaccines in the prevention of *S. pneumoniae* 6a- and *S. pneumoniae* 6b-associated invasive diseases [139,141,142]. In contrast, in regions with low immunization coverage [143], this pathogen makes a substantial contribution to the burden of invasive pneumonia.

Recently, syntheses of the spacer-armed pseudo-oligosaccharides **280**–**286** [144,145] structurally related to *S. pneumoniae* 6A phosphoglycans (Figure 14) have been reported. For the construction of a phosphodiester bond in compound **280**, Nifantiev et al. [144] used the H-phosphonate procedure. Glycoside **287** (Scheme 22) was converted into H-phosphonate **288** by treatment with $H_3PO_3$ in Py and the presence of PivCl. The condensation of H-phosphonate **288** with the primary alcohol **289**, followed by oxidation, afforded phosphodiester **290**, which, after removal of the protective groups, was converted into aminoethyl glycoside **280**.

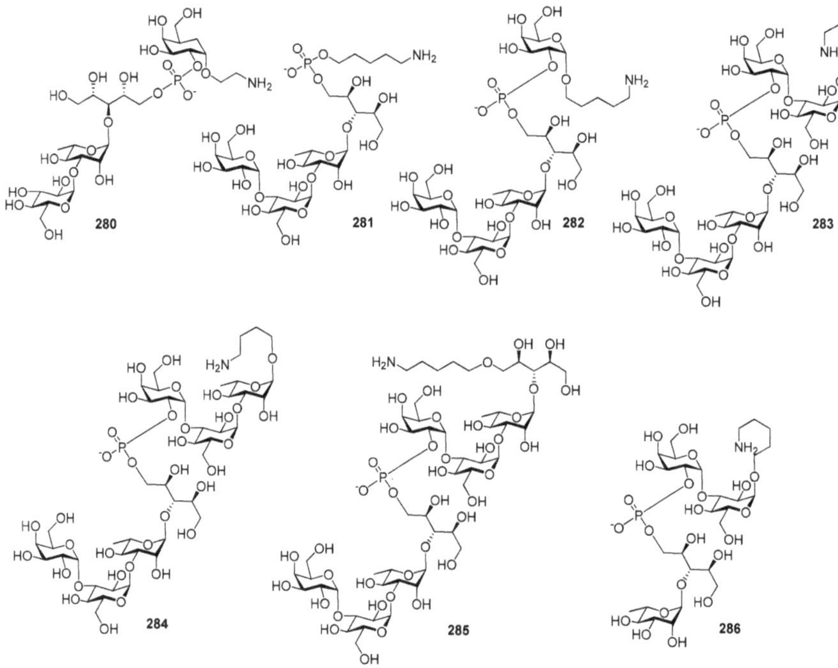

**Figure 14.** Synthetic pseudo-tetrasaccharide **280** [144] and the series of pseudo-oligosaccharides **281–286** [145] structurally related to *S. pneumoniae* 6A capsular phosphoglycans.

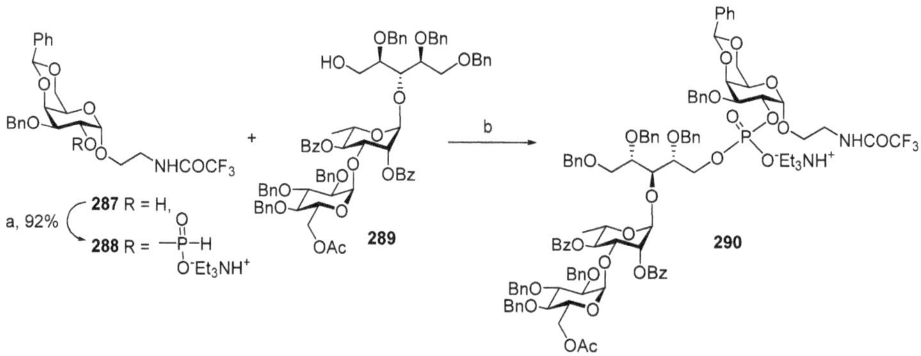

**Scheme 22.** Key building blocks for the synthesis of pseudo-oligosaccharide 280 structurally related to *S. pneumoniae* 6A capsular phosphoglycans [144]. Reagents and conditions: (**a**) PivCl, Py, $H_3PO_3$, 92%; (**b**) PivCl (4.8 eq), Py, 3 h, followed by 0.5 M $I_2$ in Py:$H_2O$ 2:1, yield over 2 steps at 73%.

Pseudo-oligosaccharides **281–286** were assembled by Taiwan scientists [145] using a phosphoramidite procedure for the establishment of phosphodiester linkage (Scheme 23). The selectively protected pseudo-tetrasaccharide **291** with a primary hydroxyl group in the ribitol residue was transformed into phosphoramidite **292** by the action of diamidite **70** and diisopropylammonium tetrazolide. In similar conditions, pseudo-disaccharide **293** was converted into phosphoramidite **294**.

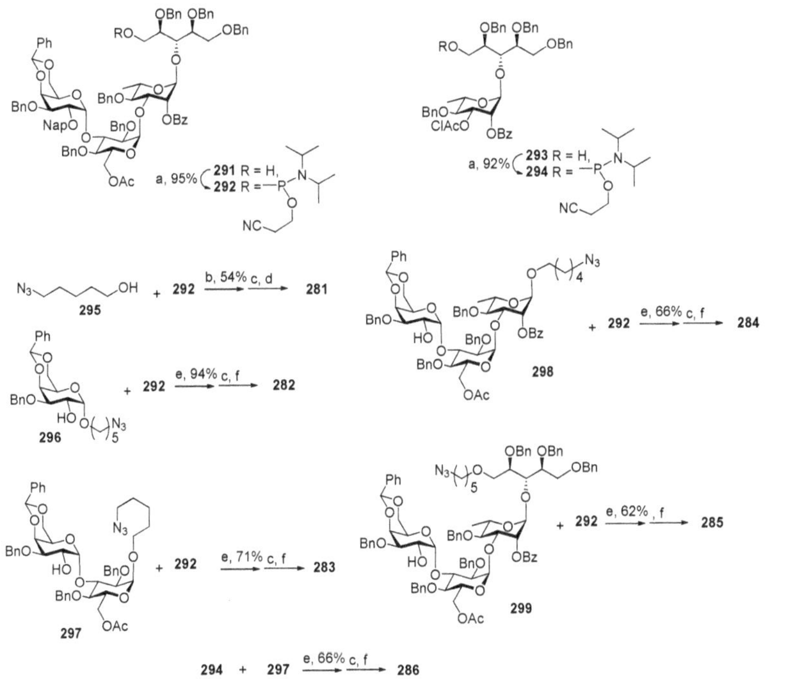

**Scheme 23.** Key building blocks for the synthesis of pseudo-oligosaccharides **281–286** structurally related to *S. pneumoniae* 6A capsular phosphoglycans [145]. Reagents and conditions: (**a**) diamidite 70, diisopropylammonium tetrazolide; (**b**) 1H-tetrazole, $CH_2Cl_2$, r.t., 1 h, followed by mCPBA, −20 °C, 30 min; (**c**) $Bu_4NOH$, $CH_2Cl_2$, $H_2O$, r.t., 4 h; (**d**) Li/$NH_3$, THF, −78 °C, 1 h, followed by MeOH, 16 h; (**e**) 5-ethylthio-1H-tetrazole, 3 Å MS, $CH_3CN$, r.t., 1 h, followed by $H_2O$, $I_2$/THF, r.t., 2 h; (**f**) 0.2 M NaOMe/MeOH, followed by $Pd(OH)_2$/C, $H_2$, MeOH, $H_2O$, AcOH.

The interaction of the key phosphoramidite **292** with 5-azidopentanol (compound **295**) in the presence of 1H-tetrazole, followed by oxidation with mCPBA, the removal of the protective groups, and the reduction of the azido group, resulted in pseudo-tetrasaccharide **281** (Scheme 23). The condensation of phosphoramidite **292** with alcohols **296–299** was carried out in the presence of 5-ethylthio-1H-tetrazole, followed by oxidation of a phosphite group to a phosphotriester with $I_2$ in THF. The efficiency of the formation of the phosphotriester linkage decreased in the sequence of monosaccharide **296** (94%), disaccharide **297** (71%), trisaccharide **298** (66%), and pseudo-tetrasaccharide **299** (62%), along with an increase in the steric hindrance of the hydroxyl group involved in condensation. The reduction and deblocking of the obtained phosphotriesters afforded the amino-spacered derivatives **282–285**. The condensation of phosphoramidite **294** with disaccharide **297**, followed by oxidation, the reduction of the azido group, and deblocking, formed pseudo-tetrasaccharide **286** [145].

Due to interest in the development of vaccine preparations based on the synthetic phosphooligosaccharides related to *S. pneumoniae* 6B, extensive libraries of the spacer-armed pseudo-oligosaccharides **300–314** [145–148] (Figure 15) related to *S. pneumoniae* 6B CPS fragments have been created. Despite their structural diversity, these compounds comprise only one phosphodiester linkage that is connected with the synthetic complexity of the target molecules.

The assembly of the spacer-armed pseudo-disaccharides **300** and **301** and pseudo-trisaccharides **302** and **303** was carried out by Vliegenthart [146] using the H-phosphonate method. For the preparation of pseudo-disaccharide **300**, the starting pseudo-disaccharide **315** was converted to the key H-phosphonate **316** by the action of chlorophosphite **129** (Scheme 24). H-phosphonate **316** was condensed with N-protected aminopropanol in the presence of PivCl in Py, the product was oxidized, and a phosphodiester **317** was obtained with a 43% yield. The deprotection of phosphodiester **317** yielded the target spacer-armed pseudo-disaccharide **300**.

Under similar conditions, alcohol **318** (Scheme 25) was transformed into H-phosphonate **319**, and the phosphonylation of alcohol **320** produced H-phosphonate **321** with an excellent yield. Phosphodiester **322** was prepared via two alternative routes. The condensation of the secondary hydroxyl group in galactoside **318** with H-phosphonate **321**, followed by oxidation, formed **322** with a 40% yield, and the interaction of a primary hydroxyl group in the selectively protected ribitol **320** with H-phosphonate **319** and oxidation of the intermediate phosphonate afforded pseudo-disaccharide **322** with a 38% yield. By the removal of the benzyl and Cbz groups, compound **322** was converted into the target pseudo-disaccharide **301** (Figure 15) [146].

Unlike the synthesis of pseudo-disaccharide **322**, the phosphitylation of alcohol **315** with H-phosphonate **319** was more efficient and readily formed pseudo-trisaccharide **323** (yield: 89%) [147], which was deprotected to yield the target spacer-armed pseudo-trisaccharide **302** (Scheme 24) [146]. The substantial difference in the yields for pseudo-disaccharide **322** (from **319** and **320**) and pseudo-trisaccharide **323** may be related to the easier accessibility of a less sterically hindered phosphodiester bond in pseudo-disaccharide **322** for the $I_2$-induced rapid rupture of the bridging P-O bonds. The total deprotection of pseudo-trisaccharide **323** resulted in the formation of the target spacer-armed pseudo-trisaccharide **302** (Figure 15) [147].

The condensation of alcohol **324** with H-phosphonate **321** (Scheme 24), followed by oxidation, afforded pseudo-tetrasaccharide **325** with a 77% yield [147]. Similar to the phosphitylation of alcohols **315** and **320** with H-phosphonate **319**, the improved efficiency of preparation of more sterically hindered pseudo-tetrasaccharide **325** compared to **322** (from **318** and **321**, 40%) [146] can be attributed to the easier destruction with $I_2$ during the oxidation step. By deprotection, pseudo-trisaccharide **325** was converted into the target spacer-armed pseudo-trisaccharide **303** (Figure 15) [147].

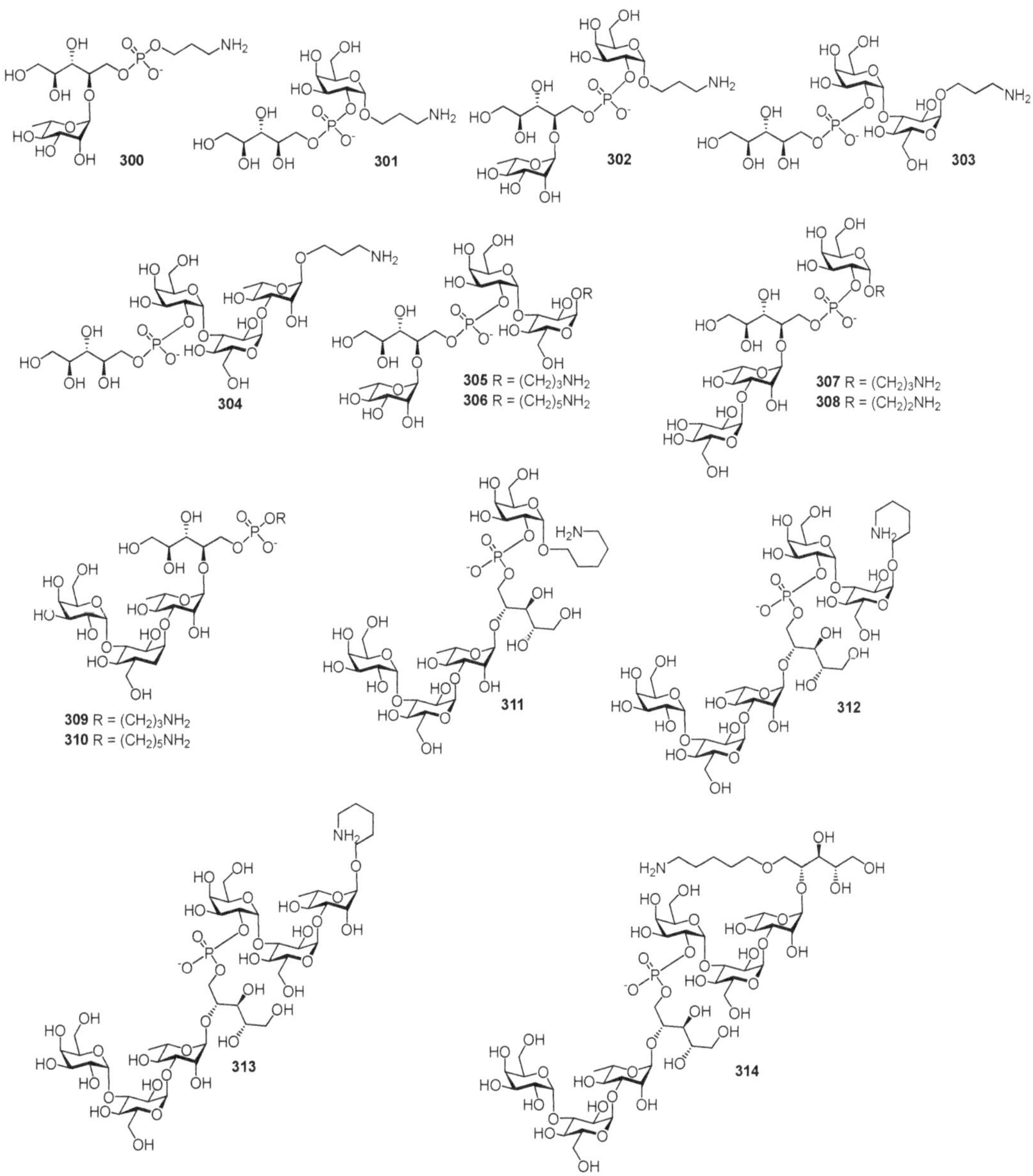

**Figure 15.** Synthetic oligosaccharides **300–303** [146], **304**, **305**, **307**, **309**, and **309** [147], **308** [148], and **306** and **310–314** [145] related to *S. pneumoniae* 6B capsular phosphoglycans.

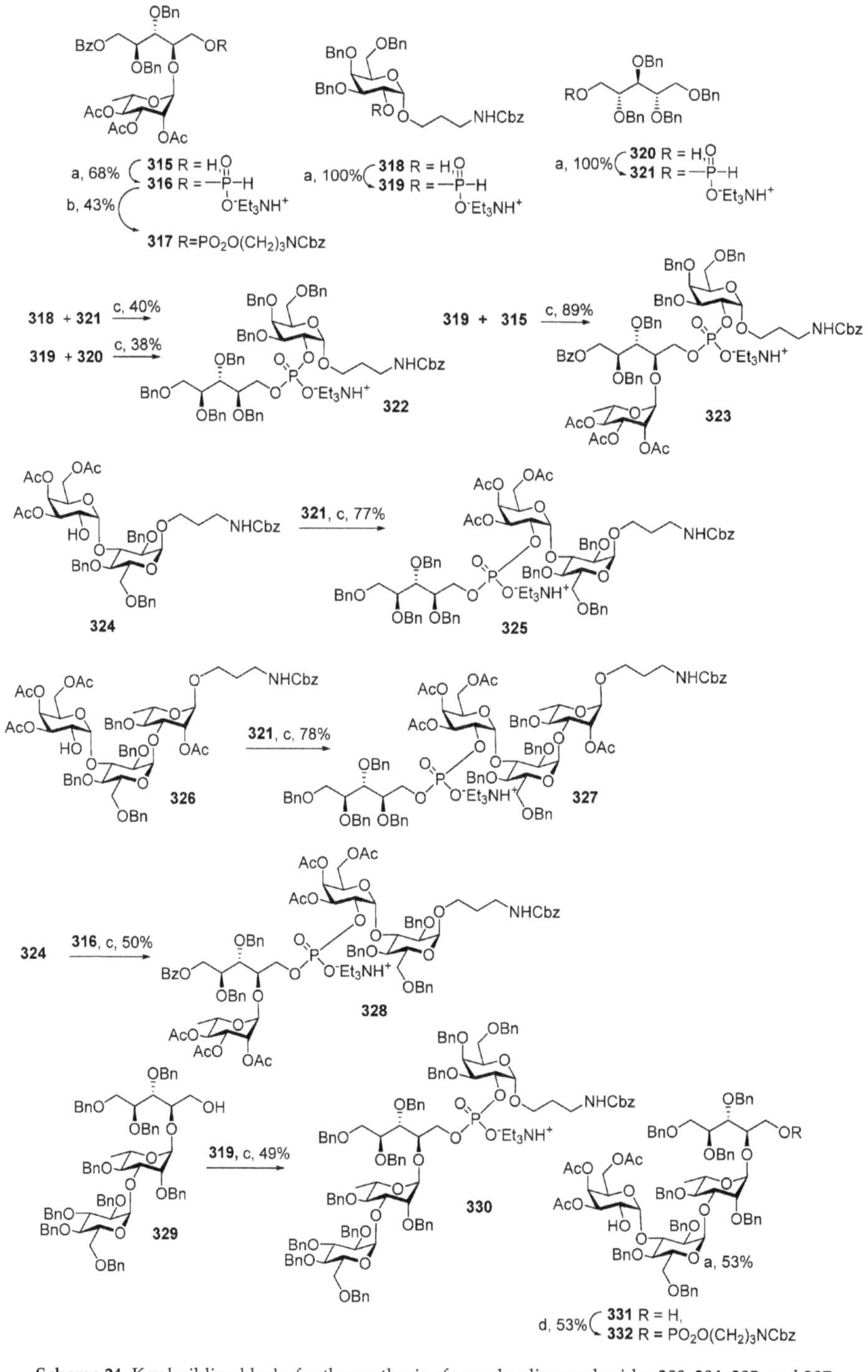

**Scheme 24.** Key building blocks for the synthesis of pseudo-oligosaccharides **300–304**, **305**, and **307** structurally related to *S. pneumoniae* 6B capsular phosphoglycans [146,147]. Reagents and conditions: (**a**) **129**, Py, 24 h; (**b**) $HOCH_2CH_2CH_2CH_2NHCbz$, PivCl, Py, followed by 0.5 M $I_2$ in Py:$H_2O$ 95:5; 20 min; (**c**) PivCl, Py, followed by 0.5 M $I_2$ in Py:$H_2O$ 95:5, 20 min; (**d**) **214**, PivCl, Py, followed by 0.5 M $I_2$ in Py:$H_2O$ 95:5, 20 min.

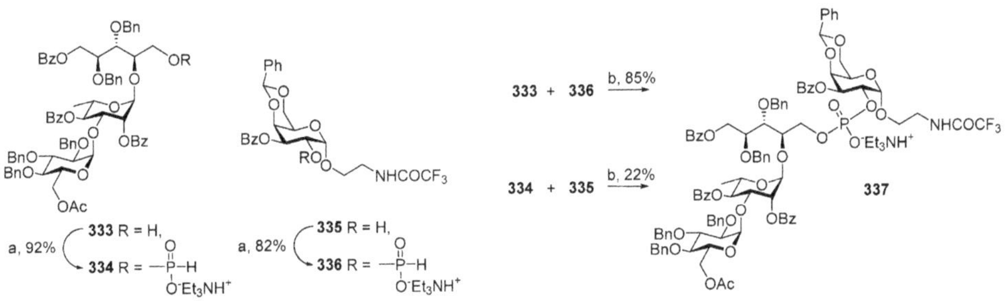

**Scheme 25.** Key building blocks for the synthesis of pseudo-oligosaccharide **308** structurally related to *S. pneumoniae* 6B phosphoglycans [148]. Reagents and conditions: (**a**) **129**, Py, 1 h; (**b**) PivCl, Py, 2 h, followed by $I_2$ in Py:$H_2O$ 2:1, 4 h.

The interaction of H-phosphonate **321** with the trisaccharide alcohol **326** (Scheme 24), followed by oxidation, formed pseudo-tetrasaccharide **327** (78%) with a phosphodiester bond between a bulky trisaccharide and a flexible ribitol residue. On the contrary, the efficiency of the formation of pseudo-tetrasaccharide **328** (from disaccharide **324**) and pseudo-tetrasaccharide **330** (from pseudo-tetrasaccharide **329**), which comprise bulky protected glycoside residues on both sides of a phosphate moiety, was much lower, and the yields did not exceed 50%. The total deprotection of pseudo-tetrasaccharides **328** and **329** resulted in the target spacer-armed compounds **304** and **305**, respectively [147].

The condensation of pseudo-tetrasaccharide **331** with H-phosphonate **214**, followed by oxidation, provided the introduction of a pre-spacer group, which is connected to the pseudo-tetrasaccharide via a phosphodiester bridge. The moderate yield of compound **332** (53%) can be attributed to the conformational flexibility of the pre-spacer group and the easier availability of the phosphate group for the action of $I_2$. Compounds **331** and **332** were deblocked to yield the target spacer-armed pseudo-tetrasaccharides **307** and **309**, respectively (Scheme 24) [147].

For the assembly of pseudo-tetrasaccharide **308**, Nifantiev et al. [148] investigated two alternative pathways according to the schemes [3 + 1], which differed in the position of the H-phosphonate group on one or the other condensing part (Scheme 25). For this purpose, alcohol **333** was converted into H-phosphonate **334** by the action of chlorophosphite **129**, and galactoside **335** was converted into H-phosphonate **336** under similar conditions. The condensation of H-phosphonate **336** with the primary hydroxyl group of pseudo-tetrasaccharide acceptor **333**, followed by oxidation, produced phosphodiester **337** in a high yield (85%), whereas the interaction of H-phosphonate **334** with the secondary hydroxyl group in galactoside **335** proceeded with low efficiency (the yield of pseudo-tetrasaccharide **337** was 22%).

For the preparation of pseudo-oligosaccharides **306** and **310–314** [145], Taiwan researchers used the universal pseudo-tetrasaccharide block **338** (Scheme 26), which was readily converted into the corresponding phosphoramidite **339** by the action of diamidite **70** in the presence of diisopropylammonium tetrazolide. In similar conditions, pseudo-disaccharide **340** was converted into phosphoramidite **341**. The condensation of alcohols **295–298** and **342** with compound **339** added a repeating unit to the chains. Thus, the phosphitylation of the primary hydroxyl group in alcohol **295** with phosphoramidite **339** in the presence of 1H-tetrazole, followed by the oxidation of phosphite to phosphate and the removal of the 2-naphthylmethyl protective group, afforded alcohol **342**, and the total deprotection and reduction of the azido group resulted in the target spacer-armed pseudo-tetrasaccharide **310**, with the spacer group connected to the pseudo-tetrasaccharide by a phosphodiester bridge.

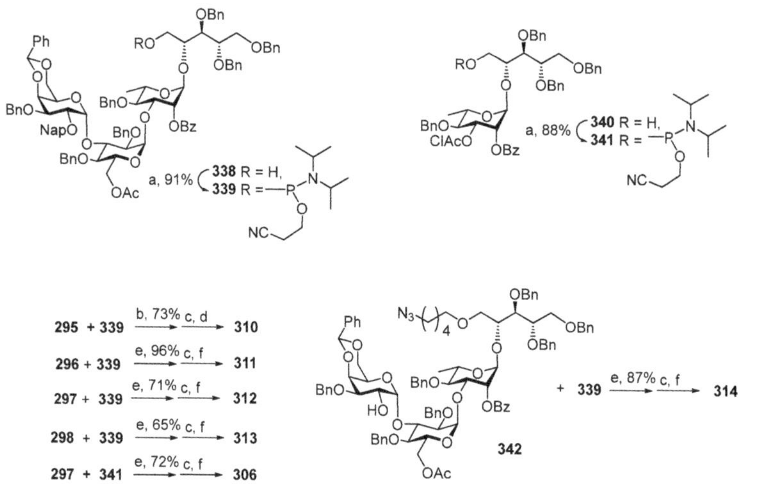


**Scheme 26.** Key building blocks for the synthesis of pseudo-oligosaccharides **281–286** structurally related to *S. pneumoniae* 6B phosphoglycans [145]. Reagents and conditions: (**a**) diamidite **70**, diisopropylammonium tetrazolide; (**b**) 1H-tetrazole, $CH_2Cl_2$, r.t., 1.5 h, followed by mCPBA, −20 °C, 30 min; (**c**) $Bu_4NOH$, $CH_2Cl_2$, $H_2O$, r.t.; (**d**) Li/$NH_3$, THF, −78 °C, 1 h, followed by MeOH, 16 h; (**e**) 5-ethylthio-1H-tetrazole, molecular sieves 3Å, $CH_3CN$, r.t., 1 h, followed by $H_2O$, $I_2$/THF, r.t., 2 h; (**f**) 0.2 M NaOMe/MeOH, followed by $Pd(OH)_2$/C, $H_2$, MeOH, $H_2O$, AcOH.

In a similar way, phosphoramidite **339** readily phosphitylated secondary hydroxyl groups in monosaccharide **296**, disaccharide **297**, and trisaccharide **298** (Scheme 26). The resulting phosphites were oxidized and deprotected to obtain target compounds **311–313**. The condensation of two pseudo-tetrasaccharides **342** and **339**, followed by oxidation, afforded a pseudo-octasaccharide (87%), with two bulky fragments connected by a phosphodiester bridge. After the deprotection and reduction of the azido group, the spacer-armed pseudo-octasaccharide **314** was obtained. Disaccharide **297** was phosphitylated with phosphoramidite **341**. The oxidation of the intermediate phosphite, removal of the protective groups, and reduction of the azido group afforded pseudo-tetrasaccharide **306**. It can be concluded that the use of phosphoramidites as phosphitylating agents in the preparation of compounds **306** and **310–314** was efficient regardless of the steric demands of the secondary hydroxyl group in the acceptor molecule. This result shows that the phosphoramidite method is preferred for the synthesis of larger structures.

The phosphoglycan structure of *S. pneumoniae* 6C differs from *S. pneumoniae* 6A in the orientation of a single hydroxyl group in one of the monosaccharide blocks (Glc in *S. pneumoniae* 6C vs. Gal in *S. pneumoniae* 6A), and a similar structural difference is observed between *S. pneumoniae* 6D and *S. pneumoniae* 6B (Figure 16). It is generally accepted that a close structural resemblance gives rise to similar antigenic properties, and one could expect cross-reactivity between all serotypes in group *S. pneumoniae* 6. However, surveillance data and clinical efficacy studies for subtypes of *S. pneumoniae* 6 obtained in connection with the use of pneumococcal vaccines are controversial. Generally, cross-protection was observed from the *S. pneumoniae* 6B conjugate in PCV7 and PCV10 against *S. pneumoniae* 6A but not against *S. pneumoniae* 6C and *S. pneumoniae* 6D. However, it was shown that PCV13, which comprises the *S. pneumoniae* 6A antigen, was effective against serotypes 6C and 6D invasive pneumococcal disease and microbial carriage [149,150]. The challenges associated with the assessment of the cross-reactivity and cross-protection of *S. pneumoniae* group 6 could be met via the identification of a glycotope for each *S. pneumoniae* 6 subtype. To this end, the series of oligosaccharides **343** and **344**, structurally related to *S. pneumoniae* 6C, and **345** and **346**, structurally related to *S. pneumoniae* 6D, were synthesized (Figure 16) [151]. The preparation of pseudo-tetrasaccharides **343** and **344**, which comprise phosphodiester linkages, was performed using phosphoramidite chemistry. To obtain pseudo-oligosaccharide **343** with an interglycosidic phosphodiester bridge, disaccharide **347** was reacted with phosphoramidite **294** in acetonitrile in the presence of 5-ethylthio-1H-tetrazole (Scheme 27). The reaction mixture was oxidized under the action of iodine in THF with the addition of water, the protective groups were removed, and the $N_3$ group was reduced to $NH_2$. The phosphitylation of the primary OH-group in pseudo-tetrasaccharide **348** with diamidite **70** resulted in the efficient formation of phosphoramidite **349**. The condensation of phosphoramidite **349** with alcohol **295**, followed by the reduction of the $N_3$ group and total deprotection, afforded pseudo-tetrasaccharide **344**, in which a phosphodiester linkage connects the ribitol part and a spacer [151].

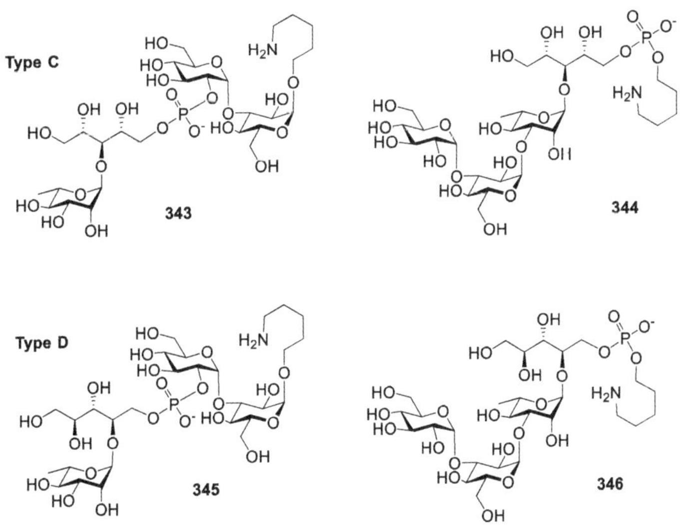


**Figure 16.** Synthetic phosphooligosaccharides related to *S. pneumoniae* 6C (compounds **343** and **344**) and *S. pneumoniae* 6D (compounds **345** and **346**) capsular phosphoglycans [151].

In a similar way, pseudo-tetrasaccharide **345** with an interglycosidic phosphodiester bridge was obtained by the condensation of disaccharide **247** and phosphoramidite **341**, oxidation, reduction of the $N_3$ group, and deprotection. For the preparation of phosphodiester **346**, pseudo-tetrasaccharide **350** was transformed into phosphoramidite **351**, which was then condensed with alcohol **295**. After oxidation, the condensation product was reduced and deprotected to yield pseudo-tetrasaccharide **346**, with a phosphodiester bridge between ribitol and a spacer. Notably, in the preparation of compounds **343**–**346**, the corresponding protected phosphodiesters were obtained in moderate yields regardless of the position of the phosphodiester bridge in the molecule. It can be concluded that the used method is not sensitive to steric factors and can be applied for the condensation of bulky counterparts.

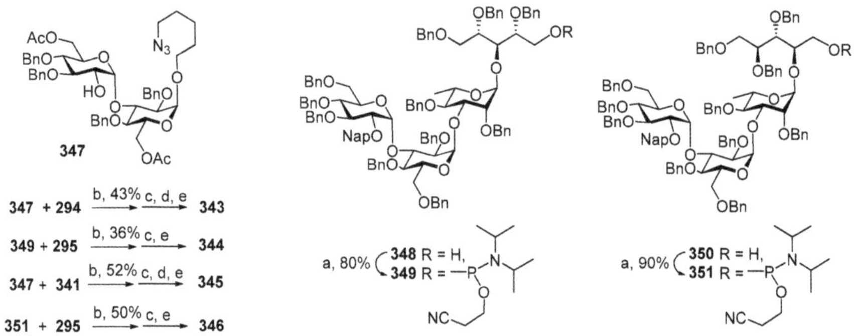

**Scheme 27.** Key building blocks for the synthesis of pseudo-oligosaccharides **343** and **344**, structurally related to *S. pneumoniae* 6C phosphoglycans, and pseudo-oligosaccharides **345** and **346**, structurally related to *S. pneumoniae* 6D phosphoglycans [151]. Reagents and conditions: (**a**) diamidite **70**, diisopropylammonium tetrazolide, 1 h; (**b**) 5-ethylthio-1H-tetrazole, molecular sieves 3Å, $CH_3CN$, r.t., 1 h, followed by $H_2O$, $I_2$/THF, r.t., 2 h; (**c**) $Bu_4NOH$, $CH_2Cl_2$, $H_2O$, r.t.; (**d**) followed by 0,3 M NaOMe, MeOH, $CH_2Cl_2$, 2 h; (**e**) $Pd(OH)_2$/C, $H_2$, MeOH, $H_2O$, AcOH, 36 h.

The cross-reactivity of *S. pneumoniae* 6A and *S. pneumoniae* 6B was evaluated in an immunological study of pseudo-tetrasaccharides **280** [144] related to *S. pneumoniae* 6A (Figure 14) and **308** [148] (Figure 15) related to *S. pneumoniae* 6B. Conjugates of these compounds with BSA (**280**-BSA and **308**-BSA, respectively) were used for the immunization of mice at a dose of 20 µg/mouse on days 0 and 14. The hyperimmune serum obtained after immunization with **280**-BSA and **308**-BSA was analyzed in ELISA experiments using the corresponding N-biotinylated conjugates **352** and **353** on streptavidin-coated plates. It was shown that the antibodies induced by the *S. pneumoniae* 6A-related conjugate **280**-BSA recognized not only the *S. pneumoniae* 6A-related N-biotinylated conjugate **352** but also the *S. pneumoniae* 6B-related N-biotinylated conjugate **353**. Inversely, the specificity of the serum obtained from mice immunized with the *S. pneumoniae* 6B-related conjugate **308**-BSA to the *S. pneumoniae* 6A-related conjugate **352** was detected. These results evidence the cross-reactivity of pseudo-tetrasaccharides **280** and **308** and indicate the presence of a common epitope in these antigens.

The immunogenic properties of synthetic phosphooligosaccharides **285** and **286** related to *S. pneumoniae* 6A phosphoglycans (Figure 14) and phosphooligosaccharides **306** and **314** related to *S. pneumoniae* 6B phosphoglycans were studied. The conjugates of these compounds with CRM197 were obtained and used for the immunization of mice at a dose of 2.2 µg of glycan per mouse, with three shots in two-week intervals. The glycan-specific IgG in the serum was analyzed using a microarray with a number of synthetic glycan antigens [145]. Mice in the negative control group were immunized with CRM197. It was found that conjugates **286**-CRM197 and **306**-CRM197, which contain pseudo-tetrasaccharides, effectively induced IgG antibodies against the majority of other synthetic antigens. On the contrary, conjugates **285**-CRM197 and **314**-CRM197 showed low immunogenicity [145].

The antigenic properties of *S. pneumoniae* 6B-related pseudo-disaccharide **301**, pseudo-trisaccharide **303**, and pseudo-tetrasaccharide **305** were investigated using their conjugates **301**-KLH–**305**-KLH with the highly immunogenic protein KLH (Figure 17) [152]. The neoglycoconjugates **301**-KLH–**305**-KLH and a conjugate obtained by the condensation of bacterial *S. pneumoniae* 6B phosphoglycans and KLH (*S. pneumoniae* 6B-KLH) were used for the immunization of mice at a dose of 2.5 µg and rabbits at a dose of 10 µg. The hyperimmune sera were analyzed in ELISA experiments with bacterial *S. pneumoniae* 6B and *S. pneumoniae* 6A phosphoglycans. Human serum samples were obtained by pooling infant antisera obtained by vaccination with PCV7-CRM197 at a dose of 4 mg of 6B PS.

It was shown that in mice conjugates, **301**-KLH with pseudo-disaccharide antigens and **303**-KLH with pseudo-trisaccharide antigens were poorly immunogenic. On the contrary, the conjugate **305**-KLH was more immunogenic than *S. pneumoniae* 6B-KLH. ELISA analysis of the rabbit hyperimmune antisera using bacterial *S. pneumoniae* 6A phosphoglycans as a coating antigen showed that conjugates **303**-KLH and **305**-KLH were able to induce anti-*S. pneumoniae* 6A antibodies. Antibody specificity was confirmed in ELISA inhibition and phagocytosis experiments. The binding of human PCV7-CRM197 antisera to the conjugates **301**-KLH–**305**-KLH was inhibited by bacterial *S. pneumoniae* 6A and *S. pneumoniae* 6B phosphooligosaccharides.

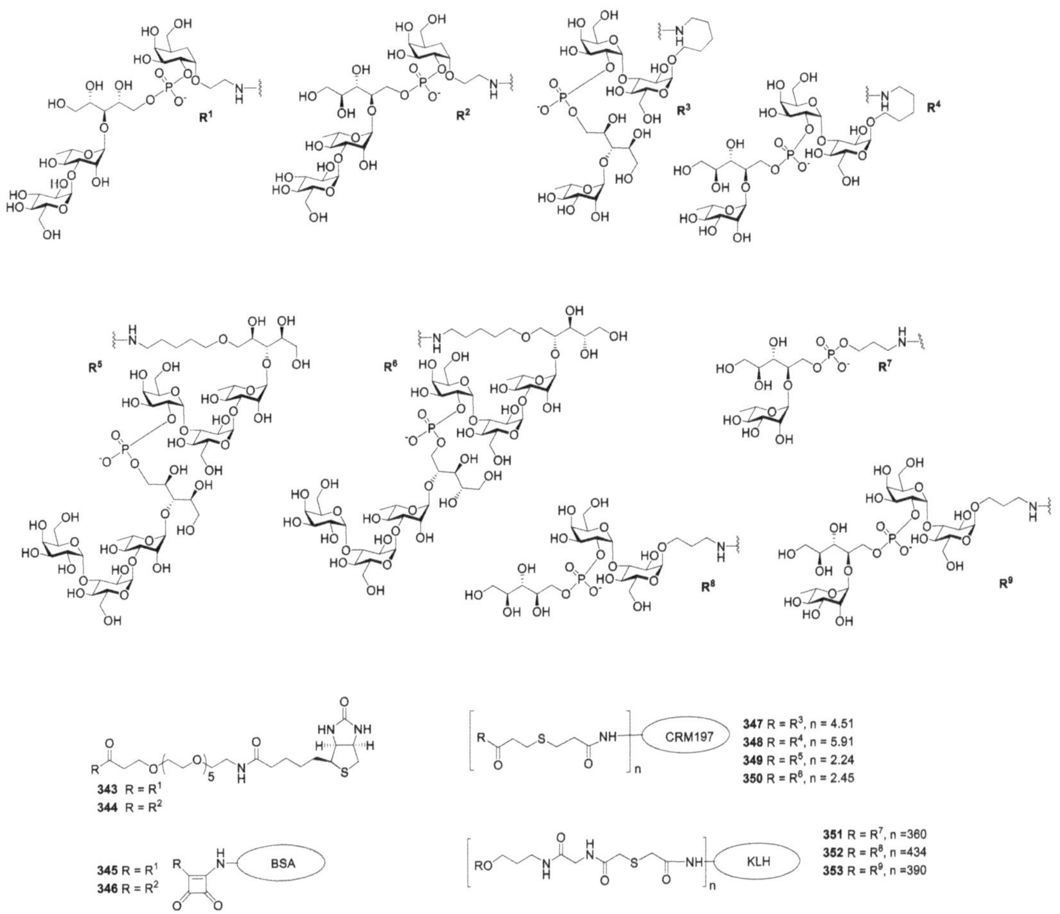


**Figure 17.** Neoglycoconjugates **343–353** obtained on the basis of synthetic pseudo-oligosaccharides structurally related to *S. pneumoniae* 6A and *S. pneumoniae* 6B capsular phosphoglycans.

*S. pneumoniae* 19F is one of the most virulent pneumococcal serotypes. Similarly to *S. pneumoniae* 6A and *S. pneumoniae* 6B, the diseases caused by *S. pneumoniae* 19F infection are associated with increased mortality and morbidity [153] despite the fact that the *S. pneumoniae* 19F capsular glycan is a component of many pneumococcal conjugated commercial vaccines. In light of this, it could be useful to develop an anti-*S. pneumoniae* 19F conjugate vaccine with a synthetic glycan antigen structurally related to the capsular glycan of *S. pneumoniae* 19F. N-trifuoroacetamidopropyl glycoside **357**, which comprises a pseudo-hexasaccharide fragment of the *S. pneumoniae* 19F capsular polysaccharide, was obtained as a mixture of isomers. In the first step, the selectively protected trisaccharide **354** with a free hydroxyl group at C-4 (ManNAc) was converted into H-phosphonate **355** under the action of chlorophosphite **129** in Py [154] (Scheme 28). The condensation of H-phosphonate **355** with hemiacetal **356**, followed by oxidation and deprotection, resulted in the formation of a diastereomeric mixture of pseudo-hexasaccharides, which contained the spacer-armed phosphooligosaccharide **357**.

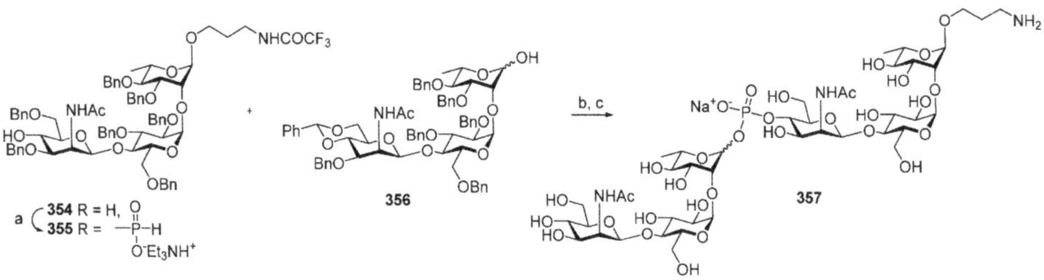

**Scheme 28.** Key building blocks for the synthesis of pseudo-oligosaccharide **357** structurally related to *S. pneumoniae* 19F capsular glycan [154]. Reagents and conditions: (**a**) **129**, acetonitrile, Py, 1 h.; (**b**) PivCl, Py, then $I_2$ B Py:$H_2O$ 19:1, yield 61% over 3 steps; (**c**) $H_2$, $Pd(OH)_2/C$, MeOH–$H_2O$.

The structure of the repeating unit of *S. pneumoniae* 19A is close to that of 19F, with the only difference being the type of linkage between Glc and Rha monosaccharide residues, which is α-D-Glc-(1→3)-Rha in *S. pneumoniae* 19A phosphoglycans and α-D-Glc-(1→2)-Rha in *S. pneumoniae* 19F phosphoglycans. Whereas *S. pneumoniae* 19F phosphoglycans are a component of all commercial PCV compositions, *S. pneumoniae* 19A phosphoglycans were not included in PCV7 and PCV10. After the introduction of PCV7, serotype replacement resulted in an increase in the number of invasive pneumococcal diseases caused by S. pneumoniae 19A [155,156]. This observation shows the absence of cross-protection between serotypes 19A and 19F. Nevertheless, the synthetic complexity of the repeating units of *S. pneumoniae* 19A and *S. pneumoniae* 19F phosphoglycans, which share a common structure, stimulated attempts to develop a universal synthetic glycoantigen for the induction of protective antibodies to both serotypes.

With a view to investigating the ability of a neoglycoconjugate with a chimeric synthetic antigen to induce a protective immune response, the Seeberger group prepared [157] phosphohexasaccharide **358**, which is a combination of *S. pneumoniae* 19A and S. pneumoniae 19F phosphoglycan repeating units in one molecule and the spacer-armed phosphotrisaccharides **359** and **360** related to repeating units of *S. pneumoniae* 19F and *S. pneumoniae* 19A, respectively (Figure 18).

Hemiacetal **361** was readily transformed into the corresponding α-H-phosphonate **362** under the action of chlorophosphite **129** and $H_3PO_3$ in Py (Scheme 29). The authors suggest that α-stereoselectivity was attained due to thermodynamic reaction control, which favored α-H-phosphonate **362** in the conditions of $S_N2$ displacement with $H_3PO_3$ at the anomeric center. In a similar manner, hemiacetal **363** was stereoselectively converted into α-H-phosphonate **364** as a precursor of the *S. pneumoniae* 19A repeating unit. α-H-Phosphonates **362** and **364** were coupled with alcohol **295**, oxidized, reduced, and deprotected to yield the spacer-armed glycoantigens **359** and **360**, which correspond to repeating units of *S. pneumoniae* 19F and *S. pneumoniae* 19A phosphoglycans [157].

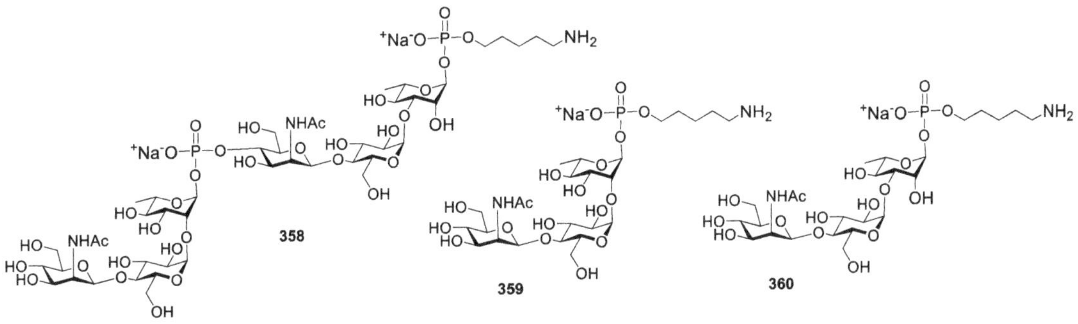

**Figure 18.** Synthetic phosphohexasaccharide hybrid antigen **358**, which comprises repeating units of *S. pneumoniae* 19F and *S. pneumoniae* 19A and phosphotrisaccharides related to *S. pneumoniae* 19F (compound **359**) and *S. pneumoniae* 19A capsular glycan (compound **360**) [157].

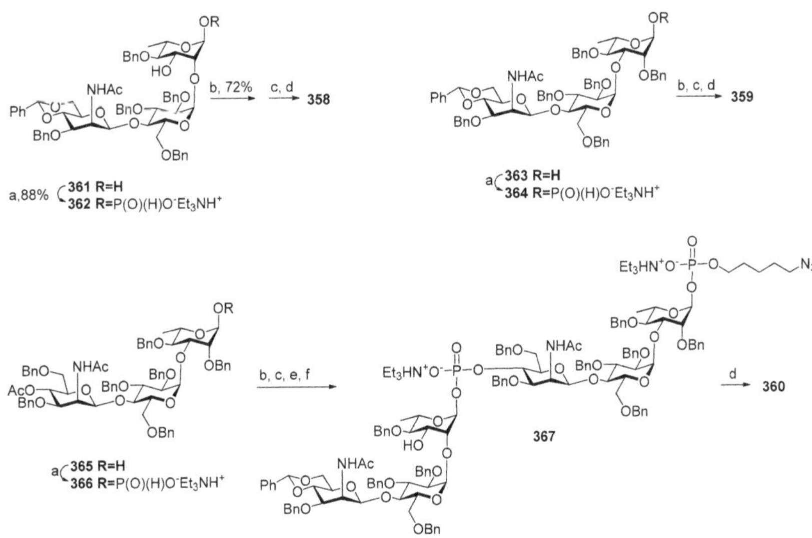

**Scheme 29.** Key building blocks for the synthesis of pseudo-oligosaccharides **358–360** [157]. Reagents and conditions: (**a**) chlorophosphite **129**, $H_3PO_3$, Py, r.t, 2 h then 50 °C, 36 h; (**b**) alcohol **295**, PivCl, pyridine, r.t., 1.5 h; (**c**) $I_2$, Py:$H_2O$ (18:2), −40 °C, 30 min, then $Et_3NHBr$; (**d**) $H_2$, Pd/C, EtOAc:MeOH:$H_2O$, 3 days, then Dowex 50W X4, $Na^+$; (**e**) 0.5 M NaOMe, MeOH, r.t., 3 h; (**f**) α-H-phosphonate **362**, PivCl, Py, r.t., 2 h.

For the preparation of the hybrid antigen 358, another precursor of the *S. pneumoniae* 19A repeating unit, hemiacetal **365** with an acetyl protecting group at O-4 (MaNAc) was phosphonylated to obtain α-H-phosphonate **366**, which was then condensed with alcohol **295** and oxidized. The following 4-O-deacetylation and phosphitylation with α-H-phosphonate **362** and oxidation yielded the protected phosphodiester **367**. The deblocking and reduction of the azido group in phosphodiester **367** afforded the conjugation-ready chimeric phosphodiester 358 [157]. Phosphooligosaccharides **358–361** were conjugated to CRM197 via a succinimidyl adipate linker, and the resulting conjugates **358**-CRM197 (average glycan loading: 5), **359**-CRM197 (average glycan loading: 5), and **360**-CRM197 (average glycan loading: 7) were adsorbed on alum and used for the vaccination of rabbits on day 0 and boosted on days 14, 28, and 133. IgG antibodies were raised in response to immunization with the semisynthetic neoglycoconjugates **358**-CRM197, **359**-CRM197, **360**-CRM197, and Prevnar 13® (PCV13).

Oligosaccharide-specific antibody titers were analyzed by glycan microarray with specific immobilized oligosaccharides. Chimeric **358**-CRM197 generated antibodies that recognized trisaccharide antigens **359** and **360** in quantities exceeding those generated by Prevnar 13®. At the same time, ELISA experiments showed that antibodies produced by vaccination with Prevnar 13® more efficiently recognized *S. pneumoniae* 19A and *S. pneumoniae* 19F phosphoglycans compared to vaccine preparations with synthetic antigens. This observation reveals that upon immunization with Prevnar 13®, the antibodies are elicited to larger glycotopes. Immunization with chimeric **358**-CRM197 produced antibodies that efficiently recognized phosphotrisaccharide antigens **359** and **360** and *S. pneumoniae* 19A and *S. pneumoniae* 19F phosphoglycans, whereas antibodies elicited by **359**-CRM197 and **360**-CRM197 did not recognize *S. pneumoniae* 19A phosphoglycans, and only moderate interaction of **359**-CRM197-induced antibodies and S. pneumoniae 19F phosphoglycans was detected. Bactericidal properties of the obtained hyperimmune sera were studied in the opsonophagocytic killing assay. Chimeric **358**-CRM197-induced opsonic antibodies were able to kill both *S. pneumoniae* 19A and *S. pneumoniae* 19F. The opsonic activity of **359**-CRM197-induced antibodies was very weak, and antibodies induced by the ST19A conjugate were not bactericidal [157].

As previously mentioned, clinically significant cross-protection between *S. pneumoniae* 19A and *S. pneumoniae* 19F serotypes has not been reported. One of the possible reasons for weak antibody cross-protection is the conformational difference between these two phosphoglycans [158]. In shorter oligosaccharides, the chain exists as another ensemble of conformers, which may be able to induce antibodies to phosphoglycans of both serotypes. The capsular phosphoglycans *S. pneumoniae* 19A and *S. pneumoniae* 19F share a common disaccharide structural element: P(1→4)ManNAc-β-(1→4)-Glc. With a view to studying the universal glycotope for these serotypes, a series of short glycans **368**–**370** (Figure 19) were prepared, which comprise the basic structural elements of the common disaccharide.

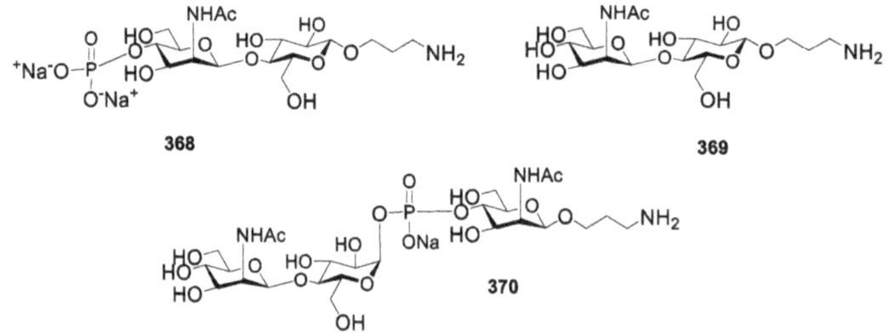

**Figure 19.** Synthetic glycans related to phosphodisaccharide P(1→4)ManNAc-β-(1→4)-Glc, which is a common structural element of *S. pneumoniae* 19A and *S. pneumoniae* 19F phosphoglycans.

The phosphoramidite approach was applied for the construction of the phosphodiester bridge in pseudotrisaccharide **370** [159]. The selectively protected glycoside **371** with a free hydroxyl group at C-4 (ManNAc) was transformed into phosphamidite **372** under the action of chlorophosphoramidite **24** (Scheme 30). The condensation of phosphamidite **372** with hemiacetal **373** resulted in the formation of a mixture of α- and β-glucopyranosides (α:β 55:45). The α-isomer was oxidized, decyanoethylated, and deprotected to obtain phosphodiester **370**.

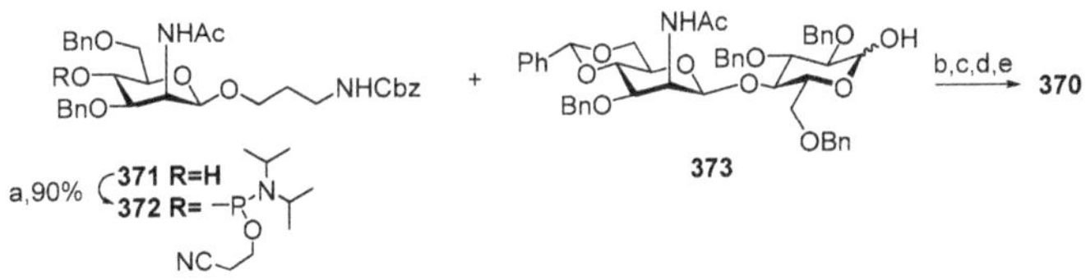


**Scheme 30.** Key building blocks for the synthesis of pseudo-trisaccharide **370** [159]. Reagents and conditions: (**a**) **24**, DIPEA, $CH_2Cl_2$, 90%; (**b**) DCI, $CH_2Cl_2$, 95%; (**c**) tBuOOH, $CH_3CN$, 0 °C to r.t., 73% (55% α); (**d**) $Et_3N$, DCM, 4 days; (**e**) Pd/C, $H_2$, MeOH/$H_2O$, 90%.

Compounds **368–370** were printed on epoxysilane-coated slides together with bacterial *S. pneumoniae* 19A and *S. pneumoniae* 19F phosphoglycans as controls. A glycan microarray was arranged, and the glycan antigens interacted with the hyperimmune sera of rabbits immunized with whole bacteria. The group's 19 sera were bound to *S. pneumoniae* 19A and *S. pneumoniae* 19F and phosphodisaccharide **368**, and the interaction with disaccharide **369** and the pseudotrisaccharide was weak. The interaction of synthetic and bacterial glycans was studied with factor reference antisera obtained by vaccination with whole cell bacteria *S. pneumoniae* 19A or *S. pneumoniae* 19F and cleared of the antibodies recognizing the common epitopes. Reference antisera to *S. pneumoniae* 19A readily bound to phosphate **368** and did not interact with compounds **369** and **370**. Reference antisera to S. pneumoniae 19F showed moderate binding to compounds **368–370**.

Two considered approaches to the search for a common epitope in *S. pneumoniae* 19A and *S. pneumoniae* 19F phosphoglycans open new horizons in the development of semisynthetic conjugated glycovaccines. The design of a universal glycotope for introduction in neoglycoconjugate vaccines is aimed at a decrease in protein loading of the vaccination dose and reduction of the production costs.

## 8. Synthesis of Pseudo-Oligosaccharides Structurally Related to Phosphoglycans of *C. Jejuni*

*C. jejuni* enteritis is one of the most common causes of intestinal infection and traveler's diarrhea. It is associated with many severe sequelae, including Guillain–Barré syndrome. Penner serotyping in combination with PCR multiplex-based typing revealed 35 capsular types [160]. Until now, no commercial vaccines for the prevention of *C. jejuni* enteritis have been developed either for humans or for cattle and poultry, which are the frequent source of human campylobacteriosis. Whole-cell vaccines are not applicable, as a number of *C. jejuni* strains present sialylated LOS that induce antibodies to the gangliosides β-D-GalNAc-(1→4)-[α-Neu5Ac-(2→3)]-β-D-Gal-(1→4)-β-D-Glc-(1→1)-N-octadecanoyl sphingosine and α-Neu5Ac-(2→3)-β-D-Gal-(1→4)-β-D-Glc-(1→1)-N-octadecanoyl sphingosine, which are associated with the onset of Guillain–Barré syndrome [161]. Despite recent advances in the understanding of the pathogenesis of *C. jejuni*, the development of subunit or glycoconjugate preventive vaccines still remains a challenge [162]. Current knowledge on *C. jejuni* infection has revealed that the key structural feature is the O-methylphosphoramidate modification of capsular glycan, which is involved in both *C. jejuni* pathogenesis and immunogenicity. In the context of the development of conjugate vaccines on the basis of synthetic glycan antigens, O-methylphosphoramidate is regarded as the immunodominant epitope [163]. However, it should be noted that this moiety is easily hydrolyzed and can be lost during the conjugation process [160].

Up to now, only one synthesis of a P-linked pseudo-oligosaccharide related to *C. jejuni* phosphoglycans has been described. Therefore, the selectively protected hemiacetal **374** was converted into α-H-phosphonate **375** under the action of chlorophosphite **129**. The condensation of phosphonate **375** with the heptabioside block **376** in the presence of PivCl/Py, followed by oxidation, afforded a phosphodiester, which was then transformed into the target pseudo-trisaccharide **377** corresponding to a repeating unit of the capsular glycan of a clinically important strain RM1221 of *C. jejuni* serotype HS:53 (Scheme 31) [164]. The prospects of conjugated vaccines for the prevention of campilobacterioses are outlined in a recent review [165].

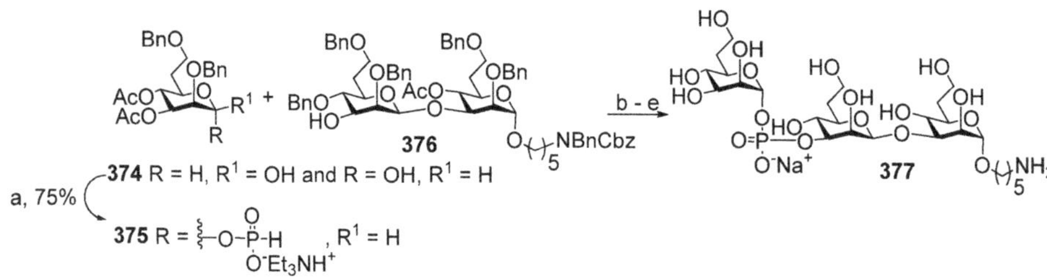

**Scheme 31.** Key building blocks for the synthesis of capsular glucan of *C. jejuni* serotype HS:53, strain RM1221 [164]. Reagents and conditions: (**a**) chlorophosphite **129**, Py, MeCN, 0 °C → r.t.; (**b**) PivCl, Py, 1 h. 74%; (**c**) $I_2$, Py/$H_2O$, 0 °C. 0,5 h., 86%; (**d**) 1.2 M MeONa/MeOH; (**e**) $H_2$/Pd-C, 48 h.

## 9. Conclusions

In this review, studies published over the last 20 years on the methods of construction of interglycosidic phosphodiester linkages used for the preparation of spacer-armed phosphooligosaccharides structurally related to the capsular phosphoglycans of the pathogenic bacteria *H. influenzae* serotypes a, b, c, and f, *N. meningitidis* serogroups a and x, *S. pneumoniae* serotypes 6a, 6b, 6c, 6f, 19a, and 19f, and the *C. jejuni* serotype HS:53, strain RM1221 are summarized.

As shown, in the syntheses of phosphoglycans, two types of intermediate compounds are widely employed: H-phosphonates and phosphoramidites. Both methods provide an effective tool for the establishment of phosphodiester bridges between different types of saccharide blocks. Due to high yields and simple procedures, these methods can be scaled up and form the basis for the manufacturing process. The H-phosphonate-based method is a one-pot polycondensation procedure used in the industrial-scale synthesis of Hib oligosaccharides for the manufacturing of the Quimi-Hib® vaccine. However, the low stability of phosphodiesters in the conditions of the oxidation of interglycosidic phosphonates poses limitations on the preparation of oligomers with two or more phosphodiester bridges. The application of phosphoramidites is, in general, more efficient and allows the preparation of longer oligomers.

The majority of the spacer-armed synthetic fragments of phosphoglycans described in this paper were converted into neoglycoconjugates, which are shown to be safe and immunogenic in laboratory animals. For *Hib*, the combination of surface plasmon resonance, saturation transfer difference nanomagnetic resonance, and X-ray crystallography evidences that the minimal epitope consists of only two repeating units [29]. In the research conducted by Seeberger et al. [69], the conjugate with a tetramer *Hib* antigen (the shortest of 4, 6, 8, and 10-mers) was shown to be the most immunogenic. For Hia, the CRM197 conjugates of 1, 2, 3, and 4-mers were almost equally immunogenic. Immunogenicity with the conjugates of the phosphono-mimetics of *MenA* was independent of the length of the pseudo-oligosaccharide antigens, and non-acetylated carba-mimetics of *MenA* failed to evoke anti-*MenA* immunity. For *MenX*, immunization with conjugates of mono-, di-, and trimer with CRM197 was inefficient. For phosphooligosaccharides related to *S. pneumoniae* capsular phosphoglycans, the correlation between the length and immunity was not analyzed, as the majority of synthetic antigens comprised only one repeating unit. It can be concluded that longer oligomers are needed for a more profound study of immunogenicity, antigenicity, and protective immunity of neoglycoconjugates on the basis of synthetic phosphooligosaccharides.

Although the first examples of polysaccharide assembly via polycondensation (for example see [166]) were published almost 40 years ago, and in the last decade, synthesis of a high-molecular weight glycan containing 25-mer related to arabinogalactan [167] and a rhamnomannan consisting of 256 monosaccharide units [168], as well as an automated glycan assembly of a 151-mer polymannoside [169] and 15-mer phosphoglycan [50] related to the LPG of *Leishmania donovani* were reported, long oligosaccharides, which relate to bacterial phosphoglycans, are yet unattainable. However, the synthetic approaches described above permit the preparation of sufficiently long oligosaccharides of this type, including spacer-armed molecules, which are applicable as indispensable tools in any glycobiology studies and glycotechnology developments.

As described in this paper, a number of *Hib* and *MenA* oligomers were synthesized using a solid support method. This method has since evolved into a flexible and universal strategy based on the use of automated synthesis platforms. In glycochemistry, the Seeberger group [170] has been developing this approach for almost 30 years, and, finally, the automated glycan assembly technology has been created, which provides quick access to homogeneous oligomers and polymers starting from the properly protected monomer blocks [171]. Recent advances in automated glycan assembly [50,172,173] offer a powerful synthetic technique for the accomplishment of key steps in oligosaccharide synthesis. Protected monomer blocks required for automated glycan assembly can be attained through HPLC-based automated synthetic facilities [174–176].

As mentioned, *MenA* and *Hib* phosphoglycans are extremely susceptible to hydrolytic cleavage. For the preparation of stable *MenA* antigens, C-phosphono-mimetics were synthesized as non-acetylated compounds, and carba-mimetics were synthesized in both non-acetylated and acetylated variants. Immunization with corresponding protein conjugates was not efficient. It can be concluded that the problem of the preparation of a stable *MenA* antigen still waits to be resolved. On the contrary, Seeberger et. al. found that the introduction of a substituent, e.g., a methyl group, at O-2 provides a stable hydrolysis-resistant synthetic polyribosylribitolphosphate, which can be used in anti-Hib vaccines [77]. Also, the stability of an interglycosidic phosphodiester linkage can be achieved by the replacement of a non-bridging oxygen atom with a borano group [177].

To date, Quimi-Hib® remains the only commercial vaccine with a fully synthetic phosphoglycan antigen. However, recent developments in glycobiology and glycochemistry provide new ways to the fundamental knowledge, which is pivotal for the design of new vaccine preparations on the basis of synthetic phosphooligosaccharides.

**Author Contributions:** Writing—original draft preparation, E.A.K., A.A.K., D.V.Y. and N.E.N.; writing—review and editing, E.A.K., D.V.Y. and N.E.N.; funding acquisition, N.E.N. All authors have read and agreed to the published version of the manuscript.

**Funding:** The study was performed with financial support from the Russian Science Foundation (project 19-73-30017-P).

**Institutional Review Board Statement:** Not applicable.

**Informed Consent Statement:** Not applicable.

**Conflicts of Interest:** The authors declare no conflicts of interest.

## Abbreviations

The following abbreviations are used in this manuscript:

| | |
|---|---|
| avDP | average degree of polymerization |
| $Ac_2O$ | acetic anhydride |
| Boc | tert-butyloxycarbonyl |
| BOM | benzyloxymethyl |
| CA | chloroacetyl |
| Cbz | benzyloxycarbonyl |
| CPS | capsular polysaccharide |
| CRM197 | cross-reacting material 197, a non-toxic mutant of the diphtheria toxin |
| DBU | 1,8-diazabicyclo(5.4.0)undec-7-ene |
| DCI | 4,5-dicyanoimidazole |
| DCSO | (+)-(10-camphorsulfonyl)oxaziridine |
| DIAD | diisopropyl azodicarboxylate |
| DIC | 1,3-diisopropylcarbodiimide |
| DIPEA | N,N-diisopropylethylamine |
| DMTr | bis(4-methoxyphenyl)-phenylmethyl |
| DMAP | 4-dimethylaminopyridine |
| DMSO | dimetylsulfoxide |
| DMTr | 4,4-dimethoxytrityl |
| EC50 | half maximal effective concentration |
| ELISA | enzyme-linked immunosorbent assay |
| EtOAc | ethyl acetate |
| EtOH | ethanol |
| HSA | human serum albumin |
| IC | inhibitory concentration |
| IgA | immunoglobulin A |
| IgG | immunoglobulin G |
| IgM | immunoglobulin M |
| KLH | keyhole limpet haemocyanin |
| mCPBA | 3-chloroperbenzoic acid |
| *MenA* | *Neisseria meningitidis* serogroup a |
| *MenX* | *Neisseria meningitidis* serogroup x |
| MMTr | monomethoxytrityl |
| MPM | 4-methoxybenzyl |
| Nap | 2-naphthylmethyl |
| MeOPN | O-methylphosphoramidate |
| OS | oligosaccharide |
| PBS | phosphate buffer saline |
| PCV | pneumococcal conjugated vaccines |
| PEG | polyethylene glycol |
| PivCl | pivaloyl chloride |
| $Ph_3P$ | triphenylphosphine |
| PhSH | thiophenol |
| PMP | 4-methoxyphenyl |
| pNP | 4-nitrophenyl |
| PRP | poly-3-β-D-ribosyl-(1→1)-D-ribitol-5-phosphate |
| Py | pyridine |
| rSBA | bactericidal assay (SBA) titers using rabbit serum |

| | |
|---|---|
| SMP | succinimidyl-3-maleimidopropionate |
| TBDPS | tert-butyldiphenylsilyl |
| TDS | thexyldimethylsilyl |
| TCA | trichloroacetic acid |
| TES | triethylsilane |
| TFA | trifluoroacetic acid |
| THF | tetrahydrofuran |
| TT | tetanus toxoid |

## References


1. Fivenson, E.M.; Dubois, L.; Bernhardt, T.G. Co-Ordinated Assembly of the Multilayered Cell Envelope of Gram-Negative Bacteria. *Curr. Opin. Microbiol.* **2024**, *79*, 102479. [CrossRef] [PubMed]
2. Ribet, D.; Cossart, P. How Bacterial Pathogens Colonize Their Hosts and Invade Deeper Tissues. *Microbes Infect.* **2015**, *17*, 173–183. [CrossRef] [PubMed]
3. Cescutti, P. Bacterial Capsular Polysaccharides and Exopolysaccharides. In *Microbial Glycobiology*; Holst, O., Brennan, P.J., von Itzstein, M., Moran, A.P., Eds.; Academic Press: San Diego, CA, USA, 2010; pp. 93–108. ISBN 978-0-12-374546-0.
4. Gao, S.; Jin, W.; Quan, Y.; Li, Y.; Shen, Y.; Yuan, S.; Yi, L.; Wang, Y.; Wang, Y. Bacterial Capsules: Occurrence, Mechanism, and Function. *npj Biofilms Microbiomes* **2024**, *10*, 21. [CrossRef] [PubMed]
5. Willis, L.M.; Whitfield, C. Structure, Biosynthesis, and Function of Bacterial Capsular Polysaccharides Synthesized by ABC Transporter-Dependent Pathways. *Carbohydr. Res.* **2013**, *378*, 35–44. [CrossRef]
6. Noel, G.J.; Hoiseth, S.K.; Edelson, P.J. Type b Capsule Inhibits Ingestion of Haemophilus Influenzae by Murine Macrophages: Studies with Isogenic Encapsulated and Unencapsulated Strains. *J. Infect. Dis.* **1992**, *166*, 178–182. [CrossRef]
7. Williams, A.E.; Maskell, D.J.; Moxon, E.R. Relationship between Intracellular Survival in Macrophages and Virulence of Haemophilus Influenzae Type b. *J. Infect. Dis.* **1991**, *163*, 1366–1369. [CrossRef]
8. Cress, B.F.; Englaender, J.A.; He, W.; Kasper, D.; Linhardt, R.J.; Koffas, M.A.G. Masquerading Microbial Pathogens: Capsular Polysaccharides Mimic Host-Tissue Molecules. *FEMS Microbiol. Rev.* **2014**, *38*, 660–697. [CrossRef]
9. Cyster, J.G.; Allen, C.D.C. B Cell Responses: Cell Interaction Dynamics and Decisions. *Cell* **2019**, *177*, 524–540. [CrossRef]
10. van der Put, R.M.F.; Metz, B.; Pieters, R.J. Carriers and Antigens: New Developments in Glycoconjugate Vaccines. *Vaccines* **2023**, *11*, 219. [CrossRef]
11. Rohokale, R.; Guo, Z. Development in the Concept of Bacterial Polysaccharide Repeating Unit-Based Antibacterial Conjugate Vaccines. *ACS Infect. Dis.* **2023**, *9*, 178–212. [CrossRef]
12. Sorieul, C.; Dolce, M.; Romano, M.R.; Codée, J.; Adamo, R. Glycoconjugate Vaccines against Antimicrobial Resistant Pathogens. *Expert Rev. Vaccines* **2023**, *22*, 1055–1078. [CrossRef]
13. Anderluh, M.; Berti, F.; Bzducha-Wróbel, A.; Chiodo, F.; Colombo, C.; Compostella, F.; Durlik, K.; Ferhati, X.; Holmdahl, R.; Jovanovic, D.; et al. Recent Advances on Smart Glycoconjugate Vaccines in Infections and Cancer. *FEBS J.* **2022**, *289*, 4251–4303. [CrossRef]
14. Berti, F.; Romano, M.R.; Micoli, F.; Adamo, R. Carbohydrate Based Meningococcal Vaccines: Past and Present Overview. *Glycoconj. J.* **2021**, *38*, 401–409. [CrossRef]
15. Khatuntseva, E.A.; Nifantiev, N.E. Glycoconjugate Vaccines for Prevention of Haemophilus Influenzae Type b Diseases. *Russ. J. Bioorganic Chem.* **2021**, *47*, 26–52. [CrossRef] [PubMed]
16. Micoli, F.; Bagnoli, F.; Rappuoli, R.; Serruto, D. The Role of Vaccines in Combatting Antimicrobial Resistance. *Nat. Rev. Microbiol.* **2021**, *19*, 287–302. [CrossRef] [PubMed]
17. Stefanetti, G.; Borriello, F.; Richichi, B.; Zanoni, I.; Lay, L. Immunobiology of Carbohydrates: Implications for Novel Vaccine and Adjuvant Design Against Infectious Diseases. *Front. Cell. Infect. Microbiol.* **2022**, *11*, 808005. [CrossRef] [PubMed]
18. Khatun, F.; Toth, I.; Stephenson, R.J. Immunology of Carbohydrate-Based Vaccines. *Adv. Drug Deliv. Rev.* **2020**, *165–166*, 117–126. [CrossRef]
19. Avci, F.; Berti, F.; Dull, P.; Hennessey, J.; Pavliak, V.; Prasad, A.K.; Vann, W.; Wacker, M.; Marcq, O. Glycoconjugates: What It Would Take to Master These Well-Known yet Little-Understood Immunogens for Vaccine Development. *mSphere* **2019**, *4*, e00520-19. [CrossRef]
20. Rappuoli, R. Glycoconjugate Vaccines: Principles and Mechanisms. *Sci. Transl. Med.* **2018**, *10*, eaat4615. [CrossRef]
21. Del Bino, L.; Østerlid, K.E.; Wu, D.-Y.; Nonne, F.; Romano, M.R.; Codée, J.; Adamo, R. Synthetic Glycans to Improve Current Glycoconjugate Vaccines and Fight Antimicrobial Resistance. *Chem. Rev.* **2022**, *122*, 15672–15716. [CrossRef]

22. Gening, M.L.; Kurbatova, E.A.; Nifantiev, N.E. Synthetic Analogs of Streptococcus Pneumoniae Capsular Polysaccharides and Immunogenic Activities of Glycoconjugates. *Russ. J. Bioorg. Chem.* **2021**, *47*, 1–25. [CrossRef]
23. Seeberger, P.H. Discovery of Semi- and Fully-Synthetic Carbohydrate Vaccines Against Bacterial Infections Using a Medicinal Chemistry Approach. *Chem. Rev.* **2021**, *121*, 3598–3626. [CrossRef]
24. Zasłona, M.E.; Downey, A.M.; Seeberger, P.H.; Moscovitz, O. Semi- and Fully Synthetic Carbohydrate Vaccines against Pathogenic Bacteria: Recent Developments. *Biochem. Soc. Trans.* **2021**, *49*, 2411–2429. [CrossRef]
25. Mettu, R.; Chen, C.-Y.; Wu, C.-Y. Synthetic Carbohydrate-Based Vaccines: Challenges and Opportunities. *J. Biomed. Sci.* **2020**, *27*, 9. [CrossRef]
26. Gening, M.L.; Kurbatova, E.A.; Tsvetkov, Y.E.; Nifantiev, N.E. Development of Approaches to a Conjugated Carbohydrate Vaccine of the Third Generation against Streptococcus Pneumoniae: The Search for Optimal Oligosaccharide Ligands. *Russ. Chem. Rev.* **2015**, *84*, 1100–1113. [CrossRef]
27. Anish, C.; Beurret, M.; Poolman, J. Combined Effects of Glycan Chain Length and Linkage Type on the Immunogenicity of Glycoconjugate Vaccines. *npj Vaccines* **2021**, *6*, 150. [CrossRef]
28. Kaplonek, P.; Seeberger, P.H. Glycan Microarrays Containing Synthetic Streptococcus Pneumoniae CPS Fragments and Their Application to Vaccine Development. *Methods Mol. Biol.* **2022**, *2460*, 193–206. [CrossRef]
29. Nonne, F.; Iacono, L.D.; Bertuzzi, S.; Unione, L.; Proietti, D.; Norais, N.; Margarit, I.; Adamo, R.; Jiménez-Barbero, J.; Carboni, F.; et al. A Multidisciplinary Structural Approach to the Identification of the Haemophilus Influenzae Type b Capsular Polysaccharide Protective Epitope. *ACS Cent. Sci.* **2024**, *10*, 978–987. [CrossRef]
30. Gerbst, A.G.; Nikolaev, A.V.; Yashunsky, D.V.; Shashkov, A.S.; Dmitrenok, A.S.; Nifantiev, N.E. Theoretical and NMR-Based Conformational Analysis of Phosphodiester-Linked Disaccharides. *Sci. Rep.* **2017**, *7*, 8934. [CrossRef]
31. Verez-Bencomo, V.; Fernández-Santana, V.; Hardy, E.; Toledo, M.E.; Rodríguez, M.C.; Heynngnezz, L.; Rodriguez, A.; Baly, A.; Herrera, L.; Izquierdo, M.; et al. A Synthetic Conjugate Polysaccharide Vaccine against Haemophilus Influenzae Type b. *Science* **2004**, *305*, 522–525. [CrossRef]
32. Rastogi, I.; Jeon, D.; Moseman, J.E.; Muralidhar, A.; Potluri, H.K.; McNeel, D.G. Role of B Cells as Antigen Presenting Cells. *Front. Immunol.* **2022**, *13*, 954936. [CrossRef]
33. Inpanathan, S.; Botelho, R.J. The Lysosome Signaling Platform: Adapting With the Times. *Front. Cell Dev. Biol.* **2019**, *7*, 113. [CrossRef]
34. Avalos, A.M.; Ploegh, H.L. Early BCR Events and Antigen Capture, Processing, and Loading on MHC Class II on B Cells. *Front. Immunol.* **2014**, *5*, 92. [CrossRef]
35. Knirel, Y.A.; Van Calsteren, M.-R. Bacterial Exopolysaccharides. In *Comprehensive Glycoscience*, 2nd ed.; Barchi, J.J., Ed.; Elsevier: Oxford, UK, 2021; pp. 21–95. ISBN 978-0-12-822244-7.
36. Whitfield, C. Structure and Assembly of Escherichia Coli Capsules. *EcoSal Plus* **2009**, *3*, 10–128. [CrossRef]
37. Tsui, F.P.; Egan, W.; Summers, M.F.; Byrd, R.A.; Schneerson, R.; Robbins, J.B. Determination of the Structure of the Escherichia Coli K100 Capsular Polysaccharide, Cross-Reactive with the Capsule from Type b Haemophilus Influenzae. *Carbohydr. Res.* **1988**, *173*, 65–74. [CrossRef]
38. Jann, K.; Jann, B.; Schmidt, M.A.; Vann, W.F. Structure of the Escherichia Coli K2 Capsular Antigen, a Teichoic Acid-like Polymer. *J. Bacteriol.* **1980**, *143*, 1108–1115. [CrossRef]
39. Nikolaev, A.V.; Botvinko, I.V.; Ross, A.J. Natural Phosphoglycans Containing Glycosyl Phosphate Units: Structural Diversity and Chemical Synthesis. *Carbohydr. Res.* **2007**, *342*, 297–344. [CrossRef]
40. Teufel, L.U.; Joosten, L.A.B.; dos Santos, J.C. Differential Structure and Immunomodulatory Functions of Lipophosphoglycan between *Leishmania* spp. *Immunol. Lett.* **2024**, *268*, 106885. [CrossRef]
41. Mikkola, S. Nucleotide Sugars in Chemistry and Biology. *Molecules* **2020**, *25*, 5755. [CrossRef]
42. Sobkowski, M.; Kraszewski, A.; Stawinski, J. Recent Advances in H-Phosphonate Chemistry. Part 1. H-Phosphonate Esters: Synthesis and Basic Reactions. *Top. Curr. Chem.* **2015**, *361*, 137–177. [CrossRef]
43. Kraszewski, A.; Stawinski, J. H-Phosphonates: Versatile Synthetic Precursors to Biologically Active Phosphorus Compounds. *Pure Appl. Chem.* **2007**, *79*, 2217–2227. [CrossRef]
44. Hall, R.H.; Alexander Todd, R.F. Webb 644. Nucleotides. Part Xli. Mixed Anhydrides as Intermediates in the Synthesis of Dinucleoside Phosphates. *J. Chem. Soc.* **1957**, 3291–3296. [CrossRef]
45. Aviñó, A.; Fàbrega, C.; Eritja, R. Nucleic Acids Chemistry: Modifications and Conjugates for Biomedicine and Nanotechnology. In *Methods for the Synthesis of Oligonucleotides*; De Gruyter: Berlin, Germany, 2021; pp. 1–44. ISBN 3-11-063579-8.
46. Garegg, P.J.; Lindh, I.; Regberg, T.; Stawinski, J.; Strömberg, R.; Henrichson, C. Nucleoside H-Phosphonates. III. Chemical Synthesis of Oligodeoxyribonucleotides by the Hydrogenphosphonate Approach. *Tetrahedron Lett.* **1986**, *34*, 4051–4054. [CrossRef]

47. Garegg, P.J.; Lindh, I.; Regberg, T.; Stawinski, J.; Strömberg, R.; Henrichson, C. Nucleoside H-Phosphonates. IV. Automated Solid Phase Synthesis of Oligoribonucleotides by the Hydrogenphosphonate Approach. *Tetrahedron Lett.* **1986**, *27*, 4055–4058. [CrossRef]
48. Nikolaev, A.V.; Shibaev, V.N.; Kochetkov, N.K. Glycosyl Phosphosugars via Glycosyl Hydrogenphosphonates. *Bioorg. Khim.* **1987**, *13*, 1591–1593.
49. Marugg, J.E.; Burik, A.; Tromp, M.; van der Marel, G.A.; van Boom, J.H. A New and Versatile Approach to the Preparation of Valuable Deoxynucleoside 3′-Phosphite Intermediates. *Tetrahedron Lett.* **1986**, *27*, 2271–2274. [CrossRef]
50. Kawato, K.; Sato, K.; Wada, T. α-Selective Solid-Phase Synthesis of Glycosyl Phosphate Repeating Structure Via the Phosphoramidite Method. *Chem. Eur. J.* **2024**, *30*, e202401226. [CrossRef]
51. Sato, K.; Hagio, T.; Sano, M.; Muramoto, K.; Yaoita, A.; Noro, M.; Hara, R.I.; Wada, T. Solid-Phase Stereocontrolled Synthesis of Oligomeric P-Modified Glycosyl Phosphate Derivatives Using the Oxazaphospholidine Method. *ACS Omega* **2021**, *6*, 20026–20041. [CrossRef]
52. Ma, Z.; Ensley, H.E.; Lowman, D.W.; Kruppa, M.D.; Williams, D.L. Recent Advances in Chemical Synthesis of Phosphodiester Linkages Found in Fungal Mannans. *Carbohydr. Res.* **2025**, *547*, 109325. [CrossRef]
53. Zhang, X.; Yang, Y.; Ding, J.; Zhao, Y.; Zhang, H.; Zhu, Y. Stereoselective Gold(I)-Catalyzed Approach to the Synthesis of Complex α-Glycosyl Phosphosaccharides. *Nat. Commun.* **2022**, *13*, 421. [CrossRef]
54. Yang, G.; Mei, G.; Shen, P.; Hong, H.; Wu, Z. Rapid Assembly of Phosphate-Bridged Tetra-Mannose by Ionic Liquid-Supported Oligosaccharide Synthesis. *Carbohydr. Res.* **2021**, *500*, 108209. [CrossRef] [PubMed]
55. Hara, R.I.; Kobayashi, S.; Noro, M.; Sato, K.; Wada, T. Synthesis and Properties of 2-Deoxy-2-Fluoromannosyl Phosphate Derivatives. *Tetrahedron* **2017**, *73*, 4560–4565. [CrossRef]
56. Oka, N.; Sato, K.; Wada, T. Recent Progress in the Synthesis of Glycosyl Phosphate Derivatives. *Trends Glycosci. Glycotechnol.* **2012**, *24*, 152–168. [CrossRef]
57. Fujita, S.; Oka, N.; Matsumura, F.; Wada, T. Synthesis of Oligo(α-D-Glycosyl Phosphate) Derivatives by a Phosphoramidite Method via Boranophosphate Intermediates. *J. Org. Chem.* **2011**, *76*, 2648–2659. [CrossRef]
58. Zamyatina, A.; Kosma, P. Synthesis of Anomeric Phosphates of Aldoses and 2-Ulosonic Acids. In *Carbohydrate Chemistry*; Academic Press: London, UK, 2009; Volume 35, pp. 71–98. ISBN 978-1-84755-970-8.
59. Egan, W.; Schneerson, R.; Werner, K.E.; Zon, G. Structural Studies and Chemistry of Bacterial Capsular Polysaccharides. Investigations of Phosphodiester-Linked Capsular Polysaccharides Isolated from Haemophilus Influenzae Types a, b, c, and f: NMR Spectroscopic Identification and Chemical Modification of End Groups and the Nature of Base-Catalyzed Hydrolytic Depolymerization. *J. Am. Chem. Soc.* **1982**, *104*, 2898–2910. [CrossRef]
60. de Oliveira Cintra, F.; Takagi, M. Study of the Chemical Stability of the Capsular Polysaccharide Produced by Haemophilus Influenzae Type b. *Carbohydr. Polym.* **2015**, *116*, 167–172. [CrossRef]
61. Lemercinier, X.; Jones, C. An NMR Spectroscopic Identity Test for the Control of the Capsular Polysaccharide from Haemophilus Influenzae Type b. *Biologicals* **2000**, *28*, 175–183. [CrossRef]
62. Sturgess, A.W.; Rush, K.; Charbonneau, R.J.; Lee, J.I.; West, D.J.; Sitrin, R.D.; Hennessy, J.P. Haemophilus Influenzae Type b Conjugate Vaccine Stability: Catalytic Depolymerization of PRP in the Presence of Aluminum Hydroxide. *Vaccine* **1999**, *17*, 1169–1178. [CrossRef]
63. Berti, F.; Romano, M.R.; Micoli, F.; Pinto, V.; Cappelletti, E.; Gavini, M.; Proietti, D.; Pluschke, G.; MacLennan, C.A.; Costantino, P. Relative Stability of Meningococcal Serogroup A and X Polysaccharides. *Vaccine* **2012**, *30*, 6409–6415. [CrossRef]
64. Hoogerhout, P.; Funke, C.W.; Mellema, J.-R.; Wagenaars, G.N.; Van Boeckel, C.A.A.; Evenberg, D.; Poolman, J.T.; Lefeber, A.W.M.; Van Der Marel, G.A.; Van Booma, J.H. Synthesis of Fragments of the Capsular Polysaccharide of Haemophilus Influnzae Type B Part II. Preparation and Structural Analysis of Fragments Comprising Two and Three Repeating Units. *J. Carbohydr. Chem.* **1988**, *7*, 399–416. [CrossRef]
65. Elie, C.J.J.; Muntendam, H.J.; van den Elst, H.; van der Marel, G.A.; van Boom, J.H.; Hoogerhout, P. Synthesis of Fragments of the Capsular Polysaccharide of Haemophilus Influenzae Type b: Part IIII-3. A Solid-phase Synthesis of a Spacer-containing Ribosylribitol Phosphate Hexamer. *Recl. Trav. Chim. Pays-Bas.* **1989**, *108*, 219–223. [CrossRef]
66. Kandil, A.A.; Chan, N.; Chong, P.; Klein, M. Synthesis of Fragments of the Capsular Polysaccharide of Haemophilus Influenzae Type b on Soluble Polymeric Support. *Synlett* **1992**, *7*, 555–557. [CrossRef]
67. Nilsson, S.; Bengtsson, M.; Norberg, T. Solid-Phase Synthesis of a Fragment of the Capsular Polysaccharide of Haemophilus Influenzae Type B Using H-Phosphonate Intermediates. *J. Carbohydr. Chem.* **1991**, *11*, 265–285. [CrossRef]
68. Fernandez Santana, V.; Peña Icart, L.; Beurret, M.; Costa, L.; Verez Bencomo, V. Glycoconjugate Vaccines against Haemophilus Influenzae Type b. *Methods Enzymol.* **2006**, *415*, 153–163. [CrossRef]

69. Baek, J.Y.; Geissner, A.; Rathwell, D.C.K.; Meierhofer, D.; Pereira, C.L.; Seeberger, P.H. A Modular Synthetic Route to Size-Defined Immunogenic Haemophilus Influenzae b Antigens Is Key to the Identification of an Octasaccharide Lead Vaccine Candidate. *Chem. Sci.* **2018**, *9*, 1279–1288. [CrossRef]
70. Hoogerhout, P.; Evenberg, D.; van Boeckel, C.A.A.; Poolman, J.T.; Beuvery, E.C.; van der Marel, G.A.; van Boom, J.H. Synthesis of Fragments of the Capsular Polysaccharide of *Haemophilus Influenzae* Type B, Comprising Two or Three Repeating Units. *Tetrahedron Lett.* **1987**, *28*, 1553–1556. [CrossRef]
71. Peeters, C.C.; Evenberg, D.; Hoogerhout, P.; Käyhty, H.; Saarinen, L.; van Boeckel, C.A.; van der Marel, G.A.; van Boom, J.H.; Poolman, J.T. Synthetic Trimer and Tetramer of 3-Beta-D-Ribose-(1-1)-D-Ribitol-5-Phosphate Conjugated to Protein Induce Antibody Responses to Haemophilus Influenzae Type b Capsular Polysaccharide in Mice and Monkeys. *Infect. Immun.* **1992**, *60*, 1826–1833. [CrossRef]
72. Kandil, A.A.; Chan, N.; Klein, M.; Chong, P. Chemical Synthesis of Haemophilus Influenzae Glycopeptide Conjugates. *Glycoconj. J.* **1997**, *14*, 13–17. [CrossRef]
73. Bacardí, D.; Cosme, K.; Aldana, L.; Merino, N.; Suárez, J.; Mosqueda, O.; Urquiza, D.; Romero, J.; Madrigal, R.; Amaya, R.; et al. Preclinical Safety Testing of the Quimi-Hib® Vaccine Adjuvanted with Aluminum Phosphate during Product Development. *Biotecnol. Apl.* **2013**, *30*, 118–124.
74. Toraño, G.; Toledo, M.E.; Baly, A.; Fernandez-Santana, V.; Rodriguez, F.; Alvarez, Y.; Serrano, T.; Musachio, A.; Hernandez, I.; Hardy, E.; et al. Phase I Clinical Evaluation of a Synthetic Oligosaccharide-Protein Conjugate Vaccine against Haemophilus Influenzae Type b in Human Adult Volunteers. *Clin. Vaccine Immunol.* **2006**, *13*, 1052–1056. [CrossRef]
75. Fernández-Santana, V.; Cardoso, F.; Rodriguez, A.; Carmenate, T.; Peña, L.; Valdés, Y.; Hardy, E.; Mawas, F.; Heynngnezz, L.; Rodríguez, M.C.; et al. Antigenicity and Immunogenicity of a Synthetic Oligosaccharide-Protein Conjugate Vaccine against Haemophilus Influenzae Type b. *Infect. Immun.* **2004**, *72*, 7115–7123. [CrossRef]
76. Käyhty, H.; Peltola, H.; Karanko, V.; Mäkelä, P.H. The Protective Level of Serum Antibodies to the Capsular Polysaccharide of Haemophilus Influenzae Type b. *J. Infect. Dis.* **1983**, *147*, 1100. [CrossRef] [PubMed]
77. Seeberger, P.H.; Pereira, C. Stable Hydrolysis-Resistant Synthetic Polyribosylribitolphosphate Derivatives as Vaccines Against Haemophilus Influenzae Type B. WO/2018/020046, 2 January 2018. Available online: https://patentscope.wipo.int/search/en/WO2018020046 (accessed on 25 June 2025).
78. Gilsdorf, J.R. Hib Vaccines: Their Impact on Haemophilus Influenzae Type b Disease. *J. Infect. Dis.* **2021**, *224*, S321–S330. [CrossRef] [PubMed]
79. Peltola, H. Worldwide Haemophilus Influenzae Type b Disease at the Beginning of the 21st Century: Global Analysis of the Disease Burden 25 Years after the Use of the Polysaccharide Vaccine and a Decade after the Advent of Conjugates. *Clin. Microbiol. Rev.* **2000**, *13*, 302–317. [CrossRef] [PubMed]
80. Bertran, M.; D'Aeth, J.C.; Hani, E.; Amin-Chowdhury, Z.; Fry, N.K.; Ramsay, M.E.; Litt, D.J.; Ladhani, S.N. Trends in Invasive Haemophilus Influenzae Serotype a Disease in England from 2008-09 to 2021-22: A Prospective National Surveillance Study. *Lancet Infect. Dis.* **2023**, *23*, 1197–1206. [CrossRef]
81. Shuel, M.; Knox, N.; Tsang, R.S.W. Global Population Structure of Haemophilus Influenzae Serotype a (Hia) and Emergence of Invasive Hia Disease: Capsule Switching or Capsule Replacement? *Can. J. Microbiol.* **2021**, *67*, 875–884. [CrossRef]
82. Slack, M.P.E.; Cripps, A.W.; Grimwood, K.; Mackenzie, G.A.; Ulanova, M. Invasive Haemophilus Influenzae Infections after 3 Decades of Hib Protein Conjugate Vaccine Use. *Clin. Microbiol. Rev.* **2021**, *34*, e0002821. [CrossRef]
83. Tsang, R.S.W.; Proulx, J.-F.; Hayden, K.; Shuel, M.; Lefebvre, B.; Boisvert, A.-A.; Moore, D. Characteristics of Invasive Haemophilus Influenzae Serotype a (Hia) from Nunavik, Canada and Comparison with Hia Strains in Other North American Arctic Regions. *Int. J. Infect. Dis.* **2017**, *57*, 104–107. [CrossRef]
84. Cox, A.D.; Barreto, L.; Ulanova, M.; Bruce, M.G.; Tsang, R. Conference contributors Developing a Vaccine for Haemophilus Influenzae Serotype a: Proceedings of a Workshop. *Can. Commun. Dis. Rep.* **2017**, *43*, 89–95. [CrossRef]
85. Boisvert, A.-A.; Moore, D. Invasive Disease Due to Haemophilus Influenzae Type A in Children in Canada's North: A Priority for Prevention. *Can. J. Infect. Dis. Med. Microbiol.* **2015**, *26*, 291–292. [CrossRef]
86. Ulanova, M.; Tsang, R.S.W. Haemophilus Influenzae Serotype a as a Cause of Serious Invasive Infections. *Lancet Infect. Dis.* **2014**, *14*, 70–82. [CrossRef] [PubMed]
87. Resman, F.; Ristovski, M.; Ahl, J.; Forsgren, A.; Gilsdorf, J.R.; Jasir, A.; Kaijser, B.; Kronvall, G.; Riesbeck, K. Invasive Disease Caused by Haemophilus Influenzae in Sweden 1997-2009; Evidence of Increasing Incidence and Clinical Burden of Non-Type b Strains. *Clin. Microbiol. Infect.* **2011**, *17*, 1638–1645. [CrossRef] [PubMed]
88. Campos, J.; Román, F.; Pérez-Vázquez, M.; Oteo, J.; Aracil, B.; Cercenado, E. Spanish Study Group for Haemophilus influenzae Type E Infections Due to Haemophilus Influenzae Serotype E: Microbiological, Clinical, and Epidemiological Features. *Clin. Infect. Dis.* **2003**, *37*, 841–845. [CrossRef] [PubMed]

89. Ribeiro, G.S.; Reis, J.N.; Cordeiro, S.M.; Lima, J.B.T.; Gouveia, E.L.; Petersen, M.; Salgado, K.; Silva, H.R.; Zanella, R.C.; Almeida, S.C.G.; et al. Prevention of Haemophilus Influenzae Type b (Hib) Meningitis and Emergence of Serotype Replacement with Type a Strains after Introduction of Hib Immunization in Brazil. *J. Infect. Dis.* **2003**, *187*, 109–116. [CrossRef]
90. Whittaker, R.; Economopoulou, A.; Dias, J.G.; Bancroft, E.; Ramliden, M.; Celentano, L.P. European Centre for Disease Prevention and Control Country Experts for Invasive Haemophilus influenzae Disease Epidemiology of Invasive Haemophilus Influenzae Disease, Europe, 2007–2014. *Emerg. Infect. Dis.* **2017**, *23*, 396–404. [CrossRef]
91. MacNeil, J.R.; Cohn, A.C.; Farley, M.; Mair, R.; Baumbach, J.; Bennett, N.; Gershman, K.; Harrison, L.H.; Lynfield, R.; Petit, S.; et al. Current Epidemiology and Trends in Invasive Haemophilus Influenzae Disease—United States, 1989–2008. *Clin. Infect. Dis.* **2011**, *53*, 1230–1236. [CrossRef]
92. Soeters, H.M.; Blain, A.; Pondo, T.; Doman, B.; Farley, M.M.; Harrison, L.H.; Lynfield, R.; Miller, L.; Petit, S.; Reingold, A.; et al. Current Epidemiology and Trends in Invasive Haemophilus Influenzae Disease-United States, 2009–2015. *Clin. Infect. Dis.* **2018**, *67*, 881–889. [CrossRef]
93. Jin, Z.; Romero-Steiner, S.; Carlone, G.M.; Robbins, J.B.; Schneerson, R. Haemophilus Influenzae Type a Infection and Its Prevention. *Infect. Immun.* **2007**, *75*, 2650–2654. [CrossRef]
94. Kohout, C.V. Synthesis of Haemophilus Influenzae Type a Oligosaccharides for Vaccine Development. Doctoral Thesis, Università degli Studi di Milano, Milan, Italy, 2021.
95. Kohout, C.V.; Del Bino, L.; Petrosilli, L.; D'Orazio, G.; Romano, M.R.; Codée, J.D.C.; Adamo, R.; Lay, L. Semisynthetic Glycoconjugates as Potential Vaccine Candidates Against Haemophilus Influenzae Type a. *Chem. Eur. J.* **2024**, *30*, e202401695. [CrossRef]
96. Kamneva, A.A.; Yashunsky, D.V.; Khatuntseva, E.A.; Nifantiev, N.E. Synthesis of Pseudooligosaccharides Related to the Capsular Phosphoglycan of Haemophilus Influenzae Type a. *Molecules* **2023**, *28*, 5688. [CrossRef]
97. Hansson, J.; Garegg, P.J.; Oscarson, S. Syntheses of Anomerically Phosphodiester-Linked Oligomers of the Repeating Units of the Haemophilus Influenzae Types c and f Capsular Polysaccharides. *J. Org. Chem.* **2001**, *66*, 6234–6243. [CrossRef] [PubMed]
98. Roberts, L. Vaccine Introduction. The Beginning of the End for Africa's Devastating Meningitis Outbreaks? *Science* **2010**, *330*, 1466–1467. [CrossRef] [PubMed]
99. Frasch, C.E.; Preziosi, M.-P.; LaForce, F.M. Development of a Group A Meningococcal Conjugate Vaccine, MenAfriVac(TM). *Hum. Vaccines Immunother.* **2012**, *8*, 715–724. [CrossRef] [PubMed]
100. Trotter, C.L.; Lingani, C.; Fernandez, K.; Cooper, L.V.; Bita, A.; Tevi-Benissan, C.; Ronveaux, O.; Préziosi, M.-P.; Stuart, J.M. Impact of MenAfriVac in Nine Countries of the African Meningitis Belt, 2010–2015: An Analysis of Surveillance Data. *Lancet Infect. Dis.* **2017**, *17*, 867–872. [CrossRef]
101. WHO. *WHO/IVB/13.04 Use of MenAfriVac™ (Meningitis A Vaccine) in a Controlled Temperature Chain (CTC) During Campaigns: Guidance for Immunization Programme Decision-Makers and Managers*; WHO Press: Geneva, Switzerland, 2013.
102. Lydon, P.; Zipursky, S.; Tevi-Benissan, C.; Djingarey, M.H.; Gbedonou, P.; Youssouf, B.O.; Zaffran, M. Economic Benefits of Keeping Vaccines at Ambient Temperature during Mass Vaccination: The Case of Meningitis A Vaccine in Chad. *Bull. World Health Organ.* **2014**, *92*, 86–92. [CrossRef]
103. Lemercinier, X.; Jones, C. Full 1H NMR Assignment and Detailed O-Acetylation Patterns of Capsular Polysaccharides from Neisseria Meningitidis Used in Vaccine Production. *Carbohydr. Res.* **1996**, *296*, 83–96. [CrossRef]
104. Gudlavalleti, S.K.; Szymanski, C.M.; Jarrell, H.C.; Stephens, D.S. In Vivo Determination of Neisseria Meningitidis Serogroup A Capsular Polysaccharide by Whole Cell High-Resolution Magic Angle Spinning NMR Spectroscopy. *Carbohydr. Res.* **2006**, *341*, 557–562. [CrossRef]
105. Berti, F.; De Ricco, R.; Rappuoli, R. Role of O-Acetylation in the Immunogenicity of Bacterial Polysaccharide Vaccines. *Molecules* **2018**, *23*, 1340. [CrossRef]
106. Berry, D.S.; Lynn, F.; Lee, C.-H.; Frasch, C.E.; Bash, M.C. Effect of O Acetylation of Neisseria Meningitidis Serogroup A Capsular Polysaccharide on Development of Functional Immune Responses. *Infect. Immun.* **2002**, *70*, 3707–3713. [CrossRef]
107. Utkina, N.S.; Eliseeva, G.I.; Nikolaev, A.V.; Shibaev, V.N. Fragmenty biopolimerov, soderzhashchikh ostatki glikozilfosfatov. 12. Sintez glikozilfosfosakharov, soderzhashchikh ostatki 2-acetamido-2-dezoksi-alpha-D-mannopiranozilfosfata, v tom chisle fragmenta kapsul'nogoantigena Neisseria meningitidis A [Biopolymer fragments, containing glycosylphosphate residues. 12. Synthesis of glycosylphosphosugars, containing 2-acetamido-2-deoxy-alpha-D-mannopyranosylphosphate residues including a fragment of the Neisseria meningitidis A capsule antigen]. *Bioorg. Khim.* **1993**, *19*, 228–235.
108. Sato, K.; Chiba, A.; Shiraishi, T.; Ogawa, Y.; Hara, R.I.; Wada, T. Solid-Phase Synthesis of N-Trichloroacetyl Mannosamine 1-Phosphate Repeating Units Mimicking Capsular Polysaccharide Derived from Neisseria Meningitidis Serotype, A. *Carbohydr. Res.* **2022**, *518*, 108585. [CrossRef]

109. Harale, K.R.; Rout, J.K.; Chhikara, M.K.; Gill, D.S.; Misra, A.K. Synthesis and Immunochemical Evaluation of a Novel Neisseria Meningitidis Serogroup A Tetrasaccharide and Its Conjugate. *Org. Chem. Front.* **2017**, *4*, 2348–2357. [CrossRef]
110. Black, A. Synthesis of Fragments of the Capsular Polysaccharide of Neisseria Meningitidis (Serogroup A) Suitable for Bioconjugation. Doctoral Thesis, University of Dundee, Dundee, Scotland, 2014.
111. Slättegård, R.; Teodorovic, P.; Kinfe, H.H.; Ravenscroft, N.; Gammon, D.W.; Oscarson, S. Synthesis of Structures Corresponding to the Capsular Polysaccharide of Neisseria Meningitidis Group, A. *Org. Biomol. Chem.* **2005**, *3*, 3782–3787. [CrossRef] [PubMed]
112. Berkin, A.; Coxon, B.; Pozsgay, V. Towards a Synthetic Glycoconjugate Vaccine against Neisseria Meningitidis, A. *Chem. Eur. J.* **2002**, *8*, 4424–4433. [CrossRef] [PubMed]
113. Torres-Sanchez, M.I.; Zaccaria, C.; Buzzi, B.; Miglio, G.; Lombardi, G.; Polito, L.; Russo, G.; Lay, L. Synthesis and Biological Evaluation of Phosphono Analogues of Capsular Polysaccharide Fragments from Neisseria Meningitidis, A. *Chem. Eur. J.* **2007**, *13*, 6623–6635. [CrossRef]
114. Teodorović, P.; Slättegård, R.; Oscarson, S. Synthesis of Stable C-Phosphonate Analogues of Neisseria Meningitidis Group A Capsular Polysaccharide Structures Using Modified Mitsunobu Reaction Conditions. *Org. Biomol. Chem.* **2006**, *4*, 4485–4490. [CrossRef]
115. Torres-Sanchez, M.I.; Draghetti, V.; Panza, L.; Lay, L.; Russo, G. Synthesis of the Phosphono Analogue of the Dimeric Subunit of Neisseria Meningitidis Type A Capsular Polysaccharide. *Synlett* **2005**, *2005*, 1147–1151. [CrossRef]
116. Enotarpi, J.; Tontini, M.; Balocchi, C.; van der Es, D.; Auberger, L.; Balducci, E.; Carboni, F.; Proietti, D.; Casini, D.; Filippov, D.V.; et al. A Stabilized Glycomimetic Conjugate Vaccine Inducing Protective Antibodies against Neisseria Meningitidis Serogroup, A. *Nat. Commun.* **2020**, *11*, 4434. [CrossRef]
117. Gao, Q.; Tontini, M.; Brogioni, G.; Nilo, A.; Filippini, S.; Harfouche, C.; Polito, L.; Romano, M.R.; Costantino, P.; Berti, F.; et al. Immunoactivity of Protein Conjugates of Carba Analogues from Neisseria Meningitidis a Capsular Polysaccharide. *ACS Chem. Biol.* **2013**, *8*, 2561–2567. [CrossRef]
118. Gao, Q.; Zaccaria, C.; Tontini, M.; Poletti, L.; Costantino, P.; Lay, L. Synthesis and Preliminary Biological Evaluation of Carba Analogues from Neisseria Meningitidis A Capsular Polysaccharide. *Org. Biomol. Chem.* **2012**, *10*, 6673–6681. [CrossRef]
119. Fiebig, T.; Freiberger, F.; Pinto, V.; Romano, M.R.; Black, A.; Litschko, C.; Bethe, A.; Yashunsky, D.; Adamo, R.; Nikolaev, A.; et al. Molecular Cloning and Functional Characterization of Components of the Capsule Biosynthesis Complex of Neisseria Meningitidis Serogroup A: Toward in Vitro Vaccine Production. *J. Biol. Chem.* **2014**, *289*, 19395–19407. [CrossRef]
120. Westerduin, P.; Veeneman, G.H.; van der Marel, G.A.; van Boom, J.H. Synthesis of the Fragment GlcNAc-α(1→p→6)-GlcNAc of the Cell Wall Polymer of Staphylococcus Lactis Having Repeating N-Acetyl-D-Glucosamine Phosphate Units. *Tetrahedron Lett.* **1986**, *27*, 6271–6274. [CrossRef]
121. Toma, L.; Legnani, L.; Rencurosi, A.; Poletti, L.; Lay, L.; Russo, G. Modeling of Synthetic Phosphono and Carba Analogues of N-Acetyl-Alpha-D-Mannosamine 1-Phosphate, the Repeating Unit of the Capsular Polysaccharide from Neisseria Meningitidis Serovar, A. *Org. Biomol. Chem.* **2009**, *7*, 3734–3740. [CrossRef] [PubMed]
122. Calloni, I.; Unione, L.; Jiménez-Osés, G.; Corzana, F.; Del Bino, L.; Corrado, A.; Pitirollo, O.; Colombo, C.; Lay, L.; Adamo, R.; et al. The Conformation of the Mannopyranosyl Phosphate Repeating Unit of the Capsular Polysaccharide of Neisseria Meningitidis Serogroup a and Its Carba-Mimetic. *Eur. J. Org. Chem.* **2018**, *2018*, 4548–4555. [CrossRef] [PubMed]
123. Hlozek, J.; Ravenscroft, N.; Kuttel, M.M. Modeling the Conformations of Neisseria Meningitidis Serogroup a CPS and a Carba-Analogue: Implications for Vaccine Development. *Carbohydr. Res.* **2019**, *486*, 107838. [CrossRef]
124. Fallarini, S.; Buzzi, B.; Giovarruscio, S.; Polito, L.; Brogioni, G.; Tontini, M.; Berti, F.; Adamo, R.; Lay, L.; Lombardi, G. A Synthetic Disaccharide Analogue from Neisseria Meningitidis A Capsular Polysaccharide Stimulates Immune Cell Responses and Induces Immunoglobulin G (IgG) Production in Mice When Protein-Conjugated. *ACS Infect. Dis.* **2015**, *1*, 487–496. [CrossRef]
125. Arjona, O.; Gómez, A.M.; López, J.C.; Plumet, J. Synthesis and Conformational and Biological Aspects of Carbasugars. *Chem. Rev.* **2007**, *107*, 1919–2036. [CrossRef]
126. Balmer, P.; Burman, C.; Serra, L.; York, L.J. Impact of Meningococcal Vaccination on Carriage and Disease Transmission: A Review of the Literature. *Hum. Vaccines Immunother.* **2018**, *14*, 1118–1130. [CrossRef]
127. Xie, O.; Pollard, A.J.; Mueller, J.E.; Norheim, G. Emergence of Serogroup X Meningococcal Disease in Africa: Need for a Vaccine. *Vaccine* **2013**, *31*, 2852–2861. [CrossRef]
128. Soumahoro, L.; Abitbol, V.; Vicic, N.; Bekkat-Berkani, R.; Safadi, M.A.P. Meningococcal Disease Outbreaks: A Moving Target and a Case for Routine Preventative Vaccination. *Infect. Dis. Ther.* **2021**, *10*, 1949–1988. [CrossRef]
129. Morelli, L.; Cancogni, D.; Tontini, M.; Nilo, A.; Filippini, S.; Costantino, P.; Romano, M.R.; Berti, F.; Adamo, R.; Lay, L. Synthesis and Immunological Evaluation of Protein Conjugates of Neisseria Meningitidis X Capsular Polysaccharide Fragments. *Beilstein J. Org. Chem.* **2014**, *10*, 2367–2376. [CrossRef] [PubMed]

130. Micoli, F.; Romano, M.R.; Tontini, M.; Cappelletti, E.; Gavini, M.; Proietti, D.; Rondini, S.; Swennen, E.; Santini, L.; Filippini, S.; et al. Development of a Glycoconjugate Vaccine to Prevent Meningitis in Africa Caused by Meningococcal Serogroup, X. *Proc. Natl. Acad. Sci. USA* **2013**, *110*, 19077–19082. [CrossRef] [PubMed]
131. Pietri, G.P.; Tontini, M.; Brogioni, B.; Oldrini, D.; Robakiewicz, S.; Henriques, P.; Calloni, I.; Abramova, V.; Santini, L.; Malić, S.; et al. Elucidating the Structural and Minimal Protective Epitope of the Serogroup X Meningococcal Capsular Polysaccharide. *Front. Mol. Biosci.* **2021**, *8*, 745360. [CrossRef] [PubMed]
132. Eliseeva, G.I.; Ivanova, I.A.; Nikolaev, A.V.; Shibaev, V.N. Fragments of Biopolymers Containing Glycosyl-Phosphate Residues. 8. Hydrogenphosphonate Synthesis of Glycosyl-Phosphosugars Containing Alpha-d-Glucopyranosyl Phosphate and 2-Acetamido-2-Deoxy-Alpha-d-Glucopyranosyl Phosphate Residues, Fragments of Bacteria Capsular Antigens and of Glycoproteins. *Bioorg. Khim.* **1991**, *17*, 1401–1411.
133. Morelli, L.; Lay, L. Synthesis of Neisseria Meningitidis X Capsular Polysaccharide Fragments. *Arkivoc* **2012**, 166–184. [CrossRef]
134. Harale, K.R.; Dumare, N.B.; Singh, D.; Misra, A.K.; Chhikara, M.K. Synthesis of a Tetrasaccharide and Its Glycoconjugate Corresponding to the Capsular Polysaccharide of Neisseria Meningitidis Serogroup X and Its Immunochemical Studies. *RSC Adv.* **2015**, *5*, 41332–41340. [CrossRef]
135. Oldrini, D.; Fiebig, T.; Romano, M.R.; Proietti, D.; Berger, M.; Tontini, M.; De Ricco, R.; Santini, L.; Morelli, L.; Lay, L.; et al. Combined Chemical Synthesis and Tailored Enzymatic Elongation Provide Fully Synthetic and Conjugation-Ready Neisseria Meningitidis Serogroup X Vaccine Antigens. *ACS Chem. Biol.* **2018**, *13*, 984–994. [CrossRef]
136. Brooks, L.R.K.; Mias, G.I. Streptococcus Pneumoniae's Virulence and Host Immunity: Aging, Diagnostics, and Prevention. *Front. Immunol.* **2018**, *9*, 1366. [CrossRef]
137. Javed; Mandal, P.K. Bacterial Surface Capsular Polysaccharides from Streptococcus Pneumoniae: A Systematic Review on Structures, Syntheses, and Glycoconjugate Vaccines. *Carbohydr. Res.* **2021**, *502*, 108277. [CrossRef]
138. Garcia Quesada, M.; Yang, Y.; Bennett, J.C.; Hayford, K.; Zeger, S.L.; Feikin, D.R.; Peterson, M.E.; Cohen, A.L.; Almeida, S.C.G.; Ampofo, K.; et al. Serotype Distribution of Remaining Pneumococcal Meningitis in the Mature PCV10/13 Period: Findings from the PSERENADE Project. *Microorganisms* **2021**, *9*, 738. [CrossRef]
139. Cleary, D.W.; Jones, J.; Gladstone, R.A.; Osman, K.L.; Devine, V.T.; Jefferies, J.M.; Bentley, S.D.; Faust, S.N.; Clarke, S.C. Changes in Serotype Prevalence of Streptococcus Pneumoniae in Southampton, UK between 2006 and 2018. *Sci. Rep.* **2022**, *12*, 13332. [CrossRef]
140. Sheppard, C.L.; Pichon, B.; George, R.C.; Hall, L.M.C. Streptococcus Pneumoniae Isolates Expressing a Capsule with Epitopes of Both Serotypes 6A and 6B. *Clin. Vaccine Immunol.* **2010**, *17*, 1820–1822. [CrossRef]
141. Pérez-García, C.; Sempere, J.; de Miguel, S.; Hita, S.; Úbeda, A.; Vidal, E.J.; Llorente, J.; Limia, A.; de Miguel, A.G.; Sanz, J.C.; et al. Surveillance of Invasive Pneumococcal Disease in Spain Exploring the Impact of the COVID-19 Pandemic (2019-2023). *J. Infect.* **2024**, *89*, 106204. [CrossRef]
142. von Gottberg, A.; Kleynhans, J.; de Gouveia, L.; Tempia, S.; Meiring, S.; Quan, V.; du Plessis, M.; von Mollendorf, C.; Crowther-Gibson, P.; Avenant, T.; et al. Long-Term Effect of Pneumococcal Conjugate Vaccines on Invasive Pneumococcal Disease Incidence among People of All Ages from National, Active, Laboratory-Based Surveillance in South Africa, 2005–2019: A Cohort Observational Study. *Lancet Glob. Health* **2024**, *12*, e1470–e1484. [CrossRef] [PubMed]
143. Singh, J.; Sundaresan, S.; Manoharan, A.; Shet, A. Serotype Distribution and Antimicrobial Susceptibility Pattern in Children$\leq$5years with Invasive Pneumococcal Disease in India—A Systematic Review. *Vaccine* **2017**, *35*, 4501–4509. [CrossRef] [PubMed]
144. Sukhova, E.V.; Yashunsky, D.V.; Kurbatova, E.A.; Akhmatova, E.A.; Tsvetkov, Y.E.; Nifantiev, N.E. Synthesis and Preliminary Immunological Evaluation of a Pseudotetrasaccharide Related to a Repeating Unit of the Streptococcus Pneumoniae Serotype 6A Capsular Polysaccharide. *Front. Mol. Biosci.* **2021**, *8*, 754753. [CrossRef]
145. Mettu, R.; Lih, Y.-H.; Vulupala, H.R.; Chen, C.-Y.; Hsu, M.-H.; Lo, H.-J.; Liao, K.-S.; Cheng, Y.-Y.; Chiu, C.-H.; Wu, C.-Y. Synthetic Library of Oligosaccharides Derived from the Capsular Polysaccharide of Streptococcus Pneumoniae Serotypes 6A and 6B and Their Immunological Studies. *ACS Infect. Dis.* **2022**, *8*, 626–634. [CrossRef]
146. Thijssen, M.J.; van Rijswijk, M.N.; Kamerling, J.P.; Vliegenthart, J.F. Synthesis of Spacer-Containing Di- and Tri-Saccharides That Represent Parts of the Capsular Polysaccharide of Streptococcus Pneumoniae Type 6B. *Carbohydr. Res.* **1998**, *306*, 93–109. [CrossRef]
147. Thijssen, M.J.; Bijkerk, M.H.; Kamerling, J.P.; Vliegenthart, J.F. Synthesis of Four Spacer-Containing "tetrasaccharides" That Represent Four Possible Repeating Units of the Capsular Polysaccharide of Streptococcus Pneumoniae Type 6B. *Carbohydr. Res.* **1998**, *306*, 111–125. [CrossRef]

148. Sukhova, E.V.; Yashunsky, D.V.; Kurbatova, E.A.; Tsvetkov, Y.E.; Nikolay, E.N. Synthesis of a Pseudotetrasaccharide Corresponding to a Repeating Unit of the Streptococcus Pneumoniae Type 6B Capsular Polysaccharide*. *J. Carbohydr. Chem.* **2018**, *37*, 1–17. [CrossRef]
149. Grant, L.R.; Hanquet, G.; Sepúlveda-Pachón, I.T.; Theilacker, C.; Baay, M.; Slack, M.P.E.; Jodar, L.; Gessner, B.D. Effects of PCV10 and PCV13 on Pneumococcal Serotype 6C Disease, Carriage, and Antimicrobial Resistance. *Vaccine* **2024**, *42*, 2983–2993. [CrossRef]
150. Feemster, K.; Hausdorff, W.P.; Banniettis, N.; Platt, H.; Velentgas, P.; Esteves-Jaramillo, A.; Burton, R.L.; Nahm, M.H.; Buchwald, U.K. Implications of Cross-Reactivity and Cross-Protection for Pneumococcal Vaccine Development. *Vaccines* **2024**, *12*, 974. [CrossRef]
151. Mettu, R.; Cheng, Y.-Y.; Vulupala, H.R.; Lih, Y.-H.; Chen, C.-Y.; Hsu, M.-H.; Lo, H.-J.; Liao, K.-S.; Chiu, C.-H.; Wu, C.-Y. Chemical Synthesis of Truncated Capsular Oligosaccharide of Serotypes 6C and 6D of Streptococcus Pneumoniae with Their Immunological Studies. *ACS Infect. Dis.* **2024**, *10*, 2161–2171. [CrossRef] [PubMed]
152. Jansen, W.T.; Hogenboom, S.; Thijssen, M.J.; Kamerling, J.P.; Vliegenthart, J.F.; Verhoef, J.; Snippe, H.; Verheul, A.F. Synthetic 6B Di-, Tri-, and Tetrasaccharide-Protein Conjugates Contain Pneumococcal Type 6A and 6B Common and 6B-Specific Epitopes That Elicit Protective Antibodies in Mice. *Infect. Immun.* **2001**, *69*, 787–793. [CrossRef] [PubMed]
153. Weinberger, D.M.; Harboe, Z.B.; Sanders, E.A.M.; Ndiritu, M.; Klugman, K.P.; Rückinger, S.; Dagan, R.; Adegbola, R.; Cutts, F.; Johnson, H.L.; et al. Association of Serotype with Risk of Death Due to Pneumococcal Pneumonia: A Meta-Analysis. *Clin. Infect. Dis.* **2010**, *51*, 692–699. [CrossRef] [PubMed]
154. Soliveri, G.; Bertolotti, A.; Panza, L.; Poletti, L.; Lay, L.; Jones, C. Synthesis of Phosphorylated Fragments of Streptococcus Pneumoniae Type 19F Capsular Polysaccharide. *J. Chem. Soc. Perkin 1* **2002**, 2174–2181. [CrossRef]
155. Desmet, S.; Theeten, H.; Laenen, L.; Cuypers, L.; Maes, P.; Bossuyt, W.; Van Heirstraeten, L.; Peetermans, W.E.; Lagrou, K. Characterization of Emerging Serotype 19A Pneumococcal Strains in Invasive Disease and Carriage, Belgium. *Emerg. Infect. Dis.* **2022**, *28*, 1606–1614. [CrossRef]
156. Isturiz, R.; Sings, H.L.; Hilton, B.; Arguedas, A.; Reinert, R.-R.; Jodar, L. Streptococcus Pneumoniae Serotype 19A: Worldwide Epidemiology. *Expert Rev. Vaccines* **2017**, *16*, 1007–1027. [CrossRef]
157. Sanapala, S.R.; Seco, B.M.S.; Baek, J.Y.; Awan, S.I.; Pereira, C.L.; Seeberger, P.H. Chimeric Oligosaccharide Conjugate Induces Opsonic Antibodies against Streptococcus Pneumoniae Serotypes 19A and 19F. *Chem. Sci.* **2020**, *11*, 7401–7407. [CrossRef]
158. Kuttel, M.M.; Jackson, G.E.; Mafata, M.; Ravenscroft, N. Capsular Polysaccharide Conformations in Pneumococcal Serotypes 19F and 19A. *Carbohydr. Res.* **2015**, *406*, 27–33. [CrossRef]
159. Morelli, L.; Lay, L.; Santana-Mederos, D.; Valdes-Balbin, Y.; Verez Bencomo, V.; van Diepen, A.; Hokke, C.H.; Chiodo, F.; Compostella, F. Glycan Array Evaluation of Synthetic Epitopes between the Capsular Polysaccharides from Streptococcus Pneumoniae 19F and 19A. *ACS Chem. Biol.* **2021**, *16*, 1671–1679. [CrossRef] [PubMed]
160. Tikhomirova, A.; McNabb, E.R.; Petterlin, L.; Bellamy, G.L.; Lin, K.H.; Santoso, C.A.; Daye, E.S.; Alhaddad, F.M.; Lee, K.P.; Roujeinikova, A. Campylobacter Jejuni Virulence Factors: Update on Emerging Issues and Trends. *J. Biomed. Sci.* **2024**, *31*, 45. [CrossRef] [PubMed]
161. Zhu, W.; Li, K.; Cui, T.; Yan, Y. Detection of Anti-Ganglioside Antibodies in Guillain-Barré Syndrome. *Ann. Transl. Med.* **2023**, *11*, 289. [CrossRef] [PubMed]
162. Poly, F.; Noll, A.J.; Riddle, M.S.; Porter, C.K. Update on Campylobacter Vaccine Development. *Hum. Vaccines Immunother.* **2019**, *15*, 1389–1400. [CrossRef]
163. Thota, V.N.; Lowary, T.L. Synthesis of 6-Deoxy-d-Ido-Heptopyranose-Containing Fragments of the Campylobacter Jejuni Strain CG8486 Capsular Polysaccharide. *Carbohydr. Res.* **2024**, *536*, 109058. [CrossRef]
164. Wang, X.; Chen, Y.; Wang, J.; Yang, Y. Total Synthesis of the Trisaccharide Antigen of the Campylobacter Jejuni RM1221 Capsular Polysaccharide via de Novo Synthesis of the 6-Deoxy-d- Manno-Heptose Building Blocks. *J. Org. Chem.* **2019**, *84*, 2393–2403. [CrossRef]
165. Cloutier, M.; Gauthier, C. Progress toward the Development of Glycan-Based Vaccines against Campylobacteriosis. *ACS Infect. Dis.* **2021**, *7*, 969–986. [CrossRef]
166. Kochetkov, N.K.; Nifant'ev, N.E.; Backinowsky, L.V. Synthesis of the Capsular Polysaccharide of Streptococcus Pneumoniae Type 14. *Tetrahedron* **1987**, *43*, 3109–3121. [CrossRef]
167. Pasari, S.; Manmode, S.; Walke, G.; Hotha, S. A Versatile Synthesis of Pentacosafuranoside Subunit Reminiscent of Mycobacterial Arabinogalactan Employing One Strategic Glycosidation Protocol. *Chem. Eur. J.* **2018**, *24*, 1128–1139. [CrossRef]
168. Zhu, Q.; Shen, Z.; Chiodo, F.; Nicolardi, S.; Molinaro, A.; Silipo, A.; Yu, B. Chemical Synthesis of Glycans up to a 128-Mer Relevant to the O-Antigen of Bacteroides Vulgatus. *Nat. Commun.* **2020**, *11*, 4142. [CrossRef]

169. Joseph, A.A.; Pardo-Vargas, A.; Seeberger, P.H. Total Synthesis of Polysaccharides by Automated Glycan Assembly. *J. Am. Chem. Soc.* **2020**, *142*, 8561–8564. [CrossRef] [PubMed]
170. Seeberger, P.H. Automated Carbohydrate Synthesis as Platform to Address Fundamental Aspects of Glycobiology--Current Status and Future Challenges. *Carbohydr. Res.* **2008**, *343*, 1889–1896. [CrossRef] [PubMed]
171. Danglad-Flores, J.; Sletten, E.T.; Reuber, E.E.; Bienert, K.; Riegler, H.; Seeberger, P.H. Optimized Platform for Automated Glycan Assembly. *Device* **2024**, *2*, 100499. [CrossRef]
172. Guberman, M.; Seeberger, P.H. Automated Glycan Assembly: A Perspective. *J. Am. Chem. Soc.* **2019**, *141*, 5581–5592. [CrossRef] [PubMed]
173. Guberman, M.; Bräutigam, M.; Seeberger, P.H. Automated Glycan Assembly of Lewis Type I and II Oligosaccharide Antigens. *Chem. Sci.* **2019**, *10*, 5634–5640. [CrossRef]
174. Kashiwagi, G.A.; Petrosilli, L.; Escopy, S.; Lay, L.; Stine, K.J.; De Meo, C.; Demchenko, A.V. HPLC-Based Automated Synthesis and Purification of Carbohydrates. *Chemistry* **2024**, *30*, e202401214. [CrossRef]
175. Lin, M.-H.; Wolf, J.B.; Sletten, E.T.; Cambié, D.; Danglad-Flores, J.; Seeberger, P.H. Enabling Technologies in Carbohydrate Chemistry: Automated Glycan Assembly, Flow Chemistry and Data Science. *Chembiochem* **2023**, *24*, e202200607. [CrossRef]
176. Escopy, S.; Singh, Y.; Stine, K.J.; Demchenko, A.V. HPLC-Based Automated Synthesis of Glycans in Solution. *Chem. Eur. J.* **2022**, *28*, e202201180. [CrossRef]
177. Sato, K.; Muramoto, K.; Hagio, T.; Hara, R.I.; Wada, T. Solid-Phase Synthesis of Glycosyl Phosphate Repeating Units via Glycosyl Boranophosphates as Stable Intermediates. *Org. Lett.* **2023**, *25*, 3927–3931. [CrossRef]

molecules

MDPI

*Review*

# Hans Paulsen: Contributions to the Investigations of Glycoprotein Biosynthesis

**Inka Brockhausen**

Department of Biomedical and Molecular Sciences, Queen's University, Kingston, ON K7L 3N6, Canada; brockhau@queensu.ca

**Abstract**

Hans Paulsen was one of the first scientists who believed that chemistry should be applied to biology and medicine. His interest in natural products and their roles solidified in the 1970s. He passed on his knowledge to hundreds of students and coworkers and advanced science with many national and international collaborators. No matter where he was, at home or travelling, he was always curious and keen to learn, from chemistry to enzymes, their roles in diseases, and the possible applications of synthetic compounds. His creative chemistry and synthesis of novel compounds made essential contributions to elucidating the mechanisms and pathways of glycoprotein biosynthesis. This review describes the biosynthetic pathways of the O- and N-glycans of glycoproteins and studies of novel substrates and inhibitors developed by Hans Paulsen's group.

**Keywords:** glycosyltransferases; glycoproteins; N-glycosylation; mucins; O-glycosylation; substrates; inhibitors; glycopeptides

## 1. Introduction

Hans Paulsen was a highly respected and successful carbohydrate chemist, valued and admired by his many coworkers, students, and collaborators. He was extremely knowledgeable in chemistry and science, but also in other cultures, politics, and history. Hans Paulsen was kind and compassionate and always interested until his passing in 2024 at the age of 102. His major strengths included an incredible memory and his ability to give informative and humorous lectures. When he developed a research plan, his excitement was contagious, and his brilliance stimulated his students to do well. He was one of the first scientists who believed that chemistry should be applied to biology and medicine. His interest in natural products and their roles solidified in the 1970s. He passed on his knowledge to hundreds of students and coworkers, and advanced science with many national and international collaborators. Paulsen visited many countries and studied their cultures and history. He enjoyed visits to Toronto, Amherst Island, and the Northern wilderness of Canada. The natural environment was conducive to making plans and discussing the value of current and future experiments and approaches to solving biochemical problems. The wilderness did not distract him from learning about enzymes, their roles in diseases, and their possible applications in solving problems through biochemistry. Always curious and full of humour, politeness, and modesty, he was a real pleasure to collaborate with.

Knowledge of the enzymes and mechanisms involved in transferring sugar residues to proteins was minimal in the 1980s. Hans Paulsen's tremendous efforts and many contributions were critical for the advancement in the field of glycoprotein biosynthesis. Accounts of Hans Paulsen's long and successful life as a carbohydrate chemist professor and emeritus have been published in journals and symposia. His main interests were carbohydrate chemistry, synthesis of unusual carbohydrate residues and linkages, selective oxidation, and specific modifications and substitutions of carbohydrate derivatives. This led to many glycan derivatives that proved to be valuable in determining enzyme specificities for their carbohydrate substrates and biosynthetic pathways.

Multiple novel chemical reactions were developed by Paulsen that established the groundwork for further chemistry of carbohydrates. While working with Kurt Heyns, Paulsen accomplished the catalytic oxidation of unusual carbohydrates [1]. He studied the role of sugar configuration and conformation in catalysis using platinum contact [2]. A major effort was made in the synthesis of carbohydrates containing nitrogen in the ring and their conformational analyses by NMR [3,4]. Many of these compounds were later shown to be potent glycosidase inhibitors [5]. Conformational analyses by NMR were developed and became standard analyses for carbohydrates [6]. He contributed to an understanding of the exo-anomeric effect and the dynamics of carbohydrate conformations. The chemistry of acyloxonium ion rearrangements was applied to many different compounds [7,8], including phosphate-containing carbohydrates [9]. These studies were extended to the synthesis and analysis of glycoprotein fragments [10]. Carbohydrates were analyzed in solution and in crystal form [11,12], contributing to our knowledge of carbohydrate structures and their biological roles.

Paulsen's efforts to advance the analyses of carbohydrate conformations were helpful in delineating the substrate binding of glycosyltransferases (GTs) and in designing specific substrate analog inhibitors. He also performed analyses of natural products, which turned out to be glucosidase and mannosidase inhibitors. These compounds could be valuable for therapy development in diseases associated with alterations of specific glycans or for controlling the immune system that relies on glycan recognition. The advancements in the structural analysis of glycans were crucial to the success of synthetic chemistry and its application in glycobiology and medicine.

The intensive efforts of the Paulsen lab succeeded in the synthesis of specific glycan derivatives for studies of glycobiology and the role of glycoproteins in cancer and immunology. Many other results of Paulsen's research led to knowledge of bacterial lipopolysaccharides and the production of specific antibodies and glycoprotein antigens. Thus, Paulsen's ideas and efforts in chemical synthesis and refined methods of structural analysis advanced our knowledge, especially in cancer cell biology. The wealth of synthetic compounds led to new knowledge of glycosyltransferase (GT) specificity, pathways, and glycoprotein functions. Compounds representing specific GT substrates were synthesized [13], allowing assays of enzyme preparation from cells or tissues expressing multiple GTs [14].

This review describes the biosynthesis of N-glycans and mucin type O-glycans by human enzymes and summarizes the work done by the Paulsen group to use synthetic chemistry for the acquisition of knowledge of biosynthetic enzymes and insight into the role of glycoproteins in health and disease.

## 2. Glycosyltransferases

The enzymes that build complex sugar chains of glycoproteins are glycosyltransferases (GTs), which have been classified into 138 families by the CAZy database (as of July 2025), based on their amino acid sequence, predicted 3-dimensional fold, mechanism, and known activity. Inverting GTs inverts the anomeric configuration of the sugar in the donor substrate to form the opposite linkage in the reaction product, while retaining GTs retain this linkage. The nucleotide sugar donor substrates for glycoproteins are UDP-αGlcNAc, UDP-αGal, UDP-αGalNAc, GDP-αMan, GDP-βFuc and CMP-βSialic acid. Dolichol (Dol) is used for membrane-bound intermediates in the endoplasmic reticulum (ER)—in particular, Dol-P-Manβ and Dol-P-Glcβ. The GT protein folds are GT-A, GT-B, and GT-C (CAZy), which have been confirmed with a limited number of crystal structures (Table 1). Table 1 lists the major GTs involved in the biosynthesis of N-glycans and mucin-type O-glycans.

**Table 1.** List of major human glycosyltransferases and glycosidases involved in glycoprotein biosynthesis. The GT names and accession numbers are derived from the Uniprot database. Included here are also the major glycoside hydrolases (GHs glucosidases and mannosidases) that are essential for the maturing of N-glycans in the ER and Golgi. *GTs with known crystal structure from human or animal sources; GALNT1,2,3,4,7,10,12 have been crystalized. CAZy GT, Carbohydrate Active Enzyme database classification; Dol, dolichol; GnT, GlcNAc-transferase; P, phosphate; Sia, sialyl; T, transferase. GT Mechanisms and Folds, and CAZy GT classification: Inverting (11,14,16,17,18,29,31,54), Inverting GT-A (2,7,12,13), Inverting GT-B (1,10,23,33), Inverting GT-C (22,57,58,59), Retaining (32), Retaining GT-A (6,27), Retaining GT-B (4).

| Enzyme Name | Uniprot | CAZy GT | Donor | Product |
|---|---|---|---|---|
| ER: N-glycan biosynthesis | | | | |
| *GlcNAc-P-T DPAGT | Q9H3H5 | - | UDP-GlcNAc | GlcNAc-PP-Dol |
| β4GlcNAcT ALG13/14 | Q9NP73 | 1 | UDP-GlcNAc | $GlcNAc_2$-PP-Dol |
| | Q96F25 | | | |
| DPM1 | O60762 | 2 | GDP-Man | Manβ-P-Dol |
| β4ManT ALG1 | Q9BT22 | 33 | GDP-Man | Man-$GlcNAc_2$-PP-Dol |
| α3/6ManT ALG2 | Q9H553 | 4 | GDP-Man | $Man_3$-$GlcNAc_2$-PP-Dol |
| α2ManT ALG11 | Q2TAA5 | 4 | GDP-Man | $Man_5$-$GlcNAc_2$-PP-Dol |
| α3ManT ALG3 | Q92685 | 58 | Dol-P-Man | $Man_7$-$GlcNAc_2$-PP-Dol |
| α2ManT ALG9 | Q9H6U8 | 22 | Dol-P-Man | $Man_9$-$GlcNAc_2$-PP-Dol |
| α2ManT ALG12 | Q9BV10 | 22 | Dol-P-Man | $Man_9$-$GlcNAc_2$-PP-Dol |

**Table 1.** *Cont.*

| Enzyme Name | Uniprot | CAZy GT | Donor | Product |
|---|---|---|---|---|
| DolPGlc synthase ALG5 | Q9Y673 | 2 | UDP-Glc | Glcβ-P-Dol |
| α3GlcT ALG6 | Q9Y672 | 57 | Dol-P-Glc | Glc-$Man_9$-$GlcNAc_2$-Dol |
| α3GlcT ALG8 | Q9BVK2 | 57 | Dol-P-Glc | $Glc_2$-$Man_9$-$GlcNAc_2$-Dol |
| α2GlcT ALG10 | Q5BKT4 | 59 | Dol-P-Glc | $Glc_3$-$Man_9$-$GlcNAc_2$-Dol |
| OST complex | - | - | Glycan-Dol | Glycan-Asn |
| α2Glc MOGS | Q13724 | GH63 | | $Glc_2$-$Man_9$-$GlcNAc_2$-Asn |
| α3Glc GANAB | Q14697 | GH31 | | $Man_9$-$GlcNAc_2$-Asn |
| α2Man MAN1B1 | Q9UKM7 | GH47 | | $Man_8$-$GlcNAc_2$-Asn |
| Golgi: N-glycan processing | | | | |
| α2Man MAN1A1 | P33908 | GH47 | | $Man_5GlcNAc_2$-Asn |
| *β2GnT I MGAT1 | P26572 | 13 | UDP-GlcNAc | GlcNAcβ2$Man_5GlcNAc_2$- |
| α3,6Man MAN2A1 | Q16706 | GH38 | | GlcNAcβ2-$Man_3GlcNAc_2$- |
| *β2GnT II MGAT2 | Q10469 | 16 | UDP-GlcNAc | GlcNAcβ$2_2Man_3GlcNAc_2$- |
| β4GnT III MGAT3 | Q09327 | 17 | UDP-GlcNAc | bisecting GlcNAcβ4Manβ- |
| β4GnT IV MGAT4A | Q9UM21 | 54 | UDP-GlcNAc | GlcNAc$β_3Man_3GlcNAc_2$- |
| *β6GnT V MGAT5 | Q09328 | 18 | UDP-GlcNAc | GlcNAc$β_4Man_3GlcNAc_2$- |
| *GNPTAB | Q3T906 | - | UDP-GlcNAc | GlcNAc-6P-Man |
| GNPTG complex | Q9UJJ9 | - | UDP-GlcNAc | GlcNAc-6P-Man |
| *α6FucT FUT8 | Q9BYC5 | 23 | GDP-Fuc | N-glycan core Fuc |
| *α6SiaT ST6Gal1 | P15907 | 29 | CMP-Sia | complex N-glycans |
| α8SiaT ST8SIA2 | Q92186 | 29 | CMP-Sia | Siaα2-8Sia- |
| *α8SiaT ST8SIA3 | O43173 | 29 | CMP-Sia | Siaα2-8Sia- |
| α8SiaT ST8SIA4 | Q92187 | 29 | CMP-Sia | Siaα2-8Sia- |
| Golgi: O-glycan biosynthesis | | | | |
| *GALNT1 | Q10472 | 27 | UDP-GalNAc | O-glycan initiation |
| *β3GalT C1GALT1 | Q9NS00 | 31 | UDP-Gal | core 1 |
| *β6GnT GCNT1 | Q02742 | 14 | UDP-GlcNAc | core 2 |
| β6GnT GCNT3 | O95395 | 14 | UDP-GlcNAc | core 2/4 |
| β6GnT GCNT4 | Q9P109 | 14 | UDP-GlcNAc | core2 |
| β3GnT B3GNT3 | Q9Y2A9 | 14 | UDP-GlcNAc | core 1,2 elongation |
| β3GnT B3GNT6 | Q6ZMB0 | 31 | UDP-GlcNAc | core 3 |
| α4GnT A4GNT | Q9UNA3 | 32 | UDP-GlcNAc | GlcNAcα4-Gal of core 2 |
| *α3SiaT ST3Gal1 | Q11201 | 29 | CMP-Sia | Sia-core 1 |
| α6SiaT ST6GalNAc1 | K7EMB6 | 29 | CMP-Sia | Sia-GalNAc-Thr |
| *α6SiaT ST6GalNAc2 | Q9UJ37 | 29 | CMP-Sia | Sia-GalNAc/Gal-GalNAc-Thr |
| α6SiaT ST6GalNAc3 | Q8NDV1 | 29 | CMP-Sia | Sia-Gal-(Sia)-GalNAc) |
| α6SiaT ST6GalNAc4 | Q9H4F1 | 29 | CMP-Sia | Sia-Gal (Sia)-GalNAc |

**Table 1.** *Cont.*

| Enzyme Name | Uniprot | CAZy GT | Donor | Product |
|---|---|---|---|---|
| Golgi: N- and O-glycan chain extension and termination | | | | |
| *iβ3GnT B3GNT2 | Q9NY97 | 31 | UDP-GlcNAc | i antigen |
| β3GnT B3GN8 | Q7Z7M8 | 31 | UDP-GlcNAc | i antigen |
| β6GnT GCNT2 | Q8N0V5 | 14 | UDP-GlcNAc | I antigen |
| β4GalNAcT B4GALNT3 | Q6L9W6 | 7 | UDP-GalNAc | Lac-diNAc |
| *α3GalNAcT, GTA, BGAT | P16442 | 6 | UDP-GalNAc | Blood group A |
| | V5ZDP0 | | | |
| β4GalNAcT B4GALNT2 | Q8NHY0 | 12 | UDP-GalNAc | Sda antigen |
| *α3GalT GTB | V9GWR7 | 6 | UDP-Gal | Blood group B |
| *β4GalT B4GALT1 | P15291 | 7 | UDP-Gal | GlcNAc extension |
| β4GalT B4GALT2 | O60909 | 7 | UDP-Gal | GlcNAc extension |
| β4GalT B4GALT3 | O60512 | 7 | UDP-Gal | GlcNAc extension |
| *β3GalT B3GALT5 | Q9Y2C3 | 31 | UDP-Gal | GlcNAc extension |
| α2FucT FUT1 | P19526 | 11 | GDP-Fuc | Blood group H/O |
| α2FucT FUT2 | Q10981 | 11 | GDP-Fuc | Blood group H/O |
| α3/4FucT FUT3 | P21217 | 10 | GDP-Fuc | Lewis antigens |
| α3FucT FUT4 | P22083 | 10 | GDP-Fuc | Lewis antigens, type 2 chains |
| α3/4FucT FUT5 | Q11128 | 10 | GDP-Fuc | Lewis antigens |
| α3FucT FUT7 | Q11130 | 10 | GDP-Fuc | sialyl-Lewis |
| *α3FucT FUT9 | Q9Y231 | 10 | GDP-Fuc | Lewis antigens |
| α3SiaT ST3Gal3 | Q11203 | 29 | CMP-Sia | Sia-Gal-GlcNAc- |
| α3SiaT ST3Gal4 | Q11206 | 29 | CMP-Sia | Sia-core 1/Gal-GlcNAc |
| α3SiaT ST3Gal6 | Q9Y274 | 29 | CMP-Sia | Sia-Gal-GlcNAc |
| α8SiaT ST8SIA6 | P61647 | 29 | CMP-Sia | Siaα2-8Sia |

Some of these GTs have broad specificities and thus can extend a number of glycoconjugates, including glycolipids and/or proteoglycans. Many GTs occur as natural variants that may have slightly different properties and specificities. Only one GT (C1GALT1) requires the coexpression of a chaperone (COSMC). Often, GTs form dimers or multi-enzyme complexes that support activities. GTs requiring Dol intermediates are localized to the membranes of the ER. The initial N-glycan structures up to $Man_5GlcNAc_2$- are made on the cytoplasmic side of the ER membrane, and after being flipped to the lumenal side, N-glycans are completed and transferred from Dol to nascent peptide. Further N-glycan processing and extension reactions occur after the transport of the glycoprotein to the Golgi.

GTA-folded GTs have one major fold with a central catalytic domain. GT-B folded GTs have 2 typical Rossman-like (nucleotide-binding) folds and the active site located in between them. GT-C folded GTs are uncommon, and Dol-binding GTs often have multiple membrane-spanning domains that are typical of GT-C folded enzymes [15]. The enzyme mechanisms of a few GTs have been suggested by their crystal structures, analysis of mutants, and enzyme kinetics. Many GTs are N-linked glycoproteins and can also be O-glycosylated.

In contrast to GTs, glycoside hydrolases are classified into 194 families (as of July 2025), with many subfamilies. These enzymes hydrolyze a glycosidic bond using a catalytic acid as proton donor and nucleophilic catalytic base. Like GTs, hydrolases act with an inverting or retaining mechanism and have a limited number of protein folds.

Most inverting GTs appear to follow a single displacement mechanism where a specific hydroxyl of the acceptor substrate forms a nucleophile that attacks carbon-1 of the sugar linked to phosphate of the nucleotide sugar donor substrate. This often involves a catalytic Asp residue in a DxD sequence [16]. The mechanisms of retaining GTs where sugar linkages in donor and product are the same remain to be clarified further. In the Golgi, GTs are commonly type II membrane proteins with a short N-terminal cytoplasmic domain followed by a transmembrane TM domain, and the folded GT domain in the lumen of the Golgi.

GlcNAc-transferases (GnT, MGAT) that synthesize N-glycan antennae are highly specific for the antennae they synthesize [17]. This specificity is likely a result of the large internal substrate-binding cavity that accommodates more than the terminal glycan. This is in contrast to certain Gal-transferases (GalTs) that can act on a single sugar residue. For example, the β1,4-GalT B4GALT1 only requires GlcNAc as a substrate, although some branch specificity for N-glycan acceptors has been observed, which may be due to steric factors [18,19]. Acceptor substrates have been designed that have all of the structures required for recognition by a specific enzyme but are as small as possible, conveniently synthesized, and handled by enzyme assays. GTs are usually highly specific for their donor substrates, with small modifications of the sugar moiety being tolerated and allowing a modified sugar to be transferred.

## 3. Measurements of Glycosyltransferase Activities

The general GT reaction is Acceptor + Donor → Product + Nucleotide. The reaction often requires a divalent metal ion to complex the nucleotide (e.g., $Mn^{2+}$ or $Mg^{2+}$), along with a buffer and detergent if a membrane-bound version of GT is used. If the enzyme source has contaminating hydrolases, specific inhibitors can increase the product yield. The GT activities can be measured accurately by tracking the transfer of a radioactive sugar residue to form the reaction product. When neutral acceptor substrates are used, the radioactive nucleotide sugar and sugar-phosphate breakdown products can be separated after incubation by anion exchange chromatography. Scintillation counting of the reaction product in the flow-through fractions will indicate the exact amount of sugar transferred, allowing the specific enzyme activity and kinetics to be determined. Most of Paulsen's compounds have been examined using the sensitive radioactive method.

If charged acceptors are used, then other assay methods could include hydrophobic chromatography, HPLC, electrophoresis, or paper chromatography. Assays using non-radioactive donor substrates require HPLC separation of assay components and mass spectrometry (MS) for estimating the product yield. TLC can also be used with a panel of standard compounds. Another method is to detect the reaction product UDP by bioluminescence (UDP-Glo TM). In order to determine the linkage synthesized, specific hydrolases, antibodies, or lectins could be employed. If sufficient amounts of reaction product are available, NMR is the best method for determining anomeric configuration and sugar linkage in the product.

For accurate and efficient measurements of GT activities, it is crucial to design a suitable acceptor substrate and select the best method to determine the amount and the new linkage in the reaction products. Paulsen's N-glycan derivatives with an octyl aglycone group were ideal for isolating reaction products by columns of silica-linked hydrophobic chains of 18 carbons. By omitting specific sugar residues, the minimum size of the active acceptor could be assessed. For GTs acting on O-glycans, aryl aglycone groups have been successful in both anion exchange and hydrophobic chromatography assays. Glycopeptide acceptors were used mainly by anion exchange chromatography and HPLC.

## 4. Biological Significance of Glycoproteins

Most eukaryotic proteins are post-translationally modified with glycans and are found inside cells, on cell surfaces, and in secretions. The Asn-linked N-glycan and Ser/Thr-linked mucin-type O-glycan structures in human glycoproteins contain the same basic sugar residues, except that these O-glycans do not have Glc and Man residues. Many of the internal and terminal carbohydrate epitopes of extended glycan structures are similar. The N- and O-glycans of glycoproteins serve to provide protein stability, conformation, folding, intracellular targeting, resistance to degradation, and dehydration. Protein glycosylation is involved in many biological functions, in the immune system, during development and fertilization, and as lubricant. The glycan moieties can interact with microbes and protect against the invasion of pathogens. Lectins and antibodies bind to glycans, which play important roles in cell–cell and cell surface receptors interactions.

## 5. Glycoprotein Biosynthesis

The initial pathways of N-glycan synthesis involve Dol lipid carriers at the cytoplasmic and lumenal sides of the ER membrane. Preformed oligosaccharides are transferred from Dol to peptides in the ER (Figure 1). In contrast, O-glycans are all added to the protein in the Golgi and do not involve Dol. During N-glycan processing, Glc and Man are removed by specific hydrolases, a process not known for O-glycan synthesis. Sulfotransferases can add sulfate esters in the Golgi to Gal or GlcNAc residues of both N- and O-glycans by transferring sulfate from 3′-phosphoadenosine-5′-phosphosulfate.

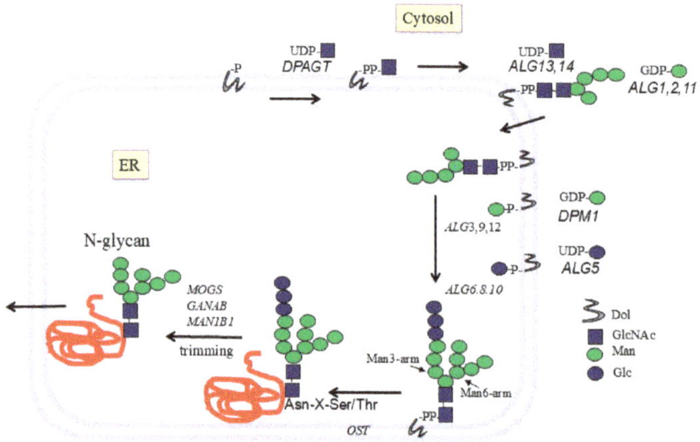


**Figure 1.** Early pathways of N-glycan biosynthesis in the ER.

Paulsen's work has contributed immensely to our knowledge of specific enzymes and pathways. The substrate binding and catalytic mechanisms of the enzymes identified in this work may differ between homologues from different species and tissues, and many enzyme variants occur. Nature has provided human mutations that unfortunately result in disease and suffering but has helped the biochemist to identify enzyme functions and the necessity for protein glycosylation. Thus, chemistry also played an important role in the discovery of the significance of glycosylation in disease.

Mucin glycoproteins have a high content of mucin-type GalNAcα-Ser/Thr-based O-glycans. Although many alterations of O-glycans in disease have been identified, there are no human mutants that lack mucin O-glycans, and 20 or more polypeptide GalNAc-transferases (GALNTs) cooperate in humans to assemble these O-glycans, indicating that they are essential for survival. The O-glycosylation of non-mucin proteins depends on the protein folding and exposure of Ser/Thr residues as potential O-glycosylation sites.

There are a number of other O-glycans that are important, e.g., O-αMan glycans of dystroglycans that may be altered in muscular dystrophies. O-αFuc is found in signalling proteins, and O-βGal in collagens. O-GlcNAcβ-Ser/Thr (O-GlcNAc), along with phosphate modifications, is common for cytoplasmic and nuclear proteins and functions in a wide range of biological activities, from apoptosis to oxidative stress responses. O-GlcNAc is not extended by additional sugars in the Golgi. However, it can be modified in vitro by β1,4-GalT. Alterations of O-GlcNAc have been implicated in protein aggregation, neurological diseases, cancer, and diabetes.

## 6. N-Glycans

N-glycans are found in all animals, as well as in many other species, and even in archaea and certain bacteria. Viruses usually carry glycoproteins with glycans that are characteristic of the cell type the virus originates from. High-mannose N-glycans have only Man residues extending the $Man_3GlcNAc_2$ core and are the least processed chains. Hybrid N-glycans have several Man residues on the Man6 arm and a GlcNAcβ1-2 antenna on the Man residue of the Man3 arm (Figure 2).

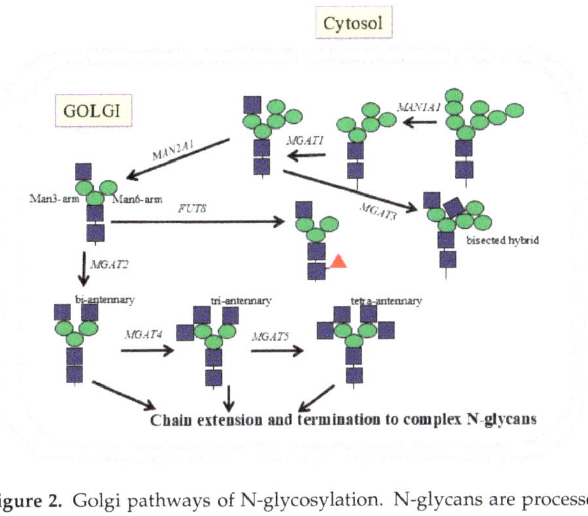


**Figure 2.** Golgi pathways of N-glycosylation. N-glycans are processed to hybrid or complex N-glycans and extension of the antennae in the Golgi. Glycans are trimmed by mannosidase, followed by the transfer of GlcNAc in β1-2 linkage to Manα1-3 of the Man3-arm by MGAT1. This facilitates further synthesis by α6-fucosylation of core GlcNAc attached to Asn, trimming of Man residues on the Man6-arm, and synthesis of further antennae by MGAT2,4,5 and bisecting structures by MGAT3. The antennae are then extended by a wealth of different enzymes that introduce repeating Gal-GlcNAc units and many possible branched and linear epitopes, Lewis antigens, blood groups, and tissue-specific structures. Glycoproteins are then transported to their preferred locations.

The antennae of complex N-glycans often consist of repeating Galβ1-4GlcNAcβ1-3- (LacNAc) sequences that can carry many possible internal and terminal sugars and epitopes, such as GlcNAc, Gal, GalNAc, Fuc, sialic acid (Sia), blood groups, i and I antigens, and LacdiNAc and Lewis determinants (Table 2). Many of these are shared with complex O-glycans and glycosphingolipids. The aberrant synthesis of N-glycans and O-glycans, leading to altered amounts of epitopes in cancer, is controlled by many enzymes. The enzymatic basis for these changes is extremely complex and usually involves the altered expression of a number of enzymes involved in the glycosylation pathways.

**Table 2.** List of mucin core structures and epitopes of both O- and N-glycans. Human glycoproteins have O-glycan structures initiated with GalNAc or core structures 1 to 4. Other core structures (5 to 8) are rare. Based on the variable tandem repeat sequences of MUC1, MUC2, MUC3, and MUC4, Paulsen's group synthesized hundreds of glycopeptides carrying one or more O-glycans with core 1 to 7. The occurrences of most of these structures have been found to be altered (increased or decreased) in cancer [20–31].

| Epitope | Structure | Alteration in Cancer |
|---|---|---|
| Tn antigen | GalNAcα-Ser/Thr- | Increased |
| Sialyl-Tn antigen | Siaα2-6GalNAcα- | Increased |
| Core 1, T antigen | Galβ1-3GalNAcα- | Increased |
| Sialyl-T antigen | Siaα2-3Galβ1-3GalNAcα- | Increased |
| Core 2 | GlcNAcβ1-6(Galβ1-3)GalNAcα- | Variable |
| Core 3 | GlcNAcβ1-3GalNAcα- | Decreased |
| Core 4 | GlcNAcβ1-6(GlcNAcβ1-3)GalNAcα- | Decreased |
| Core 5 | GalNAcα1-3GalNAcα- | Increased |
| Core 6 | GlcNAcβ1-6GalNAcα- | |
| Core 7 | GalNAcα1-6GalNAc- | |
| Core 8 | Galα1-3GalNAc- | |
| Sialyl-Lewis x | Siaα2-3Galβ1-4(Fucα1-3)GlcNAcβ- | Increased |
| Sda/Cad | GalNAcβ1-4(Siaα2-3)Galβ- | Decreased |
| i antigen | Galβ1-4GlcNAcβ1-3Galβ1- | Variable |
| I antigen | Galβ1-4GlcNAcβ1-6 | |
| | (Galβ1-4GlcNAcβ1-3)Galβ1- | |
| LacdiNAc | GalNAcβ1-4GlcNAcβ1-3- | |
| Sialyl-LacNAc | Siaα2,6Galβ1-4GlcNAc | Increased |
| Blood group O(H) | Fucα1-2Galβ- | |
| Blood group A | GalNAcα1-3(Fucα1-2)Galβ- | |
| Blood group B | Galα1-3(Fucα1-2)Galβ- | |
| Complex N-glycans | Highly branched N-glycans | Increased GnT V |
| Sialylated termini | Sialylated N-glycans | Increased ST6Gal1 |

N-glycans have gained considerable interest due to their multiple functions in protein structure, biology, and medicine. The functionally important N-glycan antennae cover a significant amount of space around glycoproteins and can be recognized by antibodies or lectins and carbohydrate-binding proteins of the immune system. N-glycan structures are often altered in cancer and in patients with congenital disorders of glycosylation (CDG), which can affect the properties and essential functions of glycoproteins [32]. For example, in cancer cells, GnT V (Table 1, Figure 2) was shown to increase in activity, thus expanding the size and complexity of branched tetra-antennary N-glycans. Thus, our focus was on designing inhibitors for N-glycan branching enzymes and controlling the functions of N-glycans.

The pathways of N-glycan synthesis are highly complex, involving many highly conserved enzymes in the ER, Golgi, cytoplasmic, and nuclear compartments and various mechanisms of transport and localization. In the early 1980s, Hans Paulsen started a fruitful collaboration with Harry Schachter (University of Toronto) to elucidate the substrate specificities and control of N-glycosylation pathways. For this purpose, Paulsen synthesized a series of N-glycans with systematic modifications of hydroxyl groups in order to explore their importance in GnT activities. Compounds included deoxy derivatives or those with specific reactive or non-reactive modifications and linkers. N-glycans were synthesized as reducing sugars or with a hydrophobic aglycone group that was useful for enzymatic reactions and isolation of enzyme products by HPLC. These derivatives helped determine the detailed recognition of substrates and enzyme specificities. Depending on the enzyme, specific hydroxyl groups were found to be critically required, and modifications in the stereochemistry, bulkiness, and hydrophobicity of substituents had variable effects on activity. These studies were complemented by our understanding of protein structures, their proposed mechanism, and the amino acids controlling the active site [33].

## 7. Early Pathways of N-Glycosylation at ER Membranes

N-glycosylation involves both nucleotide sugars and Dol-P-sugars. The early pathways to the $Man_5GlcNAc_2$-PP-dolichol structure are catalyzed by enzymes in the ER membrane. Reactions require UDP-GlcNAc and GDP-Man nucleotide sugars present in the cytosol. The N-glycan is then flipped across the membrane for further processing. In order to synthesize N-glycans, the cell must provide nucleotide sugar donor substrates and acceptor substrates that are Dol-P intermediates for the initial pathways (Figure 1) localized to the ER membrane. These early reactions have been studied mainly in yeast [34]. The biochemical importance of several of these enzymes was identified in humans with gene mutations leading to the CDG syndrome. Due to the reduction or absence of specific activities involved in the assembly of N-glycan structures, glycoproteins lack N-glycans and exhibit different electrophoretic mobility. This defect can be used as a diagnostic marker for CDG.

The initial cytoplasmic reaction is the transfer of GlcNAcα-P from UDP-GlcNAc to Dol-P by GlcNAc-1-P-transferase, named ALG7, in yeast, to form Dol-PP-αGlcNAc (Figure 1). UDP-GlcNAc: Dol-P GlcNAc phosphotransferase DPAGT1 binds to the UDP-GlcNAc analog antibiotic tunicamycin [35]. Tunicamycin also binds to GlcNAc-1-P transferases from other species, including bacteria.

An oligosaccharide consisting of 2 GlcNAc and 5 Man residues is assembled on Dol-P from UDP-αGlcNAc and GDP-αMan on the cytoplasmic side of the ER membrane. The GnT complex ALG13/ALG14 first synthesizes the di-N-acetyl-chitobiose structure GlcNAcβ1-4GlcNAc- [36]. Subsequently, the Manβ1-4 linkage to GlcNAc is synthesized by β1,4-Man-transferase (ManT) ALG1 (Table 1). ALG2 transfers Manα3 as well as Manα6 (forming the Man3- and Man6-arms) from GDP-Man to the Manβ1-4 residue. Two Man residues are then added in α1-2 linkage to the Man6 arm by α2ManT ALG11. The Dol-PP-linked $Man_5GlcNAc_2$ oligosaccharide is then transferred to the ER lumen for the further addition of Man and Glc residues. Dol is a large membrane lipid that has a significant effect on the structure and leakiness of the membrane and may be involved in the translocation process of the $Man_5$-$GlcNAc_2$ intermediate. The enzyme that transfers Man to Dol-P has been shown to require membrane phospholipids for its activity [37].

For the reactions taking place at the lumenal side of the ER membrane, both acceptor and donor substrates utilize membrane-bound Dol-P sugars that flip across the ER membrane. The donor substrate Dol-P-βMan is synthesized at the cytoplasmic side from Dol-P and GDP-Man by Dol-P-Man synthase DPM1. Dol-P-βGlc is synthesized by Dol-P-Glc synthase ALG5 using UDP-αGlc. The Glc and Man residues of these membrane-bound donor substrates are then translocated to the lumen of the ER, where they are used by GTs to transfer Man and Glc to $Man_5$-$GlcNAc_2$-PP-Dol to synthesize the large $Glc_3$-$Man_9$-$GlcNAc_2$ structure (Figure 2). The transfer of Man residues involves α3ManT ALG3 and α2ManTs ALG9 and ALG12. The final Glc residues are transferred to the Man3-arm by α3Glc-transferase (GlcT) ALG6 and ALG8, and the terminal Glc residue is transferred by α2GlcT ALG10 (Table 1).

## 8. Transfer of N-Glycans from Dolichol to Protein

After the large oligosaccharide has been assembled, it is transferred from Dol-PP to the nitrogen side chain of Asn of the nascent glycoprotein in the ER by the oligosaccharyltransferase complex OST. This inverts the α-linkage of the first GlcNAc to a β-linkage. The Asn-x-Ser/Thr sequon is absolutely required for protein N-glycosylation, and this requirement has been maintained in all eukaryotic species throughout evolution, including in archaea and several bacteria. A Pro residue between Asn and Ser/Thr or following Ser/Thr does not allow N-glycosylation, likely by preventing the formation of an essential hydrogen bond between Asn and Ser/Thr.

Processing reactions start in the ER lumen after the attachment of N-glycans to peptides. A number of chaperone and GlcT proteins are involved in protein folding and quality control. The terminal Glcα1-2 residue is cleaved by glucosidase MOGS, followed by cleavage of two Glcα1-3 residues by GANAB. Mannosidase MAN1B1 then cleaves Manα1-2, leaving a mixture of N-glycan structures from $Man_8$-$GlcNAc_2$ to $Man_5$-$GlcNAc_2$, which are precursors of hybrid and complex N-glycans. These glycoside hydrolases are type II membrane proteins, but are classified into different CAZy GH families (Table 1). The glycoproteins are then transported from the ER to the Golgi, where further glycan processing and synthesis occurs.

## 9. Golgi Reactions

All GTs in the Golgi utilize nucleotide sugar donor substrates that need to be imported into the Golgi by specific transporters. The localization of Golgi proteins is due to the gradual transport of membranous vesicles through Golgi compartments. The relatively short tm domain of GTs does not appear to be important for activity, and recombinant GTs produced in vitro usually lack the tm domain to enhance solubility. However, it is possible that the short N-terminal cytoplasmic tail of GTs plays a role in the localization of GTs within various Golgi compartments. The conserved oligomeric Golgi protein complex (COG) has been shown to control GT localization as well as retrograde Golgi to ER transport. In addition to the ER proteins, the Golgi resident protein QSOX1 also catalyzes the disulfide bonding of specific GTs, e.g., sialyltransferases (SiaTs), as well as the intermolecular bonds of large mucins [38]. However, the complex mechanisms that localize GTs and re-localize them under pathological conditions remain to be further explored. Processed and extended glycoproteins are then delivered to their final destination on the cell surface, in secretions, or in intracellular membranes.

## 10. Synthesis of N-Glycan Antennae by GlcNAc-Transferases

After transport of the glycoprotein from the ER to the Golgi, further Man residues are cleaved from N-glycans by mannosidases that are type II membrane proteins. Mannosidase 1 MAN1A1 progressively cleaves Manα1-2 residues. MGAT1 (GnT I) is the first enzyme to add an antenna and transfers GlcNAc in β1-2 linkage from UDP-GlcNAc to the Manα1-3 residue (Man3-arm) of $Man_5$-$GlcNAc_2$-Asn-R (Figure 2). The conversion of hybrid to complex N-glycans is then controlled by mannosidase 2, MAN2A1, which cleaves both the Manα1-3 and Manα1-6 residues from Manα1-6 (Man6-arm), which then exposes the substrate for GnT II.

Harry Schachter, together with Pamela Stanley [39], discovered GnT I in mutant CHO cells. It was later shown that this enzyme is essential for further modifications of N-glycans. This includes the removal of 2 Man residues by Golgi mannosidase 2, α6-fucosylation of the core GlcNAc linked to Asn, and further branching by GnT II (MGAT2) and IV (MGAT4), which then allows the formation of the fourth antenna by GnT V (MGAT5) (Figure 2). MGAT1 is also required for introducing the bisecting GlcNAc by GnT III (MGAT3) [32].

## 11. GnT I-MGAT1

The characterization of GnT enzymes was greatly facilitated by collaboration with Hans Paulsen. Paulsen's group was the first to synthesize N-glycan oligosaccharides having the Man β1-4GlcNAc linkage. The configuration and rotation of the N-acetyl group of GlcNAc represented a steric hindrance that favoured the synthesis of Manα-glycosides. Thus, the method for Manβ- synthesis was a breakthrough for the synthesis of GnT substrates of N-glycans up to 9 residues [40]. Glycan linkages and 3D conformations were extensively analyzed by NMR [41].

As an effective acceptor substrate for GnT I [39] that adds GlcNAc to the Man3-arm, Paulsen's group synthesized Manα6(Manα3)Manβ-octyl and many related compounds. It was shown that the minimum requirement for an acceptor substrate for purified rat liver GnT I is Manα1-3Manβ1-4GlcNAc, but optimal activity is observed when the Manα1-6 residue is also present. Other groups, e.g., Ole Hindsgaul (University of Alberta), Jeremy Carver (University of Toronto), and Khushi Matta (Roswell Park Institute, Buffalo, NY), also participated in the synthesis of acceptor substrates for GnTs with hydrophobic groups replacing the di-N-acetylchitobiose of the N-glycan core and helped to successfully characterize properties of GnTs.

The crystal structure of rabbit GnT I suggested its catalytic mechanism as an inverting GT-A-folded enzyme that binds UDP-GlcNAc and $Mn^{2+}$ [33] (Table 1). The protein has a $^{211}$EDD$^{213}$ motif involved in forming hydrogen bonds and interacting with UDP-GlcNAc and $Mn^{2+}$. D291 was proposed to be the catalytic base that can deprotonate the 2-hydroxyl of Manα1-3 in the acceptor, which then becomes a nucleophile that attacks carbon-1 of GlcNAc in the donor UDP-GlcNAc. The proposed acceptor binding site was shown by Manα1-3 modeled into the GnT I crystal structure, which is consistent with the specificity results using Paulsen's compounds.

Paulsen's group made significant efforts in replacing hydroxyls of the Manα1-3(Manα1-6)Manβ-octyl acceptor for GnT I with a number of small and large substituents, including O-methyl, O-propyl, O-pentyl, 4-pentenyl, 4-pentanolyl, 4-oxo-pentyl, 4,4-azo-pentyl (diazirino-pentyl), 4,5-epoxy-pentyl, 5-amino-pentyl, 5-pentanolyl, 5-iodoacetamido-pentyl. In other derivatives, the oxygen was removed or replaced with $NH_3$ [42]. The deoxy compounds indicated the requirement of the hydroxyl group, possibly as a hydrogen bond donor, while loss of activity after replacement of hydroxyl groups with larger substituents may be a result of steric hindrance.

After this extended production of new compounds, GnT I assays showed that the 2,3 and 6-hydroxyls of the Manα1-3 residue as well as the 2- and 4-hydroxyl of the Manβ residue were required for activity. The N-acetylchitobiose linked to Asn-peptide was not required and could be replaced by a number of different aglycones, including GlcNAc, hexyl, octyl, or other hydrophobic groups [14,17]. Although the Manα1-6 residue was required for optimal activity, removal of its 2-hydroxyl or replacement of the 3-, 4-, or 6-hydroxyl with an O-methyl group still allowed GnT I activity at various levels. These compounds helped to explore the substrate specificity of GnT I and the enzymatic production of acceptor substrates for the next GnTs in the pathway. This approach was also part of the design of specific competitive or irreversible inhibitors. A number of these derivatives were inhibitors with $K_i$ values between 0.76 and 10.2 mM. The most potent inhibitor was Manα6(6-O-methyl-Manα3)Manβ-octyl, where presumably the methyl group clashed with binding of the substrate to the active site of GnT I (Table 3).

**Table 3.** Acceptor substrate analog inhibitors for glycosyltransferases. Paulsen's group synthesized many acceptor substrates that are analogs of the natural N-glycans found in humans. Using these compounds, the activities of several glycosyltransferases were determined, and inhibitors were found. The compounds listed in this table showed significant inhibition of GT activity. GlcNAc-naphthyl derivatives were synthesized by W.A. Szarek, Queen's University. oct, octyl; *pnp*, *p*-nitrophenyl; -R, various hydrophobic groups.

| Enzyme | Inhibitors |
|---|---|
| GnT I | Manα6(Manα3)4-O-methyl-Manβ4GlcNAc |
| | Manα6(6-O-methyl-Manα3)Manβ-oct |
| | Manα6(6-O-4,5-epoxy-pentyl-Manα3)Manβ-oct |
| | Manα6(6-O-4,4-azo-pentyl-Manα3)Manβ-oct |
| GnT II | 2-deoxy-Manα6(GlcNAcβ2Manα3)Manβ-oct |
| | GlcNAcβ2Manα3Manβ-oct |
| | 3-O(4,4 azo)pentylManα6(GlcNAcβ2Manα3)Manβ-oct |
| | 2-deoxy-Manα6(GlcNAcβ2Manα3)Manβ-oct |
| GnT V | GlcNAcβ2(6-deoxy)Manα6Glcβ-O-R |
| | GlcNAcβ2(4-O-methyl)Manα6Glcβ-O-R |
| | GlcNAcβ2(6-deoxy, 4-O-methyl)Manα6 Glcβ-O-R |
| | GlcNAcβ2(4,6-di-O-methyl)Manα6Glcβ-*pnp* |
| O-glycan synthesis | GalNAcα-aryl |
| C1GALT1 | 6-O-(4,4-azo)pentyl-GalNAcα-Bn + UV |
| GCNT1 | Galβ3GalNAcα-*pnp* + UV |
| | Galβ3(6-deoxy) GalNAcα-Bn |
| β4GalT | GlcNAcβ-naphthyl derivatives |
| | 1-thio-N-butyryl-GlcNAcβ-2-naphthyl |

## 12. GlcNAc-Transferases (MGAT2-6)

After GnT I has acted, GnT II synthesizes the second antenna by transferring a GlcNAc residue in β1-2 linkage to the Manα1-6 residue (Figure 2). The serious human condition of CDG IIa is due to a mutation in GnT II. The synthesis of a large series of substrate analogs of the standard GnT II acceptor Manα1-6(GlcNAcβ1-2Manα1-3)Manβ-octyl allowed determination of the specificity of partially purified rat liver GnT II [43,44]. GnT II requires the 3-hydroxyl of the GlcNAc added by GnT I on the Man3-arm as well as the 3-hydroxyl of the Manα1-3 residue, which is essential for binding to GnT II. Interestingly, the trisaccharide GlcNAcβ1-2Manα1-3Manβ-octyl that lacks the Manα1-6 residue is not a substrate but is a good inhibitor of GnT II with a $K_i$ of 0.9 mM (Table 3).

Removal of the 2-hydroxyl of Manα1-6, the site of GlcNAc addition, creates a competitive inhibitor for GnT II with a $K_i$ of 0.13 mM, while replacement by a methyl group eliminates activity and does not inhibit, likely because of steric hindrance [45]. In contrast, the 3-, 4-, and 6-hydroxyls of Manα1-6 are not required for activity. In contrast to GnT I, GnT II does not need to interact with the 4-hydroxyl of Manβ-. However, substitution of the 4-hydroxyl of Manβ- by a methyl group blocked GnT II activity. This may explain why GnT II cannot act after GnT III, which introduces a bisecting GlcNAcβ1-4 residue linked to Manβ [43,44].

The structure of human MGAT2 was determined in complex with $Mn^{2+}$/UDP-GlcNAc or with the donor analog UDP and with the acceptor GlcNAcβ1-2Manα1-3(Manα1-6)Manβ1-4GlcNAc$_2$-Asn [46]. This suggested that the inverting GnT II has an overall fold resembling GT-A and that D347 is the catalytic base residue.

After the second antenna has been established, GnT IV and V can add additional GlcNAc residues that initiate tri- and tetra-antennary N-glycans, respectively (Figure 2).

GnT IV from the hen oviduct was shown to be a β1,4-GnT that can act on glycopeptides with the bi-antennary structure GlcNAcβ1-2Manα1-3 (GlcNAcβ1-2Manα1-6) Manβ1-4GlcNAc1-4GlcNAc-Asn- where both Man3- and Man6-arms had to have GlcNAc residues (added by GnT I and II) for full activity. Extension of these GlcNAc residues by Gal yielded 5% activity, while sialylation completely blocked GnT IV activity.

The tetra-antennary N-glycans are synthesized by GnT V, which is involved in cell adhesion and recognition and in cancer development [47] (Table 2). Thus, GnT V inhibitors are of importance to obtain potential tools to fight cancer and metastasis (Table 3). In contrast to GnT II, GnT V from the hamster kidney, which also acts on the Man6-arm, does not require the Man3-arm in its acceptor substrate and can act on linear trisaccharides such as GlcNAcβ1-2Manα1-6Manβ-octyl [48,49]. Paulsen's group synthesized biantennary substrate analogs that showed how specific hydroxyls of both the Man3- and Man6-arms influence GnT V activity (Figure 3). However, while the 4-hydroxyl of GlcNAcβ1-2 and the 4- and 6-hydroxyls of Manα1-6 on the Man6-arm were absolutely essential for activity, replacement of the 3-hydroxyl of GlcNAcβ1-2 with a larger pivaloyl group prevented activity, presumably due to steric interference [48]. Kanie et al. [50,51] also reported the critical role of the 3-, 4-, and 6-hydroxyls of the β1-2-linked GlcNAc in enzyme recognition. This clearly established the pathway of biosynthesis, where GnT II must add a GlcNAcβ1-2 residue to the Man6-arm before GnT V can act, and transfer of Galβ1-4 to GlcNAcβ1-2 blocks GnT V action.

GnT V is an inverting GT, but does not appear to have a DxD motif. Instead, a Glu residue forms the catalytic base, and the activity is independent of divalent metal ions. The crystal structure of the lumenal domain shows a GT-B-like fold with 2 Rossmann domains [52]. A pocket that accommodates the Man1-6 residue of the acceptor is lined by the aromatic rings of F380 and W401. GnT V requires a non-catalytic domain for activity towards glycoproteins [53,54].

GnT III requires the prior action of GnT I to add the bisecting GlcNAcβ1-4 to the internal core Manβ1-4 residue. The enzyme can also act on biantennary substrates and produce hybrid and biantennary bisected N-glycans, which causes a crowded configuration that may be the reason for the reduction of further N-glycan antennae synthesis and processing. Paulsen synthesized N-glycans with 9 sugar residues, including the bisecting GlcNAc [40]. The steric effect of the bisecting GlcNAc was examined by NMR, which showed that the Manα1-6 residue is present in a different (gg) preferred conformation [55].

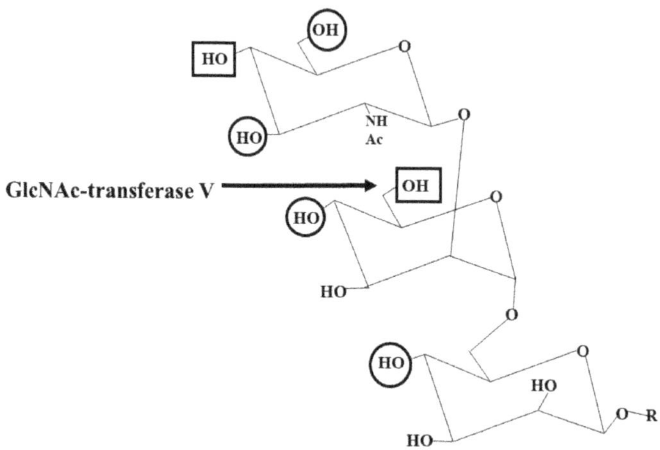


**Figure 3.** Specificity of GlcNAc-transferase V. GnT V transfers a GlcNAcβ residue to the 6-hydroxyl of the Manα1-6 residue of the N-glycan core. The acceptor substrate for purified GnT V from the hamster kidney was a derivative of the 6-arm, GlcNAcβ1-2Manα1-6Manβ-R. The 3-arm of the N-glycans is not essential for GnT V activity. R can be an oligosaccharide or a hydrophobic group such as octyl. The Manβ-residue can be replaced by Glcβ having full activity. To determine acceptor specificity, hydroxyls were deleted or replaced by a larger group. The rectangular boxes indicate an absolute requirement for the hydroxyl, while the circles indicate that the hydroxyl is important for the activity. Replacement of the 4- and 6-hydroxyls of the Manα1-6 residue with an O-methyl group yielded the GnT V inhibitor GlcNAcβ1-2(4,6-di-O-methyl)Manαl-6Glcβ-*p*np.

GnT VI synthesizes an additional antenna on the Man6-arm. This β1,4-GnT activity was found in hen oviduct [56,57] and fish. GnT VI can act on the trisaccharide acceptor GlcNAcβ1-2 (GlcNAcβ1-4)Manα-methyl, which resembles the product of GnT II and GnT V.

## 13. Modifications of N-Glycan Antennae

Lysosomal hydrolases have N-glycans with a specific Man-6-phosphate (Man6-P) marker that targets them to the lysosomes via a specific Man6-P receptor [58]. In Golgi, the GlcNAc-P-transferase GNPT ABG complex transfers GlcNAc-P from UDP-GlcNAc to the 6-hydroxyl of a Man residue on lysosomal enzymes. This is followed by hydrolysis of GlcNAc to expose the Man6-P marker that can bind to the Man6-P receptor in the Golgi, which reroutes the glycoprotein to acidic compartments and to the lysosomes.

Paulsen's compounds helped to define the association of glycoprotein biosynthesis in biological processes, such as apoptosis in porcine aortic endothelial cells [59]. GnTs I and II are usually the major GnT activities in tissues and cultured cells, with GnTs III, IV, and V activities being low in vitro. Further extension of N-glycan antennae includes the addition of Gal by β1,4-GalT B4GalT1 [18]. The inhibition of Gal extension could reduce the addition of Sia and multiple other epitopes that may play a role in the immune system and in cancer [22] (Table 3). Sialic acids, which are often found in increased amounts in cancer, form selectin (Siglec) ligands and play a role in cancer cell metastasis, immune evasion, and cancer cell survival. The expression of ST6Gal1 has been shown to be increased in cancer [24]. ST6Gal1 has been crystalized with CMP. The inverting enzyme has a GT-A-like fold and may act via an $S_N2$ mechanism [60]. However, much of the knowledge about SiaT mechanisms comes from studies of bacterial enzymes with similar activities.

## 14. Inhibition of Glycosyltransferases

GTs bind their substrates by hydrogen bonding and hydrophobic interactions and prefer specific conformations and flexibilities. Inhibitors may compete with substrates or covalently bind to GTs to cause loss of activity. Paulsen's approach to inhibitor design was to synthesize many short and long analogs of natural acceptor substrates for GTs. The synthesis of potential inhibitors included removal of hydroxyls that are essential for sugar transfer, blocking access to critical hydroxyls by introducing large groups, or preventing the folding of enzymes in the active conformation. Other strategies that could cause loss of activity could focus on the donor binding site and the binding of essential divalent metal ions.

Compounds were either free oligosaccharides or had a hydrophobic group at the reducing end of the chain. Hydroxyl groups were either deleted or replaced by methyl groups or by potentially reactive substituents such as azide and epoxide groups that can react with nucleophilic amino acids, as well as diazirino and iodoacetamido groups that are activated by UV. These modified substrates enhanced our understanding of how GTs bind to their substrates, and that small changes in the stereochemistry, conformation, or electronic, bulky, or hydrophobic character of the sugar ring substituents, depending on their positions, could block enzyme activity by affecting interactions with specific amino acids in the substrate binding and catalytic sites of enzymes. A basic requirement for understanding biosynthetic mechanisms is knowledge of the enzyme structure and substrate specificity.

Several N-glycan derivatives proved to be potent inhibitors of GTs (Table 3). However, due to their large size and hydrophilicity, it is difficult to envision their transport across the membranes to reach the GTs in the Golgi.

Since extended glycan chains carry many cancer-associated antigens, inhibition of glycoprotein extension may be useful for biological studies. Gal is added to the GlcNAc of the N-glycan antennae by a family of enzymes (Table 1). We examined the specificity and inhibition of purified bovine B4GALT1 with a large panel of acceptor substrate derivatives. While GlcNAcβ-Bn was the standard acceptor, GlcNAcα-Bn was only 5% active, and free GlcNAc showed 29% activity, indicating that it is mainly the β anomer that forms the substrate. Glycopeptide acceptors having GlcNAcβ1-3/6GalNAc-glycans were relatively less active than GlcNAcβ-Bn. Derivatives with the large 2-naphthyl as aglycone could not support activity, and most of these compounds inhibited the enzyme. 1-Thio-N-butyrylGlcNAcβ(2-naphthyl) inhibited B4GALT1 by 100% at 1 mM with a $K_i$ of 0.01 mM [18]. However, bicyclic quinolinyl compounds (having nitrogen in the ring) were active substrates. This indicates that the overall structure of the acceptor and the conformation and electronic properties of the aglycone are important for enzyme recognition. The catalytic domain of B4GALT1 has two flexible loops that change from an open to a closed conformation upon binding of UDP-Gal and metal ions [61]. This creates an acceptor binding site and seems to be a mechanism common to several GTs. After the reaction, the oligosaccharide product is released, and the loops revert to the initial conformation, releasing UDP. Thus, additional strategies are required to make inhibitory compounds applicable in biological systems. Future studies will be seeking higher potency and specificity of inhibitors, targeting strategies, and examining the biological effects in cell cultures and in animals before development into human therapeutics.

## 15. Mucin Glycoproteins

Mucins are very large glycoproteins that are heavily O-glycosylated, sometimes with 90% carbohydrates. They may also carry several N-glycans. Mucins form a viscoelastic mucus layer over the epithelia and protect the host from environmental stresses, dehydration, and infections. Humans have 20 or more genes that encode mucins and many other proteins that have mucin-like O-glycosylated domains. Human mucin-type O-glycans all start with a GalNAc residue αlinked to Ser or Thr. Four core structures are common, with several less common structures such as core 5, GalNAcα1-3GalNAc- (Table 4) [23,30]. These core structures can be sialylated or extended with structures and epitopes similar to those found on complex N-glycans. The O-glycan structures often vary between cell types and in conditions such as inflammation or in cancer and are regio-specific in the intestine. Thus, the expression of genes encoding GTs is delicately regulated, although our knowledge in this field is still marginal [23].

**Table 4.** Glycopeptides synthesized by Hans Paulsen. Paulsen synthesized hundreds of glycopeptides, some with 1 to 23 amino acids and 1 to 3 glycan chains attached to the longer peptides. Glycans were various combinations of GalNAc, core 1 to 7. The table shows one example of each series and the systematic changes to make variants of either amino acids, attachment site of glycan, type of core structure, or number of glycans attached. Bold amino acids carry glycans.

| Different Series with Examples of Glycopeptides | Variants Synthesized |
|---|---|
| AcTPPP | |
| 1. AP**T**$^{\text{GalNAc}\alpha}$ S$^{\text{GalNAc}\alpha}$ SS | Thr/Ser, position of GalNAc |
| 2. AP**T**$^{\text{Gal}\beta\text{3GalNAc}\alpha}$ SSS | Thr/Ser, position of core 1 |
| 3. Ac-**T**$^{\text{GalNAc}\alpha}$ P-*t*-Bu | protected N/C-terminal protection |
| 4. AP**T**$^{\text{GalNAc}\alpha}$-SSSTKKT | peptide length |
| 5. Ac-V**T**$^{\text{GalNAc}\alpha}$ P-$NH_2$ | peptide length, N/C-terminal protection |
| 6. Ac-PTT**T**$^{\text{GalNAc}\alpha}$ PIST-$NH_2$ | amino acids, position of GalNAc |
| 7. Ac-PTPTGTQTPT**T**$^{\text{GalNAc}\alpha}$ TPITTTTTVTPT-$NH_2$ | number of GalNAc |
| 8. AHGVT**S**$^{\text{Gal}\beta\text{3GalNAc}\alpha}$ APDTRPAPGSTAP **T**$^{\text{Gal}\beta\text{3GalNAc}\alpha}$ A | amino acids, position and number of GalNAc, core 1 |
| 9. PTTTPITTTG | amino acid sequence |
| 10. Ac-P**S**$^{\text{GalNAc}\alpha}$ **S**$^{\text{GalNAc}\alpha}$ **S**$^{\text{GalNAc}\alpha}$ PIST-$NH_2$ | amino acids, number, Ser/Thr position of glycans |
| 11. TT**T** $^{\text{GlcNAc}\beta\text{3GalNAc}\alpha}$ VTP**T** $^{\text{GlcNAc}\beta\text{3GalNAc}\alpha}$ PTG | number and position of GalNAc, core 3 |
| 12. TETTSHS**T**$^{\text{GalNAc}\alpha}$ PG | number and position of GalNAc, core 3, length of peptide |
| 13. TT**T**$^{\text{Gal}\beta\text{3GalNAc}\alpha}$ VTP**T** $^{\text{Gal}\beta\text{3GalNAc}\alpha}$ PTG | position and number of core 1 |
| 14.TTTVTPTP**T**$^{\text{GlcNAc}\beta\text{6(Gal}\beta\text{3)GalNAc}\alpha}$G | position and number of core 1, 2 |
| 15. TT**T** $^{\text{GlcNAc}\beta\text{6(Gal}\beta\text{3)GalNAc}\alpha}$ VTP**T**$^{\text{Gal}\beta\text{3GalNAc}\alpha}$ PTG | position of core 1, 2 |
| 16. TT**T**$^{\text{GlcNAc}\beta\text{6(GlcNAc}\beta\text{3)GalNAc}\alpha}$ VTP **T**$^{\text{GlcNAc}\beta\text{6(GlcNAc}\beta\text{3)GalNAc}\alpha}$ PTG | position and number of core 4 |
| 17. TTTVTP**T**$^{\text{GlcNAc}\beta\text{6GalNAc}\alpha}$ PTG | position and number of core 6 |
| 18. TT**T**$^{\text{GlcNAc}\beta\text{6(GlcNAc}\beta\text{3)GalNAc}\alpha}$ VTP**T**$^{\text{GlcNAc}\beta\text{6GalNAc}\alpha}$ PTG | position and number of core 4, 6 |
| 19. Ac-P**T**$^{\text{GalNAc}\alpha}$ **T**$^{\text{Gal}\beta\text{3GalNAc}\alpha}$ **T**$^{\text{Gal}\beta\text{3GalNAc}\alpha}$ PIST-$NH_2$ | position and number of GalNAc and core 1 |
| 20. Ac-PTT**T**$^{\text{Gal}\beta\text{3GalNAc}\alpha}$ PIST-$NH_2$ | position of core 1α or β, amino acids |
| 21. TETTSHS**T**$^{\text{Gal}\beta\text{3GalNAc}\alpha}$ PG | position and number of core 1, 2, 4, 6 |
| 22. Ac-ELS**T**$^{\text{GalNAc}\alpha}$ **T** $^{\text{GalNAc}\alpha\text{3GalNAc}\alpha}$ GPG- $NH_2$ | amino acids, number and position of GalNAc, core 5, 7 |
| 23. Ac-ELA**T**$^{\text{GalNAc}\alpha}$ VGPG-$NH_2$ | amino acids, GalNAc, core 5, 7 |

Mucins are characterized by variable number of tandem repeat (VNTR) regions that are rich in Thr/Ser/Pro residues and carry most of the O-glycans. There are three types of mucins in humans: cell surface-bound mucins, small, secreted mucins, and large, gel-forming mucins. For each mucin, several variants have been identified. Small, secreted mucins (MUC7) are relatively soluble and found in the saliva. Cell surface-bound mucins (MUC1,3,4,12,13,15,16,17,20,21,22) have a transmembrane domain and a cytoplasmic tail. These often interact with other cell surface proteins and are involved in cell signalling. Large, gel-forming mucins (MUC2,5AC, 5B,6) have a Cys-rich D domain and are polymerized to several millions of molecular weight.

## 16. Role of Mucins

The role of mucins varies from protection in mucus to cell adhesion and regulation of tumour survival and diagnostic markers in cancer. Some mucins carry hundreds of O-glycans with different structures and lengths up to more than 20 sugars, which are hydrophilic and can be negatively charged due to sialic acid and sulfate esters and bind to metal ions. The properties of mucins are dictated by the abundance of O-glycans, which cause an extended conformation of mucins and prevent protease cleavage. Mucins play an essential role in intestinal health, supporting beneficial microbiota and avoiding pathogens. O-glycans form adhesive receptors for bacteria and other microbes that can degrade them, use glycans as a source of nutrition, and compete for adhesion [62,63]. O-glycans also play a role in fertilization and sperm–egg binding. Since O-glycans often contain epitopes and antigens, e.g., Lewis-type structures, that are recognized by antibodies and lectins, and they have a controlling function in the immune system. Simple O-glycans are often associated with cancer cells, and sialylated O-glycans have been shown to increase in cancer [21]. They regulate the survival and attachment of cancer cells and can be useful biomarkers and basis for vaccines. Some growth factors and cytokine receptors are O-glycosylated. For example, death receptors require O-glycosylation for efficient apoptotic signalling in tumours [64].

There are four major mucin type O-glycan core structures (1 to 4) (Table 4), with core 1 and 2 being mostly universal, while core 3 and 4 are found in mucins from specific tissues such as intestinal or lung tissue [23]. These core structures can be extended by a variety of sugars and epitopes. Core 5 is a rare structure that is not extended but can be sialylated, and it has been found in human meconium and adenocarcinoma [30,65]. Other core structures (6 to 8) are extremely rare. The core structures are expected to have different conformations and thus display their extensions and epitopes differently. For example, cores 2 and 4 have a GlcNAcβ1-6 branch that can occupy a significant amount of space due to the rotation of the 1-6 bond. This was established by analyzing core 1 to 6 structures using NMR and energy calculations [66].

The overwhelming majority of mucins are bound to the cell membrane and have multiple roles in signalling and recognition events. These Type I membrane glycoproteins can form complexes with growth factors and tyrosine kinases; they can be receptors for mammalian lectins such as Siglecs and can interact with proteins and cells of the immune system, controlling inflammatory responses and immune responses against tumours. The cytoplasmic C-termini can be phosphorylated and interact with proteins of signal transduction and regulation of survival and cell death signals. The extracellular domains can also be cleaved and then secreted as soluble mucins with a number of additional roles. In cancer, MUC1 expression in particular has been found to be elevated in addition to alterations of their O-glycan structures [67]. For example, simple O-glycan structures Tn and T antigens (Table 2) and their sialylated forms are typically present in increased amounts in cancer, together with anti-Tn and anti-T antibodies. MUC1 and MUC4 have consistently been shown to be highly expressed in cancer. Thus, cancer vaccines based on MUC1 have been developed and tested in clinical trials, with some success in diminishing tumour load. MUC4 is a marker for pancreatic cancer and is highly immunogenic, and vaccines have been developed to target cells expressing MUC4. MUC16 is also known as ovarian cancer antigen CA125.

## 17. Glycosyltransferases That Assemble Mucin O-Glycans

O-glycans are assembled step by step in the Golgi and are directly attached to the side chain oxygen of Ser or Thr residues in the completed polypeptide chains. There is no known single amino acid sequence common for all O-glycans, but databases allow an estimate of probable O-glycosylation based on statistics of known sequences around O-glycosylation sites of glycoproteins. The first sugar residue of O-glycans (GalNAc) is added by GTs localized to Golgi membranes (Table 1). GALNT occurs as 20 isoenzymes that O-glycosylate peptides in humans. GalNAc is transferred to peptide by a family of at least 20 polypeptide GalNAc-transferases (GALNTs) that are expressed in a cell-type-specific fashion and support the essential function of mucin O-glycans. GALNTs are classified as GT27 enzymes, and most have a lectin domain connected to the catalytic domain by a flexible linker sequence. GALNTs have similar broad specificities for the amino acids near the O-glycosylation site, with a preference for nearby Pro residues that may enhance recognition of the hydroxyls of Ser/Thr. GALNT7 requires a pre-existing GalNAc-O-glycan in the peptide [68]. The crystal structures of GALNT1,2,3,4,7,10,12 have been determined and show complex structures with a GT-A fold in the GT domain. A DxH motif is found to coordinate UDP-GalNAc and a $Mn^{2+}$ cofactor [69]. Several GALNTs are aberrantly expressed in cancer. For example, the expression and activity of GALNT14 in hepatocellular and gastrointestinal cancer may be a useful disease marker [25].

GalNAcα-Ser/Thr (Tn antigen) as well as its sialylated form sialylα2-6GalNAcα-Ser/Thr (sialyl-Tn antigen) are epitopes commonly found in cancer. The α6-sialylation of GalNAc prevents further processing, with the exception of O-acetylation of Sia, which then masks the sialyl-Tn antigen.

The enzymes that synthesize the four major mucin-type O-glycan core structures (1 to 4) have been characterized and the genes cloned (Figure 4). Crystal structures are available for core 1 β1,3-GalT (C1GALT1) and core 2 (β1,6-GnT, GCNT1). The expression of C1GALT1 is uniquely dependent on the coexpression of the chaperone COSMC [70], which prevents protein degradation by ensuring proper folding of C1GALT1. In the absence of Cosmc, C1GALT1 is ubiquitinated and targeted to the proteasome for degradation. It has been observed that a number of cancer cells lack core 1 and its further metabolites, and cells express the Tn and sialyl-Tn antigen due to the lack of COSMC. The occurrence of unmodified core 1, the T antigen, is common in cancer and during early development as an oncofoetal antigen.

C1GALT1 acts on a variety of GalNAcα-R acceptor substrates, including glycoproteins and the GalNAcα-glycopeptides, synthesized by the Paulsen group. Control assays for C1GALT1 are usually carried out with GalNAcα-Bn acceptor, which is an acceptor for core 1 and core 3 synthesis and has been used as an O-glycosylation inhibitor in cells since it competes with natural glycoprotein biosynthesis. The problem is that GalNAcα-Bn is converted to a number of benzyl-terminating hydrophobic O-glycan structures, which overwhelm the storage and transport capacity of the cell and induce apoptotic cell death [71]. Since core 1 and 3 are precursors for core 2 and 4 structures (Figure 4) GalNAcα-Bn inhibits the synthesis of core 1 to 4 in glycoproteins.

Derivatives of GalNAcα-Bn were synthesized and tested as potential substrates and inhibitors [72]. Removal or substitution of the 6-hydroxyl of GalNAc slightly reduced C1GALT1 activity. 6-O-(4,4-azo)pentyl- GalNAcα-Bn was photosensitive and inhibited Gal incorporation into GalNAcα-Bn (Table 3). The results indicate that the enzyme has a relatively broad acceptor specificity and does not require the 6-hydroxyl of GalNAc, but it needs the 3- and axial 4-hydroxyl as essential requirements for binding and activity.

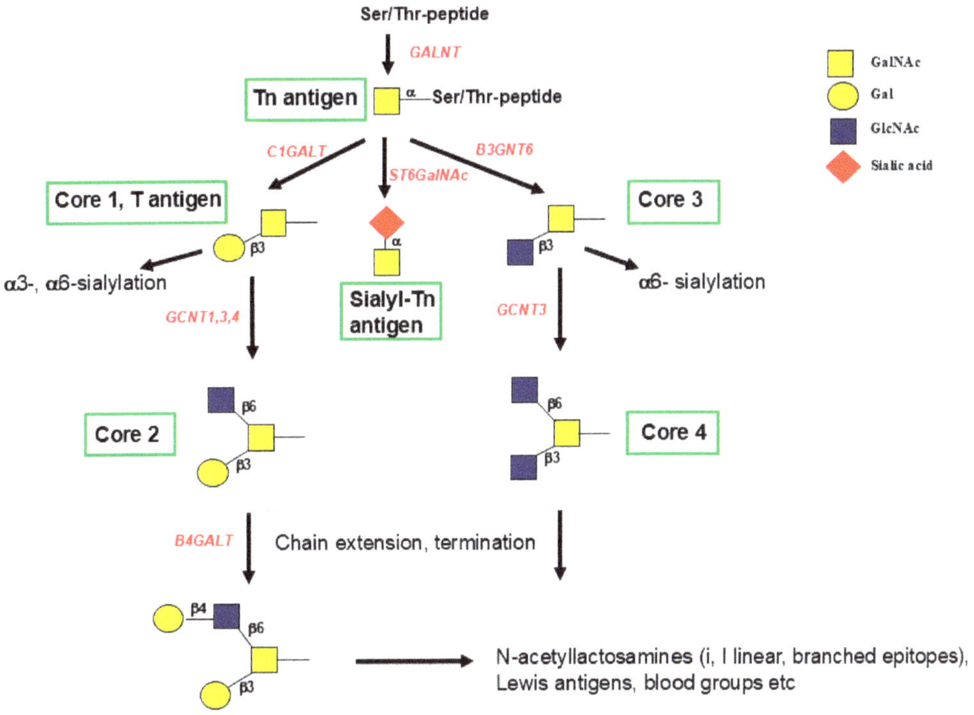


**Figure 4.** O-glycosylation pathways, synthesis of cores 1 to 4. All mucin-type O-glycans are linked through GalNAcα to Ser or Thr. This linkage is formed by one or more of 20 polypeptide GalNAc-transferases (GALNTs) that are expressed in a cell type-specific fashion. GalNAc can be converted to core 1 by β1,3-Gal-transferase C1GALT1, or to core 3 by β1,3-GlcNAc-transferase B3GNT6. Unmodified GalNAc is recognized as the cancer-associated Tn antigen. If sialic acid is added to GalNAc by ST6GalNAc1, the O-glycan becomes the Sialyl-Tn antigen, which is not further extended. Core 1 can be converted to core 2 by GCNT1, GCNT3, or GCNT4. Core 3 is converted to core 4 only by GCNT3. Unmodified core 1 is named the T antigen, which is often found in cancer. Sialylation of the T antigen also stops further extension to complex chains.

Yeh et al. [73] cloned a number of β1,6-GnTs, GCNT1, 2 and 3. GCNT1 introduces a GlcNAcβ1-6 branch onto core 1, Galβ1-3GalNAc-R, linked to various hydrophobic groups or peptides, and synthesizes core 2 (Figure 4). This activity increases in many cancer and leukaemia cells [74] and upon cancer cell differentiation. GCNT1 plays an important role in the immune system by providing the scaffold for selectin ligands sialyl-Lewis x, crucial for lymphocyte interactions [75]. In specific breast cancer cells, however, GCNT1 is not expressed [76]. The expression shows large variations among cancer cells derived from prostate cancer [31]. Although the enzyme is an inverting GT with a GT-A fold [33], it does not require $Mn^{2+}$ as a cofactor. GCNT1 is N-glycosylated and requires several conserved Cys residues for the formation of disulfide bonds. Amino acid reagents that target non-specifically His, Trp, and Cys inhibited purified GCNT1 activity up to 100%.

The groups of Hans Paulsen and Khushi Matta synthesized many derivatives of core 1, which revealed the specificity of GCNT1. Galβ1-3(6-deoxy) GalNAcα-Bn was found to be a weak inhibitor. Surprisingly, the excellent substrate Galβ1-3GalNAcα-p-nitrophenyl inhibited GCNT1 activity from acute myelogenous leukaemia cells by 87% upon UV irradiation.

Another β1,6-GnT, GCNT4, has similar activity and 44% sequence identity with GCNT1 but has a selected tissue expression. It is associated with the thymus and downregulated in a number of different cancer cells and tissues [26].

The metal ion-independent β1,6-GnT, GCNT3, synthesizes both core 2 from core 1 and core 4 from core 3. The enzyme has high activity in colonic tissue, explaining the abundance of core 4 O-glycans in colonic mucins [23]. GCNT3 has a 50.35% sequence identity with GCNT1 and is also classified as GT14, but shows very different inhibition by bis-imidazolium salts.

In many cancer cells, both core 3 and 4 synthesis by B3GNT6 and GCNT3 showed altered activity [77]. GCNT3 is not expressed in normal or cancer-derived prostate cells [31], but is variably expressed in other cancer types [29]. Huang et al. [78] suggested a role for GCNT3 in suppressing cancer.

The crystal structure of murine GCNT1 [79,80] suggests that the DxD motif is replaced by the DE sequence, with Glu320 as the catalytic base. Two positively charged amino acids (Arg378 and Lys 401) take the role of a $Mn^{2+}$ ion in stabilizing the UDP leaving group. These amino acids are conserved among these β1,6-GnTs. The other metal ion-independent β1,6-GnT, GCNT2, is 37.8% identical to GCNT1 but has a different acceptor specificity and synthesizes the branch of the I antigen [73]. It is not clear which of the Arg residues cooperate with Lys in binding UDP.

GalNAcα-Bn is also an acceptor for β1,3-GnT, B3GNT6, that synthesize core 3. The enzyme was discovered in rat and human colon tissues that are rich in mucins carrying core 3. Many colon cancer cell lines and colon cancer tissue, however, have very low enzyme activity [77]. While GnTs I, II, and V are variably active in normal and cancer prostate cells, B3GNT6 was not detected [31]. The lack of core 3 synthesis may be one factor promoting core 1 (T antigen) expression.

Rare core structures in human glycoproteins include cores 5 to 8 (Table 4). Core 5, GalNAcα1,3-GalNAc-, was found in human intestinal fetal mucins and in adenocarcinoma [30,65]. A preliminary report of core 5 synthesis using the GalNAc-mucin acceptor [81] showed that Sia may be added to GalNAc after the formation of core 5. However, the gene of this α1,3-GalNAcT has not yet been identified. It is possible that this activity is caused by a variant of a known enzyme. In contrast to O-glycan core structures 1 to 4, core 5 has not been found to be elongated, and its role is unknown. The activities or genes encoding the GTs that synthesize core 6 to 8 structures remain to be identified. It is possible that core 6 in mucins is a degradation product of core 2, due to bacterial hydrolases or due to degradation during O-glycan isolation.

## 18. Elongation of O-Glycans

A wide array of glycosyltransferases can extend core structures in complex pathways and create hundreds of different O-glycan structures with lengths of more than 20 sugar residues in mucins. The Gal and GlcNAc residues of O-glycan core structures can be elongated by enzymes that similarly elongate N-glycans. Very few of these enzymes are specific to O-glycans. This includes β1,3-GnT, B3GNT3, which adds GlcNAc to Gal of core 1 and 2 [23]. The enzyme can act on mucins or hydrophobic acceptors with core 1 or 2 structures and thus allows the attachment of complex glycans and lymphocyte homing receptors [82]. Core 2 sometimes carries an unusual GlcNAcα1-4 residue linked to Gal in gastric cells, such as pyloric glands and neck cells [83]. The retaining α1,4-GnT A4GNT may use the large mucin MUC6 as a scaffold. The expression of A4GNT together with MUC6 was found to suppress the proliferation of pancreatic tumour cells [84].

Several SiaTs are specific for mucin-type O-glycans and generally block further extension. All human SiaTs are classified in the inverting GT29 family and have common sialylmotifs that contribute to the binding of substrates [85]. ST6GalNAc1-4 transfer Sia from CMP-βSia in α2-6 linkage to GalNAc-mucins with different specificities for core 1 and sialylated core 1. ST6GalNAc1 primarily synthesizes the sialyl-Tn antigen, while ST6GalNAc2 primarily synthesizes the sialyl-T antigen. ST6GalNAc2 can act on GalNAc- and core 1- or 3-peptide acceptors [28]. The crystal structure of ST6GalNAc2 with CMP [86] shows a GT-A variant2 fold [87]. ST6GalNAc3 and 4 utilize core 1 substrates that have already been sialylated by ST3GAL1, but they do not require a peptide in the acceptor. ST6GalNAc3 has a restricted expression mainly in the brain and kidney.

ST3Gal1 specifically adds Sia in α2-3 linkage to Gal of core 1 and 2 linked to peptide or a hydrophobic group. The ST3GAL1 product, Sialyl-T antigen, is a common antigen in cancer cells. ST3Gal1 is overexpressed in breast cancer tissues [27] and in prostate cancer cells [31], where it controls proliferation, migration, and apoptosis. Porcine ST3Gal1 has been crystallized and revealed a single Rossmann domain, like other SiaTs, and a β-sheet core that contributes to binding CMP-Sia and His319 as a possible catalytic residue [88].

## 19. Synthesis of Glycopeptides

Because of the importance of mucins and O-glycosylation in human health and diseases such as cancer, the Paulsen group synthesized more than 200 peptides and glycopeptides derived from MUC1 to 4 and other glycoproteins for studies of their biological role and pathways of O-glycan biosynthesis (Table 4). It has been extremely difficult to determine the sites of O-glycan addition in complex mucins and to assign O-glycan structures to specific Thr and Ser residues. Thus, peptides and glycopeptides have been a tremendous tool to discover mechanisms of O-glycosylation, the influence of amino acids near O-glycosylation sites in acceptor substrates for GALNTs, and new cancer-associated and immunogenic epitopes.

Hundreds of glycopeptides have been produced by Paulsen's group, together with the unglycosylated control peptides as standards and controls. The development of solid-phase synthesis of GalNAcα- and Galβ1-3GalNAcα-glycopeptides using protected glycosylamino acids was a breakthrough [89,90]. Novel glycosyl amino acids were synthesized as building blocks for Fmoc-based continuous-flow synthesis using multi-column technology. This achieved synthesis of several glycopeptides simultaneously without significant degradation of the glycan-peptide bond and beta-elimination of the O-glycan. The amino acid sequences had different lengths up to 23 amino acids and were based on those from the VNTR of MUC1, 2, 3, or 4. The glycans attached to specific Thr or Ser residues included one or more GalNAc residues and selected core 1 to 7 structures (Table 4). Glycopeptides were purified, and their sequences and glycosylation were confirmed by HPLC, MS, and NMR. Computational analyses and NMR determined the conformations of glycopeptides that could potentially be recognized by individual enzymes.

To examine the preferred attachment sites for GALNTs, a series of glycopeptides were synthesized that were overall neutral with an acetyl group at the N-terminus and amide at the C-terminus and contained one or more GalNAcα-Thr/Ser or Galβ1-3GalNAcα, or in the unnatural β-linkage to Thr. The glycopeptide reaction products were analyzed by the amount of radioactive sugar transferred and by their HPLC elution times using standard glycopeptides.

Within a series, the positions of glycans and some of the Thr/Ser residues were altered in a systematic fashion to include all possibilities of glycan attachments [91]. In addition, Pro and Thr were replaced by other amino acids at +1 or -1 positions to determine their role in a GALNT substrate. The smallest peptide used as a positive control was Ac-TPPP, which showed the highest activity as an acceptor for highly purified bovine GALNT1 [92,93]. The small acceptor substrate does not have a C-terminal protection group, but clearly has easy access to the catalytic site. In glycopeptides 8 amino acids long, Pro in the +3 position (towards the C-terminus) supported the activity. None of the Thr residues are exclusive sites for GalNAc addition, but the amino acid sequences and sites of existing glycosylation clearly dictate further O-glycosylation. Substrates with core 1 in β-linkage are less active, likely due to the different conformations compared to the natural α-linkage.

A series of different lengths and GalNAc attachment to Ser did not show significant activity. It is possible that the slight hydrophobic effect of Thr plays a role in enzyme recognition. However, other members of the GALNT family may show different results and could act on Ser. Generally, site specificity is generally broad, and larger glycopeptides and those with different amino acid sequences remain to be tested as models for mucin synthesis. Glycosylated peptides are more rigid in their conformations compared to peptide alone, and large O-glycans cause crowding. Non-mucin glycoproteins often have glycosylation sites only near the N-terminus, and protein folding may block glycosylation. Mucins have large complex O-glycans, often at adjacent amino acids, and the processing enzymes only have access to their substrates in an unfolded (bottlebrush) conformation of the glycoprotein.

Although there is clearly no consensus sequence for O-glycosylation, TxxP is one of the preferred sites for mucin O-glycosylation by GALNT1. O-glycosylation is extremely complex, with substrate conformation, composition, hydrophobicity, existing glycosylation, as well as overall structure controlling recognition by GALNTs and other GTs.

Subsequently, C1GalT1, GCNT1, and B4GALT1 were examined for their ability to use glycopeptide substrates in a site-specific manner [93–95]. The synthesis of core 1 by partially purified rat liver C1GALT1 was examined with GalNAc-containing glycopeptides containing 1 to 3 GalNAc-Thr. Compared to the GalNAcα-Bn acceptor, all of the GalNAc-glycopeptides were good substrates for the enzyme that could add 1 to 3 Gal residues to each GalNAc. The enzyme was strongly influenced by the overall composition and especially amino acids at the -1 and +1 positions. These studies indicate that in addition to initial glycosylation, the second step in the glycosylation pathways of O-glycans is also controlled by the structure and glycosylation of the peptide core of substrates.

The activity of purified human GCNT1 was influenced by amino acid composition, the position of the core 1 substrate, and other glycans within the glycopeptide. Surprisingly, glycopeptides with core 1 in the unnatural β-linkage were also variably good substrates for GCNT1. The extension of core 2-glycopeptides with purified bovine B4GALT1 also showed significant variability. However, B4GALT1 has a higher activity with GlcNAcβ-Bn as a substrate than with glycopeptides.

Overall, glycopeptides helped to determine that the control of O-glycosylation is based on a multitude of complex factors, including amino acid position, properties, and charge as well as existing glycosylation. GTs that act close to the peptide (GALNT and C1GALT1) are more specific than those adding the third (GCNT1) or fourth (B4GALT1) sugar residue. Currently, Paulsen glycopeptides are still under investigation for their ability to form GT substrates, to discover novel enzymes, to enzymatically synthesize novel glycopeptides, and as potential inhibitors of bacterial biofilm formation.

## 20. Conclusions

Paulsen developed new synthetic technologies by synthesizing complex large N-glycans, modified glycans, and their fragments. This has been documented in hundreds of publications and presentations (Figure 5). His new synthesis included the discovery of methods to synthesize substituted Manβ-R compounds and to initiate solid-phase glycopeptide synthesis. Paulsen's substrates are essential for the discovery of new enzymes, for defining enzyme specificities that control the pathways of N- and O-glycosylation, and for defining the biological roles of glycans. The roles include receptors for lectins and adhesion proteins in eukaryotic systems and in bacteria that use mucins as a niche for survival and nutrition. Disease-specific alterations in GTs could be discovered. The inhibitors that Paulsen synthesized will help in the development of future therapies for cancer and other diseases, where the reduction in complex glycans could be beneficial. Many synthetic routes and analytical tools for complex glycans and glycopeptides have been developed and subsequently adopted worldwide. Future studies are expected to use this information to understand how enzymes recognize their glycan substrates and how this influences pathways of glycoprotein biosynthesis. We are grateful for the hard work of Hans Paulsen and his coworkers and students. He believed that the principles behind the contributions to glycobiology are based on the enjoyment of chemistry!

**Figure 5.** Hans Paulsen at the International Carbohydrate Symposium in Whistler, Canada (2006).

**Funding:** Funding was provided by an NSERC Discovery grant 2022-05389.

**Acknowledgments:** The author thanks the hundreds of students and postdoctoral fellows who worked with Hans Paulsen, along with many collaborators, for their hard work and dedication.

**Conflicts of Interest:** The author declares no conflicts of interest.

## References

1. Heyns, K.; Paulsen, H. Oxidative transformation of carbohydrates. VIII. Catalytic oxidation of meso-inositol to scyllo-meso-inosose. *Chem. Berichte Chem. Eur.* **1953**, *86*, 833–840. [CrossRef]
2. Heyns, K.; Paulsen, H.; Rudiger, G.; Weyer, J. Configuration and conformation selectivity in catalytic oxidation with oxygen on platinum catalysts. *Fortschritte Der Chem. Forsch.* **1969**, *11*, 285–374.

3. Paulsen, H.; Todt, K. Monosaccharides with a nitrogen-containing ring. XV. Nuclear magnetic resonance spectroscopic studies of hindered rotation of monosaccharides with a nitrogen-containing ring. *Chem. Berichte Chem. Eur.* **1967**, *100*, 3397–3404. [CrossRef]
4. Paulsen, H.; Hayauchi, Y.; Sinnwell, V. Monosaccharides containing nitrogen in the ring. XXXVII. Synthesis of 1,5-dideoxy-1,5-imino-D-galactitol. *Chem. Berichte Chem. Eur.* **1980**, *113*, 2601–2608. [CrossRef]
5. Paulsen, H.; Brockhausen, I. From imino sugars to cancer glycoproteins. *Glycoconj. J.* **2001**, *18*, 867–870. [CrossRef] [PubMed]
6. Luger, P.; Vangehr, K.; Bock, K.; Paulsen, H. Conformational analysis. Part XXIII. Determination of the conformation of 2-acetamido-1,4,6-tri-O-acetyl-2-deoxy-3-O-(2-acetamido-3,4,6-tri-O-acetyl-2-deoxy-α-D-galactopyranosyl)-2-D-galactopyranose by x-ray structural analysis, NOE measurements and HSEA calculations. *Carbohydr. Res.* **1983**, *117*, 23–38. [CrossRef]
7. Paulsen, H.; Behre, H.; Herold, C.-P. Acyloxonium ion rearrangement in carbohydrate chemistry. *Fortschritte Der Chem. Forsch.* **1970**, *14*, 472–525.
8. Paulsen, H. Cyclic acyloxonium ions in carbohydrate chemistry. *Adv. Carbohydr. Chem.* **1971**, *26*, 127–195. [CrossRef]
9. Thiem, J.; Rasch, D.; Paulsen, H. Phosphorus-containing carbohydrates, XV. Perkow reaction with α-acyloxy keto sugars for the synthesis of enol phosphates and their reactions. *Chem. Berichte Chem. Eur.* **1976**, *109*, 3588–3597. [CrossRef]
10. Schuster, O.; Klich, G.; Sinnwell, V.; Kränz, H.; Paulsen, H.; Meyer, B. "Wave-type" structure of a synthetic hexaglycosylated decapeptide: A part of the extracellular domain of human glycophorin A. *J. Biomol. NMR* **1999**, *14*, 33–45. [CrossRef]
11. Depmeier, W.; Jarchov, O.H.; Stadler, P.; Sinnwell, V.; Paulsen, H. Conformation of L-streptose reducing sugars as crystals and in solution. *Carbohydr. Res.* **1974**, *34*, 214–218. [CrossRef]
12. Strumpel, M.; Schmidt, H.-J.; Luger, P.; Paulsen, H. X-ray analysis of 3-*O*-(6-*O*-acetyl-2,4-diazido-3-*O*-benzyl-2,4-dideoxy-α-d-glucopyranosyl)-1,6-anhydro-2,4-diazido-2,4-dideoxy-β-d-glucopyranose, a disaccharide having an unusual α-d-(1→3) linkage. *Carbohydr. Res.* **1984**, *125*, 185–201. [CrossRef]
13. Paulsen, H.; Reck, F.; Brockhausen, I. Synthese von modifizierten Oligosacchariden der N-Glycoproteine als Substrate für N-Acetylglucosaminyltransferase I [Synthesis of modified oligosaccharides of the N-glycoprotein as substrate for N-acetylglucosaminyltransferase I]. *Carbohydr. Res.* **1992**, *236*, 39–71. [CrossRef] [PubMed]
14. Brockhausen, I.; Möller, G.; Yang, J.M.; Khan, S.H.; Matta, K.L.; Paulsen, H.; Grey, A.A.; Shah, R.N.; Schachter, H. Control of glycoprotein synthesis. Characterization of (1→4)-N-acetyl-beta-D-glucosaminyltransferases acting on the alpha-D-(1→3)- and alpha-D-(1→6)-linked arms of N-linked oligosaccharides. *Carbohydr. Res.* **1992**, *236*, 281–299. [CrossRef]
15. Bai, L.; Li, H. Protein N-glycosylation and O-mannosylation are catalyzed by two evolutionarily related GT-C glycosyltransferases. *Curr. Opin. Struct. Biol.* **2021**, *68*, 66–73. [CrossRef]
16. Breton, C.; Fournel-Gigleux, S.; Palcic, M.M. Recent structures, evolution and mechanisms of glycosyltransferases. *Curr. Opin. Struct. Biol.* **2012**, *22*, 540–549. [CrossRef]
17. Möller, G.; Reck, F.; Paulsen, H.; Kaur, K.J.; Sarkar, M.; Schachter, H.; Brockhausen, I. Control of glycoprotein synthesis: Substrate specificity of rat liver UDP-GlcNAc:Man alpha 3R beta 2-N-acetylglucosaminyltransferase I using synthetic substrate analogues. *Glycoconj. J.* **1992**, *9*, 180–190. [CrossRef] [PubMed]
18. Brockhausen, I.; Benn, M.; Bhat, S.; Marone, S.; Riley, J.G.; Montoya-Peleaz, P.; Vlahakis, J.Z.; Paulsen, H.; Schutzbach, J.S.; Szarek, W.A. UDP-Gal: GlcNAc-R beta1,4-galactosyltransferase--a target enzyme for drug design. Acceptor specificity and inhibition of the enzyme. *Glycoconj. J.* **2006**, *23*, 525–541. [CrossRef]
19. Ramasamy, V.; Ramakrishnan, B.; Boeggeman, E.; Ratner, D.M.; Seeberger, P.H.; Qasba, P.K. Oligosaccharide preferences of beta1,4-galactosyltransferase-I: Crystal structures of Met340His mutant of human beta1,4-galactosyltransferase-I with a pentasaccharide and trisaccharides of the N-glycan moiety. *J. Mol. Biol.* **2005**, *353*, 53–67. [CrossRef] [PubMed]
20. Dohi, T.; Yuyama, Y.; Natori, Y.; Smith, P.L.; Lowe, J.B.; Oshima, M. Detection of N-acetylgalactosaminyltransferase mRNA which determines expression of Sda blood group carbohydrate structure in human gastrointestinal mucosa and cancer. *Int. J. Cancer* **1996**, *67*, 626–631. [CrossRef]
21. Cull, J.; Pink, R.C.; Samuel, P.; Brooks, S.A. Myriad mechanisms: Factors regulating the synthesis of aberrant mucin-type O-glycosylation found on cancer cells. *Glycobiology* **2025**, *35*, cwaf023. [CrossRef] [PubMed]
22. Munkley, J. Aberrant Sialylation in Cancer: Therapeutic Opportunities. *Cancers* **2022**, *14*, 4248. [CrossRef]
23. Brockhausen, I.; Argueso, P. Mucin-type O-glycans: Biosynthesis and functions. In *Comprehensive Glycoscience*, 2nd ed.; Barchi, J., Ed.; Elsevier: Amsterdam, The Netherlands, 2021; pp. 233–252. [CrossRef]
24. Venturi, G.; Gomes Ferreira, I.; Pucci, M.; Ferracin, M.; Malagolini, N.; Chiricolo, M.; Dall'Olio, F. Impact of sialyltransferase ST6GAL1 overexpression on different colon cancer cell types. *Glycobiology* **2019**, *29*, 684–695. [CrossRef]
25. Lin, W.R.; Yeh, C.T. GALNT14: An Emerging Marker Capable of Predicting Therapeutic Outcomes in Multiple Cancers. *Int. J. Mol. Sci.* **2020**, *21*, 1491. [CrossRef]
26. Sun, H.; Chang, J.; Ye, M.; Weng, W.; Zhang, M.; Ni, S.; Tan, C.; Huang, D.; Wang, L.; Du, X.; et al. GCNT4 is Associated with Prognosis and Suppress Cell Proliferation in Gastric Cancer. *Oncol. Targets Ther.* **2020**, *13*, 8601–8613. [CrossRef]

27. Burchell, J.; Poulsom, R.; Hanby, A.; Whitehouse, C.; Cooper, L.; Clausen, H.; Miles, D.; Taylor-Papadimitriou, J. An alpha2,3 sialyltransferase (ST3Gal I) is elevated in primary breast carcinomas. *Glycobiology* **1999**, *9*, 1307–1311. [CrossRef] [PubMed]
28. Marcos, N.T.; Pinho, S.; Grandela, C.; Cruz, A.; Samyn-Petit, B.; Harduin-Lepers, A.; Almeida, R.; Silva, F.; Morais, V.; Costa, J.; et al. Role of the human ST6GalNAc-I and ST6GalNAc-II in the synthesis of the cancer-associated sialyl-Tn antigen. *Cancer Res.* **2004**, *6*, 7050–7057. [CrossRef]
29. González-Vallinas, M.; Vargas, T.; Moreno-Rubio, J.; Molina, S.; Herranz, J.; Cejas, P.; Burgos, E.; Aguayo, C.; Custodio, A.; Reglero, G.; et al. Clinical relevance of the differential expression of the glycosyltransferase gene GCNT3 in colon cancer. *Eur. J. Cancer* **2015**, *51*, 1–8. [CrossRef]
30. Kurosaka, A.; Nakajima, H.; Funakoshi, I.; Matsuyama, M.; Nagayo, T.; Yamashina, I. Structures of the major oligosaccharides from a human rectal adenocarcinoma glycoprotein. *J. Biol. Chem.* **1983**, *258*, 11594–11598. [CrossRef] [PubMed]
31. Gao, Y.; Chachadi, V.B.; Cheng, P.W.; Brockhausen, I. Glycosylation potential of human prostate cancer cell lines. *Glycoconj. J.* **2012**, *29*, 525–537. [CrossRef] [PubMed]
32. Schachter, H. Complex N-glycans: The story of the "yellow brick road". *Glycoconj. J.* **2014**, *31*, 1–5. [CrossRef]
33. Unligil, U.M.; Zhou, S.; Yuwaraj, S.; Sarkar, M.; Schachter, H.; Rini, J.M. X-ray crystal structure of rabbit N-acetylglucosaminyltransferase I: Catalytic mechanism and a new protein superfamily. *EMBO J.* **2000**, *19*, 5269–5280. [CrossRef]
34. Sharma, C.B.; Lehle, L.; Tanner, W. N-Glycosylation of yeast proteins. Characterization of the solubilized oligosaccharyl transferase. *Eur. J. Biochem.* **1981**, *116*, 101–108. [CrossRef]
35. Yoo, J.; Mashalidis, E.H.; Kuk, A.C.Y.; Yamamoto, K.; Kaeser, B.; Ichikawa, S.; Lee, S.Y. GlcNAc-1-P-transferase-tunicamycin complex structure reveals basis for inhibition of N-glycosylation. *Nature Struct. Mol. Biol.* **2018**, *25*, 217–224. [CrossRef] [PubMed]
36. Timal, S.; Hoischen, A.; Lehle, L.; Adamowicz, M.; Huijben, K.; Sykut-Cegielska, J.; Paprocka, J.; Jamroz, E.; van Spronsen, F.J.; Körner, C.; et al. Gene identification in the congenital disorders of glycosylation type I by whole-exome sequencing. *Human Mol. Gen.* **2012**, *21*, 4151–4161. [CrossRef] [PubMed]
37. Jensen, J.W.; Schutzbach, J.S. Activation of dolichyl-phospho-mannose synthase by phospholipids. *Eur. J. Biochem.* **1985**, *153*, 41–48. [CrossRef]
38. Reznik, N.; Fass, D. Disulfide bond formation and redox regulation in the Golgi apparatus. *FEBS Lett.* **2022**, *596*, 2859–2872. [CrossRef]
39. Stanley, P.; Narasimhan, S.; Siminovitch, L.; Schachter, H. Chinese hamster ovary cells selected for resistance to the cytotoxicity of phytohemagglutinin are deficient in a UDP-N-acetylglucosamine--glycoprotein N-acetylglucosaminyltransferase activity. *Proc. Nat. Acad. Sci. USA* **1975**, *72*, 3323–3327. [CrossRef]
40. Paulsen, H.; Heume, M.; Nürnberger, H. Synthese der verzweigten Nonasaccharid-Sequenz der "bisected" Struktur von N-Glycoproteinen [Synthesis of a highly branched nonasaccharide sequence of the "bisected"structure of N-glycoproteins]. *Carbohydr. Res.* **1990**, *200*, 127–166. [CrossRef] [PubMed]
41. Paulsen, H.; Peters, T.; Sinnwell, V.; Meyer, B. Konformationsanalyse modifizierter Tetrasaccharidsequenzen vom Typ der N-Glycoproteine--zum Problem der alpha-(1----6)-glycosidischen Bindung [Conformation analysis of modified tetrasaccharide sequences of the N-glycoprotein type--problem of the alpha-(1 to 6)-glycosidic bond]. *Carbohydr. Res.* **1987**, *165*, 251–266. [CrossRef]
42. Reck, F.; Springer, M.; Meinjohanns, E.; Paulsen, H.; Brockhausen, I.; Schachter, H. Synthetic substrate analogues for UDP-GlcNAc: Man alpha 1-3R beta 1-2-N-acetylglucosaminyltransferase I. Substrate specificity and inhibitors for the enzyme. *Glycoconj. J.* **1995**, *12*, 747–754. [CrossRef] [PubMed]
43. Reck, F.; Springer, M.; Paulsen, H.; Brockhausen, I.; Sarkar, M.; Schachter, H. Synthesis of tetrasaccharide analogues of the N-glycan substrate of beta-(1→2)-N-acetylglucosaminyltransferase II using trisaccharide precursors and recombinant beta-(1→2)-N-acetylglucosaminyltransferase I. *Carbohydr. Res.* **1994**, *259*, 93–101. [CrossRef]
44. Reck, F.; Meinjohanns, E.; Springer, M.; Wilkens, R.; Van Dorst, J.A.; Paulsen, H.; Möller, G.; Brockhausen, I.; Schachter, H. Synthetic substrate analogues for UDP-GlcNAc: Man alpha 1-6R beta(1-2)-N-acetylglucosaminyltransferase II. Substrate specificity and inhibitors for the enzyme. *Glycoconj. J.* **1994**, *11*, 210–216. [CrossRef] [PubMed]
45. Paulsen, H.; Springer, M.; Reck, F.; Brockhausen, I.; Schachter, H. Synthese von modifizierten Tetrasacchariden als analoge Akzeptor-Inhibitoren der N-Acetylglucosaminyltransferase II [Synthesis of modified tetrasaccharides as analog acceptor-inhibitors of N-acetylglucosaminyltransferase II]. *Carbohydr. Res.* **1995**, *275*, 403–411. [CrossRef] [PubMed]
46. Kadirvelraj, R.; Yang, J.Y.; Sanders, J.H.; Liu, L.; Ramiah, A.; Prabhakar, P.K.; Boons, G.J.; Wood, Z.A.; Moremen, K.W. Human N-acetylglucosaminyltransferase II substrate recognition uses a modular architecture that includes a convergent exosite. *Proc. Nat. Acad. Sci. USA* **2018**, *115*, 4637–4642. [CrossRef]

47. de-Souza-Ferreira, M.; Ferreira, É.E.; de-Freitas-Junior, J.C.M. Aberrant N-glycosylation in cancer: MGAT5 and β1,6-GlcNAc branched N-glycans as critical regulators of tumor development and progression. *Cell. Oncol.* **2023**, *46*, 481–501. [CrossRef]
48. Brockhausen, I.; Reck, F.; Kuhns, W.; Khan, S.; Matta, K.L.; Meinjohanns, E.; Paulsen, H.; Shah, R.N.; Baker, M.A.; Schachter, H. Substrate specificity and inhibition of UDP-GlcNAc:GlcNAc beta 1-2Man alpha 1-6R beta 1,6-N-acetylglucosaminyltransferase V using synthetic substrate analogues. *Glycoconj. J.* **1995**, *12*, 371–379. [CrossRef]
49. Khan, S.H.; Crawley, S.C.; Kanie, O.; Hindsgaul, O. A trisaccharide acceptor analog for N-acetylglucosaminyltransferase V which binds to the enzyme but sterically precludes the transfer reaction. *J. Biol. Chem.* **1993**, *268*, 2468–2473. [CrossRef] [PubMed]
50. Kanie, O.; Crawley, S.C.; Palcic, M.M.; Hindsgaul, O. Acceptor-substrate recognition by N-acetylglucosaminyltransferase-V: Critical role of the 4″-hydroxyl group in beta-D-GlcpNAc-(1→2)-alpha-D-Manp(1→6)-beta-D-Glcp-OR. *Carbohydr. Res.* **1993**, *243*, 139–164. [CrossRef]
51. Kanie, O.; Crawley, S.C.; Palcic, M.M.; Hindsgaul, O. Key involvement of all three GlcNAc hydroxyl groups in the recognition of beta-D-GlcpNAc-(1→2)-alpha-D-Manp-(1→6)-beta-D-Glcp-OR by N-acetylglucosaminyltransferase-V. *Bioorg. Med. Chem.* **1994**, *2*, 1231–1241. [CrossRef]
52. Nagae, M.; Kizuka, Y.; Mihara, E.; Kitago, Y.; Hanashima, S.; Ito, Y.; Takagi, J.; Taniguchi, N.; Yamaguchi, Y. Structure and mechanism of cancer-associated N-acetylglucosaminyltransferase-V. *Nat. Commun.* **2018**, *9*, 3380. [CrossRef]
53. Osuka, R.F.; Nagae, M.; Ohuchi, A.; Ohno, S.; Yamaguchi, Y.; Kizuka, Y. The cancer-associated glycosyltransferase GnT-V (MGAT5) recognizes the N-glycan core via residues outside its catalytic pocket. *FEBS Lett.* **2023**, *597*, 3102–3113. [CrossRef]
54. Osuka, R.F.; Yamasaki, T.; Kizuka, Y. Structure and function of N-acetylglucosaminyltransferase V (GnT-V). *Biochim. Biophys. Acta Gen. Subj.* **2024**, *1868*, 130709. [CrossRef] [PubMed]
55. Paulsen, H.; Peters, T.; Sinnwell, V.; Heume, M.; Meyer, B. Konformationsanalyse der verzweigten Pentasaccharid-sequenz der "bisected" Struktur von N-Glycoproteinen [Conformational analysis of the double pentasaccharide sequence of the "bisected" structure ov N-glycoproteins]. *Carbohydr. Res.* **1986**, *156*, 87–106. [CrossRef]
56. Brockhausen, I.; Hull, E.; Hindsgaul, O.; Schachter, H.; Shah, R.N.; Michnick, S.W.; Carver, J.P. Control of glycoprotein synthesis. Detection and characterization of a novel branching enzyme from hen oviduct, UDP-N-acetylglucosamine:GlcNAc beta 1-6 (GlcNAc beta 1-2)Man alpha-R (GlcNAc to Man) beta-4-N-acetylglucosaminyltransferase VI. *J. Biol. Chem.* **1989**, *264*, 11211–11221. [CrossRef] [PubMed]
57. Sakamoto, Y.; Taguchi, T.; Honke, K.; Korekane, H.; Watanabe, H.; Tano, Y.; Dohmae, N.; Takio, K.; Horii, A.; Taniguchi, N. Molecular cloning and expression of cDNA encoding chicken UDP-N-acetyl-D-glucosamine (GlcNAc): GlcNAcbeta 1-6(GlcNAcbeta 1-2)- manalpha 1-R[GlcNAc to man]beta 1,4N-acetylglucosaminyltransferase VI. *J. Biol. Chem.* **2000**, *275*, 36029–36034. [CrossRef]
58. Dalle Vedove, E.; Costabile, G.; Merkel, O.M. Mannose and Mannose-6-Phosphate Receptor-Targeted Drug Delivery Systems and Their Application in Cancer Therapy. *Adv. Healthc. Mater.* **2018**, *7*, e1701398. [CrossRef]
59. Brockhausen, I.; Lehotay, M.; Yang, J.M.; Qin, W.; Young, D.; Lucien, J.; Coles, J.; Paulsen, H. Glycoprotein biosynthesis in porcine aortic endothelial cells and changes in the apoptotic cell population. *Glycobiology* **2002**, *12*, 33–45. [CrossRef]
60. Kuhn, B.; Benz, J.; Greif, M.; Engel, A.M.; Sobek, H.; Rudolph, M.G. The structure of human α-2,6-sialyltransferase reveals the binding mode of complex glycans. *Acta Crystal. Sect. D Biol. Crystallogr.* **2013**, *69 Pt 9*, 1826–1838. [CrossRef] [PubMed]
61. Qasba, P.K.; Ramakrishnan, B.; Boeggeman, E. Structure and function of beta -1,4-galactosyltransferase. *Curr. Drug Targets* **2008**, *9*, 292–309. [CrossRef]
62. Inaba, R.; Vujakovic, S.; Bergstrom, K. The gut mucus network: A dynamic liaison between microbes and the immune system. *Sem. Immunol.* **2023**, *69*, 101807. [CrossRef] [PubMed]
63. Luis, A.S.; Hansson, G.C. Intestinal mucus and their glycans: A habitat for thriving microbiota. *Cell Host Microbe* **2023**, *31*, 1087–1100. [CrossRef]
64. Wagner, K.W.; Punnoose, E.A.; Januario, T.; Lawrence, D.A.; Pitti, R.M.; Lancaster, K.; Lee, D.; von Goetz, M.; Yee, S.F.; Totpal, K.; et al. Death-receptor O-glycosylation controls tumor-cell sensitivity to the proapoptotic ligand Apo2L/TRAIL. *Nature Med.* **2007**, *13*, 1070–1077. [CrossRef]
65. Hounsell, E.F.; Lawson, A.M.; Feeney, J.; Gooi, H.C.; Pickering, N.J.; Stoll, M.S.; Lui, S.C.; Feizi, T. Structural analysis of the O-glycosidically linked core-region oligosaccharides of human meconium glycoproteins which express oncofoetal antigens. *Eur. J. Biochem.* **1985**, *148*, 367–377. [CrossRef] [PubMed]
66. Pollex-Krüger, A.; Meyer, B.; Stuike-Prill, R.; Sinnwell, V.; Matta, K.L.; Brockhausen, I. Preferred conformations and dynamics of five core structures of mucin type O-glycans determined by NMR spectroscopy and force field calculations. *Glycoconj. J.* **1993**, *10*, 365–380. [CrossRef]
67. Müller, S.; Alving, K.; Peter-Katalinic, J.; Zachara, N.; Gooley, A.A.; Hanisch, F.G. High density O-glycosylation on tandem repeat peptide from secretory MUC1 of T47D breast cancer cells. *J. Biol. Chem.* **1999**, *274*, 18165–18172. [CrossRef] [PubMed]

68. Bennett, E.P.; Hassan, H.; Hollingsworth, M.A.; Clausen, H. A novel human UDP-N-acetyl-D-galactosamine:polypeptide N-acetylgalactosaminyltransferase, GalNAc-T7, with specificity for partial GalNAc-glycosylated acceptor substrates. *FEBS Lett.* **1999**, *460*, 226–230. [CrossRef]
69. Kato, K.; Hansen, L.; Clausen, H. Polypeptide N-acetylgalactosaminyltransferase-Associated Phenotypes in Mammals. *Molecules* **2021**, *26*, 5504. [CrossRef]
70. Ju, T.; Cummings, R.D. Protein glycosylation: Chaperone mutation in Tn syndrome. *Nature* **2005**, *437*, 1252. [CrossRef]
71. Patsos, G.; Hebbe-Viton, V.; Robbe-Masselot, C. O-glycan inhibitors generate aryl-glycans, induce apoptosis and lead to growth inhibition in colorectal cancer cell lines. *Glycobiology* **2009**, *19*, 382–398. [CrossRef] [PubMed]
72. Brockhausen, I.; Möller, G.; Pollex-Krüger, A.; Rutz, V.; Paulsen, H.; Matta, K.L. Control of O-glycan synthesis: Specificity and inhibition of O-glycan core 1 UDP-galactose:N-acetylgalactosamine-alpha-R beta 3-galactosyltransferase from rat liver. *Biochem. Cell Biol.* **1992**, *70*, 99–108. [CrossRef] [PubMed]
73. Yeh, J.C.; Ong, E.; Fukuda, M. Molecular cloning and expression of a novel beta-1,6-N-acetylglucosaminyltransferase that forms core 2, core 4, and I branches. *J. Biol. Chem.* **1999**, *274*, 3215–3221. [CrossRef] [PubMed]
74. Kuhns, W.; Rutz, V.; Paulsen, H.; Matta, K.L.; Baker, M.A.; Barner, M.; Granovsky, M.; Brockhausen, I. Processing O-glycan core 1, Gal beta 1-3GalNAc alpha-R. Specificities of core 2, UDP-GlcNAc: Gal beta 1-3 GalNAc-R(GlcNAc to GalNAc) beta 6-N-acetylglucosaminyltransferase and CMP-sialic acid: Gal beta 1-3GalNAc-R alpha 3-sialyltransferase. *Glycoconj. J.* **1993**, *10*, 381–394. [CrossRef]
75. Mardahl, M.; Schröter, M.F.; Engelbert, D.; Pink, M.; Sperandio, M.; Hamann, A.; Syrbe, U. Core 2 ß1,6-N-acetylglucosaminyltransferase-I, crucial for P-selectin ligand expression is controlled by a distal enhancer regulated by STAT4 and T-bet in CD4+ T helper cells 1. *Mol. Immun.* **2016**, *77*, 132–140. [CrossRef]
76. Schwientek, T.; Yeh, J.C.; Levery, S.B.; Keck, B.; Merkx, G.; van Kessel, A.G.; Fukuda, M.; Clausen, H. Control of O-glycan branch formation. Molecular cloning and characterization of a novel thymus-associated core 2 beta1, 6-n-acetylglucosaminyltransferase. *J. Biol. Chem.* **2000**, *275*, 11106–11113. [CrossRef]
77. Vavasseur, F.; Yang, J.M.; Dole, K.; Paulsen, H.; Brockhausen, I. Synthesis of O-glycan core 3: Characterization of UDP-GlcNAc: GalNAc-R beta 3-N-acetyl-glucosaminyltransferase activity from colonic mucosal tissues and lack of the activity in human cancer cell lines. *Glycobiology* **1995**, *5*, 351–357. [CrossRef]
78. Huang, M.C.; Chen, H.Y.; Huang, H.C.; Huang, J.; Liang, J.-T.; Shen, T.-L.; Lin, N.-Y.; Ho, C.-C.; Hsu, S.-M. C2GnT-M is downregulated in colorectal cancer and its re-expression causes growth inhibition of colon cancer cells. *Oncogene* **2006**, *25*, 3267–3276. [CrossRef]
79. Pak, J.E.; Arnoux, P.; Zhou, S.; Sivarajah, P.; Satkunarajah, M.; Xing, X.; Rini, J.M. X-ray crystal structure of leukocyte type core 2 beta1,6-N-acetylglucosaminyltransferase. Evidence for a convergence of metal ion-independent glycosyltransferase mechanism. *J. Biol. Chem.* **2006**, *281*, 26693–26701. [CrossRef]
80. Pak, J.E.; Satkunarajah, M.; Seetharaman, J.; Rini, J.M. Structural and mechanistic characterization of leukocyte-type core 2 β1,6-N-acetylglucosaminyltransferase: A metal-ion-independent GT-A glycosyltransferase. *J. Mol. Biol.* **2011**, *414*, 798–811. [CrossRef]
81. Kurosaka, A.; Funakoshi, I.; Matsuyama, M.; Nagayo, T.; Yamashina, I. UDP-GalNAc:GalNAc-mucin alpha-N-acetylgalactosamine transferase activity in human intestinal cancerous tissues. *FEBS Lett.* **1985**, *190*, 259–262. [CrossRef] [PubMed]
82. Yeh, J.C.; Hiraoka, N.; Petryniak, B.; Nakayama, J.; Ellies, L.G.; Rabuka, D.; Hindsgaul, O.; Marth, J.D.; Lowe, J.B.; Fukuda, M. Novel sulfated lymphocyte homing receptors and their control by a Core1 extension beta 1,3-N-acetylglucosaminyltransferase. *Cell* **2001**, *105*, 957–969. [CrossRef]
83. Nakayama, J.; Yeh, J.C.; Misra, A.K.; Ito, S.; Katsuyama, T.; Fukuda, M. Expression cloning of a human alpha1, 4-N-acetylglucosaminyltransferase that forms GlcNAcalpha1→4Galbeta→R, a glycan specifically expressed in the gastric gland mucous cell-type mucin. *Proc. Nat. Acad. Sci. USA* **1999**, *96*, 8991–8996. [CrossRef] [PubMed]
84. Yuki, A.; Fujii, C.; Yamanoi, K.; Matoba, H.; Harumiya, S.; Kawakubo, M.; Nakayama, J. Glycosylation of MUC6 by α1,4-linked N-acetylglucosamine enhances suppression of pancreatic cancer malignancy. *Cancer Sci.* **2022**, *113*, 576–586. [CrossRef]
85. Datta, A.K. Comparative sequence analysis in the sialyltransferase protein family: Analysis of motifs. *Curr. Drug Targets* **2009**, *10*, 483–498. [CrossRef] [PubMed]
86. Moremen, K.W.; Ramiah, A.; Stuart, M.; Steel, J.; Meng, L.; Forouhar, F.; Moniz, H.A.; Gahlay, G.; Gao, Z.; Chapla, D.; et al. Expression system for structural and functional studies of human glycosylation enzymes. *Nature Chem. Biol.* **2018**, *14*, 156–162. [CrossRef]

87. Grewal, R.K.; Shaikh, A.R.; Gorle, S.; Kaur, M.; Videira, P.A.; Cavallo, L.; Chawla, M. Structural Insights in Mammalian Sialyltransferases and Fucosyltransferases: We Have Come a Long Way, but It Is Still a Long Way Down. *Molecules* **2021**, *26*, 5203. [CrossRef] [PubMed]
88. Rao, F.V.; Rich, J.R.; Rakić, B.; Buddai, S.; Schwartz, M.F.; Johnson, K.; Bowe, C.; Wakarchuk, W.W.; Defrees, S.; Withers, S.G.; et al. Structural insight into mammalian sialyltransferases. *Nature Struct. Mol. Biol.* **2009**, *16*, 1186–1188. [CrossRef]
89. Meldal, M.; Bielfeldt, T.; Peters, S.; Jensen, K.J.; Paulsen, H.; Bock, K. Susceptibility of glycans to beta-elimination in Fmoc-based O-glycopeptide synthesis. *Int. J. Peptide Protein Res.* **1994**, *43*, 529–536. [CrossRef]
90. Paulsen, H.; Peters, S.; Bielfeldt, T.; Meldal, M.; Bock, K. Synthesis of the glycosyl amino acids N alpha-Fmoc-Ser[Ac4-beta-D-Galp-(1→3)-Ac2-alpha-D-GalN3p]-OPfp and N alpha-Fmoc-Thr[Ac4-beta-D-Galp-(1→3)-Ac2-alpha-D-GalN3p]-OPfp and the application in the solid-phase peptide synthesis of multiply glycosylated mucin peptides with Tn and T antigenic structures. *Carbohydr. Res.* **1995**, *268*, 17–34. [CrossRef]
91. Peters, S.; Bielfeldt, T.; Meldal, M.; Bock, K.; Paulsen, H. Multiple-column solid-phase glycopeptide synthesis. *J. Chem. Soc. Perkin Trans.* **1992**, *1*, 1163–1171. [CrossRef]
92. Brockhausen, I.; Möller, G.; Merz, G.; Adermann, K.; Paulsen, H. Control of mucin synthesis: The peptide portion of synthetic O-glycopeptide substrates influences the activity of O-glycan core 1 UDPgalactose:N-acetyl-alpha-galactosaminyl-R beta 3-galactosyltransferase. *Biochemistry* **1990**, *29*, 10206–10212. [CrossRef] [PubMed]
93. Brockhausen, I.; Toki, D.; Brockhausen, J.; Peters, S.; Bielfeldt, T.; Kleen, A.; Paulsen, H.; Meldal, M.; Hagen, F.; Tabak, L.A. Specificity of O-glycosylation by bovine colostrum UDP-GalNAc: Polypeptide alpha-N-acetylgalactosaminyltransferase using synthetic glycopeptide substrates. *Glycoconj. J.* **1996**, *13*, 849–856. [CrossRef] [PubMed]
94. Granovsky, M.; Bielfeldt, T.; Peters, S.; Paulsen, H.; Meldal, M.; Brockhausen, J.; Brockhausen, I. UDPgalactose:glycoprotein-N-acetyl-D-galactosamine 3-beta-D-galactosyltransferase activity synthesizing O-glycan core 1 is controlled by the amino acid sequence and glycosylation of glycopeptide substrates. *Eur. J. Biochem.* **1994**, *221*, 1039–1046. [CrossRef]
95. Brockhausen, I.; Dowler, T.; Paulsen, H. Site directed processing: Role of amino acid sequences and glycosylation of acceptor glycopeptides in the assembly of extended mucin type O-glycan core 2. *Biochim. Biophys. Acta* **2009**, *1790*, 1244–1257. [CrossRef] [PubMed]

molecules

MDPI

Article

# Chemical and Enzymatic Synthesis of DisialylGb5 and Other Sialosides for Glycan Array Assembly and Evaluation of Siglec-Mediated Immune Checkpoint Inhibition

Kuo-Shiang Liao [1,†], Yixuan Zhou [1,†], Cinya Chung [1,†], Chih-Chuan Kung [1], Chien-Tai Ren [1], Chung-Yi Wu [1], Yi-Wei Lou [1], Po-Kai Chuang [2], Balázs Imre [3], Yves S. Y. Hsieh [3,4] and Chi-Huey Wong [1,2,*]

1 Genomics Research Center, Academia Sinica, Taipei 115, Taiwan; shiang.cgt@gmail.com (K.-S.L.); zhouyixuan1984@outlook.com (Y.Z.); cinyachung@gmail.com (C.C.); kung061283@gmail.com (C.-C.K.); a0922720496@gmail.com (C.-T.R.); cywu@rockbiomedical.com (C.-Y.W.); lou.yi.wei@gmail.com (Y.-W.L.)
2 Department of Chemistry, The Scripps Research Institute, San Diego, CA 92037, USA; pokaipkc@gmail.com
3 School of Pharmacy, Taipei Medical University, Taipei 110, Taiwan; lalazsimre@tmu.edu.tw (B.I.); yvhsieh@tmu.edu.tw (Y.S.Y.H.)
4 Division of Glycoscience, Department of Chemistry, School of Engineering Sciences in Chemistry, Biotechnology and Health, Royal Institute of Technology (KTH), AlbaNova University Center, SE10691 Stockholm, Sweden
* Correspondence: wong@scripps.edu
† These authors contributed equally to this work.

**Abstract:** Aberrant glycosylation, especially sialylation, on cell surface is often associated with cancer progression and immunosuppression. Over-sialylation of stage-specific embryonic antigen-4 (SSEA-4) to generate disialylGb5 (DSGb5) was reported to trigger Siglec-7 recognition and suppress NK-mediated target killing. In this study, efficient chemoenzymatic and programmable one-pot methods were explored for the synthesis of DSGb5 and related sialosides for assembly of glycan microarrays and evaluation of binding specificity toward Siglecs-7, 9, 10, and 15 associated with immune checkpoint inhibition. The result showed weak binding of DSGb5 to these Siglecs; however, a truncated glycolyl glycan was identified to bind Siglec-10 strongly with a dissociation constant of 50 nM and exhibited a significant inhibition of Siglec-10 interacting with breast cancer cells.

**Keywords:** sialyl SSEA-4; programmable; chemoenzymatic synthesis; glycolyl sialic acid; immune checkpoint

## 1. Introduction

Aberrant sialylation has been recognized as a hallmark of tumorigenesis. An example of such a change is the differential expression of glycosyltransferases in cancer cells to generate tumor-associated carbohydrate antigens (TACAs) [1]. Another example is the over-sialylation of glycoproteins or glycolipids on cancer cells to enhance interaction with specific sialic acid-binding immunoglobulin-type lectins (Siglecs) on immune cells, resulting in a suppression of immune response against cancer cells, a process called immune checkpoint inhibition [2,3]. It has been demonstrated that certain sialylated glycolipids, like stage-specific embryonic antigen (SSEA-4) [4,5], exclusively expressed on cancer cells are associated with cancer progression [6–9] and the expression level correlates with poor survival of cancer patients [10]. Interestingly, SSEA-4 could be further sialylated by the enzyme ST6GalNAc6 to form disialylGb5 (DSGb5), which was first found and downregulated in renal cancer [11]. In addition, DSGb5 and ST6GalNAc6 were reported to be downregulated in colon cancer [12], resulting in the decreased expression of disialyl LeA and increased expression of sialyl LeA and SSEA-4, which was linked to the signaling pathway that maintained the survival of tyrosine kinase inhibitor (TKI)-resistant cancer cells [13–20]. Moreover, recent development of anti-SSEA-4 antibodies [10], Globo-H cancer vaccines [21–23], and studies of globo-series glycan biosynthesis [4,24] all point to SSEA-4 as a target for new cancer therapies. On the other hand, specific Siglecs on immune cells may interact with specific sialylated glycans on cancer cells to trigger immune suppression [2,3]. A recent report showed that DSGb5 on renal cancer cells interacted with Siglec-7 on NK cells and downregulated their cytotoxicity [25]. This report prompted us to further investigate the binding specificity of DSGb5 toward Siglecs and compared to that of other tumor-associated carbohydrate antigens (TACAs). Since DSGb5 is not readily available [26], we first explored methods for the synthesis of DSGb5, including the programmable one-pot method and the enzymatic method using sialyltransferases that catalyzed the sialylation of SSEA-4 to form DSGb5. We then combined the synthetic glycans and other sialylated TACAs [21,27–31] and N-glycans to construct a sialylo–glycan array to profile the binding specificity of Siglecs, especially Siglec-7, Siglec-9, Siglec-10, and Siglec-15, to assess their role in immune checkpoint inhibition.

## 2. Results and Discussion

### 2.1. The mRNA Level of ST6GALNAC6 in Brain, Breast, Colon, Lung, Pancreas, and Prostate Cancers Decreased Compared to Normal Cells

We previously showed that the elevated expression of SSEA-4 on cancer cells was caused by increased expression of β3GalT5 and ST6Gal2 [10]. Here, we compare the mRNA levels of globo-series glycotransferases in normal and cancer tissues with large-scale RNA-Seq transcriptome analyses from The Cancer Genome Atlas (TCGA) TARGET and GTEx (The Genotype-Tissue Expression) datasets. We observed a statistically significant decrease in the mRNA level of ST6GalNAc6 (Figure 1) in all cancers examined, and the decrease was significant at an early stage, while there was no obvious change in the mRNA levels of β4GalT, β3GalNT1, β3GalT5, and ST3Gal2, suggesting that downregulation of ST6GalNAc6 expression in the early stage of tumorigenesis correlated with increased expression of SSEA-4 in cancers.

### 2.2. Synthesis of DSGb5 and Other Sialylated Derivatives

We first explored the programmable one-pot method [32] for the synthesis of DSGb5 glycan. We designed two sialylated disaccharide building blocks, one with α2,3- and the other with α2,6-linkage, with distinct relative reactivity for the one-pot synthesis to give the product in moderate yield (Scheme 1a. see details in Supporting Information). We then explored the enzymatic method to sialylate SSEA-4 using the enzyme N-acetylgalactosamine α2,6-sialyltransferase VI from *Photobacterium damsela* (Pd2,6ST) [33,34]. However, Neu5Ac was transferred to the C-6 position of terminal galactose (compound **2a**) instead of the internal GalNAc (Scheme 1b). We next tested the α2,6-sialyltransferase from *Photobacterium* sp. (Psp2,6ST), and to our delight, DSGb5 glycan (compound **1a**) was obtained in 46% yield (Scheme 1b). We also designed a new enzymatic route to DSGb5 glycan that involves α2,6-sialylation of SSEA-3, followed by α2,3-sialylation at terminal galactose using a recombinant *Pasteurella multocida* α2,3-sialyltransferase (PmST1) (Scheme 1c). After purification with an ion-exchange column, the product was obtained in 10% yield, and the NMR spectrum was identical to compound **1**, except differences in the linker moiety at the reducing end. This result further confirmed that DSGb5 glycan can be obtained by α2,6-sialylation of SSEA-4 with Psp2,6ST or by α2,3-sialylation of sialyl SSEA-3 with PmST1.

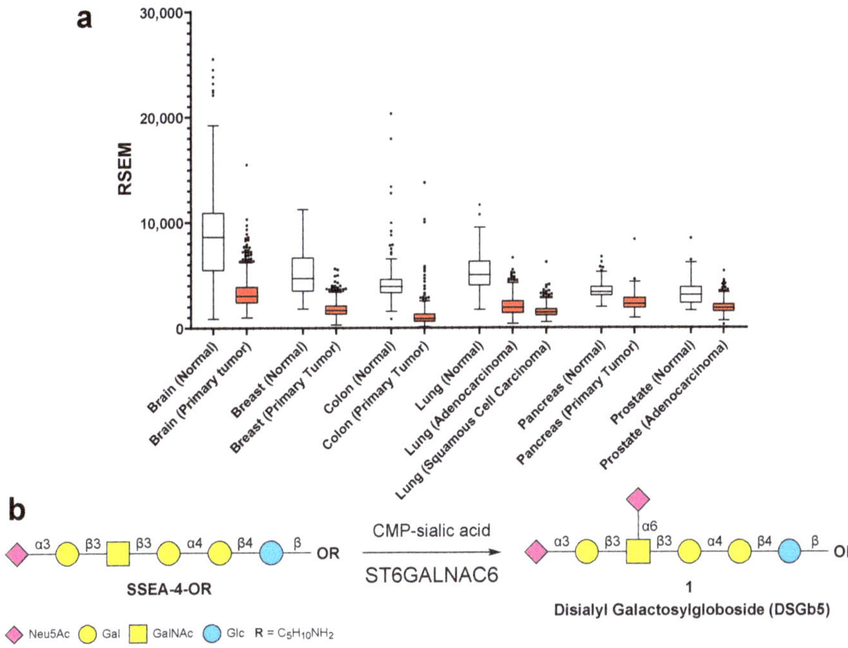


**Figure 1.** (**a**) The RSEM gene expression values of ST6GalNAc6 mRNA in brain, breast, colon, lung, pancreas, and prostate cancer cells (red box) and normal cells (white box). (**b**) Synthesis of DSGb5 by ST6GalNAc6-catalyzed sialylation of SSEA-4.

To further optimize the yield of DSGb5 glycan, we overexpressed human ST6GalNAc6 from HEK 293 cells. DSGb5 (compound **1**) was enzymatically synthesized via human ST6GalNAc6 from SSEA-4 glycan (see Materials and Methods). The result showed that human ST6GalNAc6 was more efficient and DSGb5 glycan was obtained in 82.1% yield (Figure 1).

Due to the different specificities of Pd2,6ST and Psp2,6ST, we further explored the sialylation of a range of globo-series glycan related acceptors to better understand their acceptor specificity and to expand the structural repertoire of sialylated glycans for our glycan array study of Siglecs. The obtained products and yields are summarized in Table S1. We found that Pd2,6ST accepted Gb4, Gb5, and SSEA-4 glycans with terminal Gal/GalNAc residues, whereas Gb3 and Globo-H glycans could not be sialylated by this enzyme. Previous studies also showed a lower catalytic efficiency of Pd2,6ST when the α-linked terminal GalNAc of GalNAcαAA or GalNAcαSer was used as an acceptor [33,35–37]. Similar results were found in the case of Gb3 and Globo-H glycans, in which the α-linked terminal Gal or Fuc may have hindered the activity of Pd2,6ST. Similarly, Psp2,6ST exhibited no reactivity towards Gb3 glycan, while its activity towards Gb5 glycan was lower than that of Pd2,6ST, as the former not only transferred Neu5Ac to the terminal Gal residue to afford **5a** but also sialylated GalNAc to afford disialylated oligosaccharide **16**, with 15% isolated yield. In addition, a previous study showed that Pd2,6ST catalyzed the transfer of Neu5Ac to the terminal Gal and/or GalNAc residues of the disaccharide acceptor Galβ1-3GalNAc [38]. However, only the terminal Gal was sialylated by the enzyme to afford hexasaccharide **5a** as the sole product. Interestingly, the presence of Fuc did not affect the Psp2,6ST-catalyzed sialylation of Globo-H glycan at terminal Gal to form compound **17** in 33% isolated yield. Moreover, Psp2,6ST was found to sialylate the GalNAc residue of Globo-H glycan to form compound **8** albeit in trace amounts, which was later characterized by MS/MS analysis. The sialylation of SSEA-4 glycan by Psp2,6ST showed high regioselectivity, probably due to the α2-3 linked Neu5Ac in the terminal Gal residue.

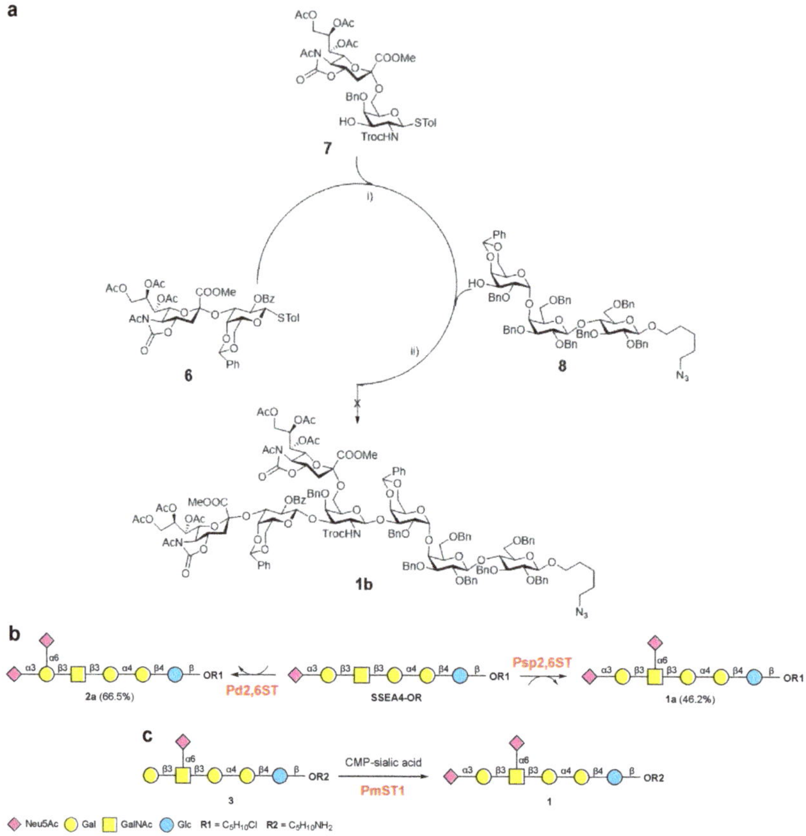


**Scheme 1.** (**a**) Programmable one-pot synthesis of DSGb5 glycan (i) NIS, TfOH, MS 4 Å, $CH_2Cl_2$, −78 °C, 2 h; (ii) NIS, TfOH, MS 4 Å, $CH_2Cl_2$, −40 °C, 2 h. (**b**) Enzymatic sialylation of SSEA-4 with Pd2,6ST and Psp2,6ST. (**c**) Synthesis of DSGb5 by enzymatic α2,3-sialylation of sialyl SSEA3. NmCSS, CMP-sialic acid synthetase from *Neisseria meningitidis*; PmST1, α2,3-sialyltransferase from *Pasteurella multocida*.

### *2.3. Glycan Array Analysis of Siglec-7, Siglec-9, Siglec-10, and Siglec-15 Binding to DSGb5 Glycan and Other Cancer-Associated Carbohydrate Antigens*

With these glycans in hand, we then created a glycan array on NHS-activated glass slides and investigated the binding specificity of several Siglecs. This array, including the glycans associated with tumor-associated glycolipids, like globo-series and RM2-related glycans, gangliosides [25], and other TACAs, was used to profile the binding specificities of Siglec-7, -9, -10, and -15. The results showed that SSEA-4 glycan (compound **GL-7**) and DSGb5 (compound **1**) lack significant binding to these four Siglecs, while Siglec-15 showed binding to the disialylated SSEA-3 glycan (Gb5), compound **2** (Figure 2d). The glycans of gangliosides GD2 (compound **GD-2**) and GD3 (compound **GD-1**) showed interaction with Siglec-7 [39–47]. We also detected a significant binding of Siglec-7 to the mono-sialylated compound **RM-2** (Figure 2a). However, Siglec-9 showed a stronger binding than Siglec-7 to the DSGb5 and **RM-4** glycans. On the other hand, Siglec-10 showed a significant binding to the unnatural **GD-5** glycan with α2,9- sialylation, and Siglec-15 recognized the SSEA-3 derivative with two sialic acids linked to the terminal galactose, compound **2** (Figure 2d).

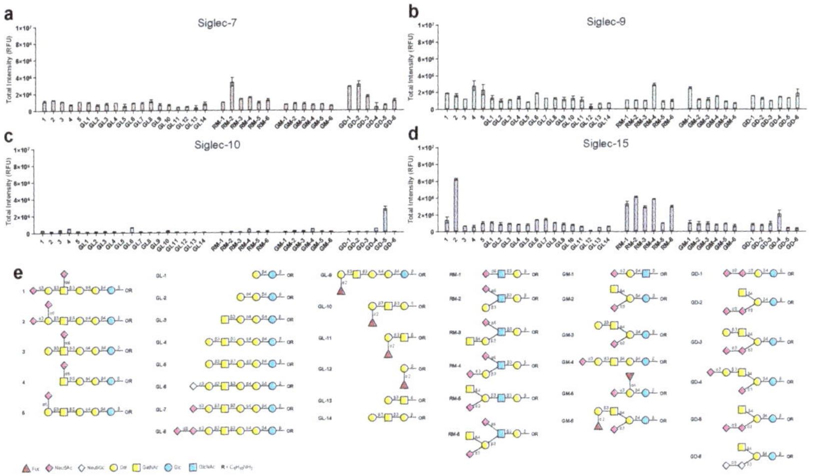

**Figure 2.** Glycan array profiling of commercially available Siglecs with TACAs. (**a**) Siglec-7. (**b**) Siglec-9. (**c**) Siglec-10. (**d**) Siglec-15. (**e**) TACAs used in this study. Siglec concentrations = 10 μg/mL.

## *2.4. Binding of Siglecs Toward Sialylated N-Glycans*

Since the glycans of cancer-associated glycolipids did not show significant binding to the Siglecs, we turned our attention to an array of sialylated *N*-glycans generated by us previously [28,29]. None of the *N*-glycans in our library showed any significant binding to Siglec-7 (Figure 3a). Interestingly, Siglec-9 recognized the glycans with terminal α2,3-sialylation (**N11**, **N14**, **N17**, and **N20**). On the other hand, the same type of *N*-glycans with α2,6-sialylaton (**N10** vs. **N11**, **N13** vs. **N14**, **N16** vs. **N17**, **N19** vs. **N20**) exhibited lower binding affinities. We also found that increasing the number of sialic acid residues in the α2,3-sialylated biantennary, triantennary, and tetraantennary *N*-glycans increased the binding affinity to Siglec-9. In contrast, no such trend was found in the case of α2,6-sialylated *N*-glycans (Figure 3b). The 2,4,2-triantennary *N*-glycan showed a stronger binding to Siglec-9 than the 2,2,6-branched one, irrespective of terminal sialic acid with α2,3- or α2,6- linkage (**N17** vs. **N16**, **N14** vs. **N13**). Interestingly, Siglec-9 did not show any significant interaction with complex type *N*-glycans with LacNAc repeats terminated with α2,3- or α2,6- siaylation. In addition, the core fucose did not affect binding to Siglec-9 as shown in the case of **N11** and **N35**. Contrary to what we observed in the case of Siglec-9, Siglec-10 preferred *N*-glycans with terminal α2,6-sialylation instead of α2,3-sialylation (**N13** vs. **N14**, **N16** vs. **N17**, and **N19** vs. **N20**). Moreover, an increasing number of terminal sialic acid groups led to a stronger binding (**N10** vs. **N11**, **N13** vs. **N14**, **N16** vs. **N17**, and **N19** vs. **N20** in Figure 3c). Siglec-15 exhibited a similar binding pattern to that of Siglec-9 toward *N*-glycans, favoring α2,3-sialosides, albeit with somewhat lower binding affinity (Figure 3d).

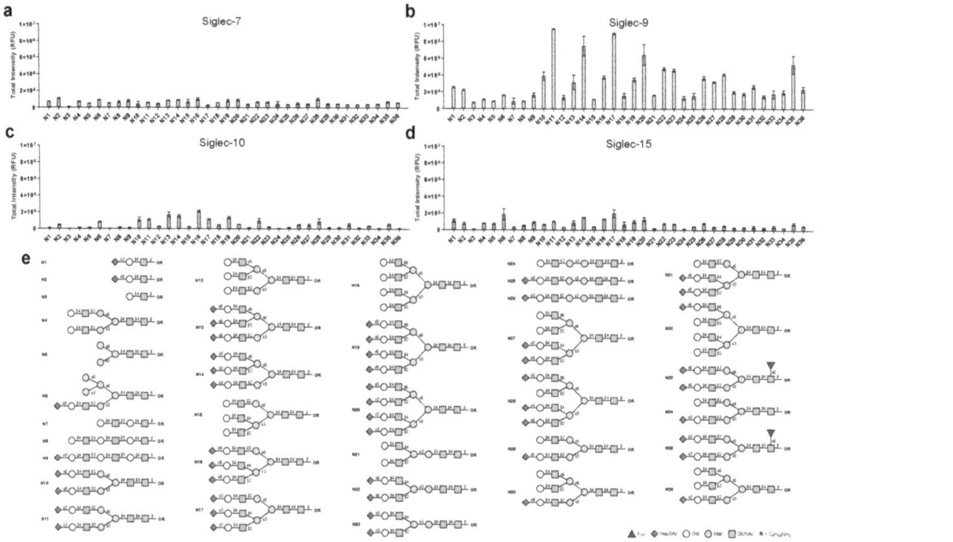

**Figure 3.** Profiling of (**a**) Siglec-7, (**b**) Siglec-9, (**c**) Siglec-10, and (**d**) Siglec-15 with *N*-glycan array. Siglec-9 significantly interacted with terminal α2,3-linked sialic acids of complex-type glycans compared to the α2,6-linked sialosides and the binding affinity increases with increase in antenna (**N11** vs. **N10**, **N14** vs. **N13**, **N17** vs. **N16**, **N20** vs. **N19**). In addition, the core fucose does not have a significant effect on binding (**N11** vs. **N35**). On the other hand, Siglec-10 prefers the *N*-sialosides with terminal α2,6-linked sialic acid. (**e**) *N*-glycans used in this study. Siglec concentrations = 10 μg/mL.

### 2.5. Inhibition of Siglec-10 Binding to CD24 on Breast Cancer Cells with Synthetic Glycans

A recent report suggested that CD24, a highly glycosylated glycosylphosphatidylinositol (GPI)-anchored protein [48,49], is overexpressed on various tumor cells [50] and is a ligand for Siglec-10. CD24 was also found to enhance the metastatic potential of malignant cells [51]. Human Siglec-10 is widely expressed in hematopoietic cell types [52], such as B and T cells, monocytes eosinophils, NK cells and macrophages [53,54]. The Siglec-10-CD24 interaction that led to immune suppression has been considered as a promising target for cancer immunotherapy [53–57]. The interaction of tumor CD24 and macrophage Siglec-10 inhibits phagocytosis and increases tumor growth and cell survival [53]. Removal of α2,3- and α2,6-linked sialic acids from CD24 by *Vibrio cholerae* neuraminidase significantly reduced binding to Siglec-10 [58]. Therefore, identification of Siglec-10 ligand may aid in the discovery of potential cancer biomarkers and the development of possible cancer therapy. Human Siglec-10 is a membrane-bound protein [59,60] that recognizes α2,6-linked sialosides on cancer cells. In this study, we identified a strong ligand (**SL-1**) of Siglec-10 using our array and the commercially available array from RayBiotech (Figure S1), a bi-antennary glycan terminated with two α2,3-glycolylneuraminic acid with a dissociation constant of 0.050 ± 0.011 μM (Figures 4a and S2). The binding affinity of Siglec-10 towards **SL-1** was better than other glycans with terminal Neu5Ac and the number of Neu5Gc-modified branches also influenced the binding affinity (**SL-1** > **SL-2**).

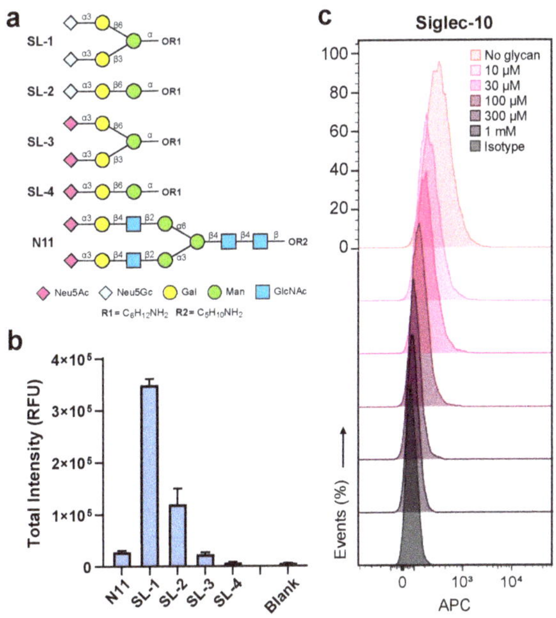


**Figure 4.** (**a**) Glycan candidates for Siglec-10 inhibition. (**b**) Siglec-10 binding signals of glycan candidates on glycan array. Binding of Siglec-10 (2.5 μg/mL) towards glycans on total glycan array (printed at 100 μM). Secondary antibody: Donkey anti-human IgG, Fcγ fragment specific, Alexa Fluor® 647-conjugated. Error bars: standard deviation. (**c**) Inhibition of Siglec-10 binding to MCF-7 at various concentrations of SL-1.

To test whether **SL-1** could compete with human breast cancer cell line MCF-7 over Siglec-10 binding, **SL-1** at different concentrations was added to a mixture of Siglec-10 and MCF-7 cells, and Siglec-10 binding signals were observed using flow cytometry. As shown in Figure 4c, binding of Siglec-10 to MCF-7 cells decreased as the concentration of glycan **SL-1** increased, suggesting that **SL-1** is a competitive inhibitor of Siglec-10 binding to MCF-7 cells. Competition analysis of Siglec-10 binding to glycans vs. MCF-7 cells was also performed with glycans **SL-2**, **SL-3**, **SL-4** and **N11**, respectively, at 100 μM. As shown in Figure S2, **SL-1** still had the strongest inhibition of Siglec-10 binding to MCF-7 cells and decreased in the order **SL-1** > **SL-2** > **SL-3** > **N11** > **SL-4**.

With **SL-2**, which contained only one of the branches of **SL-1**, Siglec-10 had a dissociation constant of 0.110 ± 0.041 μM. For **SL-3**, **SL-4**, and **N11**, the dissociation constants were similar (~0.176 μM); however, the fluorescent signals were very weak, and significantly lower than **SL-1** and **SL-2**. These results showed that Siglec-10 preferred glycolylneuraminic acid (**SL-1** and **SL-2**) over acetylneuraminic acid (**SL-3**, **SL-4** and **N11**).

The binding affinity also increased as the number of sialic acids increased (**SL-1** > **SL-2**). The multivalent glycans displayed on the microarray can be viewed as a mimic of cell-surface glycans confined in a small area. On the other hand, glycans used in the inhibition experiment in flow cytometry were dispersed in the solution as monomers. The binding affinity between glycans and Siglec-10 in solution should be lower than on microarrays. Therefore, to achieve better inhibitory effects, the interaction of **SL-1** to Siglec-10 should be increased. Since Siglec-10 is expressed on immune cells as multivalent glycoproteins, not as free form in solution, while MCF-7 cells are adherent cells, the interaction between two cells is more complex than that based on glycan array profiling. Work is in progress to identify better inhibitors of Siglec-10-mediated immune checkpoint suppression.

## 3. Materials and Methods

### *3.1. General Chemical Synthesis*

All chemicals were purchased as reagent grade and used without further purification. Anhydrous dichloromethane ($CH_2Cl_2$) was purchased from a commercial source without further distillation. Pulverized Molecular MS 4 Å (Sigma-Aldrich, St. Louis, MO, USA) for glycosylation were activated by heating at 350 °C for 10 h. Reactions were monitored by analytical thin-layer chromatography (TLC) in EM silica gel 60 F254 plates and visualized under UV (254 nm) and/or by staining with acidic ceric ammonium molybdate or p-anisaldehyde. Flash chromatography was performed on silica gel (Millipore, Burlington, MA, USA) of 40–63 μm particle size. 1H NMR spectra were recorded on a Bruker AVANCE 600 (600 MHz) spectrometer at 25 °C. Chemical shift (in ppm) was assigned according to $CDCl_3$ ($\delta$ = 7.24 ppm) and $D_2O$ ($\delta$ = 4.80 ppm). C-13 NMR spectra were obtained with Bruker AVANCE 600 spectrometer and were calibrated with $CDCl_3$ ($\delta$ = 77.00 ppm). Coupling constants (J) are reported in hertz (Hz). Chemical shifts measurements are reported in delta ($\delta$) units, and splitting patterns are described as singlet (s), doublet (d), triplet (t), quartet (q), or multiplet (m). Coupling constants (J) are reported in Hertz (Hz). High resolution ESI mass spectra were recorded on a Bruker Daltonics or Bruker Bio-TOF III spectrometer (Bruker, Billerica, MA, USA). MALDI-TOF spectra were recorded on Bruker Ultraflex II sepctrameter (Bruker).

### *3.2. Enzyme Preparations*

Codon-optimized genes for pET-15b-Pd2,6ST, pET-22b-Psp2,6ST, and pET-23a-PmST1 were synthesized by GenScript™. To construct the expression plasmid for soluble ST6GalNAc6, the pHEK293 Ultra Expression Vector I (TAKARA) was linearized with XhoI and ligated with the gene encoding soluble ST6GalNAc6 (V67–T333) fused to a C-terminal FLAG tag, using the NEBuilder® HiFi DNA Assembly Cloning Kit (NEB). The resulting plasmid was sequence-verified.

For bacterial expression of recombinant Pd2,6ST, Psp2,6ST, and PmST1, an overnight culture of BL21(DE3) was inoculated into 1 L Terrific Broth (Sigma-Aldrich) at a 1:50 (*v/v*) ratio in a 2.5 L baffled flask (Thomson, Carlsbad, CA, USA). When the culture reached an $OD_{600}$ of 0.8–1.0, protein expression was induced with 1 mM IPTG and incubated at 16 °C for 24 h. The cells were harvested, lysed using Bugbuster Protein Extraction Reagent (Millipore, Burlington, MA, USA), and the supernatant containing 6 × His-tagged proteins was purified using IMAC (Cytiva, Marlborough, MA, USA). The column was washed and eluted with eight column volumes of 50 mM MOPS, 300 mM NaCl, and either 20 mM or 300 mM imidazole (pH 7.5). The fractions were analyzed by SDS-PAGE followed by Coomassie Blue staining.

For mammalian expression of ST6GalNAc6, Expi293 cells were transfected according to the manufacturer's instructions, with a starting cell density of 1.5 million cells/mL. After 72 h, the culture supernatant was collected, filtered, and passed through anti-FLAG M2 agarose resin (Millipore). The resin was equilibrated with 10 column volumes of TBS buffer and 20 mM HEPES buffer containing 300 mM NaCl (pH 7.0) before loading. After binding, the resin was washed with 10 column volumes of the same buffer, and the protein was eluted using 8 column volumes of 0.1 M Tris-glycine (pH 3.0). The elution was immediately neutralized by adding 1/15 volume of 1 M Tris-HCl (pH 9.0). The enzyme was then buffer-exchanged and concentrated using an Amicon Ultra centrifugal filter unit (Millipore) with an appropriate molecular weight cutoff. Protein concentration was determined using the Pierce™ BCA Protein Assay Kit (Thermo Fisher Scientific, Waltham, MA, USA).

### 3.3. Enzymatic Synthesis of DSGb5 via Human ST6GalNAc6

The enzymatic reaction was subsequently performed in a reaction mixture containing 100 mM sodium cacodylate (pH 6.0), 10 mM $MgCl_2$, 1.6 mM CMP-sialic acid, 3 μg of recombinant human ST6GalNAc6, and 0.08 mM SSEA-4 glycan as acceptor substrate. The reaction mixture was incubated at 37 °C overnight with gentle shaking. The product was then isolated by solid phase extraction with HyperSep™ Hypercarb™ SPE Cartridges (Thermo Fisher Scientific) and analyzed by porous graphitic carbon liquid chromatography/tandem mass spectrometry (PGC-LC-MS/MS) using a Hypercarb™ column (Thermo Fisher Scientific).

### 3.4. Glycan Microarray

Glycan microarrays were fabricated by printing 100 μM amine-containing glycans in printing buffer (300 mM sodium phosphate buffer (pH 8.5) with 0.01% Triton X-100) onto NHS-activated glass slides (Nexterion® Slide H, SCHOTT, Mainz, Germany) as previously described [61]. Recombinant human Siglec-hFc-fusion proteins (Siglec-7, 9, 10, 15) were purchased from R&D Systems. The microarrays were blocked with SuperBlock Blocking Buffer in PBS (Thermo Fisher Scientific) at RT for 1 h and washed with 1 × PBS buffer containing 0.05% Tween-20 (PBST). Siglecs were diluted with PBST containing 30 mg/mL BSA (PBST-BSA). The Siglec samples were added to the microarray and incubated at RT for 1 h. The microarrays were washed with PBST, then added with DyLight™ 649 modified anti-human IgG antibody (donkey anti-human IgG, Fcγ fragment specific, DyLight™ 649, Jackson ImmunoResearch, West Grove, PA, USA) for 1 h in the dark. Finally, the microarrays were washed in PBST, milliQ, and dried. The microarrays were scanned at 635 nm using a GenePix 4300A Microarray Scanner (Molecular Devices, San Jose, CA, USA), and the fluorescence intensities were analyzed by the GenePix Pro 7.0 software (Molecular Devices).

The dissociation constants of Siglec-10 binding to glycans SL-1, SL-2, SL-3, SL-4 and N11 were determined using glycan microarray. Concentrations of amine-modified glycans ranging from 10 μM to 400 μM were printed onto NHS-modified glass slides [62], then treated with various concentrations of Siglec-10 coupled with Alexa Fluor® 647 modified anti-human IgG antibody (donkey anti-human IgG, Fcγ fragment specific, Alexa Fluor® 647, Jackson ImmunoResearch) at 4 °C overnight. After incubation, the microarray slides were washed with PBST, milliQ, and then dried. The microarrays were excited with 635 nm lasers and scanned using a GenePix 4300A Microarray Scanner, and the fluorescence intensities were analyzed by the GenePix Pro 7.0 software. Data analysis and Langmuir isotherms fitting was performed using PRISM (GraphPad, Boston, MA, USA) [61,62]. The microarrays were excited with 635 nm lasers and scanned using a GenePix 4300A Microarray Scanner, where the fluorescence intensities were analyzed using the GenePix Pro 7.0 software. Fluorescence intensities were plotted out against Siglec-10 concentrations to give a set of curves which were analyzed as Langmuir isotherms, assuming the reaction reached equilibrium,

$$\frac{F_{max}[P]}{[P]+K_D},$$

where $F_{max}$ is the maximum fluorescence intensity, $[P]$ is the total Siglec-10 concentration, and $K_D$ is the equilibrium dissociation constant for surface glycan and Siglec-10 binding.

### 3.5. Cell Culture

The human breast cancer cell line MCF-7 (HTB-22, the American Type Culture Collection, Manassas, VA, USA) was cultured in RPMI 1640 medium (Thermo Fisher Scientific) supplemented with 10% fetal bovine serum, nonessential amino acids (Thermo Fisher Scientific) and 1 × antibiotic-antimycotic (Thermo Fisher Scientific) [63]. The cells were incubated at 37 °C with 5% $CO_2$ and humidified atmosphere control. The culture medium was changed every 3 to 4 days.

### 3.6. Flow Cytometry

MCF-7 cells were detached from the dish surface through trypsinization, and washed with ice-cold FACS buffer (1% FBS in 1 × DPBS with 0.1% Sodium Azide) before staining with the following antibodies for respective experimental purposes: (1) For visualization of CD24 on the cell surface, cells were stained with anti-CD24 antibody, CD24 Monoclonal Antibody (eBioSN3 (SN3 A5-2H10)), PE (Invitrogen); Mouse IgG1 kappa Isotype Control (P3.6.2.8.1), PE (Thermo Fisher Scientific) was used as isotype control. (2) In Siglec-10 binding/inhibitory experiments, 5 × $10^5$ MCF-7 cells were incubated with 10 μg/mL Siglec-10, along with different concentrations of SL-1 glycan (0, 0.01 to 1 mM) in FACS buffer and incubated at 4 °C for 1 h. The cells were washed with FACS buffer before adding Alexa Flour® 647-conjugated anti-hFc antibody for Siglec-10 staining. Flow cytometry was performed on FACSCanto flow cytometer (BD Bioscience, Franklin Lakes, NJ, USA).

### 3.7. Commercial Glycan Array

Profiling of recombinant Siglec-10 was performed on commercial Glycan Array 300 (RayBiotech, Peachtree Corners, GA, USA) by the Academia Sinica Glycoscience Core Facility. Siglec-10 samples were dialyzed and labeled with biotin according to the protocol provided by the manufacturer. The glycan array slide was blocked, washed, and biotin-labeled proteins were added for incubation with the reagents provided in the kit. Cy3 equivalent dye-conjugated streptavidin was used to visualize the binding signals. The glycan arrays were excited at 532 nm laser and scanned using a GenePix 4300A Microarray Scanner.

## 4. Conclusions

It is known that cancer cells utilize the Siglec interacting pathway to evade immune cell-mediated cytotoxicity. In this study, we describe the development of facile and scalable chemo-enzymatic strategies for the synthesis of DSGb5 and sialylated derivatives. These sialylated glycans and the glycans of cancer-associated glycolipids were used to create glycan microarrays to evaluate their binding towards recombinant human Siglecs, including Siglec-7, Siglec-9, Siglec-10, and Siglec-15. It was found that SSEA-4 and DSGb5 glycans had binding to Siglec-9 and the disialyl SSEA-3 glycan with disialylated terminal Gal (compound **2**) showed binding to Siglec-9 and Siglec-15. However, these Siglecs exhibited better binding to sialylated N-glycans and the binding affinity and specificity were strongly influenced by the α2,3- or α2,6-linkage of terminal sialic acid, with Siglec-9 in favor of the α2,3-linkage and Siglec-10 the α2,6-linkage. Notably, the complex-type N-glycans are better ligands for Siglec-9 and Siglec-10 and the binding increases with increasing number of terminal sialic acids. We also found that the N-glycolylneuraminic acid derivatives are better ligands for Siglec-10, especially the truncated biantennary glycan SL-1, which exhibited a significant inhibition of macrophage-mediated phagocytosis.

**Supplementary Materials:** The following supporting information can be downloaded at https://www.mdpi.com/article/10.3390/molecules30112264/s1, detailed description of synthesis of DSGb5 and related sialosides and their characterization, including NMR data; Scheme S1–S3, Table S1–S2, Figure S1–S3, and References [1–12] are cited in the Supplementary Materials.

**Author Contributions:** Synthesis of all glycans and characterization, K.-S.L., Y.Z., and C.-T.R.; glycan array and analysis, K.-S.L., C.C., and P.-K.C.; cell culture, protein, and enzyme preparations, flow cytometry, C.C., C.-C.K., and C.-Y.W.; mRNA analysis, Y.-W.L.; consultation and draft editing, B.I. and Y.S.Y.H.; research design, review, and writing, C.-H.W. All authors have read and agreed to the published version of the manuscript.

**Funding:** This research was funded by Academia Sinica (AS-GCP-114-E01).

**Institutional Review Board Statement:** Not applicable.

**Informed Consent Statement:** Not applicable.

**Data Availability Statement:** All study data are included and available in the article and/or in the Supporting Information (SI). The raw data supporting the conclusions of this article will be made available by the authors upon request.

**Acknowledgments:** We thank the Academia Glycoscience Core Facility for their technical support. The core facility is funded by the Academia Sinica Core Facility and Innovative Instrument Project (AS-CFII-114-A13). We thank Y. P. Huang for assistance with NMR analysis and C. H. Chen and C. L. Wu for assistance with mass spectrometric analysis of synthetic compounds.

**Conflicts of Interest:** The authors declare no conflicts of interests.

## References

1. Hakomori, S. Tumor-associated carbohydrate antigens. *Annu. Rev. Immunol.* **1984**, *2*, 103–126. [CrossRef] [PubMed]
2. Duan, S.; Paulson, J.C. Siglecs as Immune Cell Checkpoints in Disease. *Annu. Rev. Immunol.* **2020**, *38*, 365–395. [CrossRef] [PubMed]
3. Smith, B.A.H.; Bertozzi, C.R. The clinical impact of glycobiology: Targeting selectins, Siglecs and mammalian glycans. *Nat. Rev. Drug Discov.* **2021**, *20*, 217–243. [CrossRef]
4. Tsai, T.I.; Lee, H.Y.; Chang, S.H.; Wang, C.H.; Tu, Y.C.; Lin, Y.C.; Hwang, D.R.; Wu, C.Y.; Wong, C.H. Effective sugar nucleotide regeneration for the large-scale enzymatic synthesis of Globo H and SSEA4. *J. Am. Chem. Soc.* **2013**, *135*, 14831–14839. [CrossRef]
5. Kung, C.C.; Lo, J.M.; Liao, K.S.; Wu, C.Y.; Cheng, L.C.; Chung, C.; Hsu, T.L.; Ma, C.; Wong, C.H. Expression of Human beta3GalT5-1 in Insect Cells as Active Glycoforms for the Efficient Synthesis of Cancer-Associated Globo-Series Glycans. *J. Am. Chem. Soc.* **2025**, *147*, 10864–10874. [CrossRef]
6. Kannagi, R.; Cochran, N.A.; Ishigami, F.; Hakomori, S.; Andrews, P.W.; Knowles, B.B.; Solter, D. Stage-specific embryonic antigens (SSEA-3 and -4) are epitopes of a unique globo-series ganglioside isolated from human teratocarcinoma cells. *EMBO J.* **1983**, *2*, 2355–2361. [CrossRef]
7. Krupnick, J.G.; Damjanov, I.; Damjanov, A.; Zhu, Z.M.; Fenderson, B.A. Globo-series carbohydrate antigens are expressed in different forms on human and murine teratocarcinoma-derived cells. *Int. J. Cancer* **1994**, *59*, 692–698. [CrossRef] [PubMed]
8. Ye, F.; Li, Y.; Hu, Y.; Zhou, C.; Hu, Y.; Chen, H. Stage-specific embryonic antigen 4 expression in epithelial ovarian carcinoma. *Int. J. Gynecol. Cancer* **2010**, *20*, 958–964. [CrossRef]
9. Lou, Y.W.; Wang, P.Y.; Yeh, S.C.; Chuang, P.K.; Li, S.T.; Wu, C.Y.; Khoo, K.H.; Hsiao, M.; Hsu, T.L.; Wong, C.H. Stage-specific embryonic antigen-4 as a potential therapeutic target in glioblastoma multiforme and other cancers. *Proc. Natl. Acad. Sci. USA* **2014**, *111*, 2482–2487. [CrossRef]
10. Lin, C.W.; Wang, Y.J.; Lai, T.Y.; Hsu, T.L.; Han, S.Y.; Wu, H.C.; Shen, C.N.; Dang, V.; Chen, M.W.; Chen, L.B.; et al. Homogeneous antibody and CAR-T cells with improved effector functions targeting SSEA-4 glycan on pancreatic cancer. *Proc. Natl. Acad. Sci. USA* **2021**, *118*, e2114774118. [CrossRef]
11. Senda, M.; Ito, A.; Tsuchida, A.; Hagiwara, T.; Kaneda, T.; Nakamura, Y.; Kasama, K.; Kiso, M.; Yoshikawa, K.; Katagiri, Y.; et al. Identification and expression of a sialyltransferase responsible for the synthesis of disialylgalactosylgloboside in normal and malignant kidney cells: Downregulation of ST6GalNAc VI in renal cancers. *Biochem. J.* **2007**, *402*, 459–470. [CrossRef] [PubMed]

12. Hugonnet, M.; Singh, P.; Haas, Q.; von Gunten, S. The Distinct Roles of Sialyltransferases in Cancer Biology and Onco-Immunology. *Front. Immunol.* **2021**, *12*, 799861. [CrossRef] [PubMed]
13. Freitas, D.P.; Teixeira, C.A.; Santos-Silva, F.; Vasconcelos, M.H.; Almeida, G.M. Therapy-induced enrichment of putative lung cancer stem-like cells. *Int. J. Cancer* **2014**, *134*, 1270–1278. [CrossRef]
14. Brune, J.C.; Tormin, A.; Johansson, M.C.; Rissler, P.; Brosjo, O.; Lofvenberg, R.; von Steyern, F.V.; Mertens, F.; Rydholm, A.; Scheding, S. Mesenchymal stromal cells from primary osteosarcoma are non-malignant and strikingly similar to their bone marrow counterparts. *Int. J. Cancer* **2011**, *129*, 319–330. [CrossRef]
15. Fonsato, V.; De Lena, M.; Tritta, S.; Brossa, A.; Calvetti, R.; Tetta, C.; Camussi, G.; Bussolati, B. Human liver stem cell-derived extracellular vesicles enhance cancer stem cell sensitivity to tyrosine kinase inhibitors through Akt/mTOR/PTEN combined modulation. *Oncotarget* **2018**, *9*, 36151–36165. [CrossRef]
16. Sourisseau, T.; Hassan, K.A.; Wistuba, I.; Penault-Llorca, F.; Adam, J.; Deutsch, E.; Soria, J.C. Lung cancer stem cell: Fancy conceptual model of tumor biology or cornerstone of a forthcoming therapeutic breakthrough? *J. Thorac. Oncol.* **2014**, *9*, 7–17. [CrossRef]
17. Bedel, A.; Pasquet, J.M.; Lippert, E.; Taillepierre, M.; Lagarde, V.; Dabernat, S.; Dubus, P.; Charaf, L.; Beliveau, F.; de Verneuil, H.; et al. Variable behavior of iPSCs derived from CML patients for response to TKI and hematopoietic differentiation. *PLoS ONE* **2013**, *8*, e71596. [CrossRef]
18. Miyauchi, M.; Koya, J.; Arai, S.; Yamazaki, S.; Honda, A.; Kataoka, K.; Yoshimi, A.; Taoka, K.; Kumano, K.; Kurokawa, M. ADAM8 Is an Antigen of Tyrosine Kinase Inhibitor-Resistant Chronic Myeloid Leukemia Cells Identified by Patient-Derived Induced Pluripotent Stem Cells. *Stem Cell Rep.* **2018**, *10*, 1115–1130. [CrossRef] [PubMed]
19. Ghosh, G.; Lian, X.; Kron, S.J.; Palecek, S.P. Properties of resistant cells generated from lung cancer cell lines treated with EGFR inhibitors. *BMC Cancer* **2012**, *12*, 95. [CrossRef]
20. Chen, N.Y.; Lin, C.W.; Lai, T.Y.; Wu, C.Y.; Liao, P.C.; Hsu, T.L.; Wong, C.H. Increased expression of SSEA-4 on TKI-resistant non-small cell lung cancer with EGFR-T790M mutation. *Proc. Natl. Acad. Sci. USA* **2024**, *121*, e2313397121. [CrossRef]
21. Huang, Y.L.; Hung, J.T.; Cheung, S.K.; Lee, H.Y.; Chu, K.C.; Li, S.T.; Lin, Y.C.; Ren, C.T.; Cheng, T.J.; Hsu, T.L.; et al. Carbohydrate-based vaccines with a glycolipid adjuvant for breast cancer. *Proc. Natl. Acad. Sci. USA* **2013**, *110*, 2517–2522. [CrossRef] [PubMed]
22. Danishefsky, S.J.; Shue, Y.K.; Chang, M.N.; Wong, C.H. Development of Globo-H cancer vaccine. *Acc. Chem. Res.* **2015**, *48*, 643–652. [CrossRef] [PubMed]
23. Wang, S.W.; Ko, Y.A.; Chen, C.Y.; Liao, K.S.; Chang, Y.H.; Lee, H.Y.; Yu, Y.H.; Lih, Y.H.; Cheng, Y.Y.; Lin, H.H.; et al. Mechanism of Antigen Presentation and Specificity of Antibody Cross-Reactivity Elicited by an Oligosaccharide-Conjugate Cancer Vaccine. *J. Am. Chem. Soc.* **2023**, *145*, 9840–9849. [CrossRef]
24. Chuang, P.K.; Hsiao, M.; Hsu, T.L.; Chang, C.F.; Wu, C.Y.; Chen, B.R.; Huang, H.W.; Liao, K.S.; Chen, C.C.; Chen, C.L.; et al. Signaling pathway of globo-series glycosphingolipids and beta1,3-galactosyltransferase V (beta3GalT5) in breast cancer. *Proc. Natl. Acad. Sci. USA* **2019**, *116*, 3518–3523. [CrossRef]
25. Kawasaki, Y.; Ito, A.; Withers, D.A.; Taima, T.; Kakoi, N.; Saito, S.; Arai, Y. Ganglioside DSGb5, preferred ligand for Siglec-7, inhibits NK cell cytotoxicity against renal cell carcinoma cells. *Glycobiology* **2010**, *20*, 1373–1379. [CrossRef]
26. Li, P.J.; Huang, S.Y.; Chiang, P.Y.; Fan, C.Y.; Guo, L.J.; Wu, D.Y.; Angata, T.; Lin, C.C. Chemoenzymatic Synthesis of DSGb5 and Sialylated Globo-series Glycans. *Angew. Chem. Int. Ed. Engl.* **2019**, *58*, 11273–11278. [CrossRef]
27. Chuang, H.Y.; Ren, C.T.; Chao, C.A.; Wu, C.Y.; Shivatare, S.S.; Cheng, T.J.; Wu, C.Y.; Wong, C.H. Synthesis and vaccine evaluation of the tumor-associated carbohydrate antigen RM2 from prostate cancer. *J. Am. Chem. Soc.* **2013**, *135*, 11140–11150. [CrossRef] [PubMed]
28. Shivatare, S.S.; Chang, S.H.; Tsai, T.I.; Ren, C.T.; Chuang, H.Y.; Hsu, L.; Lin, C.W.; Li, S.T.; Wu, C.Y.; Wong, C.H. Efficient convergent synthesis of bi-, tri-, and tetra-antennary complex type N-glycans and their HIV-1 antigenicity. *J. Am. Chem. Soc.* **2013**, *135*, 15382–15391. [CrossRef]
29. Shivatare, S.S.; Chang, S.H.; Tsai, T.I.; Tseng, S.Y.; Shivatare, V.S.; Lin, Y.S.; Cheng, Y.Y.; Ren, C.T.; Lee, C.C.; Pawar, S.; et al. Modular synthesis of N-glycans and arrays for the hetero-ligand binding analysis of HIV antibodies. *Nat. Chem.* **2016**, *8*, 338–346. [CrossRef]
30. Chu, K.C.; Ren, C.T.; Lu, C.P.; Hsu, C.H.; Sun, T.H.; Han, J.L.; Pal, B.; Chao, T.A.; Lin, Y.F.; Wu, S.H.; et al. Efficient and stereoselective synthesis of alpha(2-->9) oligosialic acids: From monomers to dodecamers. *Angew. Chem. Int. Ed. Engl.* **2011**, *50*, 9391–9395. [CrossRef]
31. Hsu, C.H.; Hung, S.C.; Wu, C.Y.; Wong, C.H. Toward automated oligosaccharide synthesis. *Angew. Chem. Int. Ed. Engl.* **2011**, *50*, 11872–11923. [CrossRef] [PubMed]

32. Hsu, C.H.; Chu, K.C.; Lin, Y.S.; Han, J.L.; Peng, Y.S.; Ren, C.T.; Wu, C.Y.; Wong, C.H. Highly alpha-selective sialyl phosphate donors for efficient preparation of natural sialosides. *Chemistry* **2010**, *16*, 1754–1760. [CrossRef]
33. Yu, H.; Huang, S.; Chokhawala, H.; Sun, M.; Zheng, H.; Chen, X. Highly efficient chemoenzymatic synthesis of naturally occurring and non-natural alpha-2,6-linked sialosides: A P. damsela alpha-2,6-sialyltransferase with extremely flexible donor-substrate specificity. *Angew. Chem. Int. Ed. Engl.* **2006**, *45*, 3938–3944. [CrossRef] [PubMed]
34. Ding, L.; Zhao, C.; Qu, J.; Li, Y.; Sugiarto, G.; Yu, H.; Wang, J.; Chen, X. A Photobacterium sp. alpha2-6-sialyltransferase (Psp2,6ST) mutant with an increased expression level and improved activities in sialylating Tn antigens. *Carbohydr. Res.* **2015**, *408*, 127–133. [CrossRef] [PubMed]
35. Ding, L.; Yu, H.; Lau, K.; Li, Y.; Muthana, S.; Wang, J.; Chen, X. Efficient chemoenzymatic synthesis of sialyl Tn-antigens and derivatives. *Chem. Commun.* **2011**, *47*, 8691–8693. [CrossRef]
36. Sun, M.; Li, Y.; Chokhawala, H.A.; Henning, R.; Chen, X. N-Terminal 112 amino acid residues are not required for the sialyltransferase activity of Photobacterium damsela alpha2,6-sialyltransferase. *Biotechnol. Lett.* **2008**, *30*, 671–676. [CrossRef]
37. Teo, C.F.; Hwang, T.S.; Chen, P.H.; Hung, C.H.; Gao, H.S.; Chang, L.S.; Lin, C.H. Synthesis of sialyl T Glycopeptides -: Enzymatic sialylation by α2,6-sialyltransferase from. *Adv. Synth. Catal.* **2005**, *347*, 967–972. [CrossRef]
38. Meng, X.; Yao, W.; Cheng, J.; Zhang, X.; Jin, L.; Yu, H.; Chen, X.; Wang, F.; Cao, H. Regioselective chemoenzymatic synthesis of ganglioside disialyl tetrasaccharide epitopes. *J. Am. Chem. Soc.* **2014**, *136*, 5205–5208. [CrossRef]
39. Ito, A.; Handa, K.; Withers, D.A.; Satoh, M.; Hakomori, S. Binding specificity of siglec7 to disialogangliosides of renal cell carcinoma: Possible role of disialogangliosides in tumor progression. *FEBS Lett.* **2001**, *498*, 116–120. [CrossRef]
40. Yamaji, T.; Teranishi, T.; Alphey, M.S.; Crocker, P.R.; Hashimoto, Y. A small region of the natural killer cell receptor, Siglec-7, is responsible for its preferred binding to alpha 2,8-disialyl and branched alpha 2,6-sialyl residues. A comparison with Siglec-9. *J. Biol. Chem.* **2002**, *277*, 6324–6332. [CrossRef]
41. Büll, C.; den Brok, M.H.; Adema, G.J. Sweet escape: Sialic acids in tumor immune evasion. *Biochim. Biophys. Acta* **2014**, *1846*, 238–246. [CrossRef] [PubMed]
42. Ravindranath, M.H.; Yesowitch, P.; Sumobay, C.; Morton, D.L. Glycoimmunomics of human cancer: Current concepts and future perspectives. *Future Oncol.* **2007**, *3*, 201–214. [CrossRef] [PubMed]
43. Krengel, U.; Bousquet, P.A. Molecular recognition of gangliosides and their potential for cancer immunotherapies. *Front. Immunol.* **2014**, *5*, 325. [CrossRef]
44. t'Hart, I.M.E.; Li, T.; Wolfert, M.A.; Wang, S.; Moremen, K.W.; Boons, G.J. Chemoenzymatic synthesis of the oligosaccharide moiety of the tumor-associated antigen disialosyl globopentaosylceramide. *Org. Biomol. Chem.* **2019**, *17*, 7304–7308. [CrossRef]
45. Hashimoto, N.; Ito, S.; Tsuchida, A.; Bhuiyan, R.H.; Okajima, T.; Yamamoto, A.; Furukawa, K.; Ohmi, Y.; Furukawa, K. The ceramide moiety of disialoganglioside (GD3) is essential for GD3 recognition by the sialic acid-binding lectin SIGLEC7 on the cell surface. *J. Biol. Chem.* **2019**, *294*, 10833–10845. [CrossRef] [PubMed]
46. Mereiter, S.; Balmaña, M.; Campos, D.; Gomes, J.; Reis, C.A. Glycosylation in the Era of Cancer-Targeted Therapy: Where Are We Heading? *Cancer Cell* **2019**, *36*, 6–16. [CrossRef]
47. Gonzalez-Gil, A.; Schnaar, R.L. Siglec Ligands. *Cells* **2021**, *10*, 1260. [CrossRef]
48. Pirruccello, S.J.; Lebien, T.W. The Human B-Cell-Associated Antigen Cd24 Is a Single Chain Sialoglycoprotein. *J. Immunol.* **1986**, *136*, 3779–3784. [CrossRef]
49. Fang, X.; Zheng, P.; Tang, J.; Liu, Y. CD24: From A to Z. *Cell Mol. Immunol.* **2010**, *7*, 100–103. [CrossRef]
50. Kristiansen, G.; Sammar, M.; Altevogt, P. Tumour biological aspects of CD24, a mucin-like adhesion molecule. *J. Mol. Histol.* **2004**, *35*, 255–262. [CrossRef]
51. Lim, S.C. CD24 and human carcinoma: Tumor biological aspects. *Biomed. Pharmacother.* **2005**, *59* (Suppl. 2), S351–S354. [CrossRef] [PubMed]
52. Munday, J.; Kerr, S.; Ni, J.; Cornish, A.L.; Zhang, J.Q.; Nicoll, G.; Floyd, H.; Mattei, M.G.; Moore, P.; Liu, D.; et al. Identification, characterization and leucocyte expression of Siglec-10, a novel human sialic acid-binding receptor. *Biochem. J.* **2001**, *355 Pt 2*, 489–497. [CrossRef]
53. Barkal, A.A.; Brewer, R.E.; Markovic, M.; Kowarsky, M.; Barkal, S.A.; Zaro, B.W.; Krishnan, V.; Hatakeyama, J.; Dorigo, O.; Barkal, L.J.; et al. CD24 signalling through macrophage Siglec-10 is a target for cancer immunotherapy. *Nature* **2019**, *572*, 392–396. [CrossRef]
54. Lim, J.; Sari-Ak, D.; Bagga, T. Siglecs as Therapeutic Targets in Cancer. *Biology* **2021**, *10*, 1178. [CrossRef]
55. Chen, G.Y.; Tang, J.; Zheng, P.; Liu, Y. CD24 and Siglec-10 selectively repress tissue damage-induced immune responses. *Science* **2009**, *323*, 1722–1725. [CrossRef]
56. Toubai, T.; Hou, G.; Mathewson, N.; Liu, C.; Wang, Y.; Oravecz-Wilson, K.; Cummings, E.; Rossi, C.; Evers, R.; Sun, Y.; et al. Siglec-G-CD24 axis controls the severity of graft-versus-host disease in mice. *Blood* **2014**, *123*, 3512–3523. [CrossRef] [PubMed]

57. Yin, S.S.; Gao, F.H. Molecular Mechanism of Tumor Cell Immune Escape Mediated by CD24/Siglec-10. *Front. Immunol.* **2020**, *11*, 1324. [CrossRef]
58. Sammar, M.; Siwetz, M.; Meiri, H.; Fleming, V.; Altevogt, P.; Huppertz, B. Expression of CD24 and Siglec-10 in first trimester placenta: Implications for immune tolerance at the fetal-maternal interface. *Histochem. Cell Biol.* **2017**, *147*, 565–574. [CrossRef] [PubMed]
59. Forgione, R.E.; Di Carluccio, C.; Guzman-Caldentey, J.; Gaglione, R.; Battista, F.; Chiodo, F.; Manabe, Y.; Arciello, A.; Del Vecchio, P.; Fukase, K.; et al. Unveiling Molecular Recognition of Sialoglycans by Human Siglec-10. *iScience* **2020**, *23*, 101231. [CrossRef]
60. Crocker, P.R.; Paulson, J.C.; Varki, A. Siglecs and their roles in the immune system. *Nat. Rev. Immunol.* **2007**, *7*, 255–266. [CrossRef]
61. Liang, P.H.; Wang, S.K.; Wong, C.H. Quantitative analysis of carbohydrate-protein interactions using glycan microarrays: Determination of surface and solution dissociation constants. *J. Am. Chem. Soc.* **2007**, *129*, 11177–11184. [CrossRef] [PubMed]
62. Shivatare, V.S.; Shivatare, S.S.; Lee, C.D.; Liang, C.H.; Liao, K.S.; Cheng, Y.Y.; Saidachary, G.; Wu, C.Y.; Lin, N.H.; Kwong, P.D.; et al. Unprecedented Role of Hybrid N-Glycans as Ligands for HIV-1 Broadly Neutralizing Antibodies. *J. Am. Chem. Soc.* **2018**, *140*, 5202–5210. [CrossRef] [PubMed]
63. Cheung, S.K.; Chuang, P.K.; Huang, H.W.; Hwang-Verslues, W.W.; Cho, C.H.; Yang, W.B.; Shen, C.N.; Hsiao, M.; Hsu, T.L.; Chang, C.F.; et al. Stage-specific embryonic antigen-3 (SSEA-3) and beta3GalT5 are cancer specific and significant markers for breast cancer stem cells. *Proc. Natl. Acad. Sci. USA* **2016**, *113*, 960–965. [CrossRef] [PubMed]

molecules

MDPI

Article

# Total Synthesis of Cardenolides Acospectoside A and Acovenoside B

**Benzhang Liu [1,2], Peng Xu [1,2,*] and Biao Yu [1,2,*]**

1 School of Chemistry and Materials Science, Hangzhou Institute for Advanced Study, University of Chinese Academy of Sciences, Hangzhou 310024, China; liubenzhang@sioc.ac.cn

2 State Key Laboratory of Chemical Biology, Shanghai Institute of Organic Chemistry, University of Chinese Academy of Sciences, Chinese Academy of Sciences, Shanghai 200032, China

* Correspondence: peterxu@sioc.ac.cn (P.X.); byu@sioc.ac.cn (B.Y.)

**Abstract:** Acospectoside A (**1**) and acovenoside B (**2**), two cytotoxic cardenolides extracted from the venomous South African bush *Acokanthera oppositifolia*, are distinguished by their unique structural motifs of the L-acovenose moiety at C-3 and a 1β-*O*-acetylated cardenolide aglycone. Here, we report the synthesis of these cardiac glycosides featuring delicate introductions of the 1-*O*-acetyl group under acid-catalyzed conditions, 14β-OH by Mukaiyama hydration, and a C17-butenolide moiety by Stille coupling.

**Keywords:** cardiac glycoside; total synthesis; cardenolides; glycosylation

## 1. Introduction

Cardiac glycosides, a structurally diverse group of natural products, are characterized by a steroid skeleton adorned with a sugar moiety at the C3 position and a lactone moiety at the C17 position [1,2]. For over two centuries, members of this family have been used in clinics for the treatment of heart failure and cardiac rhythm disorders [3–5]. Recent investigations have unveiled that diminished levels of specific endogenous cardiotonic steroids could promote tumorigenesis [6], thereby sparking significant interest in a novel hypothesis positing cardiac glycosides as potential anticancer agents [5]. In 1950, avenosides A and B, both sharing the distinctive L-acovenose moiety, were isolated from the South African poisonous bush *Acokanthera oppositifolia* [7]. Subsequent phytochemistry research paved the way for the isolation and characterization of acospectoside A and acobioside A (Figure 1) [8,9]. Notably, acospectoside A (**1**) could be metabolically converted to acovenoside B (**2**), acobioside A, and acovenoside A through enzymatic treatment with a snail (*Helix pomatia*) enzyme (Figure 1) [10,11]. These cardiac glycosides were further evaluated for their cardiotoxicity in cats and cytotoxicity against human carcinoma of the nasopharynx cells [10]. These findings indicated that the C-1 acetoxylated glycosides exhibited reduced activity compared to their hydroxylated counterparts, while monosides demonstrated greater cytotoxicity than the corresponding biosides. Amongst these cardiac glycosides, acovenoside A emerged as a potent inhibitor of the proliferation of human non-small-cell lung cancer (NSCLC) A-549 cells with an $IC_{50}$ value of 68 nM, outperforming the anticancer drug doxorubicin ($IC_{50}$ = 426 nM) [12]. However, the limited accessibility of these heterogeneous metabolites from natural sources has impeded in-depth investigations into their pharmaceutical potential and therapeutic applications. The chemical synthesis of these highly intricate cardiac glycosides presents a highly effective and transformative approach, offering a scalable and reproducible avenue for accessing these bioactive natural products and their structurally optimized derivatives [13–16]. Very recently, we developed a convenient approach to the first chemical synthesis of acovenoside A, oppofrioside, euonymuside A, and their congeners [17]. Here, we disclose the first total synthesis of acospectoside A (**1**) and acovensoide B (**2**), with 1-*O*-Ac-acovenosigenin A (**3**) as a common aglycone.

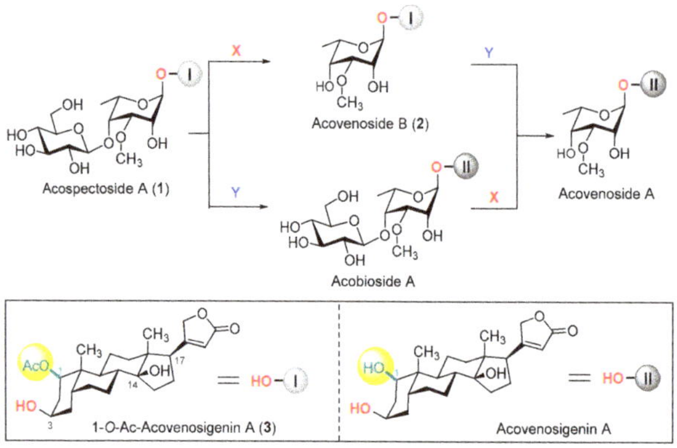

**Figure 1.** Chemical structures of acospectoside A (**1**), acovenoside B (**2**), acobioside A, acovenoside A, and their conversions. (X = removal of D-glucose with β-glucosidase; Y = deacetylation with esterase).

## 2. Results

Building upon our prior investigations into the synthesis of steroid glycosides [17–19], we envisioned the assembly of acovenosides (**1, 2**) through late-stage glycosylation, employing 1-*O*-Ac-acovenosigenin A (**3**) and glycosyl imidate donors (Scheme 1). However, our initial endeavors to achieve the selective acetylation of acovenosigenin A were unsuccessful in yielding the desired aglycone **3**. Consequently, a retrosynthetic analysis was undertaken, revealing that aglycone **3** could be elaborated on from the known intermediate (**6**) [17] via the introduction of the 1-*O*-acetyl group, modulating oxidation states at C-14, and installing the butenolide ring at C-17. A reduction of the 3-ketone group, followed by selective benzylation, led to the formation of intermediate **5**. This intermediate was then transformed into 17-ketone **4** through a sequence involving Saegusa–Ito oxidation and Mukaiyama hydration. Finally, the introduction of the C-17 butenolide moiety and selective debenzylation afforded 1-OAc-acovenosigenin A (**3**). The timing of introducing the 1-*O*-acetyl group is of strategic importance, and a successful synthetic route could be determined by trial-and-error.

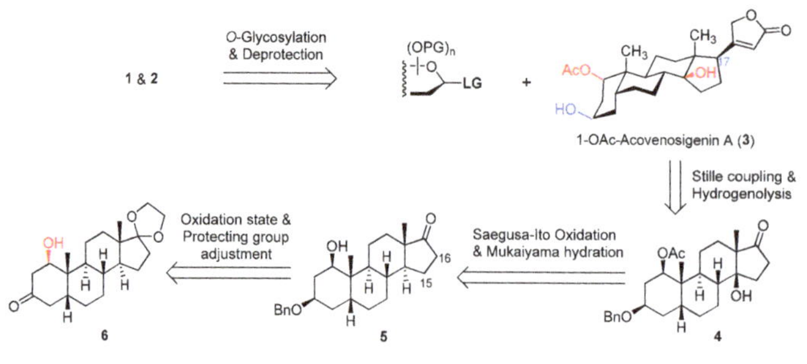

**Scheme 1.** Retrosynthetic plan for acospectoside A (**1**) and acovenoside B (**2**).

The selective reduction of 3-ketone **6** with L-selectride [20], followed by oxidative cleavage of the resulting boronic ester with a sodium hydroxide solution and hydrogen peroxide afforded 1β,3β-diol **7** (96%) (Scheme 2). Selective 3-*O*-benzylation with benzyl bromide and sodium hydride delivered the desired 1-ol **8** (76%), alongside 3-ol isomer **9** (15%). Acidic hydrolysis of the C17-ethylene glycol ketal in **8** furnished 17-ketone **5** (82%). The treatment of ketone **5** with TMSOTf and triethylamine generated the corresponding trimethylsilylenol ether. Subsequent palladium-mediated Saegusa–Ito oxidation, utilizing $O_2$ as a stoichiometric oxidant to form the 15,16-unsaturated enone **10**, required extensive optimization. The conventional Saegusa–Ito oxidation protocol typically necessitates a superstoichiometric quantity of palladium(II) acetate to achieve satisfactory yields [21]. In 1995, Larock and co-workers pioneered a catalytic variant of Saegusa–Ito oxidation using palladium(II) acetate under an oxygen atmosphere [22]. Applying Larock's protocol to the cyclic silylenol ether, with catalytic palladium(II) acetate (0.1 equiv.) and $O_2$ as the reoxidant in DMSO, afforded the desired 15,16-unsaturated enone **10** in a moderate 45% yield, alongside 12% recycled **5** (attributed to the hydrolysis of the trimethylsilyl enol ether). Concurrently, lactone **11** was isolated as a side product (26%), arising from a Baeyer–Villiger-type rearrangement of the 17-ketone followed by the unsaturation of the α,β-position of carbonyl via β-hydride elimination of palladium in one-pot [23]. Given that elevating the temperature could mitigate the formation of lactone byproducts, the Saegusa–Ito oxidation was conducted at 60 °C, yielding enone **10** (33%) and lactone **11** (5%), though nearly half of 17-ketone **5** was recycled. To enhance the conversion rate, the amount of palladium(II) acetate was further increased to 0.3 equivalents. Optimal results were achieved within 1 h, delivering enone **10** in a good 74% yield, with 15% of 17-ketone **5** recycled. Extending the reaction time to 6 h only marginally improved the conversion rate, with 7% of 17-ketone **5** recycled, while enone **10** (33%), lactone **11** (5%), and unconjugated $\Delta^{14,15}$ **12** (7%) were also obtained.

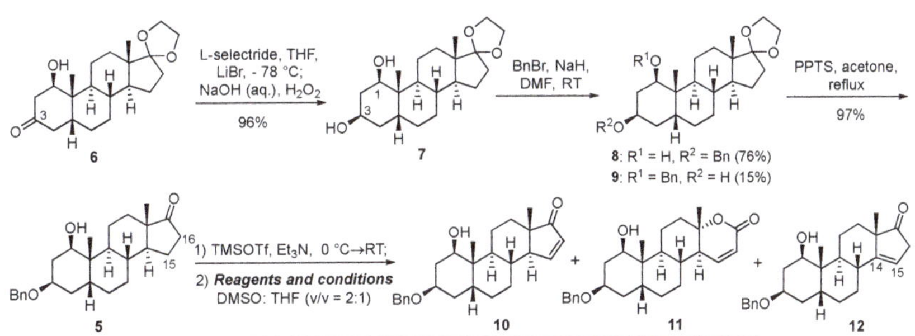


| Entry | Reagents and conditions | Results |
|---|---|---|
| 1 | $Pd(OAc)_2$ (0.1 equiv.), $O_2$ (1 atm), RT, 12 h | **10** (45%); **11** (26%); **5** (12%) |
| 2 | $Pd(OAc)_2$ (0.1 equiv.), $O_2$ (1 atm), 60 °C, 6 h | **10** (33%); **11** (5%); **5** (48%) |
| 3 | $Pd(OAc)_2$ (0.3 equiv.), $O_2$ (1 atm), 60 °C, 1 h | **10** (74%); **5** (15%) |
| 4 | $Pd(OAc)_2$ (0.3 equiv.), $O_2$ (1 atm), 60 °C, 6 h | **10** (66%); **11** (10%); **12** (7%); **5** (7%) |

**Scheme 2.** Synthesis of enone **10**.

The isomerization of **10** was carried out in the presence of *N*,*N*-diisopropylethylamine and silicon dioxide at 90 °C, yielding unconjugated $\Delta^{14,15}$ derivative **12** (82%) (Scheme 3). The strategic introduction of the 1-*O*-acetyl group was successfully achieved by employing isopropenyl acetate and *p*-toluenesulfonic acid monohydrate, resulting in the formation of the fully protected intermediate **13** (81%). Mukaiyama hydration of $\Delta^{14,15}$ **13**, utilizing $Mn(acac)_2$ and $PhSiH_3$ under aerobic conditions, furnished tertiary alcohols **4β** (68%) and **4α** (32%) on a gram scale, showcasing the efficiency and scalability of the reaction [24]. 17-Ketone **4β** was subsequently converted to vinyl iodide **14** (81%) with hydrazine hydrate and iodine. The Stille coupling reaction of vinyl iodide **14** with the known stannylated butanolide **15** catalyzed by $Pd(PPh_3)_4$ proceeded effectively, affording the desired cardiac steroid **16** in an excellent 89% yield [19]. The direct hydrogenation of the $\Delta^{16,17}$ double bond in **16** posed a risk of generating C-17 isomers. To circumvent this, the 14β-OH group in **16** was protected with TMSCl and imidazole, yielding intermediate **17** (83%). The subsequent hydrogenation of the $\Delta^{16,17}$ double bond in **17** proceeded smoothly on a gram scale, leading to the fully protected aglycone **18** (68%). Finally, the desired aglycone **3** was obtained through Pd/C-catalyzed debenzylation under atmospheric hydrogen in a mixed solvent of chloroform and methanol (78%). This sequence of reactions highlights the meticulous planning and optimization required to achieve the targeted cardiac steroid with high efficiency and selectivity.

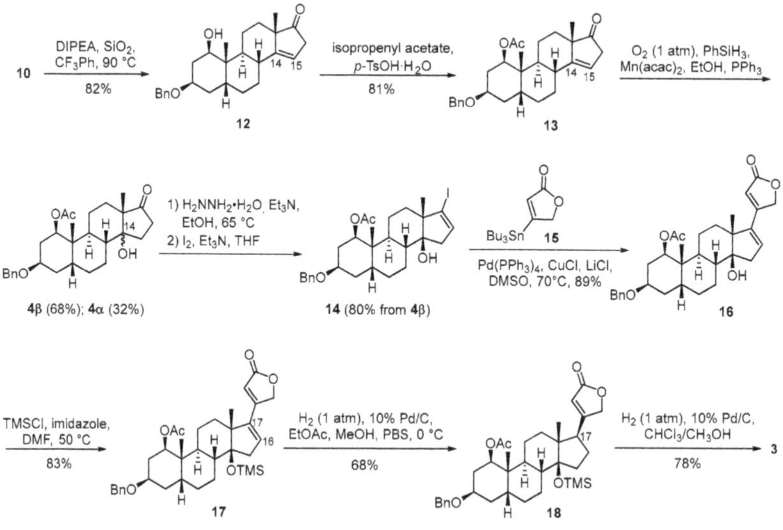

**Scheme 3.** Synthesis of 1-*O*-Ac-acovenosigenin A (**3**).

In 1969, Kapadia speculated that the observed difficulty of the saponification of 3-*O*-acetyl groups in acospectoside A (**1**) and acovenoside B (**2**) could likely be attributed to the steric hindrance imposed by an axially bound acetoxyl group [10]. The subsequent experimental transformations revealed that acetyl groups situated within the equatorial positions of the carbohydrate moieties could be selectively removed through mild saponification using potassium bicarbonate, without any discernible impact on the 3-*O*-acetyl groups in the aglycone moieties [10]. With the desired aglycone derivative **3** and monosaccharide building blocks in hand, we embarked on the assembly of the target cardiac glycosides (Scheme 4). The known lactol **19** was transformed into an imidate donor with trichloroacetonitrile and cesium carbonate. The resulting reactive intermediate was promptly subjected to glycosylation with aglycone **3**, promoted by TMSOTf, to afford glycoside **20** in a 50% yield over a two-step sequence. Finally, hydrogenative debenzylation and selective debenzoylation of the sugar residue were executed, culminating in the successful synthesis of acovenoside B (**2**) in a satisfactory 60% yield.

The known thioglycoside **21** underwent efficient debenzylation upon treatment with 2,3-dichloro-5,6-dicyano-*p*-benzoquinone (DDQ), yielding 4-ol **22** (90%). The condensation of **22** with imidate **23** was complete within 2 h catalyzed by TMSOTf at −78 °C, and the desired disaccharide **24** was obtained in a good 76% yield (Scheme 5). Recognizing that debenzoylation posed a greater synthetic challenge than deacetylation, we strategically converted the benzoyl protecting groups in **24** to acetyl groups through a meticulous two-step sequence involving saponification and subsequent acetylation with isopropenyl acetate, affording disaccharide **26**. The selective hydrolysis of the anomeric thio group was achieved using *N*-bromosuccinimide (NBS), furnishing lactol **27** in an excellent 93% yield. Lactol **27** was then condensed with trichloroacetonitrile in the presence of cesium carbonate to generate the disaccharide imidate donor, which was immediately coupled with aglycone **3** to produce bioside **28** (51% over two steps). Finally, hydrogenative debenzylation and selective deacetylation of the disaccharide residue furnished acospectoside A (**1**), with no influence on the 1-*O*-acetyl group. The NMR data of the synthetic acospectoside A (**1**) and acovenoside B (**2**) were identical to those reported for the natural products (see Tables S1 and S2 for details) [25].

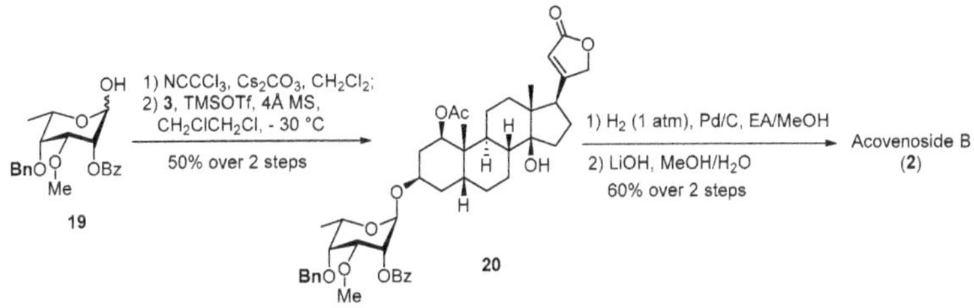


**Scheme 4.** Synthesis of acovenoside B (**2**).

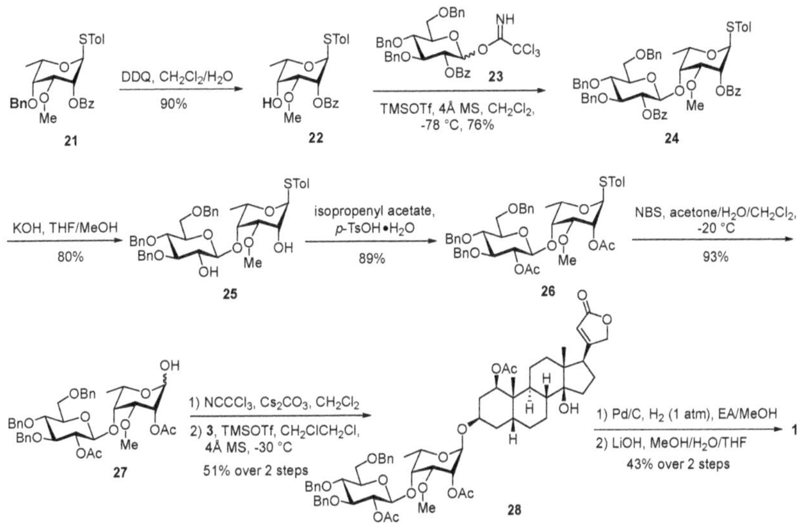


**Scheme 5.** Synthesis of acospectoside A (**1**).

## 3. Experimental Section

### 3.1. General Information

Commercial reagents were used without further purification and made in China unless specified. Crushed 4 Å molecular sieves (MSs) were activated through flame-drying under high vacuum conditions immediately prior to use. Dry $CH_2Cl_2$, DMF, and toluene were obtained by drying with activated MSs. Dry pyridine and $NEt_3$ were obtained by drying with anhydrous KOH. Anhydrous THF was obtained by refluxing with Na under an argon atmosphere. Thin layer chromatography (TLC) was performed on TLC silica gel 60 F254 (Merck, Darmstadt, Germany). The TLC plates were visualized with UV light and/or by staining with $EtOH/H_2SO_4$ (10%, *v*/*v*). Flash column chromatography was performed on silica gel, SiliaFlash P60 (40–63 µm, Silicycle, Quebec, QC, Canada). The NMR spectra were measured on a Bruker AM 400 (Zurich, Switzerland), Agilent 500 MHz (Santa Clara, CA, USA) NMR spectrometer at 25 °C. $^1H$ and $^{13}C$ NMR signals were calibrated to the residual proton and carbon resonance of the solvent ($CDCl_3$: $\delta_H$ = 7.26 ppm, $\delta_C$ = 77.16 ppm; $CD_3OD$: $\delta_H$ = 3.31 ppm, $\delta_C$ = 49.00 ppm; $C_5D_5N$: $\delta_H$ =8.74 ppm, $\delta_C$ = 155.35 ppm). High-resolution mass spectra were recorded with Shimadzu Biotech Axima Performance FTMS (Kyoto, Japan), maXis 4G FTMS, Thermo Scientific Q Exactive HF Orbitrap-FTMS (Waltham, MA, USA), or Agilent-TOF/LC-MS 1260-6230 FTMS. Optical rotations were measured on an Anton Paar MCP5500 polarimeter (Graz, Austria).

### 3.2. Synthesis of Compounds **1–28**

#### 3.2.1. Synthesis of Compound **7**

To a mixture of compound **6** (5.2 g, 14.9 mmol, 1.0 equiv.) and LiBr (7.8 g, 89.5 mmol, 6.0 equiv.), dry THF (150 mL) was added under an atmosphere of Ar. L-selectride (22.4 mL, 1 M, 1.5 equiv., 0.5 mL/min) was added dropwise at −78 °C to the resulting mixture under an Ar atmosphere and stirred for 2 h at the same temperature. TLC indicated the reaction was complete (petroleum ether/EtOAc = 2/1). The reaction mixture was quenched with a 2 N NaOH aqueous solution (75 mL) and 30% $H_2O_2$ (7.5 mL). The mixture was extracted with EtOAc (150 mL × 3), and the organic phase was washed with brine, and dried with $Na_2SO_4$. After filtration, the solution was concentrated under reduced pressure. The residue was purified by silica gel column chromatography (petroleum ether/EtOAc/$CH_2Cl_2$ = 2/1/0.6) to give compound **7** (5.0 g, 96%) as a white amorphous solid. $[\alpha]_D^{25}$ = −19.6 (*c* 1.0, $CHCl_3$); $^1H$ NMR (600 MHz, $CDCl_3$) δ 4.13 (s, 1H), 3.95–3.79 (m, 5H), 3.42 (s, 2H), 2.09–1.89 (m, 4H), 1.86–1.69 (m, 3H), 1.68–1.60 (m, 1H), 1.53–1.31 (m, 7H), 1.28–1.18 (m, 4H), 1.15–1.10 (m, 1H), 1.09 (s, 3H), 0.82 (d, *J* = 1.7 Hz, 3H); $^{13}C$ NMR (125 MHz, $CDCl_3$) δ 119.6, 74.0, 68.5, 65.3, 64.6, 50.5, 45.9, 41.9, 39.8, 36.2, 34.2, 33.7, 32.3, 31.0, 30.7, 26.0, 25.3, 22.7, 20.5, 19.0, 14.6; HR-ESI-MS (*m*/*z*) calcd for $C_{21}H_{34}O_4Na$ $[M+Na]^+$: 373.2349, found: 373.2347.

#### 3.2.2. Synthesis of Compound **8** and **9**

To a stirring solution of compound **7** (5.0 g, 14.4 mmol, 1.0 equiv.) and BnBr (1.7 mL, 14.4 mmol, 1.0 equiv.) in dry DMF (140 mL), 60% NaH (1.4 g, 36.0 mmol, 2.5 equiv.) was added in portions at room temperature. And the reaction mixture was stirred for 2 h at the same temperature. TLC indicated the reaction was complete. The reaction mixture was quenched with a saturated $NH_4Cl$ solution (20 mL). $H_2O$ (100 mL) was added to the mixture; the mixture extracted with EtOAc (300 mL), and the organic phase was washed with brine, and dried with $Na_2SO_4$. After filtration, the solution was concentrated under reduced pressure. The residue was purified by silica gel column chromatography (petroleum ether/EtOAc = 10/1 → 4/1) to give compound **8** (4.8 g, 76%) and to give compound **9** (923 mg, 15%) as white amorphous solids.

**8**: $[\alpha]_D^{25}$ = −19.0 (*c* 1.0, $CHCl_3$); $^1$H NMR (500 MHz, $CDCl_3$) δ 7.37–7.26 (m, 5H), 4.56–4.48 (m, 2H), 3.99 (d, *J* = 9.7 Hz, 1H), 3.95–3.82 (m, 6H), 3.71 (dt, *J* = 9.7, 3.0 Hz, 1H), 2.10 (dq, *J* = 14.8, 2.8 Hz, 1H), 2.02–1.92 (m, 2H), 1.91–1.86 (m, 1H), 1.86–1.74 (m, 2H), 1.70 (dt, *J* = 15.2, 3.1 Hz, 1H), 1.67–1.59 (m, 2H), 1.55–1.36 (m, 6H), 1.34–1.19 (m, 4H), 1.10 (s, 3H), 0.84 (s, 3H); $^{13}$C NMR (125 MHz, $CDCl_3$) δ 138.1, 128.6, 127.8, 127.7, 119.6, 76.0, 73.1, 70.7, 65.3, 64.6, 50.6, 45.9, 42.1, 40.0, 36.2, 34.2, 31.1, 31.0, 30.1, 29.9, 26.2, 25.3, 22.8, 20.4, 18.9, 14.6; HR-ESI-MS (*m*/*z*) calcd for $C_{28}H_{40}O_4Na$ $[M+Na]^+$: 463.2819, found: 463.2823.

**9**: $[\alpha]_D^{25}$ = −42.1 (*c* 1.0, $CHCl_3$); $^1$H NMR (500 MHz, $CDCl_3$) δ 7.36–7.31 (m, 4H), 7.31–7.27 (m, 1H), 4.69 (d, *J* = 11.0 Hz, 1H), 4.40 (d, *J* = 11.0 Hz, 1H), 4.33 (d, *J* = 10.4 Hz, 1H), 3.98 (dt, *J* = 10.6, 3.0 Hz, 1H), 3.94–3.81 (m, 4H), 3.60 (s, 1H), 2.16–2.04 (m, 2H), 2.02–1.91 (m, 2H), 1.86–1.73 (m, 2H), 1.70–1.59 (m, 2H), 1.55–1.35 (m, 6H), 1.32–1.20 (m, 5H), 1.11 (s, 3H), 0.84 (s, 3H); $^{13}$C NMR (125 MHz, $CDCl_3$) δ 138.0, 128.6, 128.1, 127.9, 119.5, 82.8, 73.0, 68.0, 65.3, 64.6, 50.6, 45.9, 41.6, 40.2, 36.0, 34.2, 33.9, 31.3, 30.9, 28.6, 25.9, 25.2, 22.7, 20.7, 18.9, 14.6; HR-ESI-MS (*m*/*z*) calcd for $C_{28}H_{40}O_4Na$ $[M+Na]^+$: 463.2819, found: 463.2822.

#### 3.2.3. Synthesis of Compound **5**

To a stirring solution of **8** (4.7 g, 10.7 mmol, 1.0 equiv.) in acetone (11 mL), pyridinium *p*-toluenesulfonate (4.0 g, 16.1 mmol, 1.5 equiv.) was added. The resulting mixture was heated to reflux for 3 h. TLC indicated the reaction was complete. The reaction mixture was concentrated under reduced pressure and added to a 1 N HCl aqueous solution (100 mL). The resulting mixture was extracted with $CH_2Cl_2$ (100 mL × 3); the combined organic phase was washed with brine, and dried with $Na_2SO_4$. After filtration, the solution was concentrated under reduced pressure. The residue was purified by silica gel column chromatography (petroleum ether/EtOAc = 10/1) to give **5** (4.1 g, 97%) as a white amorphous solid. $[\alpha]_D^{25}$ = +55.6 (*c* 1.0, $CHCl_3$); $^1$H NMR (500 MHz, $CDCl_3$) δ 7.36–7.26 (m, 5H), 4.58–4.49 (m, 2H), 4.03 (d, *J* = 9.7 Hz, 1H), 3.93–3.87 (m, 1H), 3.71 (dt, *J* = 9.8, 3.1 Hz, 1H), 2.43 (ddd, *J* = 19.1, 8.9, 1.1 Hz, 1H), 2.12 (dq, *J* = 14.9, 2.9 Hz, 1H), 2.07–2.00 (m, 2H), 1.96–1.85 (m, 3H), 1.81 (dt, *J* = 12.8, 3.0 Hz, 1H), 1.73–1.62 (m, 3H), 1.60–1.55 (m, 1H), 1.51 (ddd, *J* = 12.5, 9.1, 3.4 Hz, 1H), 1.48–1.42 (m, 1H), 1.40–1.29 (m, 3H), 1.28–1.15 (m, 3H), 1.13 (s, 3H), 0.86 (s, 3H); $^{13}$C NMR (125 MHz, $CDCl_3$) δ 221.3, 138.0, 128.6, 128.6, 127.9, 127.7, 75.8, 72.9, 70.8, 51.5, 47.7, 42.4, 40.2, 35.9, 35.6, 31.9, 31.1, 30.0, 26.1, 25.0, 21.9, 20.2, 18.9, 14.0; HR-ESI-MS (*m*/*z*) calcd for $C_{26}H_{36}O_3Na$ $[M+Na]^+$: 419.2557, found: 419.2553.

#### 3.2.4. Synthesis of Compounds **10** and **11**

To a stirring solution of **5** (115 mg, 0.290 mmol, 1.0 equiv.) in dry $CH_2Cl_2$ (6 mL), $NEt_3$ (0.161 mL, 1.02 mmol, 4.0 equiv.), and TMSOTf (0.185 mL, 0.185 mmol, 3.5 equiv.) were added successively at 0 °C under an Ar atmosphere. The mixture was gradually warmed up to room temperature and stirred for 1 h. TLC indicated the reaction was complete. The reaction mixture was quenched with a saturated $NaHCO_3$ solution (15 mL) at 0 °C and extracted with $CH_2Cl_2$ (15 mL × 3); the combined organic phase was washed with brine, and dried with $Na_2SO_4$. After filtration, the solution was concentrated under reduced pressure. The resulting residue was dissolved in dry DMSO (4 mL) and THF (2 mL). $Pd(OAc)_2$ (6.5 mg, 0.029 mmol, 0.1 equiv.) was added to the solution. The mixture was heated to 60 °C for 12 h under an $O_2$ atmosphere. TLC indicated the reaction was complete (petroleum ether/EtOAc = 2/1). The reaction mixture was added to a saturated $NH_4Cl$ solution (20 mL) and extracted with EtOAc (20 mL × 3); the combined organic phase was washed with an aqueous $NH_4Cl$, brine, and dried with $Na_2SO_4$. After filtration, the solution was concentrated under reduced pressure. The residue was dissolved with acetonitrile (20 mL). To the solution, $FeCl_3$ (49 mg, 0.304 mmol, 1.05 equiv.) was added at 0 °C and stirred for 10 min. The mixture was neutralized by $K_2CO_3$ (120 mg, 0.870 mmol, 3.0 equiv.) and filtered; the filtrate was concentrated under reduced pressure and purified by silica gel column chromatography (petroleum ether/EtOAc = 6/1 → 4/1) to give **10** (51 mg, 45%), **11** (31 mg, 26%), and **5** (14 mg, 12%) as white amorphous solids.

**10:** $[\alpha]_D^{25}$ = −55.2 (*c* 1.0, $CHCl_3$); $^1$H NMR (400 MHz, $CDCl_3$) δ 7.52 (dd, *J* = 6.0, 1.8 Hz, 1H), 7.38–7.27 (m, 5H), 6.02 (dd, *J* = 5.9, 3.1 Hz, 1H), 4.59–4.50 (m, 2H), 4.09 (d, *J* = 9.6 Hz, 1H), 3.91 (q, *J* = 3.0 Hz, 1H), 3.72 (dt, *J* = 9.7, 3.0 Hz, 1H), 2.32 (ddd, *J* = 11.4, 3.1, 1.9 Hz, 1H), 2.15 (dq, *J* = 14.9, 2.9 Hz, 1H), 2.11–2.03 (m, 1H), 1.99–1.85 (m, 4H), 1.81–1.73 (m, 1H), 1.71–1.64 (m, 2H), 1.57–1.45 (m, 3H), 1.42–1.24 (m, 3H), 1.18 (s, 3H), 1.07 (s, 3H); $^{13}$C NMR (100 MHz, $CDCl_3$) δ 213.3, 158.6, 137.9, 131.8, 128.6, 127.9, 127.7, 75.7, 72.7, 70.8, 57.0, 51.0, 43.7, 40.4, 33.0, 31.0, 29.9, 29.8, 29.4, 26.0, 25.1, 20.8, 19.9, 18.8; HR-ESI-MS (*m*/*z*) calcd for $C_{26}H_{34}O_3Na$ $[M+Na]^+$: 417.2400, found: 417.2387.

**11:** $[\alpha]_D^{25}$ = −24.2 (*c* 1.0, $CHCl_3$); $^1$H NMR (500 MHz, $CDCl_3$) δ 7.38–7.27 (m, 5H), 6.81 (dd, *J* = 9.9, 2.2 Hz, 1H), 6.02 (dd, *J* = 9.8, 3.0 Hz, 1H), 4.59–4.49 (m, 2H), 3.95–3.90 (m, 1H), 3.72 (d, *J* = 3.1 Hz, 1H), 2.33 (dt, *J* = 11.4, 2.5 Hz, 1H), 2.18–2.12 (m, 1H), 2.09–2.02 (m, 1H), 1.99 (dt, *J* = 12.6, 3.5 Hz, 1H), 1.93–1.82 (m, 3H), 1.76 (dd, *J* = 13.2, 4.4 Hz, 1H), 1.72 (s, 2H), 1.64–1.58 (m, 1H), 1.59–1.48 (m, 1H), 1.49–1.40 (m, 1H), 1.41–1.35 (m, 1H), 1.32 (s, 3H), 1.28–1.21 (m, 1H), 1.18 (dd, *J* = 12.2, 4.0 Hz, 1H), 1.09 (s, 3H); $^{13}$C NMR (125 MHz, $CDCl_3$) δ 164.0, 145.4, 137.9, 128.6, 127.9, 127.7, 121.8, 83.3, 75.6, 72.5, 70.9, 48.6, 41.9, 40.2, 38.4, 35.8, 30.6, 30.0, 29.8, 25.9, 24.6, 21.7, 18.7, 18.6; HR-ESI-MS (*m*/*z*) calcd for $C_{26}H_{35}O_4$ $[M+H]^+$: 411.2530, found: 411.2523.

To a stirring solution of **5** (15.0 g, 37.8 mmol, 1.0 equiv.) in dry $CH_2Cl_2$ (120 mL), $NEt_3$ (21.0 mL, 151 mmol, 4.0 equiv.), and TMSOTf (23.9 mL, 132 mmol, 3.5 equiv.) were added successively at 0 °C under an Ar atmosphere. The mixture was gradually warmed up to room temperature and stirred for 1 h. TLC indicated the reaction was complete. The reaction mixture was quenched with a saturated $NaHCO_3$ solution (100 mL) at 0 °C and the aqueous phase was extracted with $CH_2Cl_2$ (100 mL × 2). The combined organic phase was washed with brine, and dried with $Na_2SO_4$. After filtration, the solution was concentrated under reduced pressure. The resulting residue was dissolved in dry DMSO (280 mL) and THF (140 mL). $Pd(OAc)_2$ (2.5 g, 11.3 mmol, 0.3 equiv.) was added to the solution. The mixture was heated to 60 °C for 12 h under an $O_2$ atmosphere. TLC indicated the reaction was complete (petroleum ether/EtOAc = 2/1). $FeCl_3$ (6.4 g, 39.7 mmol, 1.05 equiv.) was added to the reaction mixture at 0 °C and stirred for 10 min. The mixture was neutralized by $K_2CO_3$ (16.4 g, 119.0 mmol) and filtered; the filtrate was concentrated under reduced pressure and purified by silica gel column chromatography (petroleum ether/EtOAc/$CH_2Cl_2$ = 8/1/0.9 → 4/1/0.9) to give **10** (11.0 g, 74%) and **5** (2.3 g, 15%) as white amorphous solids.

3.2.5. Synthesis of Compound **12**

To a mixture of **10** (2.9 g, 7.40 mmol, 1.0 equiv.) and $SiO_2$ (8.0 g), benzotrifluoride (80 mL) and DIPEA (67.3 mL, 407 mmol, 55.0 equiv.) were added successively under an atmosphere of Ar. The resulting mixture was stirred for 2 h at 90 °C under an Ar atmosphere. TLC indicated the reaction was complete (petroleum ether/EtOA = 2/1). The reaction mixture was filtered, and the filter cake was washed with EtOA (150 mL). The filtrate was concentrated under reduced pressure. The residue was dissolved with $CH_2Cl_2$ (100 mL), and the organic phase was washed with an aqueous 5 wt% HCl solution, brine, and dried with $Na_2SO_4$. After filtration, the solution was concentrated under reduced pressure. The residue was purified by silica gel column chromatography (petroleum ether/EtOAc = 10:1) to give **12** (2.4 g, 82%) as a white amorphous solid. $[\alpha]_D^{25}$ = +84.3 (*c* 1.0, $CHCl_3$); $^1$H NMR (400 MHz, $CDCl_3$) δ 7.39–7.27 (m, 5H), 5.50 (q, *J* = 2.1 Hz, 1H), 4.59–4.48 (m, 2H), 4.03 (d, *J* = 9.7 Hz, 1H), 3.89 (t, *J* = 2.9 Hz, 1H), 3.73 (d, *J* = 9.3 Hz, 1H), 3.00 (ddd, *J* = 23.1, 3.9, 1.9 Hz, 1H), 2.82 (dt, *J* = 23.1, 2.3 Hz, 1H), 2.26 (ddd, *J* = 13.6, 9.0, 4.9 Hz, 1H), 2.16–2.09 (m, 1H), 2.07–2.01 (m, 1H), 1.99–1.83 (m, 2H), 1.78 (dt, *J* = 12.9, 3.2 Hz, 1H), 1.70–1.58 (m, 3H), 1.55–1.48 (m, 1H), 1.48–1.42 (m, 1H), 1.41–1.34 (m, 1H), 1.29–1.19 (m, 2H), 1.15 (s, 3H), 1.11 (s, 3H); $^{13}$C NMR (100 MHz, $CDCl_3$) δ 153.6, 138.0, 128.6, 127.9, 127.7, 113.1, 75.7, 72.7, 70.8, 50.9, 43.0, 41.5, 40.4, 35.8, 33.5, 31.1, 30.3, 29.8, 25.8, 22.7, 20.8, 20.1, 18.8; HR-ESI-MS (*m/z*) calcd for $C_{26}H_{34}O_3Na$ $[M+Na]^+$: 417.2400, found: 417.2411.

### 3.2.6. Synthesis of Compound **13**

To a stirring solution of compound **12** (2.4 g, 6.03 mmol, 1.0 equiv.) in isopropenyl acetate (12 mL), $TsOH{\cdot}H_2O$ (4.6 g, 24.1 mmol, 4.0 equiv.) was added at room temperature. After, the reaction mixture was stirred for 10 min at the same temperature. TLC indicated the reaction was complete. $CH_2Cl_2$ (100 mL) was added, and the reaction mixture was quenched with a saturated $NaHCO_3$ solution (100 mL) at 0 °C. The organic phase was washed with brine, and dried with $Na_2SO_4$. After filtration, the solution was concentrated under reduced pressure. The residue was purified by silica gel column chromatography (petroleum ether/EtOAc = 10/1) to give compound **13** (2.14 g, 81%) as a white amorphous solid. $[\alpha]_D^{25}$ = +51.9 (*c* 1.0, $CHCl_3$); $^1$H NMR (500 MHz, $CDCl_3$) δ 7.34–7.29 (m, 4H), 7.26–7.22 (m, 1H), 5.52 (q, *J* = 2.1 Hz, 1H), 4.97 (t, *J* = 3.0 Hz, 1H), 4.47–4.39 (m, 2H), 3.73 (q, *J* = 3.0 Hz, 1H), 3.00 (ddd, *J* = 23.0, 3.8, 1.8 Hz, 1H), 2.82 (dt, *J* = 23.1, 2.2 Hz, 1H), 2.32–2.19 (m, 3H), 1.99–1.95 (m, 1H), 1.94 (s, 3H), 1.93–1.87 (m, 2H), 1.80 (dt, *J* = 13.1, 3.1 Hz, 1H), 1.67–1.61 (m, 3H), 1.55 (dt, *J* = 15.7, 3.3 Hz, 1H), 1.50–1.31 (m, 3H), 1.23 (td, *J* = 13.1, 3.9 Hz, 1H), 1.11 (s, 3H), 1.05 (s, 3H); $^{13}$C NMR (125 MHz, $CDCl_3$) δ 222.3, 171.5, 153.4, 139.2, 128.4, 127.4, 127.3, 113.4, 74.2, 72.6, 69.9, 50.9, 42.3, 41.5, 39.0, 35.7, 33.4, 32.1, 30.5, 27.4, 25.6, 22.3, 21.4, 21.3, 20.2, 18.4; HR-ESI-MS (*m/z*) calcd for $C_{28}H_{36}O_4Na$ $[M+Na]^+$: 459.2506, found: 459.2512.

### 3.2.7. Synthesis of Compound **4α** and **4β**

To the solution of **13** (5.0 g, 11.5 mmol, 1.0 equiv.) in EtOH (230 mL), $PPh_3$ (6.03 g, 23.0 mmol, 2.0 equiv.), and $Mn(acac)_2$ (1.46 g, 5.75 mmol, 0.5 equiv.) were added. The reaction mixture was bubbled with $O_2$ for 40 min. Next, to the solution, $PhSiH_3$ (4.25 mL, 34.5 mmol, 3.0 equiv.) was added. The resulting mixture was stirred for 3 h at room temperature under an $O_2$ atmosphere. TLC indicated the reaction was complete (petroleum ether/acetone = 4/1). The reaction mixture was quenched with a saturated $Na_2S_2O_3$ solution (25 mL) at 0 °C. $H_2O$ (400 mL) was carefully added to the reaction mixture at room temperature. The resulting mixture was extracted with EtOAc (400 mL × 2); the combined organic phase was washed with brine, and dried with $Na_2SO_4$. After filtration, the solution was concentrated under reduced pressure. The residue was purified by silica gel column chromatography (petroleum ether/acetone = 8/1) to give **4β** (3.6 g, 68%) and **4α** (1.7 g, 32%) as white amorphous solids.

**4β**: $[\alpha]_D^{25}$ = −4.7 (*c* 1.0, $CHCl_3$); $^1$H NMR (500 MHz, $CDCl_3$) δ 7.35–7.30 (m, 4H), 7.26–7.23 (m, 1H), 4.95 (t, *J* = 3.1 Hz, 1H), 4.48–4.41 (m, 2H), 3.79–3.74 (m, 1H), 2.44–2.39 (m, 2H), 2.30–2.24 (m, 1H), 2.23–2.10 (m, 3H), 1.94 (s, 2H), 1.93–1.80 (m, 3H), 1.78–1.66 (m, 3H), 1.64–1.50 (m, 4H), 1.44–1.34 (m, 3H), 1.30–1.21 (m, 2H), 1.05 (s, 2H), 1.00 (s, 3H); $^{13}$C NMR (125 MHz, $CDCl_3$) δ 221.0, 171.5, 139.1, 128.4, 127.4, 127.3, 82.5, 74.2, 72.4, 70.0, 41.6, 41.5, 38.8, 37.4, 33.1, 32.0, 31.7, 30.5, 27.5, 27.5, 25.8, 21.4, 20.4, 18.9, 18.5, 13.0; HR-ESI-MS (*m/z*) calcd for $C_{28}H_{42}O_5N$ $[M+NH_4]^+$: 472.3057, found: 472.3050.

**4α**: $[\alpha]_D^{25}$ = +11.2 (*c* 1.0, $CHCl_3$); $^1H$ NMR (600 MHz, $CDCl_3$) δ 7.32–7.30 (m, 3H), 7.27–7.22 (m, 2H), 4.98–4.94 (m, 1H), 4.51–4.34 (m, 2H), 3.77–3.73 (m, 1H), 2.41 (ddd, *J* = 18.7, 9.5, 2.2 Hz, 1H), 2.33 (dt, *J* = 18.6, 8.6 Hz, 1H), 2.29–2.22 (m, 1H), 2.22–2.17 (m, 1H), 2.02–1.94 (m, 2H), 1.94 (s, 3H), 1.92–1.82 (m, 4H), 1.76 (td, *J* = 13.4, 4.5 Hz, 1H), 1.68–1.62 (m, 2H), 1.60–1.49 (m, 2H), 1.48–1.26 (m, 4H), 1.03 (s, 3H), 0.99 (s, 3H); $^{13}C$ NMR (150 MHz, $CDCl_3$) δ 218.7, 171.5, 139.2, 128.4, 127.4, 127.3, 81.4, 74.6, 72.6, 69.9, 52.6, 38.8, 38.2, 34.6, 33.1, 32.0, 30.5, 30.5, 27.2, 25.6, 25.1, 21.4, 19.5, 19.3, 18.5, 18.2; HR-ESI-MS (*m*/*z*) calcd for $C_{28}H_{38}O_5Na$ $[M+Na]^+$: 477.2611, found: 477.2612.

### 3.2.8. Synthesis of Compound **14**

To a stirring solution of **4β** (4.9 g, 10.7 mmol, 1.0 equiv.) in EtOH (110 mL), $Et_3N$ (4.5 mL, 32.1 mmol, 3.0 equiv.) and $H_2NNH_2{\cdot}H_2O$ (1.6 mL, 32.1 mmol, 3.0 equiv.) were added. The mixture was stirred at 65 °C for 5 h. TLC indicated the reaction was complete ($CH_2Cl_2$/MeOH = 20/1). The reaction mixture was concentrated under reduced pressure. The residue was purified by silica gel column chromatography ($CH_2Cl_2$/MeOH = 60/1 → 20/1) to give intermediate hydrazone (4.4 g, 88%) as a white amorphous solid.

To the intermediate hydrazone (4.4 g, 9.5 mmol, 1.0 equiv.) in dry THF (100 mL), $Et_3N$ (13.1 mL, 94.5 mmol, 10.0 equiv.) and a solution of $I_2$ (6.0 g, 23.6 mmol, 2.5 equiv.) in THF (100 mL) were added successively. The resulting mixture was stirred at 25 °C for 30 min under an Ar atmosphere. TLC indicated the reaction was complete ($CH_2Cl_2$/MeOH = 20/1). The reaction mixture was quenched with a saturated $NaHCO_3$ solution (100 mL) and a $Na_2SO_3$ solution (100 mL) at 0 °C, and EtOAc (200 mL) was added. The mixture was extracted with EtOAc (100 mL × 2); the combined organic phase was washed with brine, and dried with $Na_2SO_4$. After filtration, the solution was concentrated under reduced pressure. The residue was purified by silica gel column chromatography (petroleum ether/EtOAc = 8/1) to give **14** (4.9 g, 91%) as a white foamy solid. $[\alpha]_D^{25}$ = −9.54 (*c* 1.0, $CHCl_3$); $^1H$ NMR (500 MHz, $CDCl_3$) $^1H$ NMR (500 MHz, $CDCl_3$) δ 7.34–7.29 (m, 4H), 7.26–7.22 (m, 1H), 6.11 (dd, *J* = 3.2, 1.8 Hz, 1H), 5.00–4.96 (m, 1H), 4.49–4.39 (m, 2H), 3.78–3.72 (m, 1H), 2.54 (dd, *J* = 16.5, 1.9 Hz, 1H), 2.25 (ddd, *J* = 16.5, 7.1, 3.0 Hz, 2H), 2.18 (dd, *J* = 13.4, 3.6 Hz, 1H), 1.95 (s, 3H), 1.91 (dd, *J* = 13.8, 3.6 Hz, 1H), 1.88–1.70 (m, 4H), 1.67–1.59 (m, 2H), 1.51–1.39 (m, 2H), 1.38–1.24 (m, 2H), 1.14–1.07 (m, 1H)), 1.05 (s, 3H), 0.99 (s, 3H), 0.97–0.90 (m, 1H); $^{13}C$ NMR (125 MHz, $CDCl_3$) δ 171.5, 139.2, 133.7, 128.4, 127.4, 127.3, 111.4, 82.4, 74.3, 72.5, 69.9, 54.8, 42.8, 41.3, 38.8, 37.9, 37.5, 31.7, 30.5, 27.6, 25.8, 21.5, 20.4, 20.2, 18.7, 18.1; HR-ESI-MS (*m*/*z*) calcd for $C_{28}H_{37}O_4INa$ $[M+Na]^+$: 587.1629, found: 587.1633.

### 3.2.9. Synthesis of Compound **16**

A mixture of **14** (4.0 g, 7.1 mmol, 1.0 equiv.) and **16** (8.0 g, 21.3 mmol, 3.0 equiv.) was co-evaporated with toluene twice and recharged with Ar. CuCl (10.5 g, 106.0 mmol, 15.0 equiv.), LiCl (6.02 g, 142 mmol, 20.0 equiv.), and $Pd(Ph_3P)_4$ (819 mg, 0.709 mmol, 0.1 equiv.) were added. The resulting mixture was recharged with Ar and dry DMSO (150 mL) was added. The mixture was stirred for 2 h at 70 °C under an Ar atmosphere after Freeze–Pump–Thaw Degassing was performed twice. TLC indicated the reaction was complete (PE/EA = 2/1). The reaction mixture was quenched with a PBS (350 mL) at 0 °C. The mixture was filtered. The filter cake was washed with EtOAc (200 mL × 5). The aqueous phase was extracted with EtOAc (300 mL × 2); the combined organic phase was washed with a saturated $NH_4Cl$ solution (800 mL × 2) and brine, and dried with $Na_2SO_4$. After filtration, the solution was concentrated under reduced pressure. The residue was purified by silica gel column chromatography (containing 10% $K_2CO_3$, and petroleum ether/EtOAc/$CH_2Cl_2$ = 2/1/0.2) to give compound **16** (3.3 g, 89%) as a pale yellow amorphous solid. $[\alpha]_D^{25}$ = +37.5 (*c* 1.0, $CHCl_3$); $^1H$ NMR (400 MHz, $CDCl_3$) δ 7.36–7.29 (m, 4H), 7.24 (d, *J* = 6.1 Hz, 1H), 6.10 (d, *J* = 2.8 Hz, 1H), 5.96 (s, 1H), 5.02–4.89 (m, 3H), 4.49–4.39 (m, 2H), 3.75 (s, 1H), 2.73–2.64 (m, 1H), 2.39 (dd, *J* = 18.4, 3.4 Hz, 1H), 2.26 (dd, *J* = 15.9, 2.8 Hz, 1H), 2.20 (d, *J* = 13.3 Hz, 1H), 2.06–1.98 (m, 1H), 1.95 (s, 3H), 1.91–1.72 (m, 3H), 1.69–1.59 (m, 3H), 1.51–1.33 (m, 4H), 1.28 (s, 3H), 1.18–1.05 (m, 1H), 1.01 (s, 3H), 0.95–0.88 (m, 1H); $^{13}C$ NMR (100 MHz, $CDCl_3$) δ 174.5, 171.5, 158.3, 144.0, 139.1, 132.2, 128.4, 127.4, 127.2, 112.6, 85.5, 74.2, 72.4, 71.8, 69.9, 52.2, 41.0, 40.6, 38.7, 38.4, 37.8, 31.6, 30.5, 27.6, 25.8, 21.4, 20.4, 20.3, 18.7, 16.7; HR-ESI-MS (*m*/*z*) calcd for $C_{32}H_{40}O_6Na$ $[M+Na]^+$: 543.2717, found: 543.2715.

3.2.10. Synthesis of Compound **17**

To a mixture of **16** (2.0 g, 3.8 mmol, 1.0 equiv.) and imidazole (1.6 g, 23.0 mmol, 6.0 equiv.), dry DMF (40 mL) was added under an atmosphere of Ar. TMSCl (1.5 mL, 11.5 mmol, 3.0 equiv.) was added to the resulting mixture under an Ar atmosphere and stirred for 12 h at 50 °C. TLC indicated the reaction was complete (petroleum ether/EtOAc = 1/1.5). The reaction mixture was quenched with MeOH (5 mL) at 0 °C. The mixture was extracted with $CH_2Cl_2$ (200 mL), washed with brine (200 mL × 3), and dried with $Na_2SO_4$. After filtration, the solution was concentrated under reduced pressure. The residue was purified by silica gel column chromatography (petroleum ether/EtOAc = 3/1) to give **17** (1.9 g, 83%) as a white foamy solid. $[\alpha]_D^{25}$ = +39.6 (*c* 1.0, $CHCl_3$); $^1H$ NMR (400 MHz, $CDCl_3$) δ 7.34–7.26 (m, 5H), 6.03 (d, *J* = 2.8 Hz, 1H), 5.96 (s, 1H), 5.03–4.86 (m, 3H), 4.49–4.39 (m, 2H), 3.74 (d, *J* = 3.9 Hz, 1H), 2.59 (d, *J* = 18.5 Hz, 1H), 2.44 (dd, *J* = 18.6, 3.4 Hz, 1H), 2.25 (dd, *J* = 15.8, 2.8 Hz, 1H), 2.19 (d, *J* = 14.2 Hz, 1H), 1.98 (d, *J* = 3.6 Hz, 1H), 1.94 (s, 3H), 1.92–1.79 (m, 2H), 1.78–1.68 (m, 2H), 1.67–1.55 (m, 3H), 1.42–1.34 (m, 2H), 1.28–1.23 (m, 1H), 1.20 (s, 3H), 1.17–1.11 (m, 1H), 1.05 (d, *J* = 13.7 Hz, 1H), 1.00 (s, 3H), 0.00 (s, 9H); $^{13}C$ NMR (100 MHz, $CDCl_3$) δ 174.5, 171.5, 158.4, 145.0, 139.2, 132.0, 128.4, 127.4, 127.3, 112.6, 89.8, 74.3, 72.5, 71.7, 69.9, 52.9, 42.0, 39.4, 38.7, 38.7, 37.8, 31.7, 30.5, 27.6, 26.0, 21.5, 20.7, 20.3, 18.8, 17.4, 2.8; HR-ESI-MS (*m*/*z*) calcd for $C_{35}H_{48}O_6SiNa$ $[M+Na]^+$: 615.3112, found: 615.3120.

3.2.11. Synthesis of Compound **18**

To a solution of **17** (1.9 g, 3.2 mmol, 1.0 equiv.) in a mixture of EtOAc/MeOH/PBS (60 mL/30 mL/60 mL), Pd/C (1.0 g) was added. The reaction mixture was recharged with $H_2$ thrice, and then stirred for 3 h at 0 °C under an atmosphere of $H_2$. TLC indicated the reaction was complete (petroleum ether/EtOAc = 2/1). The reaction mixture was filtered, and the filter cake was washed with EtOAc (100 mL × 3). The filtrate was concentrated under reduced pressure. The resulting residue was purified by silica gel column chromatography (petroleum ether/EtOAc = 2/1 → 1/1.5) to give **18** (1.3 g, 68%) as a colorless amorphous solid. $[\alpha]_D^{25}$ = −4.35 (*c* 1.0, $CHCl_3$); $^1H$ NMR (500 MHz, $CDCl_3$) δ 7.34–7.29 (m, 4H), 7.26–7.24 (m, 1H), 5.84 (t, *J* = 1.5 Hz, 1H), 4.95–4.90 (m, 1H), 4.79–4.67 (m, 2H), 4.49–4.39 (m, 2H), 3.76 (t, *J* = 3.1 Hz, 1H), 2.56 (t, *J* = 7.6 Hz, 1H), 2.30–2.22 (m, 1H), 2.17 (d, *J* = 13.4 Hz, 1H), 2.10–2.02 (m, 1H), 1.98–1.95 (m, 1H), 1.94 (s, 3H), 1.93–1.85 (m, 2H), 1.84–1.69 (m, 4H), 1.67–1.55 (m, 3H), 1.50 (td, *J* = 11.8, 3.5 Hz, 1H), 1.43–1.33 (m, 3H), 1.32–1.18 (m, 2H), 0.97 (s, 3H), 0.88 (s, 3H), 0.13 (s, 9H); $^{13}C$ NMR (125 MHz, $CDCl_3$) δ 174.4, 173.9, 171.5, 139.1, 128.4, 127.4, 127.3, 117.2, 91.3, 74.3, 74.0, 72.5, 70.0, 50.9, 50.6, 41.5, 40.9, 39.2, 38.2, 34.1, 31.7, 30.7, 27.4(2C), 26.1, 22.5, 21.5, 21.2, 18.4, 18.3, 3.1; HR-ESI-MS (*m*/*z*) calcd for $C_{35}H_{50}O_6SiNa$ $[M+Na]^+$: 617.3269, found: 617.3275.

### 3.2.12. Synthesis of Compound **3**

To a solution of **18** (1.5 g, 2.6 mmol, 1.0 equiv.) in a mixture of $CHCl_3$/MeOH (60 mL/ 60 mL, Pd/C (0.8 g) was added. The reaction mixture was recharged with $H_2$ thrice, and then stirred for 2 h at room temperature under an atmosphere of $H_2$. TLC indicated the reaction was complete (petroleum ether/EtOAc = 1/1.5). The reaction mixture was filtered. The filtrate was concentrated under reduced pressure. The resulting residue was purified by silica gel column chromatography ($CH_2Cl_2$/MeOH = 30/1) to give **3** (869 mg, 78%) as a colorless amorphous solid. $[\alpha]_D^{25}$ = +9.4 (*c* 1.0, $CHCl_3$); $^1$H NMR (500 MHz, $CDCl_3$) δ 5.88 (s, 1H), 5.26 (d, *J* = 2.9 Hz, 1H), 4.97 (d, *J* = 18.0 Hz, 1H), 4.80 (d, *J* = 17.6 Hz, 1H), 4.04 (s, 1H), 2.80–2.75 (m, 1H), 2.23–2.11 (m, 1H), 2.09 (s, 3H), 2.06–1.79 (m, 6H), 1.77–1.68 (m, 2H), 1.66–1.59 (m, 2H), 1.58–1.49 (m, 2H), 1.47–1.26 (m, 6H), 0.96 (s, 3H), 0.87 (s, 3H); $^{13}$C NMR (125 MHz, $CDCl_3$) δ 174.6, 174.4, 169.9, 118.0, 85.4, 75.8, 73.6, 66.9, 50.8, 49.5, 41.8, 39.8, 39.1, 37.1, 33.3(2 C), 31.5, 31.1, 27.0, 25.8, 21.6, 21.4, 20.8, 18.5, 15.9; HR-ESI-MS (*m/z*) calcd for $C_{25}H_{36}O_6Na$ $[M+Na]^+$: 455.2404, found: 455.2411.

### 3.2.13. Synthesis of Compound **20**

To a stirring solution of **19** (10.0 mg, 0.027 mmol, 1.0 equiv.) in dry $CH_2Cl_2$ (7 mL), $NCCCl_3$ (11 μL, 0.108 mmol, 4.0 equiv.) and $Cs_2CO_3$ (0.87 mg, 2.69 μmol, 0.1 equiv.) were added. The resulting mixture was stirred for 3 h at room temperature. TLC indicated the reaction was complete. The reaction mixture was filtered. The filtrate was concentrated in vacuo. The residue and aglycone **3** (9.7 mg, 0.022 mmol, 1.0 equiv.) were co-evaporated with toluene three times and dissolved in dry $CH_2Cl_2$ (1.5 mL); a 4Å molecular sieve (150 mg) was added. The mixture was stirred for 15 min at 25 °C under an Ar atmosphere. Next, TMSOTf (0.81 μL, 4.48 μmol, 0.2 equiv.) was added at −78 °C. The resulting mixture was stirred for 10 min at −78 °C. TLC indicated the reaction was complete (petroleum ether/EtOAc = 1/2). The reaction mixture was quenched with a saturated $NaHCO_3$ aqueous solution (2 mL) and filtered. The filtrate was added to the saturated $NaHCO_3$ solution (20 mL). The mixture was extracted with $CH_2Cl_2$ (20 mL × 3), washed with brine, and dried with $Na_2SO_4$. After filtration, the solution was concentrated under reduced pressure. The residue was purified by silica gel column chromatography (petroleum ether/EtOAc = 4/1) to give compound **20** (8.9 mg, 50%) as a white amorphous solid. $[\alpha]_D^{25}$ = −34.5 (*c* 1.0, $CHCl_3$); $^1$H NMR (600 MHz, $CDCl_3$) δ 8.01–7.98 (m, 2H), 7.46–7.38 (m, 3H), 7.34–7.27 (m, 3H), 7.17–7.11 (m, 2H), 5.87 (t, *J* = 1.9 Hz, 1H), 5.43–5.34 (m, 1H), 5.00–4.94 (m, 3H), 4.91 (t, *J* = 3.0 Hz, 1H), 4.80 (dd, *J* = 18.1, 1.8 Hz, 1H), 4.62 (d, *J* = 11.6 Hz, 1H), 4.04 (s, 1H), 3.95–3.90 (m, 1H), 3.68–3.63 (m, 2H), 3.43 (s, 3H), 2.77 (dd, *J* = 9.5, 5.5 Hz, 1H), 2.19–2.10 (m, 2H), 2.08 (s, 3H), 2.07–1.99 (m, 2H), 1.91–1.76 (m, 4H), 1.74–1.65 (m, 3H), 1.64–1.57 (m, 3H), 1.57–1.50 (m, 2H), 1.48–1.42 (m, 1H), 1.39 (d, *J* = 15.3 Hz, 1H), 1.35 (d, *J* = 6.5 Hz, 3H), 1.30 (dt, *J* = 13.1, 3.4 Hz, 1H), 0.99 (s, 3H), 0.88 (s, 3H); δ $^{13}$C NMR (150 MHz, $CDCl_3$) δ 174.5, 174.3, 171.0, 166.7, 139.3, 133.0, 130.4, 129.9, 128.3, 128.3, 128.0, 127.4, 118.0, 95.7, 85.5, 77.9, 76.4, 74.7, 74.3, 73.5, 69.3, 67.9, 67.3, 57.1, 50.9, 49.6, 41.7, 39.9, 38.8, 37.0, 33.3, 31.2, 30.0, 29.8, 28.0, 27.0, 25.8, 21.7, 20.7, 18.6, 17.1, 15.9; HR-ESI-MS (*m/z*) calcd for $C_{46}H_{58}O_{11}Na$ $[M+Na]^+$: 809.3871, found: 809.3875.

### 3.2.14. Synthesis of Compound **2**

To a solution of **20** (8.9 mg, 11.3 μmol, 1.0 equiv.) in EtOAc/MeOH (1.0 mL/1.0 mL), 10% Pd/C (8.9 mg) was added. The reaction mixture was recharged with $H_2$ 3 times, and then stirred for 18 h at 25 °C under an atmosphere of $H_2$. TLC indicated the reaction was complete (petroleum ether/EtOAc = 1/2). The reaction mixture was filtered. The filtrate was concentrated under reduced pressure. The residue was dissolved in MeOH/$H_2O$/THF

(1.5 mL/0.5 mL/0.5 mL); LiOH (2.0 mg, 83.5 μmol, 7.4 equiv.) was added and then the mixture was stirred for 0.5 h at 25 °C. TLC indicated the reaction was complete (petroleum ether/EtOAc = 1/2). The reaction mixture was quenched with HOAc (4.8 μL, 83.5 μmol, 7.4 equiv.) and concentrated in vacuo. The residue was purified by reverse-phase columns ($H_2O$/MeOH = 1/2) to afford acovensoide B (**2**) (4.0 mg, 60% over 2 steps) as a white amorphous solid. $[\alpha]_D^{25} = -53.0$ (*c* 0.4, $CHCl_3$); $^1$H NMR (400 MHz, $CDCl_3$) δ 5.88 (s, 1H), 5.01–4.93 (m, 2H), 4.88 (t, *J* = 3.2 Hz, 1H), 4.80 (d, *J* = 18.1 Hz, 1H), 4.04 (s, 1H), 3.86 (s, 1H), 3.84–3.76 (m, 2H), 3.47 (s, 3H), 3.34 (t, *J* = 3.2 Hz, 1H), 2.78 (dd, *J* = 9.3, 5.6 Hz, 1H), 2.22–2.03 (m, 3H), 2.01 (s, 3H), 1.97 (s, 1H), 1.88–1.68 (m, 7H), 1.58–1.49 (m, 4H), 1.46–1.39 (m, 1H), 1.38–1.32 (m, 2H), 1.29 (d, *J* = 6.5 Hz, 3H), 1.25–1.22 (m, 1H), 0.97 (s, 3H), 0.87 (s, 3H); $^{13}$C NMR (150 MHz, $CDCl_3$) δ 174.5, 174.3, 170.8, 117.9, 97.4, 85.4, 75.4, 74.2, 73.5, 70.1, 68.7, 68.6, 66.2, 55.6, 50.8, 49.5, 41.7, 39.8, 38.8, 36.9, 33.2, 31.1, 29.9, 27.8, 26.9, 25.8, 21.6, 21.5, 20.6, 18.5, 16.6, 15.8; HR-ESI-MS (*m*/*z*) calcd for $C_{32}H_{48}O_{10}Na$ [M+Na]$^+$: 615.3140, found: 615.3142.

3.2.15. Synthesis of Compound **22**

To a stirring solution of **21** (200 mg, 0.418 mmol, 1.0 equiv.) in a mixture of $CH_2Cl_2$ /$H_2O$ (4 mL/0.4 mL), DDQ (285 mg, 1.3 mmol, 3.0 equiv.) was added. The resulting mixture was stirred for 12 h at room temperature. TLC indicated the reaction was complete (petroleum ether/EtOAc = 4/1). The reaction mixture was quenched with a saturated $NaHCO_3$ solution (20 mL). The mixture was extracted with $CH_2Cl_2$ (20 mL × 3), washed with brine, and dried with $Na_2SO_4$. After filtration, the solution was concentrated under reduced pressure. The residue was purified by silica gel column chromatography (petroleum ether/EtOAc = 4/1) to give compound **22** (146 mg, 90%) as a pale yellow amorphous solid. $[\alpha]_D^{25} = -106.8$ (*c* 1.0, $CHCl_3$); $^1$H NMR (500 MHz, $CDCl_3$) δ 8.02–7.99 (m, 2H), 7.59–7.54 (m, 1H), 7.44 (t, *J* = 7.8 Hz, 2H), 7.40–7.36 (m, 2H), 7.16–7.10 (m, 2H), 5.70 (dt, *J* = 3.5, 1.3 Hz, 1H), 5.56 (d, *J* = 1.4 Hz, 1H), 4.50–4.45 (m, 1H), 3.92 (ddt, *J* = 7.9, 3.1, 1.3 Hz, 1H), 3.66 (t, *J* = 3.5 Hz, 1H), 3.48 (s, 3H), 2.55 (d, *J* = 7.9 Hz, 1H), 2.33 (s, 3H), 1.40 (d, *J* = 6.5 Hz, 3H); $^{13}$C NMR (125 MHz, $CDCl_3$) δ 165.6, 138.2, 133.6, 132.5, 130.1, 129.9, 129.8, 129.5, 128.7, 87.0, 75.2, 70.2, 69.6, 68.2, 56.5, 21.3, 16.5; HR-ESI-MS (*m*/*z*) calcd for $C_{21}H_{24}O_5SNa$ [M+Na]$^+$: 411.1237, found: 411.1242.

3.2.16. Synthesis of Compound **24**

To a stirring solution of 2-*O*-benzoyl-3,4,6-tri-*O*-benzyl-D-glucopyranose (223 mg, 0.402 mmol, 1.0 equiv.) in dry $CH_2Cl_2$ (4 mL), $NCCCl_3$ (0.164 mL, 1.6 mmol, 4.0 equiv.) and $Cs_2CO_3$ (13.1 mg, 40.2 μmol, 0.1 equiv.) were added. The resulting mixture was stirred for 4 h at room temperature. TLC indicated the reaction was complete (petroleum ether/EtOAc = 4/1). The reaction mixture was filtered. The filtrate was concentrated in vacuo.

The residue and acceptor **22** (130 mg, 0.335 mmol, 1.0 equiv.) were co-evaporated with toluene thrice and dissolved in dry $CH_2Cl_2$ (3.5 mL); a 4 Å molecular sieve (350 mg) was added. The mixture was stirred for 15 min at 25 °C under an Ar atmosphere. Next, TMSOTf (0.81 μL, 4.48 μmol, 0.2 equiv.) was added at −78 °C. The resulting mixture was stirred for 2 h at −78 °C. TLC indicated the reaction was complete (petroleum ether/EtOAc = 3/1). The reaction mixture was quenched with a saturated $NaHCO_3$ solution (2 mL) and filtered. The filtrate was added to the saturated $NaHCO_3$ solution (20 mL). The mixture was extracted with $CH_2Cl_2$ (20 mL × 3), washed with brine, and dried with $Na_2SO_4$. After filtration, the solution was concentrated under reduced pressure. The residue was purified by silica gel column chromatography (petroleum ether/EtOAc = 6/1) to give compound **24** (234 mg, 76%) as a pale yellow foamy solid. $[\alpha]_D^{25} = -57.0$ (*c* 1.0, $CHCl_3$); $^1$H NMR (500 MHz, $CDCl_3$) δ 8.31–8.25 (m, 2H), 8.17–8.11 (m, 2H), 7.75–7.65 (m, 4H), 7.55 (t, *J* = 7.8 z,

2H), 7.49–7.37 (m, 11H), 7.34 (dd, $J$ = 7.3, 2.1 Hz, 2H), 7.26 (d, $J$ = 3.0 Hz, 4H), 7.18 (d, $J$ = 7.9 Hz, 2H), 5.58 (dd, $J$ = 9.4, 8.0 Hz, 1H), 5.49–5.42 (m, 2H), 4.95 (d, $J$ = 10.9 Hz, 1H), 4.88 (d, $J$ = 11.1 Hz, 1H), 4.78 (d, $J$ = 9.2 Hz, 1H), 4.76 (d, $J$ = 6.0 Hz, 1H), 4.71 (d, $J$ = 11.0 Hz, 1H), 4.62 (s, 2H), 4.42–4.36 (m, 1H), 4.09 (t, $J$ = 3.4 Hz, 1H), 3.95 (t, $J$ = 9.1 Hz, 1H), 3.91 (s, 1H), 3.86–3.76 (m, 3H), 3.74–3.69 (m, 1H), 3.58 (s, 3H), 2.42 (s, 3H), 1.38 (d, $J$ = 6.7 Hz, 3H); $^{13}$C NMR (125 MHz, $CDCl_3$) δ 166.2, 165.0, 138.1, 137.9(2), 137.9(0), 137.8, 133.2, 133.1, 132.7, 130.5, 130.1, 129.9(2), 129.8(6), 129.8, 129.5, 128.6(4), 128.5(9), 128.5(1), 128.4(8), 128.4, 128.3, 128.2, 128.0, 127.9, 127.8(0), 127.7(7), 102.2, 83.2, 78.2, 76.0, 75.6, 75.2, 75.1, 74.0, 73.7, 69.9, 69.2, 60.5, 58.8, 21.2, 15.4; HR-ESI-MS ($m/z$) calcd for $C_{55}H_{56}O_{11}SNa$ [M+Na]$^+$: 947.3436, found: 947.3429.

### 3.2.17. Synthesis of Compound **25**

To a stirring solution of **24** (60.0 mg, 64.9 μmol, 1.0 equiv.) in THF/MeOH (1 mL/1 mL), KOH (10.0 mg, 0.178 mmol, 2.7 equiv.) was added at room temperature. After, the reaction mixture was stirred for 12 h at the same temperature. TLC indicated the reaction was complete (petroleum ether/EtOAc = 2/1). The reaction mixture was quenched with HOAc (10.5 μL, 183 μmol, 2.7 equiv.). $H_2O$ (30 mL) was added to the mixture; the mixture was extracted with $CH_2Cl_2$ (30 mL × 3), washed with brine, and dried with $Na_2SO_4$. After filtration, the solution was concentrated under reduced pressure. The residue was purified by silica gel column chromatography (petroleum ether/EtOAc = 2/1) to give compound **25** (37.4 mg, 80%) as a white amorphous solid. $[\alpha]_D^{25} = -128.8$ ($c$ 1.0, $CHCl_3$); $^1$H NMR (500 MHz, $CDCl_3$) δ 7.41–7.27 (m, 15H), 7.22–7.19 (m, 2H), 7.13 (dt, $J$ = 8.4, 1.3 Hz, 2H), 5.55 (d, $J$ = 1.4 Hz, 1H), 5.03 (d, $J$ = 11.2 Hz, 1H), 4.86 (d, $J$ = 10.9 Hz, 1H), 4.83 (d, $J$ = 11.2 Hz, 1H), 4.57 (d, $J$ = 10.8 Hz, 1H), 4.53 (d, $J$ = 12.0 Hz, 1H), 4.51–4.47 (m, 2H), 4.36 (q, $J$ = 6.5 Hz, 1H), 4.17 (d, $J$ = 6.4 Hz, 1H), 4.12 (dt, $J$ = 2.7, 1.2 Hz, 1H), 4.03 (s, 1H), 3.71–3.66 (m, 2H), 3.65–3.55 (m, 3H), 3.52–3.49 (m, 1H), 3.52–3.49 (m, 1H), 3.51 (s, 3H), 2.34 (s, 3H), 1.31 (d, $J$ = 6.5 Hz, 3H); $^{13}$C NMR (125 MHz, $CDCl_3$) δ 138.9, 138.2, 138.2, 137.8, 132.1, 130.3, 130.0, 128.5, 128.5, 128.5, 128.1, 128.1, 127.9, 127.8, 127.8, 127.7, 102.0, 90.0, 84.4, 77.3, 76.0, 75.5, 75.4, 75.1, 73.8, 73.6, 73.5, 69.3, 69.2, 67.9, 56.3, 21.2, 17.3; HR-ESI-MS ($m/z$) calcd for $C_{41}H_{48}O_9SNa$ [M+Na]$^+$: 739.2911, found: 739.2918.

### 3.2.18. Synthesis of Compound **26**

To a stirring solution of compound **25** (37.4 mg, 52.2 μmol, 1.0 equiv.) in isopropenyl acetate (1 mL), TsOH·$H_2O$ (30 mg, 158 μmol, 3.0 equiv.) was added at room temperature. After, the reaction mixture was stirred for 1 h at the same temperature. TLC indicated the reaction was complete (petroleum ether/EtOAc = 2/1). $CH_2Cl_2$ (30 mL) was added, and the reaction mixture was quenched with a saturated $NaHCO_3$ solution (30 mL) and extracted with $CH_2Cl_2$ (30 mL × 2). The organic phase was washed with brine, and dried with $Na_2SO_4$. After filtration, the solution was concentrated under reduced pressure. The residue was purified by silica gel column chromatography (petroleum ether/EtOAc = 4/1) to give compound **26** (37.3 mg, 89%) as a white amorphous solid. $[\alpha]_D^{25} = -49.0$ ($c$ 1.0, $CHCl_3$); $^1$H NMR (500 MHz, $CDCl_3$) δ 7.37–7.26 (m, 15H), 7.19 (dd, $J$ = 7.3, 2.2 Hz, 2H), 7.11 (d, $J$ = 7.9 Hz, 2H), 5.31 (d, $J$ = 3.8 Hz, 1H), 5.15–5.10 (m, 1H), 5.10–5.06 (m, 1H), 4.79 (dd, $J$ = 11.2, 2.4 Hz, 2H), 4.67 (d, $J$ = 11.4 Hz, 1H), 4.57–4.48 (m, 3H), 4.42 (d, $J$ = 7.9 Hz, 1H), 4.28 (qd, $J$ = 6.7, 2.6 Hz, 1H), 3.91 (t, $J$ = 3.0 Hz, 1H), 3.72 (dd, $J$ = 10.6, 2.0 Hz, 1H), 3.68–3.62 (m, 3H), 3.61–3.57 (m, 1H), 3.53–3.47 (m, 1H), 3.44 (s, 3H), 2.32 (s, 3H), 2.14 (s, 3H), 1.95 (s, 3H), 1.31 (d, $J$ = 6.7 Hz, 3H); $^{13}$C NMR (125 MHz, $CDCl_3$) δ 171.2, 169.3, 138.3, 138.0, 138.0, 137.9, 132.7, 129.9, 129.7, 128.6, 128.5, 128.2, 128.0, 128.0, 127.9, 127.9, 127.8, 101.2, 83.3, 78.1,

76.6, 75.2, 75.1, 75.1, 74.4, 73.7, 73.5, 68.8, 58.1, 21.2(2C), 21.1, 15.9; HR-ESI-MS (*m*/*z*) calcd for $C_{45}H_{52}O_{11}SNa$ [M+Na]$^+$: 823.3125, found: 823.3123.

3.2.19. Synthesis of Compound **27**

To a stirring solution of **26** (37.3 mg, 0.047 mmol, 1.0 equiv.) in a mixture of acetone/$H_2O$/$CH_2Cl_2$ (1 mL/0.1 mL/1 mL), NBS (24.9 mg, 0.140 mmol, 3.0 equiv.) was added at −20 °C. The resulting mixture was stirred for 0.5 h at the same temperature. TLC indicated the reaction was complete (petroleum ether/EtOAc = 3/1). The reaction mixture was quenched with $NEt_3$ (0.5 mL). A saturated $NaHCO_3$ solution (30 mL) was added. The mixture was extracted with $CH_2Cl_2$ (30 mL × 3), washed with brine, and dried with $Na_2SO_4$. After filtration, the solution was concentrated under reduced pressure. The residue was purified by silica gel column chromatography (petroleum ether/EtOAc = 2/1) to give compound **27** (30.0 mg, 93%) as a white foamy solid. The solid was used in the next step without further characterization.

3.2.20. Synthesis of Compound **28**

To a stirring solution of **27** (30.0 mg, 0.043 mmol, 1.0 equiv.) in dry $CH_2Cl_2$ (2 mL), $NCCCl_3$ (17.7 µL, 0.173 mmol, 4.0 equiv.) and $Cs_2CO_3$ (1.4 mg, 2.69 µmol, 0.1 equiv.) were added. The resulting mixture was stirred for 2 h at room temperature. TLC indicated the reaction was complete (petroleum ether/EtOAc = 1/2). The reaction mixture was filtered. The filtrate was concentrated in vacuo.

The residue and aglycone **3** (12.5 mg, 0.029 mmol, 1.5 equiv.) were co-evaporated with toluene three times and dissolved in dry $CH_2Cl_2$ (2.5 mL); a 4 Å molecular sieve (250 mg) was added. The mixture was stirred for 10 min at 25 °C under an Ar atmosphere. Next, TMSOTf (1.1 µL, 5.76 µmol, 0.2 equiv.) was added at −78 °C. The resulting mixture was stirred for 1 h at −78 °C. TLC indicated the reaction was complete (petroleum ether/EtOAc = 1/2). The reaction mixture was quenched with a saturated $NaHCO_3$ solution (2 mL) and filtered. The filtrate was added to the saturated $NaHCO_3$ solution (20 mL). The mixture was extracted with $CH_2Cl_2$ (20 mL × 3), and the organic phase was washed with brine, and dried with $Na_2SO_4$. After filtration, the solution was concentrated under reduced pressure. The residue was purified by silica gel column chromatography (petroleum ether/EtOAc = 1/1) to give compound **28** (16.4 mg, 51% over 2 steps) as a pale yellow foamy solid. $[\alpha]_D^{25} = -34.5$ (*c* 1.0, $CHCl_3$); $^1$H NMR (400 MHz, $CDCl_3$) δ 7.34–7.27 (m, 13H), 7.20–7.15 (m, 2H), 5.87 (t, *J* = 1.8 Hz, 1H), 5.14–5.07 (m, 1H), 4.97 (dd, *J* = 18.3, 1.8 Hz, 1H), 4.91 (t, *J* = 3.0 Hz, 1H), 4.87 (t, *J* = 2.9 Hz, 1H), 4.83–4.74 (m, 4H), 4.66 (d, *J* = 11.4 Hz, 1H), 4.56–4.48 (m, 3H), 4.39 (d, *J* = 7.9 Hz, 1H), 3.95 (s, 1H), 3.86–3.83 (m, 1H), 3.81 (d, *J* = 6.7 Hz, 1H), 3.72–3.58 (m, 5H), 3.52–3.46 (m, 2H), 3.38 (s, 3H), 2.77 (dd, *J* = 9.5, 5.5 Hz, 1H), 2.17 (s, 1H), 2.13 (s, 3H), 2.11–2.06 (m, 2H), 2.01 (s, 3H), 1.98–1.96 (m, 3H), 1.94 (d, *J* = 1.4 Hz, 1H), 1.87–1.66 (m, 10H), 1.60–1.50 (m, 2H), 1.43–1.34 (m, 3H), 1.25 (d, *J* = 6.7 Hz, 3H), 0.96 (s, 3H), 0.87 (s, 3H); $^{13}$C NMR (150 MHz, $CDCl_3$) δ 174.5, 174.3, 171.7, 171.1, 169.3, 138.3, 138.0, 137.9, 128.6, 128.6, 128.5, 128.3, 128.1, 128.0, 127.9, 127.9, 118.0, 101.4, 95.0, 85.5, 83.4, 78.2, 75.2, 75.1, 75.1, 74.3, 74.1, 73.6, 73.6, 73.5, 70.0, 69.4, 68.9, 67.1, 60.5, 57.5, 50.9, 49.6, 41.7, 39.9, 38.8, 37.0, 33.3, 31.1, 29.9, 28.5, 28.4, 27.0, 25.8, 21.7, 21.6, 21.3, 21.1, 20.6, 18.5, 16.5, 15.9; HR-ESI-MS (*m*/*z*) calcd for $C_{63}H_{80}O_{17}Na$ [M+Na]$^+$: 1131.5288, found: 1131.5281.

3.2.21. Synthesis of Compound **1**

To a solution of **28** (8.6 mg, 7.8 µmol, 1.0 equiv.) in EtOAc/MeOH (1.0 mL/1.0 mL), 10% Pd/C (8.6 mg) was added. The reaction mixture was recharged with $H_2$ thrice, and then stirred for 5 h at 25 °C under an atmosphere of $H_2$. TLC indicated the reaction was complete ($CH_2Cl_2$/MeOH = 20/1). The reaction mixture was filtered. The filtrate was concentrated under reduced pressure.

The residue was dissolved in MeOH/$H_2O$/THF (1.5 mL/0.5 mL/0.5 mL); LiOH (2.0 mg, 83.5 μmol, 10.8 equiv.) was added and then the mixture was stirred for 2 h at 25 °C. TLC indicated the reaction was complete ($CH_2Cl_2$ / MeOH = 10/1). The reaction mixture was quenched with HOAc (4.8 μL, 83.5 μmol, 10.8 equiv.) and concentrated in vacuo. The residue was purified by reverse-phase columns ($H_2O$/MeOH = 2/1 → 1/2) to afford acospectoside A (**1**) (2.5 mg, 43% over 2 steps) as a white amorphous solid. $[\alpha]_D^{25} = -27.0$ (*c* 0.3, $CH_3OH$); $^1H$ NMR (400 MHz, $C_5D_5N$) δ 6.16 (d, *J* = 1.9 Hz, 1H), 5.50 (s, 1H), 5.35 (d, *J* = 18.4 Hz, 1H), 5.31 (d, *J* = 1.7 Hz, 1H), 5.18 (s, 1H), 5.09 (s, 1H), 4.57 (dd, *J* = 11.5, 2.6 Hz, 1H), 4.49 (d, *J* = 2.6 Hz, 1H), 4.44–4.37 (m, 1H), 4.29–4.18 (m, 3H), 4.14 (s, 1H), 4.10 (d, *J* = 6.6 Hz, 1H), 3.99 (q, *J* = 8.5, 7.2 Hz, 2H), 3.78 (t, *J* = 3.1 Hz, 1H), 3.70 (s, 3H), 2.86–2.76 (m, 1H), 2.29–2.22 (m, 1H), 2.21 (s, 3H), 2.17–2.06 (m, 4H), 2.02–1.89 (m, 3H), 1.87–1.79 (m, 2H), 1.74 (d, *J* = 6.6 Hz, 3H), 1.68 (d, *J* = 12.5 Hz, 1H), 1.64–1.54 (m, 1H), 1.47–1.19 (m, 8H), 1.04 (s, 3H), 1.00 (s, 3H); $^{13}C$ NMR (150 MHz, $C_5D_5N$) δ 175.7, 174.4, 170.6, 117.6, 105.1, 99.0, 84.3, 78.4, 78.3, 76.5, 76.4, 75.3, 74.3, 73.5, 71.3, 69.8, 69.4, 67.3, 62.6, 55.9, 51.1, 49.8, 41.5, 39.4, 38.9, 36.7, 32.9, 31.4, 29.9, 28.1, 27.1, 26.1, 21.7, 21.3, 21.0, 18.3, 17.2, 16.0; HR-ESI-MS (*m*/*z*) calcd for $C_{38}H_{58}O_{15}Na$ $[M+Na]^+$: 777.3668, found: 777.3676.

## 4. Conclusions

Here, we report the total synthesis of cytotoxic cardenolides acospectoside A (**1**) and acovensoide B (**2**), which were isolated from the South African poisonous bush *Acokanthera oppositifolia* and relevant plants. The synthesis features the selective introductions of a 15,16-unsaturated enone via the Larock protocol of Saegusa–Ito oxidation, 14β-OH by Mukaiyama hydration, and a C17-butenolide moiety by Stille coupling. Leveraging the strategic 1-*O*-acetylation of the aglycone moiety, glycosylation with imidate donors proceeds with remarkable selectivity, yielding glycosylated products in satisfactory amounts. Given the structural conservation inherent in the acovenoside family, the present synthetic approach should offer a promising gateway to accessing a broader array of the acovenoside congeners. This advancement stands to facilitate comprehensive investigations into the biological and pharmacological activities of these components, which have long been esteemed in traditional medicinal practices.

**Supplementary Materials:** The following supporting information can be downloaded at https://www.mdpi.com/article/10.3390/molecules30112297/s1: Copies of the NMR spectra of compounds **1–28**; Figure S1: $^1H$ NMR spectrum of compound **7** ($CDCl_3$, 500 MHz); and Table S1: Comparison of the Spectroscopic Data of Natural and Synthetic Acovenoside B (**2**).

**Author Contributions:** Conceptualization, P.X. and B.Y.; methodology, B.L.; formal analysis, B.L.; investigation, B.L.; resources, B.Y.; data curation, B.L.; writing—original draft preparation, B.L. and P.X.; writing—review and editing, B.Y.; visualization, B.L.; supervision, P.X. and B.Y.; project administration, P.X. and B.Y.; funding acquisition, B.Y. All authors have read and agreed to the published version of the manuscript.

**Funding:** This research was funded by the National Natural Science Foundation of China (22031011), the Strategic Priority Research Program of the Chinese Academy of Sciences (XDB1060000), the Shanghai Municipal Science and Technology Major Project, and the Hangzhou leading innovation and entrepreneurship team project (TD2022002).

**Institutional Review Board Statement:** This study does not involve humans or animals.

**Informed Consent Statement:** Informed consent was obtained from all subjects involved in this study.

**Data Availability Statement:** The data presented in this study are available in the article and Supplementary Material.

**Conflicts of Interest:** The authors declare no conflicts of interest.

## References


1. Radford, D.J.; Gillies, A.D.; Hinds, J.A.; Duffy, P. Naturally occurring cardiac glycosides. *Med. J. Aust.* **1986**, *144*, 540–544. [CrossRef] [PubMed]
2. Albrecht, H.P. Cardiac glycosides. In *Naturally Occurring Glycosides*; Ikan, R., Ed.; John Wiley & Sons Ltd.: New York, NY, USA, 1999; p. 83.
3. Prassas, I.; Diamandis, E.P. Novel therapeutic applications of cardiac glycosides. *Nat. Rev. Drug Discov.* **2008**, *7*, 926–935. [CrossRef] [PubMed]
4. Botelho, A.F.M.; Pierezan, F.; Soto-Blanco, B.; Melo, M.M. A review of cardiac glycosides: Structure, toxicokinetics, clinical signs, diagnosis and antineoplastic potential. *Toxicon* **2019**, *158*, 63–68. [CrossRef] [PubMed]
5. Hou, Y.; Shang, C.; Meng, T.; Lou, W. Anticancer potential of cardiac glycosides and steroid-azole hybrids. *Steroids* **2021**, *171*, 108852. [CrossRef]
6. Diederich, M.; Muller, F.; Cerella, C. Cardiac glycosides: From molecular targets to immunogenic cell death. *Biochem. Pharmacol.* **2017**, *125*, 1–11. [CrossRef]
7. von Euw, J.; Reichstein, T. Acovenosid A und acovensoid B, zwei glykoside aus den samen von acokanthera venenata G. Don. Erste Mitteilung. Glykoside und aglykone, 53. Mitteilung. *Helv. Chim. Acta* **1950**, *33*, 485–502.
8. Kapadia, G.J. Acospectoside A, a new cardenolide glycoside. *J. Pharm. Sci.* **1965**, *54*, 1834–1835. [CrossRef]
9. Kapadia, G.J. Structural elucidation of acospectoside A. *Lloydia* **1964**, *27*, 272.
10. Kapadia, G.J. Acospectoside A II: The structure of the cardenolide glycoside. *J. Pharm. Sci.* **1969**, *58*, 1555–1557. [CrossRef]
11. Kapadia, G.J. Acospectoside A III: Selective conversion into acovenoside B using snail enzyme with inhibited esterase activity. *J. Pharm. Sci.* **1970**, *59*, 723–724. [CrossRef]
12. Gaafary, M.E.; Ezzat, S.M.; Sayed, A.M.E.; Sabry, O.M.; Hafner, S.; Lang, S.; Schmiech, M.; Syrovets, T.; Simmet, T. Acovenoside A induces mitotic catastrophe followed by apoptosis in non-small-cell lung cancer cells. *J. Nat. Prod.* **2017**, *80*, 3203–3210. [CrossRef] [PubMed]
13. Heasley, B. Chemical synthesis of the cardiotonic steroid glycosides and related natural products. *Chem. Eur. J.* **2012**, *18*, 3092–3120. [CrossRef] [PubMed]
14. Urabe, D.; Nakagawa, Y.; Mukai, K.; Fukushima, K.-i.; Aoki, N.; Itoh, H.; Nagatomo, M.; Inoue, M. Total synthesis and biological evaluation of 19-hydroxysarmentogenin-3-*O*-α-L-rhamnoside, trewianin, and their aglycons. *J. Org. Chem.* **2018**, *83*, 13888–13910. [CrossRef] [PubMed]
15. Bhattarai, B.; Nagorny, P. Enantioselective Total synthesis of cannogenol-3-*O*-α-L-rhamnoside via sequential Cu(II)-catalyzed michael addition/intramolecular aldol cyclization reactions. *Org. Lett.* **2018**, *20*, 154–157. [CrossRef]
16. Carney, N.; Perry, N.; Garabedian, J.; Nagorny, P. Development of α-selective glycosylation with L-oleandral and its application to the total synthesis of oleandrin. *Org. Lett.* **2023**, *25*, 966–971. [CrossRef]
17. Liu, B.; Bi, S.; Wang, J.; Xu, P.; Yu, B. Synthesis of acovenosides: Cardiac glycosides with potent antitumor activities. *Org. Lett.* **2024**, *26*, 8725–8729. [CrossRef]
18. Wang, R.; Xiao, L.; Pan, J.; Bao, G.; Zhu, Y.; Zhu, D.; Wang, J.; Pei, C.; Ma, Q.; Fu, X.; et al. Natural product P57 induces hypothermia through targeting pyridoxal kinase. *Nat. Commun.* **2023**, *14*, 5984. [CrossRef]
19. Zhang, Z.; Wu, R.; Cao, S.; Li, J.; Huang, G.; Wang, H.; Yang, T.; Tang, W.; Xu, P.; Yu, B. Merging total synthesis and NMR technology for deciphering the realistic structure of natural 2,6-dideoxyglycosides. *Sci. Adv.* **2024**, *10*, eadn1305. [CrossRef]
20. Tückmantel, W.; Kozikowski, A.P.; Romanczyk, L.J. Studies in polyphenol chemistry and bioactivity. 1. Preparation of building blocks from (+)-Catechin. Procyanidin formation. Synthesis of the cancer cell growth inhibitor, 3-*O*-galloyl-(2*R*,3*R*)-epicatechin-4*β*,8-[3-*O*-galloyl-(2*R*,3*R*)-epicatechin]. *J. Am. Chem. Soc.* **1999**, *121*, 12073–12081. [CrossRef]
21. Ito, Y.; Hirao, T.; Saegusa, T. Synthesis of α,β-unsaturated carbonyl compounds by palladium(II)-catalyzed dehydrosilylation of silyl enol ethers. *J. Org. Chem.* **1978**, *43*, 1011–1013. [CrossRef]
22. Larock, R.C.; Hightower, T.R.; Kraus, G.A.; Hahn, P.; Zhang, D. A simple, effective, new, palladium-catalyzed conversion of enol silanes to enones and enals. *Tetrahedron Lett.* **1995**, *36*, 2423–2426. [CrossRef]

23. Tamiya, M.; Takada, F.; Isaka, N.; Iimura, N.; Ishiguro, M. Modification of D-ring moiety of steroids—A novel palladium catalyzed Bayer–Villiger type rearrangement of cyclic silylenol ether derivatives. *Hetrocycles* **2011**, *82*, 1119–1125.
24. Isayama, S.; Mukaiyama, T. A new method for preparation of alcohols from olefins with molecular oxygen and phenylsilane by the use of bis(acetylacetonato)cobalt(II). *Chem. Lett.* **1989**, *18*, 1071–1074. [CrossRef]
25. Hanna, A.G.; Elgamal, M.H.A.; Hassan, A.Z.; Duddeck, H.; Simon, A.; Kovács, J.; Tóth, G. Complete $^{1}$H and $^{13}$C signal assignments of 5β-cardenolides isolated from *Acokanthera spectabilis* Hook F. *Magn. Reson. Chem.* **1998**, *36*, 936–942. [CrossRef]

*molecules*

MDPI

*Article*

# Synthesis of Azide-Labeled β-Lactosylceramide Analogs Containing Different Lipid Chains as Useful Glycosphingolipid Probes †

**Basant Mohamed [1,2], Rajendra Rohokale [1], Xin Yan [1], Amany M. Ghanim [2], Nermine A. Osman [2], Hanan A. Abdel-Fattah [2] and Zhongwu Guo [1,3,*]**

1 Department of Chemistry, University of Florida, Gainesville, FL 32611, USA; bmhusin@zu.edu.eg (B.M.); rajendra.rohokal@chem.ufl.edu (R.R.); x.yan@chem.ufl.edu (X.Y.)
2 Department of Pharmaceutical Organic Chemistry, Faculty of Pharmacy, Zagazig University, Zagazig 44519, Egypt; ammettwalli@pharmacy.zu.edu.eg (A.M.G.); namohamd@pharmacy.zu.edu.eg (N.A.O.); hahamied@pharmacy.zu.edu.eg (H.A.A.-F.)
3 UF Health Cancer Center, University of Florida, Gainesville, FL 32611, USA
* Correspondence: zguo@chem.ufl.edu

† This article is dedicated to the late Professor Hans Paulsen who has made seminal contributions to the development of carbohydrate chemistry.

**Abstract**

β-Lactosylceramide (β-LacCer) is not only a key intermediate in the biosynthesis of complex glycosphingolipids (GSLs) but also an important regulator of many biological processes. To facilitate the investigation of β-LacCer and other GSLs, a series of β-LacCer analogs with an azido group at the 6-C-position of the D-galactose in lactose and varied forms of the ceramide moiety were synthesized from commercially available lactose in sixteen linear steps by a versatile and diversity-oriented strategy, which engaged lipid remodeling and glycan functionalization at the final stage. These azide-labeled β-LacCer analogs are flexible and universal platforms that are suitable for further functionalization with other molecular tags via straightforward and biocompatible click chemistry, thereby paving the way for their application to various biological studies.

**Keywords:** carbohydrates; glycolipids; glycosphingolipids; β-Lactosylceramide; lipids; lactose; azide; synthesis

## 1. Introduction

Glycosphingolipids (GSLs) are ubiquitous and abundant glycolipids in the cell membrane [1] that are composed of a hydrophilic glycan head and a lipophilic ceramide (Cer) tail coupled together by a glycosidic bond [2,3]. The amphiphilic property of GSLs enables them to adhere to the cell membrane outer leaflet and traffic through the lipid bilayer [4,5]. Inside the cell membrane, GSLs tend to self-aggregate in specific microdomains of the lipid rafts, where signaling molecules are localized and enriched [6,7]. As a result, GSLs play an important role in numerous biophysiological processes, such as cellular recognition and signaling, cell differentiation and proliferation, etc. [8]. GSL dysregulation is also associated with many human diseases, including cancer and Alzheimer's disease [9,10]. For example, a number of GSLs have been identified as tumor-associated carbohydrate antigens (TACAs) and utilized as molecular targets for the development of therapeutic cancer vaccines [9,11].

Natural GSLs are structurally diverse, with potential structural variations in both the glycan and the Cer moiety [1]. β-Lactosylceramide (β-LacCer) is a simple GSL having the D-glucose (Glc) residue of lactose, a disaccharide, β-linked to the Cer moiety. In mammals, it is the key intermediate during the biosynthesis of the majority of complex GSLs, e.g., *globo-*, *isoglobo-*, *ganglio-*, *lacto-*, and *neolecto*-series GSLs (Figure 1A) [12,13]. In the meantime, it has been demonstrated that β-LacCer is involved in the human central nervous system, immune system, and diseases like cancer, inflammation, and neurodegeneration [14–16]. For instance, β-LacCer overexpression is found in leukemia, renal cancer, and cholangiocarcinoma [17], and LacCer accumulation is also related to atherosclerosis and aberrant autophagy [18,19]. Thus, β-LacCer is identified as a drug target [20]. Additionally, similar to other GSLs, the lipids in β-LacCer usually contain 16–20 carbons (C16–20) (such as C18, Figure 1B), while the forms of β-LacCer with longer (>C24) lipid chains are also found in neutrophils, which become molecular targets for the treatment of immunological disorders [21–23].

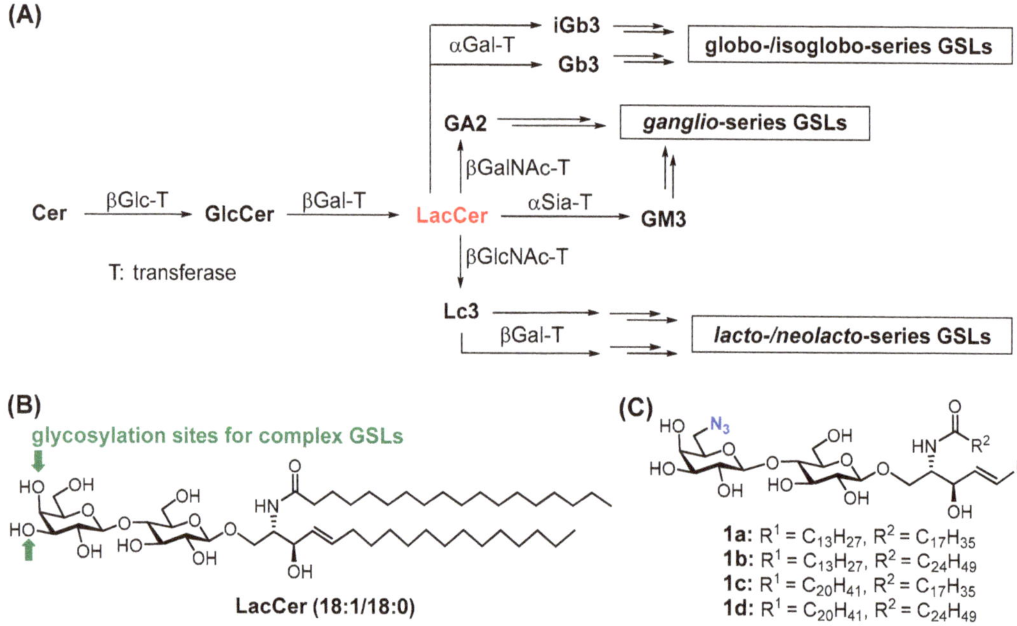


**Figure 1.** (**A**) The biosynthetic pathways for *globo-*, *isoglobo-*, *ganglio-*, *lacto-*, and *neolecto*-series GSLs, utilizing β-LacCer as the key intermediate, and some key enzymes involved, and structures of (**B**) a representative β-LacCer with the most common (18:1/18:0) lipid form found in mammals and (**C**) designed β-LacCer analogs with different lipid forms as useful probes to study β-LacCer and other GSLs.

However, detailed investigation of β-LacCer, especially within complex matrices like the lipid membrane, is challenging due to the lack of fluorophores and other visible functionalities in its structure. To address the problem, it is necessary to functionalize β-LacCer with molecular tags (e.g., affinity and fluorescent tags) to facilitate various modern analytical technologies. In this context, we designed a series of β-LacCer analogs **1a–d** (Figure 1C) containing an azido group and diverse lipid chains and synthesized them by a diversity-oriented strategy. These β-LacCer analogs should be useful molecular tools.

## 2. Results and Discussion

In the designed β-LacCer probes **1a–d** (Figure 1C), we planned to attach an azido group to the 6-*C*-position of D-galactose (Gal) in the lactose moiety for several reasons. First, the azido group is a flexible platform to facilitate further functionalization of **1a–d** with various molecular labels through straightforward and biocompatible click chemistry [24,25]. Next, the azido group is small—not significantly larger than a hydroxyl group, and it is expected to have a minimal impact on the properties of β-LacCer. Consequently, if **1a–d** are used as biosynthetic precursors for metabolic glycoengineering, they are likely to be acceptable by enzymes. Moreover, the 6-*C*-position of the Gal residue in lactose is distinct, because it is a primary carbon at the non-reducing end and is probably the least sterically hindered position, making it relatively efficient and straightforward to functionalize this position and subsequently modify the synthetic probes. The designed probes also contain different forms of lipids, which will facilitate the investigation of how lipid structures of β-LacCer influence its chemical, biophysical, and biological properties.

Our synthesis of the target molecules **1a–d** (Scheme 1) started with commercially available and inexpensive lactose. It was first converted into the *O*-acetyl and 4′,6′-*O*-benzylidene protected glycoside of *p*-toluenethiol (TolSH) **3** as an α,β-isomeric mixture via a series of well-established reactions reported in the literature [26–29]. Next, the 4′,6′-*O*-benzylidene group in **3** was selectively removed upon treatment with 80% acetic acid in water at 85 °C to afford diol **4** as an anomeric mixture. Selective tosylation of the 6′-OH group in **4** was achieved by its reaction with *p*-toluenesulfonylchloride (*p*-TsCl) and 4-dimethylaminopyridine (DMAP). The regioselective attack of *p*-TsCl to the 6′-OH group is probably because this primary hydroxyl group is less sterically hindered than the secondary 4′-OH group in **4**. The equivalence and concentration of *p*-TsCl, along with the order to add reagents, are also substantial in the control of this regioselectivity. The formation of **5** was confirmed by the downfield shifts of its 6′-H NMR signals, compared to those of **4** (from δ 4.06 and 3.72 to 4.41 and 4.07 ppm), and the appearance of $^1$H signals of an additional toluene moiety at δ 7.40–7.35 and 2.47 ppm, respectively, in the $^1$H NMR spectrum of **5**. The free hydroxyl group remaining in **5** was then acetylated using acetic anhydride, pyridine, and DMAP. To reveal the reducing-end anomeric position for glycosidation, the resultant **6** was applied to oxidative hydrolysis with *N*-bromosuccinimide (NBS) in water and acetone (1:4) to afford hemiacetal **7** (63%) as an anomeric mixture, together with unreacted **6** (34%). Next, **7** was converted into imidate **8** as a glycosyl donor upon reacting with trichloroacetonitrile catalyzed by 1,8-diazabicyclo-[5.4.0]-undec-7-ene (DBU). After brief purification, the reactive glycosyl imidate **8** was immediately submitted to the glycosylation reaction with Cer precursor **9** [30,31], utilizing trimethylsilyltrifluoromethanesulfonate (TMSOTf) as the promotor. This reaction produced **10** as the key synthetic intermediate in excellent stereoselectivity, and the β-configuration of its newly formed glycosidic bond is confirmed by the large coupling constant ($J$ = 7.8 Hz) of its anomeric proton at δ 4.43 in the $^1$H NMR spectrum. The common intermediate **10** was then used for diversity-oriented synthesis of all β-LacCer probes **1a–d**.

The final steps for **1a–d** synthesis include lipid remodeling, glycan functionalization, and global deprotection, as outlined in Scheme 2. Olefin cross-metathesis of pentadecene **11a** or docosene **11b** with **10** in the presence of the 2nd generation Hoveyda−Grubbs catalyst [32] allowed on-site generation of the sphingosine moiety of varied chain lengths in very good yields (80–83%), although the reaction was rather slow, taking six days to complete [31]. The *E*-configuration of the C=C bond in **12a** and **12b** was verified by the olefinic $^1$H-$^1$H coupling constant ($J$ = 14.6–14.8 Hz) in their $^1$H NMR spectra. To attach the *N*-fatty acyl chain, the *tert*-butoxycarbonyl (Boc) group in **12a,b** was selectively removed with 10% trifluoroacetic acid (TFA) in dichloromethane (DCM) to afford the intermediate free amines, which were directly subjected to *N*-acylation employing stearic acid or pentacosanoic acid, with *N*-(3-dimethylaminopropyl)-*N′*-ethylcarbodiimide hydrochloride (EDC) and DMAP as the condensation reagents. These reactions gave 6′-*O*-tosylated β-LacCer derivatives **13a–d** containing different lipid chains in good overall yields (63–65%, two steps). Subsequently, nucleophilic substitution of the tosylate in **13a–d** with sodium azide in dimethylformamide (DMF) proceeded smoothly to provide fully protected LacCer analogs **14a–d** in high yields (69–80%). Eventually, all *O*-acyl groups in **14a–d** were removed with sodium methoxide in methanol and DCM (1:1) to produce the synthetic targets **1a–d** in 80–90% yields. The final products, as well as all new intermediates involved in these syntheses, have been fully characterized with NMR and high-resolution mass spectrometry (HR-MS) data (see data and Figures in Supporting Information).

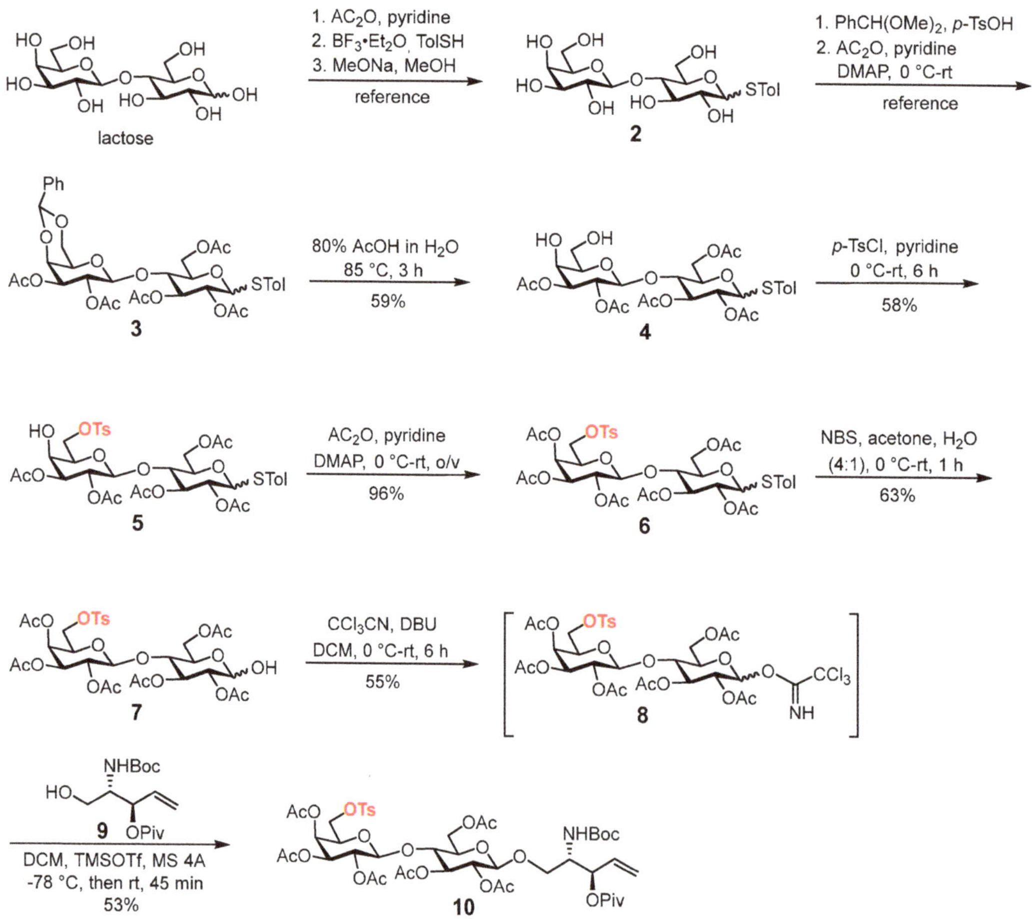


**Scheme 1.** Preparation of **10**—the key and common building block for the synthesis of all designed probes in the present work.

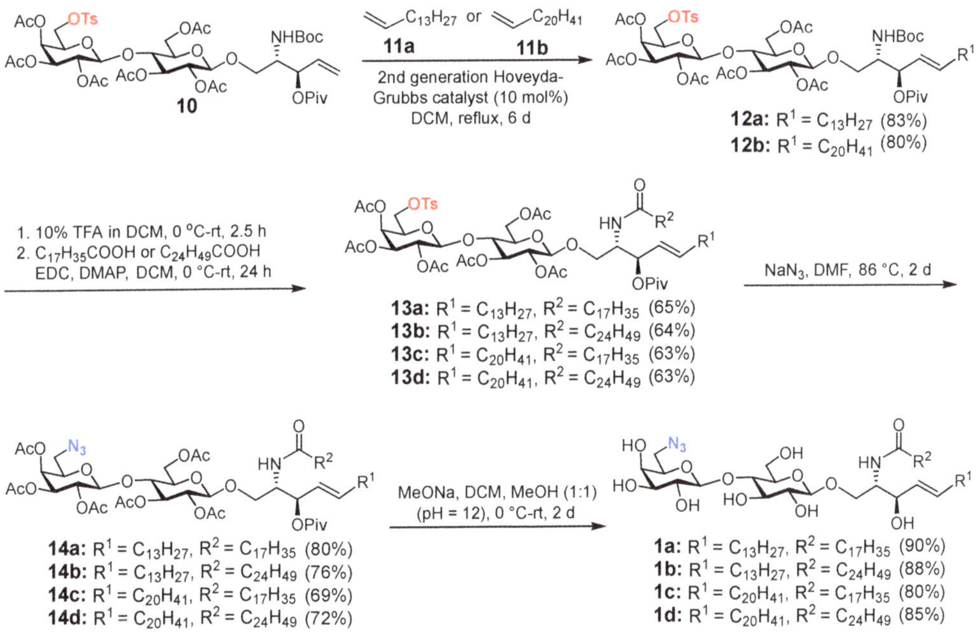

**Scheme 2.** Assembly of the designed β-LacCer probes **1a–d**.

## 3. Experimental Section

### 3.1. General Methods

Chemicals, reagents, and solvents were commercial and used without further purification. Molecular sieves 4 Å (MS 4Å) were flame-dried under a high vacuum and used after being cooled to room temperature (rt) under an $N_2$ atmosphere. Thin-layer chromatography (TLC) was performed on silica gel 60 Å F254 plates, with detection utilizing a UV lamp at a wavelength of 254 nm followed by charring with 10% (*v*/*v*) $H_2SO_4$ in ethanol or ninhydrin and anisaldehyde staining. Flash column chromatography was performed using silica gel 60 (230–400 mesh). NMR spectra were acquired on a 400, 500, or 600 MHz spectrometer, with chemical shifts (δ) reported in ppm as referenced to internal tetramethylsilane (TMS) ($^1$H NMR: δ 0.00 ppm), $CDCl_3$ ($^1$H NMR: δ 7.26 ppm; $^{13}$C{$^1$H} NMR: δ 77.2 ppm), or $CD_3OD$ ($^1$H NMR: δ 3.31 ppm; $^{13}$C{$^1$H} NMR: δ 49.0 ppm). Coupling constants (*J*) were reported in Hz. Peak, and coupling constant assignments were based upon $^1$H NMR, $^1$H–$^1$H COSY, and $^1$H–$^{13}$C{$^1$H} HSQC experiments. HR-MS spectra were recorded on the XEVO-G2-XS Q-TOF-ESI instrument. An aluminum heating block was used for heating reaction mixtures. Intermediates **2**, **3**, and **9** were prepared according to the reported protocols, and their $^1$H NMR spectra were the same as those in the literature [26–29].

### 3.2. *p-Methylphenyl 2,3-di-O-acetyl-β-D-galactopyranosyl-(1→4)-2,3,6-tri-O-acetyl-1-thio-β-D-glucopyranoside (**4**)*

A solution of **3** (51.5 g, 68.96 mmol) in 80% aqueous AcOH was heated at 85 °C using an aluminum block for 3 h. After the reaction was complete, as indicated by the consumption of all starting material on TLC, the solvent was removed under reduced pressure followed by co-evaporation with toluene three times. Thereafter, ethyl acetate (EtOAc) was added, and the solution was washed with $NaHCO_3$, water, and brine multiple times. The organic layer was dried over $MgSO_4$, filtered, and concentrated under vacuum.

The crude product was purified using silica gel column chromatography to afford **4** (59%, 26.8 g) as a white solid. TLC: $R_f$ = 0.61 (EtOAc:Hex 80:20). $^1$H NMR (500 MHz, $CDCl_3$): δ 7.35 (d, *J* = 8.0 Hz, 2H), 7.09 (d, *J* = 8.0 Hz, 2H), 5.31–5.02 (m, 2H), 4.95–4.79 (m, 2H), 4.61 (d, *J* = 10.1 Hz, 1H, anomeric), 4.53 (dd, *J* = 11.7, 1.9 Hz, 1H), 4.46 (d, *J* = 7.9 Hz, 1H, anomeric), 4.08 (dt, *J* = 11.9, 8.5 Hz, 2H), 3.93–3.72 (m, 3H), 3.68–3.48 (m, 2H), 3.45–3.30 (m, 1H), 2.79 (dd, *J* = 7.1, 4.7 Hz, 1H), 2.33 (s, 3H), 2.09 (s, 3H), 2.07 (s, 3H), 2.05 (s, 6H, 2 × $CH_3$), 2.02 (s, 3H). $^{13}$C NMR (126 MHz, $CDCl_3$): δ 170.8, 170.6, 170.4, 169.71, 169.7, 138.7, 133.7, 129.7, 127.8, 101.1, 85.6, 76.7, 76.3, 74.5, 73.6, 70.4, 70.3, 69.8, 67.8, 62.4, 62.1, 21.3, 21.01, 21.0, 20.9, 20.8. HR-ESI-MS *m/z*: $[M + H_2O + H]^+$ Calcd for $C_{29}H_{41}O_{16}S$ 677.2115; Found 677.2106.

### 3.3. *p-Methylphenyl 2,3-di-O-acetyl-6-O-(p-toluenesulfonyl)-β-D-galactopyranosyl-(1→4)-2,3,6-tri-O-acetyl-1-thio-β-D-glucopyranoside* (**5**)

To a solution of **4** (20.5 g, 31.12 mmol) dissolved in pyridine (93.2 mL) was added *p*-TsCl (5.93 g, 31.12 mmol) portionwise under an $N_2$ atmosphere in an ice bath. The mixture was stirred at rt for 6 h, at which point the reaction did not show further progress. Next, MeOH was added dropwise to the reaction mixture, and the solvent was removed under vacuo. The residue was extracted with EtOAc and washed twice with brine. The organic layer was dried over $MgSO_4$ and concentrated under reduced pressure. Purification of the product by silica gel column chromatography afforded **5** (58%, 14.67 g) as a white waxy solid. TLC: $R_f$ = 0.95 (EtOAc:Hex 80:20). $^1$H NMR (400 MHz, $CDCl_3$): δ 7.78 (d, *J* = 8.3 Hz, 2H), 7.40–7.35 (m, 4H), 7.10 (d, *J* = 8.0 Hz, 2H), 5.15 (t, *J* = 9.1 Hz, 1H), 5.09 (dd, *J* = 10.2, 7.9 Hz, 1H), 4.92–4.86 (m, 1H), 4.83 (d, *J* = 9.4 Hz, 1H), 4.58 (d, *J* = 10.1 Hz, 1H, anomeric), 4.50 (dd, *J* = 11.9, 1.9 Hz, 1H), 4.41 (d, *J* = 7.9 Hz, 1H, anomeric), 4.24 (dd, *J* = 10.4, 6.5 Hz, 1H), 4.12–4.05 (m, 2H), 4.03 (t, *J* = 4.4 Hz, 1H), 3.76 (t, *J* = 6.4 Hz, 1H), 3.68 (t, *J* = 9.5 Hz, 1H), 3.58 (ddd, *J* = 10.0, 5.3, 1.9 Hz, 1H), 2.47 (s, 3H), 2.33 (s, 3H), 2.10 (s, 3H), 2.08 (s, 3H), 2.06 (s, 3H), 2.01 (s, 3H), 1.90 (s, 3H). $^{13}$C NMR (101 MHz, $CDCl_3$): δ 170.5, 170.2, 170.1, 169.6, 169.5, 145.5, 138.7, 133.7, 132.6, 130.2, 129.7, 128.0, 127.9, 100.8, 85.6, 76.7, 75.9, 74.0, 73.2, 72.3, 70.3, 69.5, 67.1, 66.5, 62.2, 21.8, 21.3, 20.92, 20.91, 20.9, 20.8, 20.7. HR-ESI-MS *m/z*: $[M + NH_4]^+$ Calcd for $C_{36}H_{48}NO_{17}S_2$ 830.2364; Found 830.2354

### 3.4. *p-Methylphenyl 2,3,4-tri-O-acetyl-6-O-(p-toluenesulfonyl)-β-D-galactopyranosyl-(1→4)-2,3,6-tri-O-acetyl-1-thio-β-D-glucopyranoside* (**6**)

After **5** (14.5 g, 17.84 mmol) was dissolved in pyridine (90 mL), $Ac_2O$ (2.53 mL, 26.76 mmol) and DMAP (0.4 g, 3.57 mmol) were added at 0 °C under an $N_2$ atmosphere. The mixture was stirred at rt overnight. After the complete consumption of **5** as indicated by TLC, MeOH was added dropwise to quench the reaction. The solvent was evaporated under vacuo. The resulting residue was extracted with EtOAc, and the organic layer was washed three times with $H_2O$ and brine, dried over $MgSO_4$, and finally, concentrated under vaccum. The residue was purified by silica gel column chromatography to afford **6** (96%, 14.6 g) as a white waxy solid. TLC: $R_f$ = 0.66 (DCM:MeOH 95:5). $^1$H NMR (400 MHz, $CDCl_3$): δ 7.75 (d, *J* = 8.3 Hz, 2H), 7.36 (dd, *J* = 8.1, 1.7 Hz, 4H), 7.10 (d, *J* = 7.9 Hz, 2H), 5.32 (dd, *J* = 3.4, 0.8 Hz, 1H), 5.18 (t, *J* = 9.1 Hz, 1H), 5.04 (dd, *J* = 10.4, 7.9 Hz, 1H), 4.91 (dd, *J* = 10.4, 3.4 Hz, 1H), 4.83 (t, *J* = 9.6 Hz, 1H), 4.59 (d, *J* = 10.1 Hz, 1H, anomeric), 4.50 (dd, *J* = 11.9, 2.0 Hz, 1H), 4.44 (d, *J* = 7.9 Hz, 1H, anomeric), 4.08–3.95 (m, 3H), 3.88 (dd, *J* = 7.5, 6.5 Hz, 1H), 3.70 (t, *J* = 9.5 Hz, 1H), 3.59 (ddd, *J* = 10.0, 5.2, 1.9 Hz, 1H), 2.46 (s, 3H), 2.33 (s, 3H), 2.10 (s, 3H), 2.09 (s, 3H), 2.04 (s, 3H), 2.01 (s, 3H), 1.95 (s, 3H), 1.94 (s, 3H). $^{13}$C NMR (101 MHz, $CDCl_3$): δ 170.4, 170.01, 170.0, 169.8, 169.6, 169.1, 145.6, 138.8, 133.8, 132.3, 130.2, 129.8, 128.1, 127.8, 100.9, 85.6, 76.7, 76.1, 74.0, 70.91, 70.9, 70.4, 69.1, 66.6, 65.5, 62.1, 21.8, 21.3,

21.01, 21.0, 20.9, 20.7, 20.61, 20.6. HR-ESI-MS *m/z*: $[M + HCOO]^-$ Calcd for $C_{39}H_{47}O_{20}S_2$ 899.2103; Found 899.2074.

### *3.5. 2,3,4-Tri-O-acetyl-6-O-(p-toluenesulfonyl)-β-D-galactopyranosyl-(1→4)-2,3,6-tri-O-acetyl-D-glucopyranose (**7**)*

To a solution of **6** (8.0 g, 9.36 mmol) dissolved in acetone (160 mL) and $H_2O$ (40 mL) was added NBS (3.33 g, 18.72 mmol) at 0 °C. The mixture was stirred at rt for 1 h, while the reaction was monitored by TLC. Once no further progress was observed with TLC, the reaction was quenched by adding solutions of $Na_2S_2O_3$ and $NaHCO_3$ dropwise. Acetone was removed under vacuum and then saturated aq. $NaHCO_3$ solution was added to the residue. The mixture was extracted with DCM three times. The organic layers were combined, dried over $MgSO_4$, and concentrated under vacuum. The residue was purified by silica gel column chromatography to afford **7** (63%, 4.40 g) as a white solid, along with recovery of **6** (34%, 2.70 g). TLC: $R_f = 0.25$ (EtOAc:Hex 60:40). $^1$H NMR (400 MHz, $CDCl_3$): δ 7.77 (d, *J* = 8.3 Hz, 2H), 7.37 (d, *J* = 8.4 Hz, 2H), 5.54 (t, *J* = 9.7 Hz, 1H), 5.38 (t, *J* = 3.4 Hz, 1H), 5.32 (d, *J* = 2.5 Hz, 1H), 5.09 (dd, *J* = 10.4, 7.9 Hz, 1H), 4.92 (dd, *J* = 10.4, 3.4 Hz, 1H), 4.83 (dd, *J* = 10.1, 3.4 Hz, 1H, anomeric), 4.53 (d, *J* = 7.9 Hz, 1H, anomeric), 4.48 (ddd, *J* = 10.4, 7.3, 2.3 Hz, 2H), 4.23 (ddd, *J* = 10.1, 4.1, 1.9 Hz, 1H), 4.12 (dd, *J* = 12.0, 4.1 Hz, 1H), 4.07 (dd, *J* = 10.6, 4.2 Hz, 2H), 3.96 (t, *J* = 3.3 Hz, 1H), 3.94–3.90 (m, 1H), 3.79 (t, *J* = 9.7 Hz, 1H), 2.13 (s, 3H), 2.09 (s, 3H), 2.06 (s, 3H), 2.03 (s, 3H), 2.01 (s, 3H), 1.94 (s, 3H). $^{13}$C NMR (101 MHz, $CDCl_3$): δ 170.7, 170.4, 170.1, 169.8, 169.2, 169.1, 132.3, 130.2, 129.8, 128.2, 100.6, 90.3, 77.4, 76.3, 71.6, 71.0, 69.3, 69.1, 68.3, 66.8, 66.0, 62.0, 21.8, 21.1, 21.0, 20.9, 20.8, 20.7, 20.6. HR-ESI-MS *m/z*: $[M-H_2O + H]^+$ Calcd for $C_{31}H_{39}O_{18}S$ 731.1857; Found: 731.1833.

### *3.6. 2,3,4-Tri-O-acetyl-6-O-(p-toluenesulfonyl)-β-D-galactopyranosyl-(1→4)-2,3,6-tri-O-acetyl-D-glucopyranosyl trichloroimidate (**8**)*

To a solution of **7** (990 mg, 1.322 mmol) in dry DCM (9.5 mL) were added DBU (0.132 mmol, 20 μL) and trichloroacetonitrile (2.12 mL, 21.15 mmol) dropwise at 0 °C under an $N_2$ atmosphere. The mixture was stirred at rt for 3 h. After the reaction was completed as indicated by TLC, the solvent was removed under reduced pressure, and the residue was purified by flash column chromatography using $Et_3N$-deactivated silica to give **8** (55%, 649 mg). TLC: $R_f = 0.45$ (EtOAc:Hex 50:50). $^1$H NMR (400 MHz, $CDCl_3$): δ 8.66 (s, 1H, NH), 7.77 (d, *J* = 8.2 Hz, 2H), 7.38 (d, *J* = 8.1 Hz, 2H), 6.49 (d, *J* = 3.7 Hz, 1H, anomeric), 5.55 (t, *J* = 9.6 Hz, 1H), 5.33 (d, *J* = 3.3 Hz, 1H), 5.13–5.08 (m, 1H), 5.05 (dd, *J* = 10.2, 3.9 Hz, 1H), 4.93 (dd, *J* = 10.4, 3.4 Hz, 1H), 4.51 (d, *J* = 8.0 Hz, 1H, anomeric), 4.49–4.44 (m, 1H), 4.14–4.05 (m, 1H), 3.99 (dd, *J* = 10.0, 6.6 Hz, 1H), 3.95–3.81 (m, 1H), 2.11 (s, 3H), 2.06 (s, 3H), 2.03 (s, 6H, 2 × $CH_3$), 2.02 (s, 3H), 1.95 (s, 3H). $^{13}$C NMR (101 MHz, $CDCl_3$): δ 170.4, 170.2, 170.1, 170.0, 169.5, 169.2, 161.1, 145.7, 132.3, 130.2, 128.2, 101.3, 93.0, 77.4, 76.0, 71.1, 71.0, 70.2, 69.7, 69.1, 66.7, 65.6, 61.6, 21.8, 21.2, 21.05, 21.0, 20.8, 20.7, 20.6.

### *3.7. (2S,3R)-2-(tert-Butoxycarbonyl)amino-3-pivaloyloxy-pent-4-en-1-yl 2,3,4-tri-O-acetyl-6-O-(p-toluenesulfonyl)-β-D-galactopyranosyl-(1→4)-2,3,6-tri-O-acetyl-D-glucopyranoside (**10**)*

The mixture of glycosyl donor **8** (600 mg, 0.671 mmol), acceptor **9** (168 mg, 0.559 mmol), and flame-dried MS 4Å (1.5 g) in dry DCM (12 mL) was stirred at rt for 15 min under an $N_2$ atmosphere and then cooled to -78 °C. Thereafter, TMSOTf (4.0 μL, 22.4 μmol) was added dropwise followed by stirring at the same temperature (−78 °C) for 15 min. The mixture was gradually warmed to rt and stirred for another 45 min. When TLS indicated the complete consumption of **9**, the reaction was quenched with $Et_3N$ at 0 °C, and the mixture was filtered through a Celite pad. The filtrate was washed with saturated $NaHCO_3$solution, and the organic layer was dried over $MgSO_4$ and concentrated under vacuo. Silica gel column chromatography of the residue afforded **10** (53% yield, 306 mg) as a white solid. TLC: $R_f$ = 0.50 (EtOAc:Hex 50:50).$^1$H NMR (600 MHz, $CDCl_3$): δ 7.77 (d, *J* = 8.3 Hz, 2H), 7.38 (d, *J* = 8.1 Hz, 2H), 5.83–5.73 (m, 1H, =CH), 5.32 (dd, *J* = 11.2, 5.1 Hz, 2H), 5.29 (d, *J* = 1.2 Hz, 1H), 5.26 (d, *J* = 10.7 Hz, 1H), 5.17 (t, *J* = 9.3 Hz, 1H), 5.06 (dd, *J* = 10.3, 7.9 Hz, 1H), 4.92 (dd, *J* = 10.4, 3.4 Hz, 1H), 4.88 (dd, *J* = 9.4, 8.0 Hz, 1H), 4.70 (d, *J* = 8.9 Hz, 1H, -NH), 4.46 (d, *J* = 7.8 Hz, 1H, anomeric), 4.45–4.44 (m, 1H), 4.43 (d, *J* = 7.8 Hz, 1H, anomeric), 4.08–3.97 (m, 4H), 3.94 (dd, *J* = 10.0, 3.4 Hz, 1H), 3.90 (t, *J* = 6.5 Hz, 1H), 3.77 (t, *J* = 9.5 Hz, 1H), 3.63–3.57 (m, 1H), 3.51 (dd, *J* = 10.1, 4.9 Hz, 1H), 2.46 (s, 1H), 2.11 (s, 1H), 2.06 (s, 1H), 2.04 (s, 1H), 2.02 (s, 1H), 1.99 (s, 1H), 1.94 (s, 1H), 1.42 (s, 9H), 1.19 (s, 9H). $^{13}$C NMR (151 MHz, $CDCl_3$): δ 177.1, 170.5, 170.01, 170.0, 169.9, 169.8, 169.7, 169.2, 155.3, 145.7, 132.4, 130.2, 128.2, 118.7, 100.91, 100.9, 79.8, 76.1, 73.3, 72.8, 72.6, 71.8, 70.91, 70.9, 69.1, 68.5, 66.7, 65.6, 62.1, 60.5, 52.2, 39.0, 28.5, 27.2, 21.8, 21.2, 21.0, 20.9, 20.71, 20.7, 20.6. HR-ESI-MS *m*/*z*: $[M + HCOO]^-$ Calcd for $C_{47}H_{66}NO_{25}S$ 1076.3645; Found 1076.3657.

### *3.8. (2S,3R,E)-2-(tert-Butoxycarbonyl)amino-3-pivaloyloxy-octadec-4-en-1-yl 2,3,4-tri-O-acetyl-6-O-(p-toluenesulfonyl)-β-D-galactopyranosyl-(1→4)-2,3,6-tri-O-acetyl-D-glucopyranoside (**12a**)*

To an $N_2$-bubbled mixture of **10** (150 mg, 0.145 mmol) and 1-pentadecene **11a** (183.5 mg, 0.237 mL, 0.872 mmol) in dry DCM (28 mL) was added 2nd generation Hoveyda—Grubbs catalyst (9 mg, 10 mol%). The reaction mixture was refluxed at 40 °C for 6 d. Each day, another batch of **11a** (183.5 mg, 0.237 mL, 0.873 mmol) and the catalyst (4.53 mg, 5 mol%) was added to the mixture. Upon the completion of reaction, DMSO (2 drops) was added, and the mixture was stirred at rt for 40 min. The solvent was removed under vacuo, and the residue was purified by silica gel column chromatography to give **12a** (83%, 146.5 mg) as a brownish solid. TLC: $R_f$ = 0.83 (EtOAc:Hex 80:20). $^1$H NMR (400 MHz, $CDCl_3$): δ 7.78 (d, *J* = 8.3 Hz, 2H), 7.39 (d, *J* = 8.0 Hz, 2H), 5.75 (dt, *J* = 14.6, 7.6 Hz, 1H, =CH-), 5.40–5.30 (m, 2H, =CH- and 4′-H), 5.18 (dd, *J* = 17.4, 8.2 Hz, 2H), 5.06 (dd, *J* = 10.4, 7.9 Hz, 1H), 4.90 (ddd, *J* = 17.3, 9.9, 5.7 Hz, 2H), 4.67 (d, *J* = 9.2 Hz, 1H, NH), 4.46 (d, *J* = 7.8 Hz, 1H, anomeric), 4.43 (d, *J* = 7.7 Hz, 1H, anomeric), 4.10–3.96 (m, 4H), 3.91 (dd, *J* = 13.5, 7.8 Hz, 2H), 3.77 (t, *J* = 9.4 Hz, 1H), 3.59 (ddd, *J* = 9.8, 4.7, 1.7 Hz, 1H), 3.49 (dd, *J* = 9.7, 4.4 Hz, 1H), 2.47 (s, 3H), 2.11 (s, 3H), 2.07 (s, 3H), 2.06 (s, 3H), 2.03 (s, 3H), 1.99 (s, 3H), 1.95 (s, 3H), 1.42 (s, 9H), 1.30–1.21 (m, 24H), 1.17 (s, 9H), 0.88 (t, *J* = 6.8 Hz, 3H). $^{13}$C NMR (101 MHz, $CDCl_3$): δ 177.0, 171.3, 170.5, 170.01, 170.0, 169.9, 169.8, 169.2, 145.6, 137.0, 132.4, 130.2, 128.2, 124.6, 100.91, 100.9, 77.4, 76.1, 73.3, 72.71, 72.7, 71.8, 70.9, 69.1, 68.61, 68.6, 66.6, 65.5, 62.1, 60.5, 52.3, 38.9, 32.5, 32.1, 29.81, 29.8, 29.7, 29.5, 29.3, 29.1, 28.5, 27.2, 22.8, 21.8, 21.2, 21.0, 20.9, 20.8, 20.71, 20.7, 20.6, 14.31, 14.3. HR-ESI-MS *m*/*z*: $[M + HCOO]^-$ Calcd for $C_{60}H_{92}NO_{25}S$ 1258.5680; found: 1258.5686

### *3.9. (2S,3R,E)-2-(tert-Butoxycarbonyl)amino-3-pivaloyloxy-pentacos-4-en-1-yl 2,3,4-tri-O-acetyl-6-O-(p-toluenesulfonyl)-β-D-galactopyranosyl-(1→4)-2,3,6-tri-O-acetyl-D-glucopyranoside (**12b**)*

Compound **12b** (80%, 152.6 mg) was prepared as a blackish brown solid from **11b** by the same method and conditions utilized to prepare **12a**. TLC: $R_f$ = 0.92 (EtOAc:Hex 80:20). $^1$H NMR (600 MHz, $CDCl_3$): δ 7.78 (d, *J* = 8.3 Hz, 2H), 7.38 (d, *J* = 8.1 Hz, 2H), 5.75 (dt, *J* = 14.6, 7.6 Hz, 1H, =CH-), 5.40–5.30 (m, 2H, =CH- and 4′-H), 5.18 (dd, *J* = 17.3, 8.1 Hz, 2H), 5.06 (dd, *J* = 10.4, 7.9 Hz, 1H), 4.89 (ddd, *J* = 17.3, 9.9, 5.7 Hz, 2H), 4.66 (d, *J* = 8.9 Hz, 1H, NH), 4.46 (d, *J* = 7.8 Hz, 1H, anomeric), 4.45–4.44 (m, 1H), 4.43 (d, *J* = 7.6 Hz, 1H, anomeric),

4.05 (dd, $J$ = 11.1, 5.0 Hz, 2H), 3.99 (dd, $J$ = 10.2, 6.6 Hz, 2H), 3.91 (dd, $J$ = 13.4, 7.5 Hz, 2H), 3.77 (t, $J$ = 9.4 Hz, 1H), 3.59 (ddd, $J$ = 9.6, 4.6, 1.5 Hz, 1H), 3.48 (dd, $J$ = 9.7, 4.4 Hz, 1H), 2.47 (s, 3H), 2.10 (s, 3H), 2.06 (s, 6H), 2.02 (s, 3H), 1.99 (s, 3H), 1.94 (s, 3H), 1.42 (s, 9H), 1.36–1.20 (m, 38H), 1.17 (s, 9H), 0.88 (t, $J$ = 6.8 Hz, 3H). $^{13}$C NMR (151 MHz, $CDCl_3$): δ 177.0, 170.5, 170.01, 170.0, 169.9, 169.81, 169.8, 169.2, 145.6, 137.0, 132.4, 130.2, 128.1, 124.6, 100.91, 100.9, 76.1, 73.3, 72.71, 72.7, 71.8, 70.9, 69.0, 68.61, 68.6, 66.6, 65.5, 62.1, 52.3, 38.9, 38.4, 35.9, 32.5, 32.1, 29.82, 29.81, 29.8, 29.6, 29.5, 29.3, 29.1, 28.5, 27.21, 27.2, 22.8, 21.8, 21.0, 20.9, 20.8, 20.7, 20.61, 20.6, 14.3. HR-ESI-MS $m/z$: [M + $CH_3OH$–H]$^-$ Calcd for $C_{67}H_{108}NO_{24}S$ 1342.6982; Found: 1342.7032.

### *3.10. (2S,3R,E)-2-Octadecanamido-3-pivaloyloxy-octadec-4-en-1-yl 2,3,4-tri-O-acetyl-6-O-(p-toluenesulfonyl)-β-D-galactopyranosyl-(1→4)-2,3,6-tri-O-acetyl-D-glucopyranoside (**13a**)*

To a stirred solution of **12a** (59 mg, 48.58 μmol) in dry DCM (10 mL) was added TFA (1.0 mL) dropwise over 5 min in an ice bath. After the ice bath was removed, the mixture was stirred at rt for 2.5 h. Toluene was added, and the organic layer was washed with $NaHCO_3$ solution three times, dried over $Na_2SO_4$, and filtered. The filtrate was concentrated under reduced pressure, affording the free amine derivative. To the solution of stearic acid (20.2 mg, 70.9 μmol) dissolved in DCM (8.7 mL) in an ice-bath was added EDC (13.6 mg, 70.9 μmol) under an $N_2$ atmosphere. After the mixture was stirred for 15 min, the above-obtained free amine (39.5 mg, 35.45 μmol) in DCM (4.3 mL) was added dropwise followed by the addition of DMAP (0.8 mg, 7.1 μmol). The reaction mixture was stirred at rt for 24 h. $NaHCO_3$ solution was added dropwise to the mixture at 0 °C, and the aqueous layer was extracted with DCM three times. The combined organic layer was concentrated under vacuo, and the residue was purified using silica gel column chromatography to afford **13a** (65%, 43.6 mg) as an off-white solid. TLC: $R_f$ = 0.55 (EtOAc:Hex 50:50). $^1$H NMR (600 MHz, $CDCl_3$): δ 7.78 (d, $J$ = 8.3 Hz, 2H), 7.38 (d, $J$ = 8.1 Hz, 2H), 5.75 (dt, $J$ = 14.8, 7.4 Hz, 1H, =CH-), 5.61 (d, $J$ = 9.3 Hz, 1H, −NHCO−), 5.36–5.33 (m, 2H, =CH- and 4′-H), 5.23 (t, $J$ = 7.1 Hz, 1H), 5.19 (t, $J$ = 9.3 Hz, 1H), 5.06 (dd, $J$ = 10.3, 8.0 Hz, 1H), 4.92 (dd, $J$ = 10.4, 3.4 Hz, 1H), 4.86 (dd, $J$ = 9.4, 7.9 Hz, 1H), 4.47 (d, $J$ = 7.9 Hz, 1H, anomeric), 4.45–4.44 (m, 1H), 4.43 (d, $J$ = 7.8 Hz, 1H, anomeric), 4.33 (ddd, $J$ = 11.3, 8.7, 4.1 Hz, 1H), 4.06 (ddd, $J$ = 16.9, 11.1, 5.6 Hz, 1H), 3.99 (dd, $J$ = 10.2, 6.5 Hz, 1H), 3.90 (dd, $J$ = 12.6, 5.6 Hz, 1H), 3.78 (t, $J$ = 9.5 Hz, 1H), 3.60 (ddd, $J$ = 9.9, 4.7, 2.0 Hz, 1H), 3.52 (dd, $J$ = 10.0, 4.4 Hz, 1H), 2.47 (s, 3H), 2.11 (s, 3H), 2.06 (s, 3H), 2.04 (s, 3H), 2.03 (s, 3H), 2.00 (s, 3H), 1.95 (s, 3H), 1.59–1.56 (m, 2H), 1.25 (s, 56H), 1.17 (s, 9H), 0.88 (t, $J$ = 7.0 Hz, 6H, 2 × Me). $^{13}$C NMR (151 MHz, $CDCl_3$): δ 177.1, 172.7, 170.5, 170.01, 170.0, 169.81, 169.8, 169.2, 145.7, 137.1, 132.4, 130.2, 128.2, 124.9, 100.9, 100.6, 76.0, 73.2, 72.8, 72.5, 72.0, 71.0, 70.9, 69.1, 67.8, 66.7, 65.6, 62.0, 50.6, 38.9, 37.0, 32.5, 32.1, 29.9, 29.8, 29.71, 29.7, 29.6, 29.5, 29.4, 29.2, 27.2, 25.9, 22.9, 21.8, 21.0, 20.91, 20.9, 20.8, 20.7, 20.6, 14.3. HR-ESI-MS $m/z$: [M + Na]$^+$ Calcd for $C_{72}H_{117}NNaO_{22}S$: 1402.7686; Found: 1402.7698.

### *3.11. (2S,3R,E)-2-Pentacosanamido-3-pivaloyloxy-octadec-4-en-1-yl 2,3,4-tri-O-acetyl-6-O-(p-toluenesulfonyl)-β-D-galactopyranosyl-(1→4)-2,3,6-tri-O-acetyl-D-glucopyranoside (**13b**)*

Compound **13b** (64%, 44.4 mg) was prepared as a white solid from **12b** (57 mg, 46.9 μmol) by the same method and conditions employed to prepare **13a**. TLC: $R_f$ = 0.5 (EtOAc:Hex 50:50). $^1$H NMR (600 MHz, $CDCl_3$): δ 7.78 (d, $J$ = 8.3 Hz, 2H), 7.38 (d, $J$ = 8.1 Hz, 2H), 5.75 (dt, $J$ = 14.7, 7.3 Hz, 1H, =CH-), 5.61 (d, $J$ = 9.2 Hz, 1H, -NHCO-), 5.36–5.33 (m, 2H, =CH- and 4′-H), 5.23 (t, $J$ = 7.1 Hz, 1H), 5.18 (t, $J$ = 9.3 Hz, 1H), 5.06 (dd, $J$ = 10.3, 8.0 Hz, 1H), 4.92 (dd, $J$ = 10.4, 3.4 Hz, 1H), 4.86 (dd, $J$ = 9.4, 7.9 Hz, 1H), 4.47 (d, $J$ = 7.9 Hz, 1H, anomeric), 4.45 (d, *J* = 1.6 Hz, 1H), 4.43 (d, *J* = 7.8 Hz, 1H, anomeric), 4.33 (ddd, *J* = 11.3, 8.8, 4.1 Hz, 1H), 4.06 (ddd, *J* = 16.8, 11.1, 5.6 Hz, 2H), 3.99 (dd, *J* = 10.2, 6.6 Hz, 1H), 3.90 (dd, *J* = 12.0, 5.5 Hz, 2H), 3.78 (t, *J* = 9.5 Hz, 1H), 3.60 (ddd, *J* = 9.8, 4.7, 1.9 Hz, 1H), 3.52 (dd, *J* = 10.0, 4.4 Hz, 1H), 2.47 (s, 3H), 2.11 (s, 3H), 2.06 (s, 3H), 2.04 (s, 3H), 2.03 (s, 3H), 1.99 (s, 3H), 1.95 (s, 3H), 1.59–1.56 (m, 2H), 1.32–1.22 (m, 66H), 1.17 (s, 9H), 0.88 (t, *J* = 7.0 Hz, 6H, 2 × Me). $^{13}$C NMR (151 MHz, $CDCl_3$): δ 177.1, 172.7, 170.5, 170.01, 170.0, 169.81, 169.8, 169.2, 145.6, 137.1, 132.4, 130.2, 128.2, 124.9, 100.9, 100.6, 76.0, 73.2, 72.8, 72.5, 72.0, 71.0, 70.9, 69.1, 67.8, 66.7, 65.6, 62.1, 50.6, 38.9, 37.0, 32.5, 32.1, 29.91, 29.9, 29.8, 29.71, 29.7, 29.6, 29.51, 29.5, 29.4, 29.2, 27.2, 25.9, 22.9, 21.8, 21.0, 20.91, 20.9, 20.8, 20.7, 20.6, 14.3. HR-ESI-MS *m*/*z*: [M + HCOO]$^-$ Calcd for $C_{80}H_{132}NO_{24}S$ 1522.8860; found: 1522.8896.

### *3.12. (2S,3R,E)-2-Octadecanamido-3-pivaloyloxy-pentacos-4-en-1-yl 2,3,4-tri-O-acetyl-6-O-(p-toluenesulfonyl)-β-D-galactopyranosyl-(1→4)-2,3,6-tri-O-acetyl-D-glucopyranoside (**13c**)*

Compound **13c** (63%, 36.2 mg) as a white solid was prepared from **12c** (51 mg, 38.9 μmol) by the same method and conditions employed to prepare **13a**. TLC: $R_f$ = 0.5 (EtOAc:Hex 50:50). $^1$H NMR (600 MHz, $CDCl_3$): δ 7.78 (d, *J* = 8.3 Hz, 2H), 7.38 (d, *J* = 8.0 Hz, 2H), 5.75 (dt, *J* = 14.7, 7.3 Hz, 1H, =CH-), 5.61 (d, *J* = 9.2 Hz, 1H, -NHCO-), 5.36–5.33 (m, 2H, =CH- and 4′-H), 5.23 (t, *J* = 7.2 Hz, 1H), 5.18 (t, *J* = 9.3 Hz, 1H), 5.06 (dd, *J* = 10.4, 7.9 Hz, 1H), 4.92 (dd, *J* = 10.4, 3.4 Hz, 1H), 4.86 (dd, *J* = 9.4, 7.9 Hz, 1H), 4.47 (d, *J* = 7.9 Hz, 1H, anomeric), 4.45 (d, *J* = 1.6 Hz, 1H), 4.43 (d, *J* = 7.8 Hz, 1H, anomeric), 4.33 (ddd, *J* = 11.3, 8.8, 4.2 Hz, 1H), 4.06 (ddd, *J* = 16.9, 11.1, 5.6 Hz, 2H), 3.99 (dd, *J* = 10.2, 6.6 Hz, 1H), 3.93–3.88 (m, 2H), 3.78 (t, *J* = 9.5 Hz, 1H), 3.60 (ddd, *J* = 9.9, 4.7, 2.0 Hz, 1H), 3.52 (dd, *J* = 10.0, 4.4 Hz, 1H), 2.47 (s, 3H), 2.11 (s, 3H), 2.06 (s, 3H), 2.04 (s, 3H), 2.03 (s, 3H), 1.99 (s, 3H), 1.95 (s, 3H), 1.57 (b, 2H), 1.35–1.22 (m, 70H), 1.17 (s, 9H), 0.88 (t, *J* = 7.0 Hz, 6H, 2 × Me). $^{13}$C NMR (151 MHz, $CDCl_3$): δ 177.1, 172.7, 170.5, 170.01, 170.0, 169.81, 169.8, 169.2, 145.6, 137.2, 132.4, 130.2, 128.2, 124.9, 100.9, 100.6, 76.0, 73.1, 72.8, 72.5, 72.0, 71.0, 70.9, 69.1, 67.8, 66.7, 65.6, 62.0, 50.6, 38.9, 37.0, 32.5, 32.1, 29.9, 29.8, 29.71, 29.7, 29.6, 29.5, 29.4, 29.2, 27.2, 25.9, 22.9, 21.8, 21.0, 20.91, 20.9, 20.8, 20.7, 20.6, 14.3. HR-ESI-MS *m*/*z*: [M + $CH_3OH$-H]$^-$ Calcd for $C_{80}H_{134}NO_{23}S$ 1508.9067; found: 1508.9098.

### *3.13. (2S,3R,E)-2-Pentacosanamido-3-pivaloyloxy-pentacos-4-en-1-yl 2,3,4-tri-O-acetyl-6-O-(p-toluenesulfonyl)-β-D-galactopyranosyl-(1→4)-2,3,6-tri-O-acetyl-D-glucopyranoside (**13d**)*

Compound **13d** (63%, 37.8 mg) was prepared as a white solid from **12d** (50 mg, 38.1 μmol) by the same method and conditions employed to prepare **13a**. TLC: $R_f$ = 0.5 (EtOAc:Hex 50:50). $^1$H NMR (600 MHz, $CDCl_3$): δ 7.77 (d, *J* = 8.3 Hz, 2H), 7.38 (d, *J* = 8.1 Hz, 2H), 5.75 (dt, *J* = 14.8, 7.3 Hz, 1H, =CH-), 5.61 (d, *J* = 9.2 Hz, 1H, -NHCO-), 5.36–5.33 (m, 2H, =CH- and 4′-H), 5.23 (t, *J* = 7.1 Hz, 1H), 5.18 (t, *J* = 9.3 Hz, 1H), 5.06 (dd, *J* = 10.3, 8.0 Hz, 1H), 4.92 (dd, *J* = 10.4, 3.4 Hz, 1H), 4.86 (dd, *J* = 9.4, 7.9 Hz, 1H), 4.47 (d, *J* = 7.9 Hz, 1H, anomeric), 4.45 (d, J = 1.4 Hz, 1H), 4.43 (d, *J* = 7.8 Hz, 1H, anomeric), 4.33 (ddd, *J* = 11.3, 8.7, 4.1 Hz, 1H), 4.05 (ddd, *J* = 16.9, 11.1, 5.6 Hz, 2H), 3.99 (dd, *J* = 10.2, 6.5 Hz, 1H), 3.90 (dd, *J* = 11.9, 5.5 Hz, 2H), 3.78 (t, *J* = 9.5 Hz, 1H), 3.60 (ddd, *J* = 9.9, 4.7, 2.0 Hz, 1H), 3.52 (dd, *J* = 10.0, 4.4 Hz, 1H), 2.47 (s, 3H), 2.11 (s, 3H), 2.06 (s, 3H), 2.04 (s, 3H), 2.03 (s, 3H), 1.99 (s, 3H), 1.95 (s, 3H), 1.64 (dd, *J* = 14.7, 7.2 Hz, 2H), 1.59–1.56 (m, 2H), 1.34–1.23 (m, 80H), 1.17 (s, 9H), 0.88 (t, *J* = 7.0 Hz, 6H, 2 × Me). $^{13}$C NMR (151 MHz, $CDCl_3$): δ 177.1, 172.7, 170.5, 170.01, 170.0, 169.81, 169.8, 169.2, 145.6, 137.2, 132.4, 130.2, 128.2, 124.9, 100.9, 100.6, 76.0, 73.1, 72.8, 72.5, 72.0, 71.0, 70.9, 69.1, 67.8, 66.7, 65.6, 62.0, 50.6, 38.9, 37.0, 32.5, 32.1, 29.91, 29.9, 29.8, 29.71, 29.7, 29.6, 29.5, 29.4, 29.2, 27.2, 25.9, 22.9, 21.8, 21.0, 20.91, 20.9, 20.8, 20.7, 20.6, 14.3. HR-ESI-MS *m*/*z*: [M-H]$^-$ Calcd for $C_{86}H_{144}NO_{22}S$ 1574.9901; found: 1574.9834.

*3.14. (2S,3R,E)-2-Octadecanamido-3-pivaloyloxy-octadec-4-en-1-yl 2,3,4-tri-O-acetyl-6-azido-6-deoxy-β-D-galactopyranosyl-(1→4)-2,3,6-tri-O-acetyl-D-glucopyranoside (**14a**)*

To a solution of **13a** (30 mg, 21.7 μmol) in dry DMF (6.0 mL) was added $NaN_3$ (28 mg, 434.5 μmol), and the mixture was heated at 86 °C for 2 d. When the starting material **13a** disappeared in TLC, the reaction was allowed to cool to rt. The mixture was diluted with EtOAc and cooled at 0 °C. After cold water was added, the mixture was extracted with EtOAc three times. The combined organic layers were washed with brine, filtered, dried over $Na_2SO_4$, and condensed under vacuum. The residue was purified through silica gel column chromatography to afford **14a** as a white solid (80%, 21.8 mg). TLC: $R_f$ = 0.45 (EtOAc:Hex 40:60). $^1$H NMR (600 MHz, $CDCl_3$): δ 5.75 (dt, *J* = 14.8, 7.4 Hz, 1H, =CH-), 5.62 (d, *J* = 9.2 Hz, 1H, −NHCO−), 5.38–5.30 (m, 2H, =CH- and 4′-H), 5.25–5.17 (m, 2H), 5.09 (dd, *J* = 10.4, 7.9 Hz, 1H), 4.96 (dd, *J* = 10.4, 3.5 Hz, 1H), 4.88 (dd, *J* = 9.5, 7.8 Hz, 1H), 4.49–4.44 (m, 1H), 4.50 (d, *J* = 7.9 Hz, 1H, anomeric), 4.42 (d, *J* = 7.7 Hz, 1H, anomeric), 4.32 (ddd, *J* = 11.4, 8.6, 4.1 Hz, 1H), 4.05 (dd, *J* = 12.0, 4.9 Hz, 1H), 3.90 (dd, *J* = 10.0, 3.7 Hz, 1H), 3.85 (t, *J* = 9.5 Hz, 1H), 3.72 (t, *J* = 6.7 Hz, 1H), 3.60 (ddd, *J* = 9.9, 4.7, 2.0 Hz, 1H), 3.52 (dd, *J* = 10.0, 4.4 Hz, 1H), 3.47 (dd, *J* = 12.8, 7.3 Hz, 1H), 3.26 (dd, *J* = 12.8, 5.6 Hz, 1H), 2.17 (s, 3H), 2.12 (s, 3H), 2.07 (s, 3H), 2.04 (s, 3H), 2.04 (s, 3H), 1.97 (s, 3H), 1.59–1.56 (b, 2H), 1.35–1.22 (m, 56H), 1.17 (s, 9H), 0.88 (t, *J* = 7.0 Hz, 6H, 2 × Me). $^{13}$C NMR (151 MHz, $CDCl_3$): δ 177.1, 172.7, 170.5, 170.2, 170.1, 169.81, 169.8, 169.2, 137.2, 125.0, 100.8, 100.7, 75.4, 73.2, 72.9, 72.5, 72.4, 72.0, 71.1, 69.3, 69.2, 67.8, 67.6, 62.0, 50.6, 50.4, 38.9, 37.0, 32.5, 32.1, 29.9, 29.8, 29.71, 29.7, 29.6, 29.5, 29.4, 29.2, 27.2, 25.9, 22.9, 21.1, 21.0, 20.9, 20.8, 20.7, 14.3. HR-ESI-MS *m*/*z*: $[M + HCOO]^-$ Calcd for $C_{66}H_{111}N_4O_{21}$: 1295.7741; Found: 1295.7753.

*3.15. (2S,3R,E)-2-Pentacosanamido-3-pivaloyloxy-octadec-4-en-1-yl 2,3,4-tri-O-acetyl-6-azido-6-deoxy-β-D-galactopyranosyl-(1→4)-2,3,6-tri-O-acetyl-D-glucopyranoside (**14b**)*

Compound **14b** (76%, 20.8 mg) was prepared as a white solid from **13b** (30 mg, 20.3 μmol) by the same method and conditions employed to prepare **14a**. TLC: $R_f$ = 0.65 (EtOAc:Hex 40:60). $^1$H NMR (600 MHz, $CDCl_3$): δ 5.75 (dt, *J* = 14.7, 7.4 Hz, 1H, =CH-), 5.61 (d, *J* = 9.2 Hz, 1H, -NHCO-), 5.37–5.31 (m, 2H, =CH- and 4′-H), 5.21 (dt, *J* = 15.5, 8.3 Hz, 2H), 5.09 (dd, *J* = 10.4, 7.9 Hz, 1H), 4.96 (dd, *J* = 10.4, 3.5 Hz, 1H), 4.88 (dd, *J* = 9.5, 7.8 Hz, 1H), 4.51 (d, *J* = 7.8 Hz, 1H, anomeric), 4.48 (d, *J* = 1.7 Hz, 1H), 4.42 (d, *J* = 7.7 Hz, 1H, anomeric), 4.32 (ddd, *J* = 11.4, 8.4, 4.1 Hz, 1H), 4.05 (dd, *J* = 12.0, 4.9 Hz, 1H), 3.91 (dd, *J* = 9.9, 3.7 Hz, 1H), 3.85 (t, *J* = 9.5 Hz, 1H), 3.72 (t, *J* = 6.6 Hz, 1H), 3.60 (ddd, *J* = 9.9, 4.7, 2.0 Hz, 1H), 3.52 (dd, *J* = 10.0, 4.4 Hz, 1H), 3.47 (dd, *J* = 12.8, 7.3 Hz, 1H), 3.26 (dd, *J* = 12.8, 5.6 Hz, 1H), 2.17 (s, 3H), 2.12 (s, 3H), 2.07 (s, 3H), 2.04 (s, 3H), 2.04 (s, 3H), 1.97 (s, 3H), 1.59–1.56 (b, 2H), 1.37–1.21 (m, 70H), 1.17 (s, 9H), 0.88 (t, *J* = 7.0 Hz, 6H, 2 × Me). $^{13}$C NMR (151 MHz, $CDCl_3$): δ 177.1, 172.7, 170.4, 170.2, 170.1, 169.81, 169.8, 169.2, 137.2, 125.0, 100.8, 100.7, 75.4, 73.2, 72.9, 72.5, 72.4, 72.0, 71.1, 69.2, 67.8, 67.6, 62.4, 62.2, 50.6, 50.4, 38.9, 37.0, 32.5, 32.1, 29.9, 29.8, 29.71, 29.7, 29.6, 29.5, 29.4, 29.2, 27.2, 25.9, 22.9, 21.1, 21.0, 20.9, 20.8, 20.7, 14.3. HR-ESI-MS *m*/*z*: $[M + CH_3OH\text{-}H]^-$ Calcd for $C_{73}H_{127}N_4O_{20}$ 1379.9044; Found: 1379.9058.

*3.16. (2S,3R,E)-2-Octadecanamido-3-pivaloyloxy-pentacos-4-en-1-yl 2,3,4-tri-O-acetyl-6-azido-6-deoxy-β-D-galactopyranosyl-(1→4)-2,3,6-tri-O-acetyl-D-glucopyranoside (**14c**)*

Compound **14c** (69%, 18.9 mg) was prepared as a white solid from **13c** (30 mg, 20.3 μmol) by the same method and conditions employed to prepare **14a**. TLC: $R_f$ = 0.45 (EtOAc:Hex 40:60). $^1$H NMR (600 MHz, $CDCl_3$) δ 5.75 (dt, *J* = 14.7, 7.4 Hz, 1H, =CH-), 5.62 (d, *J* = 9.3 Hz, 1H, -NHCO-), 5.35–5.32 (m, 2H, =CH- and 4′-H), 5.24–5.16 (m, 2H), 5.09 (dd,

$J$ = 10.4, 7.9 Hz, 1H), 4.96 (dd, $J$ = 10.4, 3.5 Hz, 1H), 4.87 (dd, $J$ = 9.5, 7.8 Hz, 1H), 4.50 (d, $J$ = 7.8 Hz, 1H, anomeric), 4.48 (d, $J$ = 1.8 Hz, 1H), 4.42 (d, $J$ = 7.7 Hz, 1H, anomeric), 4.32 (ddd, $J$ = 11.4, 8.4, 4.1 Hz, 1H), 4.05 (dd, $J$ = 12.0, 4.9 Hz, 1H), 3.90 (dd, $J$ = 10.0, 3.7 Hz, 1H), 3.85 (t, $J$ = 9.5 Hz, 1H), 3.72 (t, $J$ = 6.6 Hz, 1H), 3.60 (ddd, $J$ = 9.8, 4.7, 2.0 Hz, 1H), 3.51 (dd, $J$ = 10.0, 4.4 Hz, 1H), 3.46 (dd, $J$ = 12.8, 7.3 Hz, 1H), 3.26 (dd, $J$ = 12.8, 5.6 Hz, 1H), 2.16 (s, 3H), 2.11 (s, 3H), 2.07 (s, 3H), 2.04 (s, 6H), 1.97 (s, 3H), 1.59–1.56 (b, 2H), 1.34–1.23 (m, 70H), 1.17 (s, 9H), 0.88 (t, $J$ = 7.0 Hz, 6H, 2 × Me). $^{13}$C NMR (151 MHz, $CDCl_3$): δ 177.1, 172.7, 170.4, 170.2, 170.1, 169.81, 169.8, 169.2, 137.2, 125.0, 100.71, 100.7, 75.4, 73.2, 72.9, 72.5, 72.4, 71.9, 71.1, 69.2, 67.8, 67.6, 62.1, 62.0, 50.6, 50.4, 38.9, 37.0, 32.5, 32.1, 29.9, 29.81, 29.8, 29.71, 29.7, 29.6, 29.5, 29.4, 29.2, 27.2, 25.9, 22.9, 21.1, 21.0, 20.9, 20.8, 20.7, 14.3. HR-ESI-MS $m/z$: $[M + HCOO]^-$ Calcd for $C_{73}H_{125}N_4O_{21}$ 1393.8837; Found: 1393.8878.

### *3.17. (2S,3R,E)-2-Pentacosanamido-3-pivaloyloxy-pentacos-4-en-1-yl 2,3,4-tri-O-acetyl-6-azido-6-deoxy-β-D-galactopyranosyl-(1→4)-2,3,6-tri-O-acetyl-D-glucopyranoside (**14d**)*

Compound **14d** (72%, 18.5 mg) was prepared as a white solid from **13d** (28 mg, 17.8 µmol) by the same method and conditions employed to prepare **14a**. TLC: $R_f$ = 0.66 (EtOAc:Hex 40:60). $^1$H NMR (600 MHz, $CDCl_3$): δ 5.75 (dt, $J$ = 14.7, 7.4 Hz, 1H, =CH-), 5.61 (d, $J$ = 9.2 Hz, 1H, -NHCO-), 5.36–5.32 (m, 2H, =CH- and 4′-H), 5.24–5.17 (m, 2H), 5.09 (dd, $J$ = 10.4, 7.9 Hz, 1H), 4.96 (dd, $J$ = 10.4, 3.5 Hz, 1H), 4.87 (dd, $J$ = 9.5, 7.8 Hz, 1H), 4.50 (d, $J$ = 7.9 Hz, 1H, anomeric), 4.48 (d, $J$ = 1.7 Hz, 1H), 4.42 (d, $J$ = 7.7 Hz, 1H, anomeric), 4.32 (ddd, $J$ = 11.4, 8.2, 4.1 Hz, 1H), 4.05 (dd, $J$ = 12.0, 4.9 Hz, 1H), 3.90 (dd, $J$ = 9.9, 3.7 Hz, 1H), 3.85 (t, $J$ = 9.5 Hz, 1H), 3.72 (t, $J$ = 6.5 Hz, 1H), 3.60 (ddd, $J$ = 9.8, 4.7, 2.0 Hz, 1H), 3.51 (dd, $J$ = 10.0, 4.4 Hz, 1H), 3.47 (dd, $J$ = 12.8, 7.3 Hz, 1H), 3.26 (dd, $J$ = 12.8, 5.7 Hz, 1H), 2.17 (s, 3H), 2.12 (s, 3H), 2.07 (s, 3H), 2.04 (s, 6H, 2 × $CH_3$), 1.97 (s, 3H), 1.59–1.56 (b, 2H), 1.32–1.23 (m, 84H), 1.17 (s, 9H), 0.88 (t, $J$ = 7.0 Hz, 6H, 2 × Me). $^{13}$C NMR (151 MHz, $CDCl_3$): δ 177.1, 172.7, 170.4, 170.2, 170.1, 169.81, 169.8, 169.2, 137.2, 125.0, 100.8, 100.7, 75.4, 73.2, 72.9, 72.5, 72.4, 72.0, 71.1, 69.2, 67.8, 67.6, 63.2, 62.0, 50.6, 50.4, 38.9, 37.0, 32.5, 32.1, 29.9, 29.89, 29.8, 29.71, 29.7, 29.6, 29.5, 29.4, 29.2, 27.2, 25.9, 22.9, 21.1, 20.91, 20.9, 20.8, 20.7, 14.3. HR-ESI-MS $m/z$: $[M + CH_3OH-H]^-$ Calcd for $C_{80}H_{141}N_4O_{20}$ 1478.0139; Found: 1478.0077.

### *3.18. (2S,3R,E)-2-Octadecanamido-3-hydroxy-octadec-4-en-1-yl 6-azido-6-deoxy-β-D-galactopyranosyl-(1→4)-β-D-glucopyranoside (**1a**)*

To a solution of **14a** (19.5 mg, 15.6 µmol) in a mixture of dry DCM and MeOH (1:1, 2 mL) was added NaOMe in MeOH (5 M, 43.6 µL, 0.2 mmol) dropwise at 0 °C. The ice bath was removed, and the mixture was stirred at rt under $N_2$ for 2 d. After the starting material was completely consumed as indicated by TLC, Amberlyst $H^+$ resin was added with stirring to neutralize the reaction mixture. The solvent was removed under vacuum, and the residue was subjected to silica gel column chromatography to afford **1a** as an off-white solid (90%, 12.8 mg). TLC: $R_f$ = 0.51 (DCM:MeOH 85:15). $^1$H NMR (600 MHz, $CDCl_3$:$CD_3OD$ 2:1): δ 5.65 (dt, $J$ = 15.0, 7.4 Hz, 1H, =CH-), 5.41 (dd, $J$ = 15.0, 7.6 Hz, 1H, =CH-), 4.32 (d, $J$ = 7.8 Hz, 1H anomeric), 4.25 (d, $J$ = 7.8 Hz, 1H anomeric), 4.17 (dd, $J$ = 10.0, 3.9 Hz, 1H), 4.05 (t, $J$ = 7.6 Hz, 1H), 3.96–3.93 (m, 1H), 3.82 (d, $J$ = 3.0 Hz, 2H), 3.77 (d, $J$ = 2.9 Hz, 1H), 3.65–3.61 (m, 2H), 3.60–3.58 (m, 1H), 3.56–3.52 (m, 4H), 3.50 (dd, $J$ = 9.8, 3.0 Hz, 2H), 3.47 (dd, $J$ = 9.7, 3.2 Hz, 1H), 3.38–3.35 (m, 1H), 2.13 (t, $J$ = 7.7 Hz, 2H), 2.00–1.96 (m, 2H), 1.56–1.52 (m, 2H), 1.34–1.30 (m, 2H), 1.28–1.20 (m, 48H), 0.84 (t, $J$ = 7.0 Hz, 6H, 2 × Me). $^{13}$C NMR (151 MHz, $CDCl_3$:$CD_3OD$ 2:1): δ 175.0, 134.8, 129.7, 104.3, 103.3, 80.4, 75.3, 75.2, 74.0, 73.8, 73.7, 72.4, 71.3, 69.2, 69.0, 61.2, 53.6, 51.6, 36.9, 32.8, 32.3, 30.12, 30.11, 30.1, 30.03, 30.02, 30.01, 30.0, 29.9, 29.8, 29.72, 29.71, 29.7, 26.4, 23.0, 14.3. HR-ESI-MS $m/z$: $[M + HCOO]^-$ Calcd for $C_{49}H_{91}N_4O_{14}$ 959.6532; Found: 959.6563.

### *3.19. (2S,3R,E)-2-Pentacosanamido-3-(hydroxy)-octadec-4-en-1-yl 6-azido-6-deoxy-β-D-galactopyranosyl-(1→4)-β-D-glucopyranoside* (**1b**)

Compound **1b** (88%, 11.3 mg) was prepared as a white solid from **14b** (17 mg, 12.6 μmol) by the same method and conditions utilized to prepare **1a**. TLC: $R_f$ = 0.66 (DCM:MeOH 85:15). $^1$H NMR (600 MHz, $CDCl_3$:$CD_3OD$ 2:1): δ 5.63 (dt, *J* = 14.9, 7.2 Hz, 1H, =CH-), 5.40 (dd, *J* = 14.9, 7.5 Hz, 1H, =CH-), 4.30 (d, *J* = 7.8 Hz, 1H anomeric), 4.23 (d, *J* = 7.8 Hz, 1H anomeric), 4.14 (dd, *J* = 10.1, 3.9 Hz, 1H), 4.04 (t, *J* = 7.4 Hz, 1H), 3.80 (d, *J* = 3.0 Hz, 2H), 3.76 (d, *J* = 3.0 Hz, 1H), 3.61–3.59 (m, 2H), 3.58–3.56 (m, 4H), 3.54–3.51 (m, 4H), 3.49 (dd, *J* = 10.4, 1.9 Hz, 1H), 3.46 (dd, *J* = 9.7, 3.2 Hz, 1H), 2.11 (t, *J* = 7.7 Hz, 2H), 1.98–1.94 (m, 2H), 1.54–1.50 (m, 2H), 1.31–1.28 (m, 2H), 1.26–1.19 (m, 62H), 0.83 (t, *J* = 7.0 Hz, 6H, 2 × Me). $^{13}$C NMR (151 MHz, $CDCl_3$:$CD_3OD$ 2:1): δ 174.6, 134.6, 129.2, 104.0, 103.0, 80.2, 75.0, 74.9, 73.7, 73.4, 72.3, 71.0, 68.8, 68.7, 63.6, 61.0, 53.5, 53.3, 51.3, 36.6, 32.5, 32.0, 29.82, 29.81, 29.8, 29.71, 29.7, 29.6, 29.52, 29.51, 29.5, 29.4, 26.0, 22.8, 14.1. HR-ESI-MS *m*/*z*: [M + HCOO]$^-$ Calcd for $C_{56}H_{105}N_4O_{14}$ 1057.7628; Found: 1057.7659.

### *3.20. (2S,3R,E)-2-Octadecanamido-3-(hydroxy)-pentacos-4-en-1-yl 6-azido-6-deoxy-β-D-galactopyranosyl-(1→4)-β-D-glucopyranoside* (**1c**)

Compound **1c** (80%, 10.8 mg) was prepared as an off-white solid from **14c** (18 mg, 13.3 μmol) by the same method and conditions utilized to prepare **1a**. TLC: $R_f$ = 0.70 (DCM:MeOH 85:15). $^1$H NMR (600 MHz, $CDCl_3$:$CD_3OD$ 2:1): δ 5.64 (dt, *J* = 14.9, 7.3 Hz, 1H, =CH-), 5.41 (dd, *J* = 14.9, 7.6 Hz, 1H, =CH-), 4.30 (1H, anomeric, overlapped with DHO signal), 4.25 (d, *J* = 7.8 Hz, 1H, anomeric), 4.16 (dd, *J* = 10.0, 3.9 Hz, 1H), 4.05 (t, *J* = 7.7 Hz, 1H), 3.96–3.92 (m, 2H), 3.92–3.89 (m, 3H), 3.82 (d, *J* = 3.3 Hz, 2H), 3.77 (d, *J* = 3.3 Hz, 1H), 3.63–3.58 (m, 2H), 3.55–3.52 (m, 3H), 3.50 (dd, *J* = 10.5, 3.3 Hz, 1H), 3.47 (dd, *J* = 9.7, 3.2 Hz, 1H), 2.15–2.10 (m, 2H), 1.99–1.96 (m, 2H), 1.56–1.52 (m, 2H), 1.34–1.29 (m, 2H), 1.27–1.20 (m, 62H), 0.84 (t, *J* = 7.0 Hz, 6H, 2 × Me). $^{13}$C NMR (151 MHz, $CDCl_3$:$CD_3OD$ 2:1): δ 174.9, 134.8, 129.6, 104.2, 103.2, 80.4, 75.2, 75.1, 74.0, 73.7, 73.6, 72.4, 71.2, 70.6, 69.1, 61.1, 55.3, 53.5, 51.5, 49.9, 49.6, 36.8, 32.7, 32.2, 30.1, 30.03, 30.02, 30.01, 30.0, 29.91, 29.9, 29.8, 29.73, 29.72, 29.71, 29.7, 29.6, 26.3, 23.0, 14.2. HR-ESI-MS *m*/*z*: [M + HCOO]$^-$ Calcd for $C_{56}H_{105}N_4O_{14}$ 1057.7628; Found: 1057.7659.

### *3.21. (2S,3R,E)-2-Pentacosanamido-3-(hydroxy)-pentacos-4-en-1-yl 6-azido-6-deoxy-β-D-galactopyranosyl-(1→4)-β-D-glucopyranoside* (**1d**)

Compound **1d** (85%, 11.1 mg) was prepared as a white solid from **14d** (17 mg, 11.7 μmol) by the same method and conditions used to prepare **1a**. TLC: $R_f$ = 0.75 (DCM:MeOH 85:15). $^1$H NMR (600 MHz, $CDCl_3$:$CD_3OD$ 2:1): δ 5.63 (dt, *J* = 14.9, 7.3 Hz, 1H, =CH-), 5.39 (dd, *J* = 14.9, 7.5 Hz, 1H, =CH-), 4.30 (d, *J* = 7.8 Hz, 1H, anomeric), 4.23 (d, *J* = 7.8 Hz, 1H, anomeric), 4.13 (dd, *J* = 10.0, 4.0 Hz, 1H), 4.04 (t, *J* = 7.4 Hz, 1H), 3.94–3.90 (m, 1H), 3.79 (d, *J* = 2.7 Hz, 3H), 3.61–3.56 (m, 3H), 3.54–3.52 (m, 2H), 3.51 (dd, *J* = 7.8, 3.1 Hz, 2H), 3.48 (dd, *J* = 10.7, 3.2 Hz, 1H), 3.44–3.44 (m, 3H), 2.11 (t, *J* = 7.6 Hz, 2H), 1.94–1.97 (m, 2H), 1.59–1.47 (m, 4H), 1.31–1.16 (m, 76H), 0.82 (t, *J* = 7.0 Hz, 6H). $^{13}$C NMR (151 MHz, $CDCl_3$:$CD_3OD$ 2:1): δ 177.6, 134.5, 129.0, 103.8, 102.8, 79.9, 74.8, 73.6, 73.21, 73.2, 72.2, 70.8, 68.61, 68.6, 63.5, 60.9, 53.1, 51.1, 36.5, 32.3, 31.9, 29.71, 29.7, 29.61, 29.6, 29.51, 29.5, 29.42, 29.41, 29.4, 29.31, 29.3, 29.21, 29.2, 25.9, 22.6, 14.0. HR-ESI-MS *m*/*z*: [M + HCOO]$^-$ Calcd for $C_{63}H_{119}N_4O_{14}$ 1155.8723; Found: 1155.87622.1.

## 4. Conclusions

β-LacCer is not only a key intermediate in complex GSL biosynthesis but also an important player in many biological events and diseases. However, in-depth investigation of β-LacCer is hindered by its lack of fluorophores or other useful functionalities. To overcome this problem, we designed and synthesized a series of β-LacCer analogs carrying an azido group in its glycan and varied Cer moieties. In the literature, there are many reports [33–46] (just cite a few) about the synthesis of β-LacCer and its analogs, but none of these reports have described azide-functionalized β-LacCer analogs like **1a–d** in this work. To facilitate rapid access to **1a–d**, we have employed a diversity-oriented synthetic strategy that is highlighted by the late-stage lipid remodeling for on-site assembly of the Cer moiety. Therefore, all the target molecules **1a–d** could be readily derived from the same intermediate **10** within five robust steps. This strategy and its protocols are streamlined so that they will be applicable to the synthesis of various lipid forms of β-LacCer and its analogs. The azide in **1a–d** is suitable for further and direct functionalization by means of straightforward and biocompatible click chemistry to attach a diversity of molecular tags, such as affinity and fluorescent tags that enable different modern analytical technologies. Alternatively, the azido group in **1a–d** can be reduced to generate a free primary amino group that is suitable for further introduction of molecular labels via chemoselective *N*-acylation. Moreover, additional functionalization of **1a–d** through biocompatible click chemistry can be conducted either before or after their introduction to cells/tissues [24,25], thereby expanding their application scopes. The flexibility to introduce fluorophores or other large molecular labels later also makes **1a–d** especially valuable. Because having a hydroxyl group substituted with a small azido group will have a minimal impact on the interaction of these probes with target molecules in/on cells and probably make the probes more easily acceptable by enzymes, these probes can better mimic the behaviors and bioactivities of β-LacCer than those containing large functional groups. Consequently, this work not only has established a novel and efficient method for the synthesis of β-LacCer and its analogs but also has provided **1a–d** as powerful probes or tools for in-depth investigation of β-LacCer biology, including the influences of its lipid structures on the biological functions and metabolisms of β-LacCer and other GSLs, as well as their related diseases.

**Supplementary Materials:** The following supporting information can be downloaded at: https://www.mdpi.com/article/10.3390/molecules30132667/s1, the NMR and MS spectra of all new compounds, including **4**, **5**, **6**, **7**, **8** (only NMR), **10**, **12a,b**, **13a–d**, **14a–d**, and **1a–d**.

**Author Contributions:** B.M. performed the experiments and data interpretation and drafted the manuscript; R.R. and X.Y. were engaged in experimental design and revision, provided advice, and verified data; A.M.G., N.A.O., and H.A.A.-F. provided research supervision and advice; Z.G. is overall responsible for this project, including experimental designs, data analysis, and research supervision. All authors have read and agreed to the published version of the manuscript.

**Funding:** We acknowledge the Egyptian Government, represented by the Egyptian Cultural and Educational Bureau at Washington DC, and the Mission Sector of Egyptian Ministry of Higher Education for supporting B.M. This work is partly supported by an NIH grant (R21 AI170129) to Z.G.

**Data Availability Statement:** The original contributions presented in this study are included in the article/Supplementary Material. Further inquiries can be directed to the corresponding author.

**Acknowledgments:** Z.G. is grateful to Steven and Rebecca Scott for their endowment to support our research.

**Conflicts of Interest:** The authors declare no conflict of interest.

## References


1. Guo, Z. The structural diversity of natural glycosphingolipids (GSLs). *J. Carbohydr. Chem.* **2022**, *41*, 63–154. [CrossRef] [PubMed]
2. Hakomori, S.I. Structure and function of glycosphingolipids and sphingolipids: Recollections and future trends. *Biochim. Biophys. Acta* **2008**, *1780*, 325–346. [CrossRef]
3. Ando, H.; Komura, N. Recent progress in the synthesis of glycosphingolipids. *Curr. Opin. Chem. Biol.* **2024**, *78*, 102423. [CrossRef] [PubMed]
4. Sonnino, S.; Prinetti, A. Gangliosides as regulators of cell membrane organization and functions. *Adv. Exp. Med. Biol.* **2010**, *688*, 165–184. [PubMed]
5. Saxena, K.; Zimmermann, P.; Schmidt, R.R.; Shipley, G.G. Bilayer properties of totally synthetic C16:0-lactosyl-ceramide. *Biophys. J.* **2000**, *78*, 306–312. [CrossRef]
6. Sarmento, M.J.; Ricardo, J.C.; Amaro, M.; Šachl, R. Organization of gangliosides into membrane nanodomains. *FEBS Lett.* **2020**, *594*, 3668–3697. [CrossRef]
7. Nakayama, H.; Ogawa, H.; Takamori, K.; Iwabuchi, K. GSL-enriched membrane microdomains in innate immune responses. *Arch. Immunol. Ther. Exp.* **2013**, *61*, 217–228. [CrossRef]
8. Wennekes, T.; van den Berg, R.J.; Boot, R.G.; van der Marel, G.A.; Overkleeft, H.S.; Aerts, J.M. Glycosphingolipids—Nature, function, and pharmacological modulation. *Angew. Chem. Int. Ed. Engl.* **2009**, *48*, 8848–8869. [CrossRef] [PubMed]
9. Hakomori, S. Cancer-associated glycosphingolipid antigens: Their structure, organization, and function. *Acta Anat.* **1998**, *161*, 79–90. [CrossRef]
10. Mutoh, T.; Hirabayashi, Y.; Mihara, T.; Ueda, M.; Koga, H.; Ueda, A.; Kokura, T.; Yamamoto, H. Role of glycosphingolipids and therapeutic perspectives on Alzheimer's disease. CNS Neurol. *Disord. Drug Targets* **2006**, *5*, 375–380. [CrossRef]
11. Hakomori, S.; Zhang, Y. Glycosphingolipid antigens and cancer therapy. *Chem. Biol.* **1997**, *3*, 97–104. [CrossRef] [PubMed]
12. Sandhoff, K.; Kolter, T. Biosynthesis and degradation of mammalian glycosphingolipids. *Philos. Trans. R. Soc. Lond. B Biol. Sci.* **2003**, *358*, 847–861. [CrossRef] [PubMed]
13. Yu, R.K.; Tsai, Y.T.; Ariga, T.; Yanagisawa, M. Structures, biosynthesis, and functions of gangliosides—An overview. *J. Oleo. Sci.* **2011**, *60*, 537–544. [CrossRef] [PubMed]
14. Yu, W.; Ying, J.; Wang, X.; Liu, X.; Zhao, T.; Yoon, S.; Zheng, Q.; Fang, Y.; Yang, D.; Hua, F. The Involvement of Lactosylceramide in Central Nervous System Inflammation Related to Neurodegenerative Disease. *Front. Aging Neurosci.* **2021**, *13*, 691230. [CrossRef]
15. Nakayama, H.; Nagafuku, M.; Suzuki, A.; Iwabuchi, K.; Inokuchi, J.I. The regulatory roles of glycosphingolipid-enriched lipid rafts in immune systems. *FEBS Lett.* **2018**, *592*, 3921–3942. [CrossRef]
16. Furukawa, K.; Ohmi, Y.; Kondo, Y.; Ohkawa, Y.; Tajima, O.; Furukawa, K. Regulatory function of glycosphingolipids in the inflammation and degeneration. *Arch. Biochem. Biophys.* **2015**, *571*, 58–65. [CrossRef]
17. Silsirivanit, A.; Phoomak, C.; Teeravirote, K.; Wattanavises, S.; Seubwai, W.; Saengboonmee, C.; Zhan, Z.; Inokuchi, J.I.; Suzuki, A.; Wongkham, S. Overexpression of HexCer and LacCer containing 2-hydroxylated fatty acids in cholangiocarcinoma and the association of the increase of LacCer (d18:1-h23:0) with shorter survival of the patients. *Glycoconj. J.* **2019**, *36*, 103–111. [CrossRef]
18. Bodas, M.; Min, T.; Vij, N. Lactosylceramide-accumulation in lipid-rafts mediate aberrant-autophagy, inflammation and apoptosis in cigarette smoke induced emphysema. *Apoptosis* **2015**, *20*, 725–739. [CrossRef]
19. Chatterjee, S. Sphingolipids in atherosclerosis and vascular biology. *Arter. Thromb. Vasc. Biol.* **1998**, *18*, 1523–1533. [CrossRef]
20. Chatterjee, S.; Kolmakova, A.; Rajesh, M. Regulation of lactosylceramide synthase (glucosylceramide beta1-->4 galactosyltransferase); implication as a drug target. *Curr. Drug Targets* **2008**, *9*, 272–281. [CrossRef]
21. Ekyalongo, R.C.; Nakayama, H.; Kina, K.; Kaga, N.; Iwabuchi, K. Organization and functions of glycolipid-enriched microdomains in phagocytes. *Biochim. Biophys. Acta* **2015**, *1851*, 90–97. [CrossRef]
22. Yoshizaki, F.; Nakayama, H.; Iwahara, C.; Takamori, K.; Ogawa, H.; Iwabuchi, K. Role of glycosphingolipid-enriched microdomains in innate immunity: Microdomain-dependent phagocytic cell functions. *Biochim. Biophys. Acta* **2008**, *1780*, 383–392. [CrossRef]
23. Chiricozzi, E.; Ciampa, M.G.; Brasile, G.; Compostella, F.; Prinetti, A.; Nakayama, H.; Ekyalongo, R.C.; Iwabuchi, K.; Sonnino, S.; Mauri, L. Direct interaction, instrumental for signaling processes, between LacCer and Lyn in the lipid rafts of neutrophil-like cells. *J. Lipid Res.* **2015**, *56*, 129–141. [CrossRef] [PubMed]
24. Kolb, H.C.; Finn, M.G.; Sharpless, K.B. Click chemistry: Diverse chemical function from a few good reactions. *Angew. Chem. Int. Ed.* **2001**, *40*, 2004–2021. [CrossRef]
25. McKay, C.S.; Finn, M.G. Click chemistry in complex mixtures: Bioorthogonal bioconjugation. *Chem. Biol.* **2014**, *21*, 1075–1101. [CrossRef] [PubMed]
26. Chao, C.S.; Li, C.W.; Chen, M.C.; Chang, S.S.; Mong, K.K. Low-concentration 1,2-trans beta-selective glycosylation strategy and its applications in oligosaccharide synthesis. *Chem. Eur. J.* **2009**, *15*, 10972–10982. [CrossRef]

27. Mong, K.T.; Chao, C.; Chen, M.; Lin, C. Tandem one-pot acetalation-acetylation for direct access to differentially protected thioglycosides and O-glycosides with p-toluenesulfonic acid. *Synlett* **2009**, *2009*, 603–606. [CrossRef]
28. D'Angelo, K.A.; Taylor, M.S. Borinic acid catalyzed stereo- and regioselective couplings of glycosyl methanesulfonates. *J. Am. Chem. Soc.* **2016**, *138*, 11058–11066. [CrossRef]
29. Larsen, K.; Olsen, C.E.; Motawia, M.S. A facile protocol for direct conversion of unprotected sugars into phenyl 4,6-O-benzylidene-per-O-acetylated-1,2-trans-thioglycosides. *Carbohydr. Res.* **2003**, *338*, 199–202. [CrossRef]
30. Li, Q.; Jaiswal, M.; Rohokale, R.S.; Guo, Z. A diversity-oriented strategy for chemoenzymatic synthesis of glycosphingolipids and related derivatives. *Org. Lett.* **2020**, *22*, 8245–8249. [CrossRef]
31. Rohokale, R.S.; Li, Q.; Guo, Z. A diversity-oriented strategy for chemical synthesis of glycosphingolipids: Synthesis of glycosphingolipid LcGg4 and its analogues and derivatives. *J. Org. Chem.* **2021**, *86*, 1633–1648. [CrossRef] [PubMed]
32. Meek, S.J.; O'Brien, R.V.; Llaveria, J.; Schrock, R.R.; Hoveyda, A.H. Catalytic Z-selective olefin cross-metathesis for natural product synthesis. *Nature* **2011**, *471*, 461–466. [CrossRef] [PubMed]
33. Hasegawa, A.; Morita, M.; Kojima, Y.; Ishida, H.; Kiso, M. Synthesis of cerebroside, lactosyl ceramide, and ganglioside GM3 analogs containing beta-thioglycosidically linked ceramide. *Carbohydr. Res.* **1991**, *214*, 43–53. [CrossRef] [PubMed]
34. Albrecht, B.; Pütz, U.; Schwarzmann, G. Synthesis of fluorescent and radioactive analogues of two lactosylceramides and glucosylceramide containing beta-thioglycosidic bonds that are resistant to enzymatic degradation. *Carbohydr. Res.* **1995**, *276*, 289–308. [CrossRef] [PubMed]
35. Miura, Y.; Arai, T.; Yamagata, T. Synthesis of amphiphilic lactosides that possess a lactosylceramide-mimicking N-acyl structure: Alternative universal substrates for endo-type glycosylceramidases. *Carbohydr. Res.* **1996**, *289*, 193–199. [CrossRef]
36. Ohlsson, J.; Magnusson, G. Analogues of glycosphingolipids and glycerolipids suitable for conjugation to gold- and amino-functionalised surfaces. *Tetrahedron* **2000**, *56*, 9975–9984. [CrossRef]
37. Zhang, Y.; Toyokuni, T.; Ruan, F.; Hakomori, S.I. A one pot synthesis of mono- and di-lactosyl sphingosines. *Glycoconj. J.* **2001**, *18*, 557–563. [CrossRef]
38. Liu, Y.; Bittman, R. Synthesis of fluorescent lactosylceramide stereoisomers. *Chem. Phys. Lipids* **2006**, *142*, 58–69. [CrossRef]
39. Liu, Y.; Ding, N.; Xiao, H.; Li, Y. Efficient syntheses of a series of glycosphingolipids with 1,2-trans-glycosidic linkages. *J. Carbohydr. Chem.* **2006**, *25*, 471–489. [CrossRef]
40. Morales-Serna, J.A.; Boutureira, O.; Díaz, Y.; Matheu, M.I.; Castillón, S. Highly efficient and stereoselective synthesis of beta-glycolipids. *Org. Biomol. Chem.* **2008**, *6*, 443–446. [CrossRef]
41. Morales-Serna, J.A.; Díaz, Y.; Matheu, M.I.; Castillón, S. Efficient synthesis of β-glycosphingolipids by reaction of stannylceramides with glycosyl Iodides promoted by TBAI/AW 300 molecular sieves. *Eur. J. Org. Chem.* **2009**, *2009*, 3849–3852. [CrossRef]
42. Tanaka, H.N.; Yoshimura, Y.; Dovichi, N.J.; Palcic, M.M.; Hindsgaul, O. A concise chemical synthesis of a fluorescent βGal-(1,4)-S-βGlc-Cer derivative and its enzymatic elongation by glycosyltransferases. *Tetrahedron Lett.* **2012**, *53*, 1812–1815. [CrossRef]
43. Nagatsuka, T.; Uzawa, H.; Sato, K.; Kondo, S.; Izumi, M.; Yokoyama, K.; Ohsawa, I.; Seto, Y.; Neri, P.; Mori, H.; et al. Localized surface plasmon resonance detection of biological toxins using cell surface oligosaccharides on glyco chips. *ACS Appl. Mater. Interfaces* **2013**, *5*, 4173–4180. [CrossRef] [PubMed]
44. Cheng, J.M.; Dangerfield, E.M.; Timmer, M.S.; Stocker, B.L. A divergent approach to the synthesis of iGb3 sugar and lipid analogues via a lactosyl 2-azido-sphingosine intermediate. *Org. Biomol. Chem.* **2014**, *12*, 2729–2736. [CrossRef]
45. Son, S.H.; Daikoku, S.; Ohtake, A.; Suzuki, K.; Kabayama, K.; Ito, Y.; Kanie, O. Syntheses of lactosyl ceramide analogues carrying novel bifunctional BODIPY dyes directed towards the differential analysis of multiplexed glycosphingolipids by MS/MS using iTRAQ. *Chem. Commun.* **2014**, *50*, 3010–3013. [CrossRef]
46. D'Angelo, K.A.; Taylor, M.S. Borinic acid-catalyzed stereo- and site-selective synthesis of β-glycosylceramides. *Chem. Commun.* **2017**, *53*, 5978–5980. [CrossRef]

*molecules*

MDPI

Article

# Towards Understanding the Basis of Brucella Antigen–Antibody Specificity

**Amika Sood [1], David R. Bundle [2,*] and Robert J. Woods [1,*]**

[1] Complex Carbohydrate Research Center, University of Georgia, Athens, GA 30602, USA; amika@uga.edu
[2] Department of Chemistry, University of Alberta, Edmonton, AB T6G 2R3, Canada
* Correspondence: dave.bundle@ualberta.ca (D.R.B.); rwoods@ccrc.uga.edu (R.J.W.)

**Abstract**

Brucellosis continues to be a significant global zoonotic infection, with diagnosis largely relying on the detection of antibodies against the Brucella O-polysaccharide (O-PS) A and M antigens. In this study, computational methods, including homology modeling, molecular docking, and molecular dynamics simulations, were applied to investigate the interaction of the four murine monoclonal antibodies (mAbs) YsT9.1, YsT9.2, Bm10, and Bm28 with hexasaccharide fragments of the A and M epitopes. Through stringent stability criteria, based on interaction energies and mobility of the antigens, high-affinity binding of A antigen with YsT9.1 antibody and M antigen with Bm10 antibody was predicted. In both the complexes hydrophobic interactions dominate the antigen–antibody binding. These findings align well with experimental epitope mapping, indicating YsT9.1's preference for internal sequences of the A epitope and Bm10's preference for internal elements of the M epitope. Interestingly, no stable complexes were identified for YsT9.2 or Bm28 interacting with A or M antigen. This study provides valuable insights into the mechanism of molecular recognition of Brucella O-antigens that can be applied for the development of improved diagnostics, synthetic glycomimetics, and improved vaccine strategies.

**Keywords:** *Brucella* A and M antigens; *Brucella* monoclonal antibodies; homology modeling; Vina-Carb docking; MD simulations; MM-GBSA binding energy; bound complexes; GLYCAM; AMBER

## 1. Introduction

Brucellosis is regarded by the World Health Organization as one of the most serious zoonotic bacterial diseases [1]. It is a costly, highly contagious disease that affects cattle, sheep, goats, pigs, camels, and other productive animals worldwide [2]. It is endemic in a number of countries and costs of the disease worldwide amount annually to several billion dollars. The detection of antibodies against *Brucella* antigens has been at the forefront of the understanding, control, and eradication of brucellosis for over 120 years. Presumptive diagnosis of brucellosis primarily depends on detection, in animal or human sera, of antibodies to the O-polysaccharide (O-PS) component of *Brucella* lipopolysaccharide [2].

Wilson and Miles established in 1932 that the major serological antigens of pathogenic Brucella species were two surface antigens A and M simultaneously expressed on all smooth strains of the bacterium albeit in varying quantitative amounts within a single molecule [3]. Strains designated $A^{+}M^{-}$ possessed a preponderance of A over M antigen and $M^+A^-$ strains higher M over A. It is the detection of antibodies to these A and M antigens that forms the basis for the diagnosis of brucellosis [2,4].

In 1939, the A and M antigens were found to contain a polyhydroxyamino compound and formate [5,6]. This insight was confirmed when 4-formamido-4,6-dideoxy-α-D-mannopyrannose (Rha4NFo) was isolated as the sole monosaccharide component of the repeating portion of O-PS from *B. abortus* lipopolysaccharide [7]. *Brucella* A dominant strains were originally suggested to be composed of exclusively α1,2-linked D-Rha4NFo residues [7]. *Brucella* M dominant strains were proposed to contain one α1,3-linked D-Rha4NFo for every four α1,2-linked D-Rha4NFo residues [8]. However, it was soon recognized that this model was not complete, and M epitopes were always present in all smooth strains of the pathogen [9].

Recently the fine structure of *Brucella* O-PS has been re-examined and a revised structure proposed [10] (Figure 1). This work confirmed the original single molecule paradigm of Wilson and Miles. All *Brucella* strains with a smooth LPS (except *B. suis* biovar 2 [11]) essentially consist of two polymeric elements, a 1,2-linked homopolymer that is capped by at least one tetrasaccharide repeating unit containing a α1,3-linked D-Rha4NFo (Figure 1). This latter tetrasaccharide defines the M antigen, while the exclusively α1,2-linked polysaccharide component defines the A antigen. An epitope referred to as C/Y requires from 2 to 4 α1,2-linked D-Rha4NFo residues [12]. It can be appreciated from the general *Brucella* O-antigen structure (Figure 1) that the A antigen is universally present in all smooth *Brucella* with variable expression of the M antigen (Figure 1, $m \geq 2$) [10]. The simultaneous expression of two antigenic determinants in a single homopolysaccharide and the recognition of these by antibodies has considerable practical implications in recognition of the antigens by antibodies.

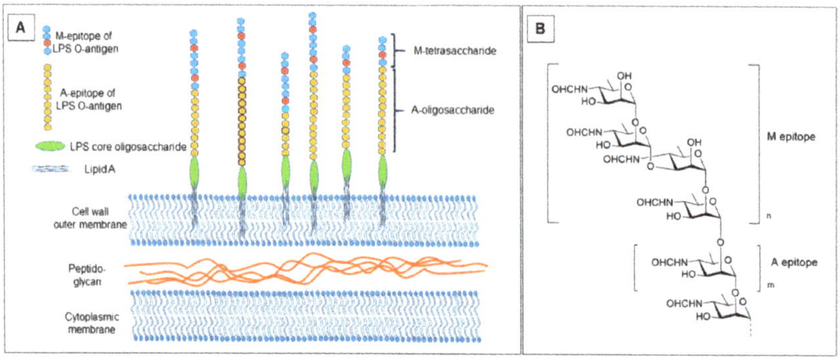


**Figure 1.** (**A**) Schematic representation of LPS molecules anchored by their Lipid A component in the outer membrane of the Gram-negative coccobacilli, Brucella. The majority of Brucella strains expressing O-polysaccharide have an internal sequence of 1,2 linked 4-formamido-4-deoxy D mannopyranose residues of variable length typically 12–16 residues capped at their reducing end by 1 or 2 copies of M tetrasaccharide. *B. melitensis* 16M may contain as many as 34 4-formamido-4-deoxy-D-mannopyranose residues and include more than 2 copies of capping tetrasaccharide. O-polysaccharides are attached through the core oligosaccharide to Lipid A. (**B**) Definitive structural studies established the structure of the *Brucella* O-antigen with a capping M epitope terminating an A type α1,2-linked D-Rha4NFo polymer, where usually $m \leq 16$ [10]. The capping tetrasaccharide occurs a minimum of once, $n = 1$ or may be repeated several times, $n > 1$. Epitope mapping suggests the A epitope likely comprises four or more α1,2-linked D-Rha4NFo residues, e.g., [→2)αRha4NFo(1-]$_4$ [13,14]. The M epitope is minimally the disaccharide Rha4NFo(1→3)Rha4NFo but could be as large as the capping tetrasaccharide, [Rha4NFo(1→2)Rha4NFo(1→3)Rha4NFo(1→2) Rha4NFo(1-] [13,14].

During studies that helped establish the structural basis of the A and M antigens, several monoclonal antibodies, YsT9.1, YsT9.2 Bm10, and Bm28 were developed that proved crucial in linking antibody binding profiles to the fine structures of the two antigens [7–9,13,15]. Subsequently the binding sites of these have been subjected to extensive epitope mapping by a large panel of purified polysaccharide antigens and synthetic oligosaccharides [14–17]. To gain insight into the potential binding modes of these antigens with these distinct monoclonal antibodies, the antibodies have been sequenced and attempts were made to obtain crystal structures of their antigen bound forms. Only one of the four (YsT9.1), an IgG2b [13] with binding profile A >> M yielded a crystal structure without a bound antigen [18] and a rudimentary binding mode was proposed for the interaction with the A antigen [19]. The remaining three antibodies all belonged to the IgG3 subclass, which in our hands failed to provide Fab fragments that crystallized with or without an antigen [Bundle unpublished results]. Two of the three, (Bm10 and Bm28), had binding profiles M >> A [15] and one (YsT9.2) [13] bound A and M antigens with equal affinity [15]. Previously, we have employed a range of experimental approaches, including point mutagenesis [20], chemical modification of the ligands [21], saturation transfer difference NMR spectroscopy [20], x-ray crystallography [21], and deuterium exchange mass spectrometry [22] to characterize antibody paratopes and ligand epitopes, many of which leverage the power of MD simulations to guide experimental design and data interpretation. However homogeneous samples of the A and M epitopes of Brucella O-antigen (Figure 1) are currently unavailable for such experimental work.

This study computationally explores the binding of four murine monoclonal antibodies (mAbs), namely YsT9.1, YsT9.2, Bm10, and Bm28, to hexasaccharide fragments of the A and M epitopes of Brucella O-antigen. By integrating homology modeling, Vina-Carb molecular docking, and molecular dynamics simulations, we attempted to clarify the molecular basis of their differential binding specificities. The docked poses can be further interrogated using extensive experimental epitope mapping data published for these antibodies [14,16,17]. This computational modeling approach aims to overcome a number of unique hurdles to the A and M antigens, including frame shifting between overlapping epitopes and the occurrence of two energetically accessible rotamers of the formamido group [23].

## 2. Results

### 2.1. Docking Analysis

To identify the preferred binding sites of the antibody, AutoDock Vina-Carb was used to perform docking of fragments of the A and M epitopes of the O-antigens to the murine antibodies YsT9.1, YsT9.2, Bm10, and Bm28. The docked poses were then divided into clusters using a partitional k-means clustering algorithm. Each docking simulation resulted in a different number of clusters (Table S1) each with a different number of poses. The top ranked pose from each cluster was selected for MD simulation (Figure 2).

### 2.2. Ligand Stability

To assess the stability of the docked antigen poses when bound to the antibodies, each complex was subjected to independent molecular dynamics (MD) simulations (100 ns) in triplicate. The Root Mean Square Fluctuation (RMSF) of the antigen atoms was calculated from the final 25 ns of simulation to quantify their mobility within the binding site, with lower RMSF values indicating more stable complexes. The average RMSF values across the three replicates for each antibody–antigen cluster are presented in the top graphs in each panel of Figure 3. Significant variation in antigen mobility was observed across the different antibody–antigen complexes and even among the different docking clusters for the same antibody and antigen.

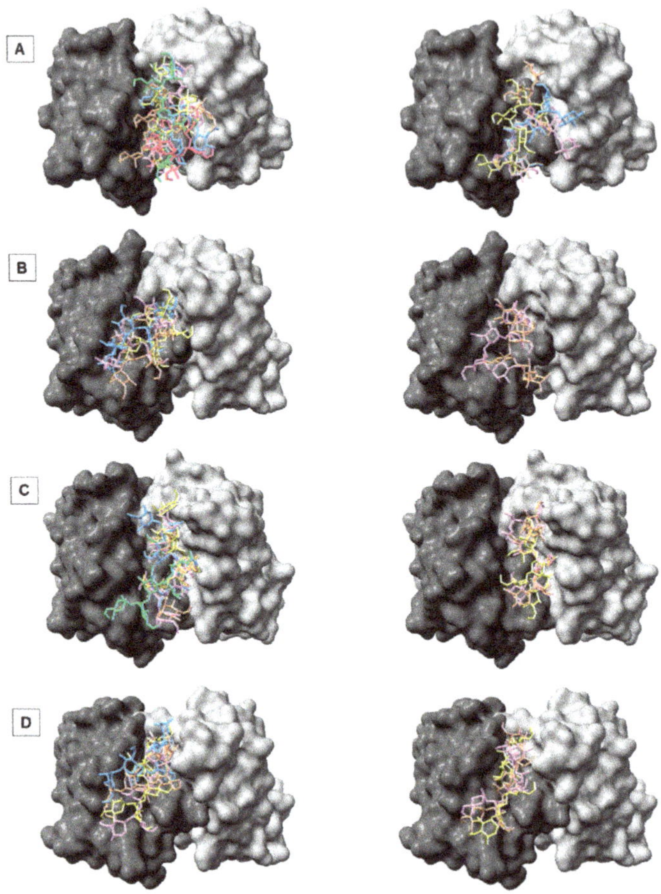

**Figure 2.** The top docked poses from each cluster (orange, pink, yellow, blue, green, and red) obtained after docking for the A antigen (**left**) and M antigen (**right**) to murine *Brucella* antibodies YsT9.1 (**A**), YsT9.2 (**B**), Bm10 (**C**), and Bm28 (**D**). The antigens are shown as sticks and the antibodies as solvent-accessible surface. The heavy chain of the antibodies is colored dark gray, while the light chain of the antibodies is colored light gray.

For example, in panel A representing YsT9.1, most A antigen clusters had relatively low RMSF values (<2 Å), while cluster 'f' has an average RMSF of 6.7 Å, breaking away from this trend. Overall, the M antigen clusters exhibited higher fluctuations compared to the A antigen interacting with YsT9.1. Similarly, panel B representing YsT9.2, panel C representing Bm10, and panel D representing Bm28 from Figure 3 showed varied RMSF values across their numerous clusters. The overall RMSF values across all calculated ligand RMSF values ranged from a minimum of 0.6 Å to a maximum of 13.3 Å, with a median of 1.8 Å.

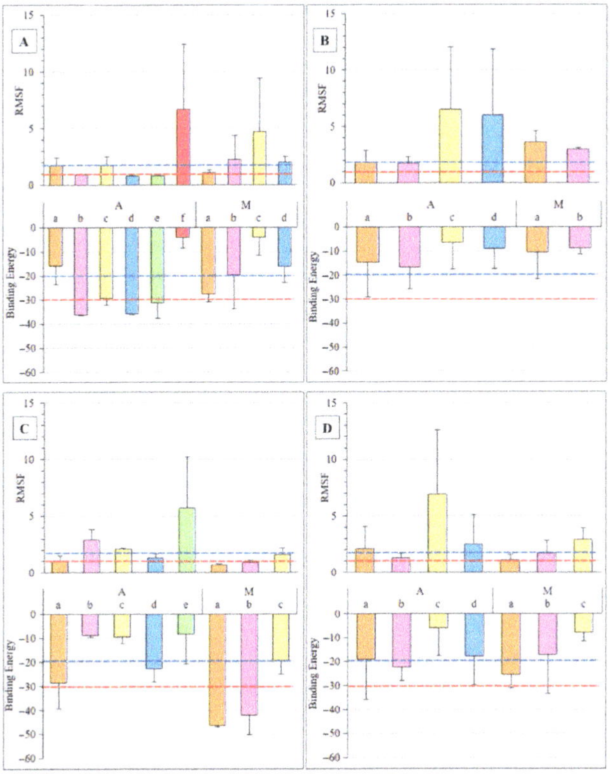


**Figure 3.** The RMSF (Å) and Binding Energy (kcal/mol) values computed using the last 25 ns of the MD simulations for YsT9.1 (**A**), YsT9.2 (**B**), Bm10 (**C**), and Bm28 (**D**). The A and M antigens are labeled on the *x*-axis and the different clusters (shown structurally in Figure 3) are also indicated alphabetically labeled here (a: orange, b: pink, c: yellow, d: blue, e: green, f: red). The blue lines indicate the median values, and the red lines indicate the cut-off values.

### 2.3. Interaction Energies

The interaction energy for each complex was quantified by calculating the MM-GBSA binding energy with IGB2 parameters, as advised previously [24], from snapshots (2500) taken from the last 25 ns of the MD simulations. Lower (more negative) binding energy values indicate a stronger, more favorable interaction. The average binding energies from the triplicate MD simulation runs for each cluster are presented in the bottom graphs of each panel of Figure 3. Similarly to the RMSF data, the binding energy values displayed considerable variation from complex to complex and across different clusters. For instance, in the bottom graph of panel A of Figure 3 (YsT9.1), several A antigen clusters show favorable (negative) binding energies, while M antigen clusters have less favorable values, consistent with YsT9.1's preference for the A antigen. In the bottom graph of panel C of Figure 3 (Bm10), some M antigen clusters exhibit highly favorable binding energies compared to A antigens. Across all calculated binding energy values, the range extended from a minimum (most favorable) of −50.8 kcal/mol to a maximum (least favorable) of 1.3 kcal/mol, with a median of −19.2 kcal/mol.

Given that both lower RMSF (indicating higher stability) and lower binding energy (indicating stronger affinity) are indicative of a more favorable interaction, a direct correlation between these two parameters was anticipated. As illustrated in Figure 4, a positive correlation was observed between RMSF and binding energy within a specific range, where increasing stability (decreasing RMSF from approximately 2 Å down to 1 Å) corresponded to stronger binding (decreasing binding energy from approximately 0 kcal/mol down to −30 kcal/mol). Beyond these values, the relationship became asymptotic: very low RMSF values (below ~1 Å) were associated with strongly favorable binding energies (below approximately −30 kcal/mol), while high RMSF values (above ~2 Å) consistently corresponded to binding energies close to 0 kcal/mol, indicative of unstable or non-binding poses. Based on this observed correlation, and to focus on stable interactions, only complexes with ligand RMSF values below 1 Å and binding affinities below −30 kcal/mol in each of the three independent MD runs were considered for further analysis. Clearly, the interaction energies will depend on the calculation method and on the system; however, the RMSF–energy correlation is a useful general feature for developing acceptance criteria. Applying these criteria, the only stable complexes identified were the A antigen bound to YsT9.1 through clusters 'b' and 'd', and the M antigen bound to Bm10 through cluster 'a'. These stable interactions are visually evident in Figure 3 as specific bars within Panels A and C falling below both the red dotted line in the RMSF plot (~1 Å) and the red dotted line in the Binding Energy plot (~−30 kcal/mol).

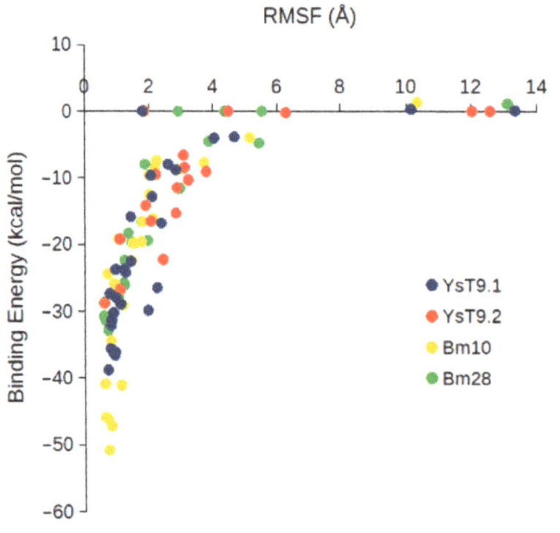


**Figure 4.** RMSF (Å) versus binding energy (kcal/mol) for all mAb-ligand poses (YsT9.1, YsT9.2, Bm10, and Bm28 shown as blue, red, yellow, and green circles, respectively) simulated in triplicate by MD indicating the strong correlation between low RMSF and strong binding. Complexes with either RMSF below 1 Å or binding energy below −30 kcal/mol were considered stable.

### 2.4. Binding Modes

To gain a better understanding of how the antigens interact within the identified stable complexes, we analyzed the MD trajectories to identify distinct binding modes. For the stable YsT9.1 antibody complex with A antigen (derived from initial docking clusters 'b' and 'd'), all frames (2500 frames from the last 25 ns) from the three independent MD simulations for each of the two clusters (total six simulations) were merged. Similarly, for the stable Bm10 antibody complex with M antigen (derived from initial docking cluster 'a'), all frames (2500 frames from the last 25 ns) from its three independent MD simulations were merged. These merged trajectories were then subjected to clustering based on the 3D coordinates of the antigen atoms. This analysis yielded three representative structures for both the YsT9.1-A complex and the Bm10-M complex, representing the most populated conformational states of the antigen within the binding site (Figure 5).

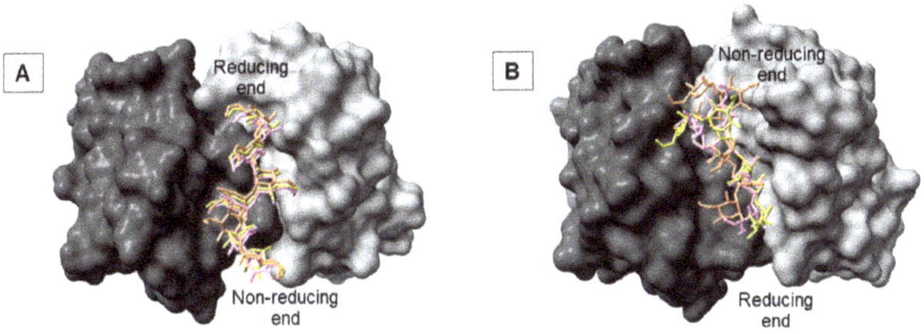


**Figure 5.** Representative stable-binding modes (orange, pink, yellow) of mAb Ys9.1 in complex with A antigen (**A**) and mAb Bm10 in complex with M antigen (**B**). The antigens are shown as sticks and the antibodies as solvent-accessible surface. The heavy chain of the antibodies is colored dark gray, and the light chain of the antibodies is colored light gray.

The structural variations among these representative binding modes were assessed by calculating the Root Mean Square Deviation (RMSD) of the ring atoms (C1, C2, C3, C4, C5, O5) of the hexasaccharide antigens. For the YsT9.1-A complex, the three representative structures showed RMSDs ranging from 0.8 Å to 2.1 Å relative to each other (Table S2). As visualized in Figure 5A, these representative structures occupied largely overlapping 3D space within the YsT9.1 binding site. Therefore, these variations were considered to represent minor conformational adjustments around a singular, dominant binding mode. Consequently, for subsequent analyses of the YsT9.1-A interaction, the binding energy values were averaged across all six independent MD simulation runs (three for cluster 'b' and three for cluster 'd').

For the Bm10-M complex, the three representative structures exhibited larger RMSDs, ranging from 1.4 Å to 3.9 Å (Table S3). Visual inspection (Figure 5B) revealed that two of these representative structures (depicted in yellow and pink) occupied a similar region of the binding site, defining what was termed Binding Mode 1. The third representative structure (depicted in orange) occupied a distinct spatial location, which was designated as Binding Mode 2. Accordingly, for the Bm10-M interaction, binding energy values for Binding Mode 1 were averaged over the two MD simulations whose trajectories were closest to the centroids of the yellow and pink representative structures. The binding energy for Binding Mode 2 was derived from the single MD simulation whose trajectory was closest to the centroid of the third representative structure. This approach allowed us to analyze the energetic contributions separately for these distinct binding orientations observed in the stable Bm10-M complex.

### 2.5. *Contribution to Binding from the Perspective of the Antibody*

To identify the amino acid residues of the antibody that are critical for interaction with the antigens (the paratope), the per-residue contributions to the total MM-GBSA binding energy were analyzed. Residues exhibiting favorable contributions stronger than -1.0 kcal/mol were considered. For the YsT9.1 antibody interacting with the A antigen, key residues contributing significantly to the binding were primarily located in the heavy chain, including Tyr32H, Tyr33H, Asp101H, Tyr103H, Pro105H, and Ala106H, whereas the residues making contributions to the binding from the light chain were Tyr32L, Tyr50L, and Gly91L (Table S4, Figure 6A left).

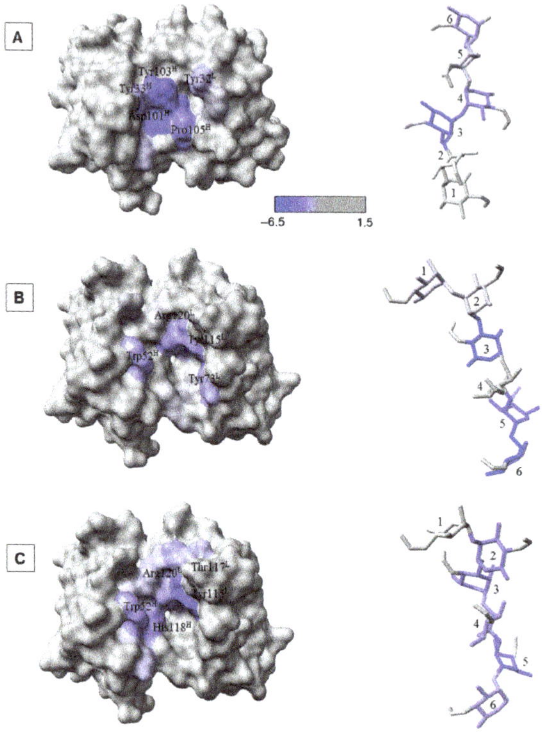


**Figure 6.** Per-residue binding energy contributions (blue color scale) for the antibody and antigen in the YsT9.1-A complex (**A**), Bm10-M complex in binding mode 1 (**B**), and Bm10-M complex in binding mode 2 (**C**). The antibodies are shown with a solvent-accessible surface (**left**) and the antigen as sticks (**right**). The H and L superscripts indicate heavy chain and light chains, respectively. Antigen residues are numbered 1 through 6, corresponding from the non-reducing to the reducing end.

Analysis of the per-residue contributions for the Bm10 antibody's interaction with the M antigen revealed distinct sets of contributing residues for its two binding modes. In Binding Mode 1, the amino acids making significant contributions from the heavy chain were Trp52$^{H}$, Arg78$^{H}$, Gly119$^{H}$, and His120$^{H}$, and from the light chain were Leu70$^{L}$, Tyr73$^{L}$, Gln113$^{L}$, Tyr115$^{L}$, Tyr118$^{L}$, and Arg120$^{L}$ (Table S5, Figure 6B left). Whereas, for Binding Mode 2, a somewhat different set of residues contributed to the binding, including heavy chain residues Tyr51$^{H}$, Trp52$^{H}$, Trp66$^{H}$, Ile69$^{H}$, Arg78$^{H}$, and His118$^{H}$, and light chain residues Tyr49$^{L}$, Tyr56$^{L}$, Gln113$^{L}$, Tyr115$^{L}$, Tyr116$^{L}$, Thr117$^{L}$, Tyr118$^{L}$, and Arg120$^{L}$ (Table S6, Figure 6C left). Both the binding modes shared five residues that are involved in the binding of the M antigen (Trp52$^{H}$, Arg78$^{H}$, Gln113$^{L}$, Tyr115$^{L}$, and Arg120$^{L}$), which implied that these residues formed an integral part of the Bm10 binding site, independent of the specific orientation of the antigen.

The nature of the interactions between the antigen and antibodies can be deduced by studying the energetic decomposition of MM-GBSA binding energies. The magnitude of the hydrophobic contributions is indicated by the sum of van der Waals and nonpolar desolvation energies, while that of the polar contribution is indicated by the sum of electrostatic and polar desolvation energies. The energetic decomposition revealed that the interactions in all three complexes were predominantly hydrophobic. For the YsT9.1-A interaction, the hydrophobic contribution was $-19.6 \pm 0.3$ kcal/mol, which was significantly larger in magnitude than the electrostatic contribution of −1.7 ± 0.8 kcal/mol. Similarly, for the Bm10-M complex in Binding Mode 1, hydrophobic interactions contributed −20.1 ± 0.6 kcal/mol, while electrostatic contributions were a less favorable 2.3 ± 4.3 kcal/mol. In Binding Mode 2 of Bm10-M, the hydrophobic contribution was even more substantial at −28.7 kcal/mol, compared to an unfavorable electrostatic contribution of 2.1 kcal/mol. This strongly indicates that the burial of hydrophobic surfaces upon complex formation is a major driving force for stable binding in these systems.

The difference in contributions based on the antibody chains (heavy and light) were also analyzed and indicated a notable difference between the behavior of YsT9.1 and Bm10 antibodies. For the YsT9.1-A complex, the heavy chain (−16.7 ± 0.3 kcal/mol) provided a substantially larger contribution to the overall binding energy compared to the light chain (−4.6 ± 0.1 kcal/mol). Alternately, for the Bm10-M complex, the light chain contributed more favorably to binding (−11.7 ± 0.5 kcal/mol in Binding Mode 1 and −16.1 kcal/mol for Binding Mode 2) compared to the heavy chain (−6.1 ± 0.8 kcal/mol in Binding Mode 1 and −10.1 kcal/mol in Binding Mode 2). This implies distinct roles for the heavy and light chains in shaping the binding interfaces.

### 2.6. Contribution to Binding from the Perspective of the Antigen

To identify the monosaccharide residues within the oligosaccharide ligands that contribute most to the stability of the interaction (the epitope residues) the per-residue contributions of the antigens to the binding energy in the stable complexes were examined. The residues of the carbohydrate antigen are numbered 1 through 6, corresponding to the non-reducing end to the reducing end (Figure 6 right). The binding orientation of the antigen was notably different between the YsT9.1-A and Bm10-M complexes. In the Bm10 binding site (Figure 6B,C right), the antigen was bound in a flipped orientation relative to its position in YsT9.1 (Figure 6A right). Despite this difference in overall orientation, it was observed that across all three stable binding cases the residue at the non-reducing end of the antigen (Residue 1) consistently made the weakest contribution to the overall binding energy (Tables S7–S9). Rather, the significant interactions for stable binding primarily involved the internal and reducing-end residues of the hexasaccharide antigens.

### 2.7. Comparison of Preferred Binding Modes with Experimental Epitope Binding Data

Since the YsT9.1 and YsT9.2 mAbs were generated after immunization with killed cells of *Yersinia enterocolitica* O:9 that possesses an exclusively α1,2-linked Rha4NFo O-polysaccharide [13], it would be expected that their binding sites would not readily accept 1,3 linked elements characteristic of the M antigen. This is confirmed by the modeling data. Also, a 1,2-linked pentasaccharide was the most active oligosaccharide inhibitor of YsT9.1 binding to the A polysaccharide [15] and a hexasaccharide conjugated to protein was shown to be a better binder to YsT9.1 than either 1,2-linked penta- or tetrasaccharides [14]. This is fully consistent with the extended internal 1,2-linked internal sequences predicted by computational modeling reported here.

The internal trisaccharide Rha4NFo(1→3)Rha4NFo(1→2)Rha4NFo sequence of the M tetrasaccharide was a more effective inhibitor of Bm10 binding with the M-polysaccharide, while the Rha4NFo(1→2)Rha4NFo(1→3)Rha4NFo trisaccharide was weakly bound by it [17]. Consistent with calculations and predicted binding poses, the terminal non-reducing Rha4NFo of the M tetrasaccharide was not essential for binding Bm10, a fact supported when a mannose residue replaced the terminal non-reducing Rha4NFo residue [17]. Furthermore, a pentasaccharide (Rha4NFo(1→3)Rha4NFo(1→2)Rha4NFo(1→2)Rha4NFo(1→2)Rha4NFo) was a potent inhibitor of Bm10 [15]. This supports the prediction of calculations that a

1,3 linked disaccharide provides the primary focus of protein ligand binding and that additional Bm10- oligosaccharide interactions at the reducing end are possible.

## 3. Discussion

The study of antibody–antigen interactions is crucial for understanding the molecular basis of immune recognition and for developing targeted therapeutic strategies. In this work, we performed an extensive computational analysis of the binding of the four murine mAbs, YsT9.1, YsT9.2, Bm10, and Bm28, against the A and M epitopes of Brucella O-antigen. First, the 3D structures of the YsT9.2, Bm10, and Bm28 antibodies were generated by homology modeling. This was followed by docking analysis performed by using Vina-Carb, which generated a diverse set of binding poses for each antibody–antigen pair. Subsequently, the clustering of these poses provided distinct binding orientations for further investigation by MD simulations.

The MD simulations were performed in triplicates (100 ns each) and provided critical insights into the stability of the docked complexes through the Root Mean Square Fluctuation (RMSF) analysis. Significant variations were observed in the mobility of the antigens interacting with different antibodies and even between replicates of the same complex.

To further understand the basis of antibody–antigen preferences, the strength of their interactions was quantified through MM-GBSA binding energy calculations. There was a large variation in the binding strength for different poses ranging from −50.8 kcal/mol to 1.3 kcal/mol, reflecting the diversity of the binding interactions. A strong positive correlation was observed within a specific range of stable complexes (binding energy below −30 kcal/mol and RMSF below 2 Å), where lower RMSF (greater stability) corresponded to more favorable binding energies. However, beyond these values, the relationship became asymptotic, where poses with high mobility or RMSF values had low binding affinities, while stable or low RMSF poses (<1 Å) also had high binding affinities (<−30 kcal/mol). This correlation provided a key criterion for identifying stable binding poses from a large number of trajectories.

On the basis of this stringent stability criteria, only two complexes were deemed to be high affinity, the A antigen bound to YsT9.1 antibody (clusters 'b' and 'd') and the M antigen bound to Bm10 antibody (cluster 'a'). This finding aligns well with experimental observations indicating YsT9.1's preference for the A epitope and Bm10's preference for the M epitope. However, no stable complexes were identified for YsT9.2 or for Bm28 antibodies with either A or M antigens. This could be a result of the lack of the antibody flexibility while docking, as rigid protein docking may not fully capture the conformational changes that can happen in the antibody upon antigen binding. Additionally, although the antibody homology models have high GMQE scores, they may not represent the true 3D structures of the antibodies. These shortcomings could potentially be addressed through flexible antibody docking or enhanced sampling techniques.

Detailed analysis of the stable YsT9.1-A and Bm10-M complexes through trajectory clustering revealed distinct, yet in the case of Bm10-M, multiple binding modes. For YsT9.1-A, the representative structures from the two stable clusters ('b' and 'd') occupied similar 3D space, suggesting a predominant binding mode for this interaction. On the other hand, the Bm10-M complex presented with two different binding modes from a singular stable cluster 'a'. This observation suggests the potential for a single antibody to recognize an oligosaccharide epitope in more than one orientations.

Analysis of the per-residue contributions to binding energy shed light on the molecular determinants of specificity. In both the stable YsT9.1-A and Bm10-M complexes, a significant number of hydrophobic amino acids were found to contribute favorably to binding.

The energetic decomposition confirmed the highly hydrophobic nature of these interactions, with van der Waals and nonpolar desolvation energies being the dominant driving forces, while electrostatic contributions were considerably smaller and even unfavorable in some cases for Bm10-M. The importance of hydrophobic contacts in antibody-carbohydrate antigen interactions have also been observed previously [25] and found in crystal structures of antibody–carbohydrate complexes [26–28].

The differences in the contribution to the binding energy from the heavy and the light chain of the antibodies were also observed. For the YsT9.1-A antigen complex, the heavy chain contributed more significantly to the binding energy with key contributions coming from residues Tyr32$^{H}$, Tyr33$^{H}$, Asp101$^{H}$, and Tyr103$^{H}$. On the contrary, in the case of Bm10-M antigen interactions the light chains were favored, but significant contributions came from both heavy and light chains. Interestingly, the two binding modes of Bm10-M complexes have five common residues (Trp52$^{H}$, Arg78$^{H}$, Gln113$^{L}$, Tyr115$^{L}$, and Arg120$^{L}$) with significant contribution to binding. This suggests that there might be a conserved binding site present on the Bm10 antibody that can accommodate the M antigen in multiple orientations.

Analysis of the antigen's contribution revealed an interesting pattern: in all identified stable complexes, the residue at the non-reducing end of the hexasaccharide had one of the lowest contributions to the overall binding energy. This implies that the antibodies have evolved to recognize an internal epitope rather than the terminal sequence of the polysaccharide antigen. Additionally, the antigens interacted with the antibodies in a completely different orientation, where the direction of binding was flipped between the YsT9.1-A and Bm10-M complexes. This suggests that while both antibodies recognize their preferred antigens, they do so by orienting the antigen differently within their binding pockets.

## 4. Methods

### *4.1. Brucella A and M Specific mAbs*

#### 4.1.1. Monoclonal Antibodies

The four monoclonal antibodies studied here were generated in two separate experiments. YsT9.1 and YsT9.2 resulted from hybridomas generated from the spleens of mice immunized with the killed cells of *Yersinia enterocolitica* O:9 [13]. The cell wall of this bacterium contains an O-antigen of exclusively α1,2 linked D-Rha4NFo residues [29] and should therefore result in antibodies specific for α1,2 linked D-Rha4NFo sequences. Monoclonal antibodies Bm10 and Bm28 were generated from mice immunized with *Brucella melitensis* 16M [15], the prototypical *Brucella* variant expressing the M antigen. This variant expresses at least two repeating units of the M tetrasaccharide (Figure 1) [10].

#### 4.1.2. Sequencing of mAbs

The sequence of YsT9.1 Fab was determined by conventional protein sequencing methods [18]. The sequences of YsT9.2, Bm10, and Bm10 Fv domains were determined from mRNA extracted from hybridoma cells performed under contract (Fusion Antibodies, Belfast, UK). Protein sequences are provided in Supplementary Material (Figures S1–S6). Accession numbers for all sequence data deposited with GenBank and ABCD databases are listed at the end of the paper.

### *4.2. Modeling*

#### 4.2.1. Structure Preparation

Murine mAbs Bm10, Bm28, YsT9.1, and YsT9.2 were considered for the analysis. The 3D structure of YsT9.1 as apo-protein is available in the PDB database with PDB

ID 1MAM [18]. The 3D structures of the light and heavy chains of Bm10, Bm28, and YsT9.2 were generated using AlphaFold DB models through the SWISS-MODEL homology modeling server (swissmodel.expasy.org) [30]. All of the AlphaFold DB model structures had Global Model Quality Estimate (GMQE) quality scores greater than 0.83. GMQE is a quality estimate that integrates properties from target-template alignment and template structure to assess model quality. It aids in selecting optimal templates for modeling and is updated after model construction to enhance reliability. A homology structure with a GMQE score value above 0.7 is considered reliable. The Carbohydrate Builder at GLYCAM-Web (https://glycam.org/cb/ (accessed on 20 November 2023)) [31] was used to generate 3D structures of hexasaccharide fragments of the A and M epitopes of the Brucella O antigen (Figure 7). Chimera [32] was used to modify the hydroxyl group at O4 position to a formamido group in Z conformation.

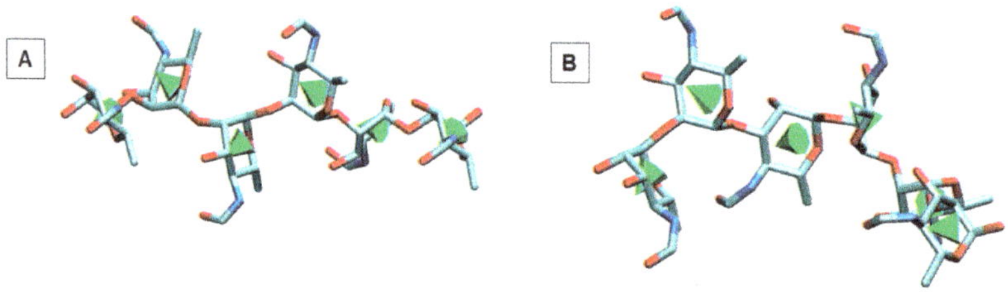


**Figure 7.** (**A**). The 3D structures of the A antigen: Rha4NFoα(1→2)Rha4NFoα(1→2)Rha4NFoα (1→2) Rha4NFoα(1→2)Rha4NFoα(1→2)Rha4NFo (top). (**B**). The 3D structures of the M antigen: Rha4NFoα(1→2)Rha4NFoα(1→3) Rha4NFoα(1→2)Rha4NFoα(1→2)Rha4NFoα(1→2) Rha4NFo (bottom), generated using the Carbohydrate Builder at GLYCAM-Web [31]. All the formamido groups in both the antigens are in Z conformation.

#### 4.2.2. Docking and Clustering

All the input files for docking were prepared using AutoDock Tools [33] and docking was performed using Vina-Carb [34] with a grid box of dimensions x = 26, y = 26, and z = 38 Å, which was placed at the center of the complementarity determining regions (CDRs) of the antibody [35]. Default values were used, with the following exceptions: exhaustiveness = 80, num_modes = 100, chi_cutoff = 2, and chi_coeff = 1. All of the hydroxyl groups and glycosidic torsion angles were treated as flexible, while the protein was held rigid; the formamido group was also fixed in the Z (s-cis) conformation. Subsequently, the docked poses were clustered using the partitional k-means clustering algorithm provided by the Multi-scale modeling tools in the structural biology (MMTSB) [36] toolset. Only the highest-ranked pose from each cluster, as determined by Vina-Carb, was selected for subsequent MD simulation analysis.

#### 4.2.3. Molecular Dynamics (MD) Simulations

Antechamber was used to develop GAFF [37] partial atomic charges and force field parameters for the formamido group, ff14SB [38] parameters were employed for the amino acids, and GLYCAM06j [39] parameters for the sugar. All the complexes were solvated in a truncated octahedral box of TIP5P [40] water molecules with counter ions (Cl-) added to neutralize the charge, using the tLEAP module of AMBER. Previously [41], we have shown that the TIP5P model performs well with the GLYCAM/AMBER force field. All histidine residues were considered neutral with a hydrogen atom at the ε-position. The ionization states of the ionizable side chains (ASP, GLU, ARG, LYS) were set appropriately for a neutral pH, and kept in that state throughout the simulation. All simulations were performed using the Amber14 software suite [42]. The geometry was initially relaxed, followed by heating and equilibration following a recommended nine-step protocol [43]. This was followed by a 100 ns production run in triplicates for each complex. The atom coordinates were saved to the trajectory file every 10 ps or every 5000 simulation steps.

#### 4.2.4. Simulation Data Analysis

Root mean square deviation (RMSD) (Figure S7) and Root mean square fluctuation (RMSF) values were calculated for the antigen residues from the last 25 ns of the 100 ns of MD simulation, by extracting all of the last 2500 poses, using the cpptraj [44]. To identify different binding modes within the MD simulation, the trajectories were analyzed using MDAnalysis [45]. First, the independent MD simulations were merged and the interactions between antigen and antibody were quantified by computing distances between all atom pairs (interaction threshold <4.0 Å). For each frame, key features including mean and standard deviation of distances, interaction count, hydrogen bond count (distance < 3.5 Å and angle > 150°), and center of mass of the antigen were extracted. These features were normalized and subsequently clustered with K-Means ($n = 3$). Representative structures for each cluster were identified by finding the frame closest to the centroids of the clusters. Absolute binding affinity and per-residue contributions were estimated with the MMPBSA.py.MPI [46] module of AMBER employing the molecular mechanical (MM) interaction energies with a generalized Born approximation (MM-GBSA) for the desolvation free energies (igb = 2) [47] for 2500 snapshots extracted evenly from the last 25 ns of the 100 ns of MD simulation using the single trajectory method. The net binding energies were computed as the difference between those for the complex minus, those for the protein, and the ligand.

## 5. Conclusions

This study provides unique insights into the molecular basis of YsT9.1 and Bm10 antibody recognition of the *Brucella* O-antigen A and M epitopes. Stable binding modes were identified, interaction strengths were quantified, and crucial antibody and antigen residues involved in these interactions were objectively located. The observed dominance of hydrophobic forces and the differential contributions from the heavy and the light chains of the antibodies offers a deeper understanding of the recognition mechanism. The predicted binding modes and identified interaction hot spots may help to guide the design of synthetic glycomimetics, the engineering of epitope-specific antibodies, and possibly inform vaccine design.

**Supplementary Materials:** The following supporting information can be downloaded at: https://www.mdpi.com/article/10.3390/molecules30142906/s1, Figure S1: Graphical depiction of YsT9.2 Heavy Chain and CDR Loops; Figure S2: Graphical depiction of YsT9.2 Light Chain and CDR Loops; Figure S3: Graphical representation of the Bm10 VH chain and CDR Loops; Figure S4: Graphical representation of the Bm10 VL and CDR Loops; Figure S5: Graphical representation of the Bm10 VH and CDR Loops; Figure S6: Graphical representation of the VL Chain and CDR loops; Table S1: The number of clusters from each Antigen–Antibody docking simulation; Figure S7: The RMSD values, averaged over three independent MD simulations, of the top docked poses (a–f) from each cluster, obtained after docking the A antigen (left panels) or the M antigen (right panels) to murine Brucella antibodies YsT9.1 (A), YsT9.2 (B), Bm10 (C), and Bm28 (D); Table S2: RMSD between the ring atoms of representative structures of YsT9.1-A complex; Table S3: RMSD between the ring atoms of representative structures of Bm10-M complex; Table S4: Per-residue contribution of YsT9.1 antibody in complex with A antigen; Table S5: Per-residue contribution of Bm10 antibody in complex with M antigen in binding mode 1; Table S6: Per-residue contribution of Bm10 antibody in complex with M antigen in binding mode 2; Table S7: Per-residue contribution of the A antigen in complex with YsT9.1 antibody; Table S8: Per-residue contribution of the M antigen in complex with Bm10 antibody in binding mode 1; Table S9: Per-residue contribution of the M antigen in complex with Bm10 antibody in binding mode 2.

**Author Contributions:** All authors contributed to the conceptualization of this work, including the original draft preparation, writing, review, and editing. D.R.B. generated, characterized, and sequenced the mAbs and also synthesized the oligosaccharides and performed epitope mapping studies with the synthetic oligosaccharides and natural polysaccharides. A.S. and R.J.W. designed, executed, and analyzed the modeling studies. All authors have read and agreed to the published version of the manuscript.

**Funding:** D.R.B. thanks the National Research Council of Canada, the Natural Science and Engineering Research Council of Canada (NSERC Discovery Grant #138355), and Bill and Melinda Gates Grand Challenges Round 11 for support (OPP1096883). R.J.W. thanks the National Institutes of Health (NIH R01 GM135473) for support.

**Institutional Review Board Statement:** Not applicable.

**Informed Consent Statement:** Not applicable.

**Data Availability Statement:** All the homology models and top docked poses from each cluster in this article can be freely accessed as PDB files at https://zenodo.org/records/15800977 (accessed on 29 June 2025) The DNA sequencing data are openly available at GenBank [https://www.ncbi.nlm.nih.gov/], accession numbers as follows: YsT9.2 VH Chain; PV749939-49944: VL Chain; PV749945-PV749951; Bm10 VH Chain; PV749910-PV749917: VL Chain; PV749918-PV749924; Bm28 VH Chain; PV749925-PV749932: VL Chain; PV749933- PV749938. Fv consensus amino acid sequences are available at the ABCD (AntiBodies Chemically Defined) database [https://web.expasy.org/abcd/ (accessed on 29 June 2025)], accession numbers as follows: YsT9.2 ABCD_BF026; Bm10 ABCD_BF027; Bm28 ABCD_BF028.

**Conflicts of Interest:** The authors declare no conflicts of interest.

## Abbreviations

The following abbreviations are used in this manuscript:

| | |
|---|---|
| PS | Polysaccharide |
| mAb | Monoclonal Antibody |
| Rha4NFo | 4-formamido-4,6-dideoxy-α-D-mannopyrannose |
| GMQE | Global Model Quality Estimate |
| CDR | Complementarity-Determining Region |
| MMTSB | Multiscale Modeling Tools for Structural Biology |
| RMSF | Root Mean Square Fluctuation |
| RMSD | Root Mean Square Deviation |
| MD | Molecular Dynamics |
| MM | Molecular Mechanics |
| GBSA | Generalized Born and Surface Area |

## References

1. Maudlin, I.; Weber, S. *The Control of Neglected Zoonotic Diseases: A Route to Poverty Alleviation*; WHO/SDE/FOS/2006.1; World Health Organization: Geneva, Switzerland, 2006; p. 83.
2. Corbel, M.J. *Brucellosis in Humans and Animals*; World Health Organization: Geneva, Switzerland, 2006.
3. Wilson, G.; Miles, A. The serological differentiation of smooth strains of the Brucella group. *Br. J. Exp. Pathol.* **1932**, *13*, 1–13.
4. Nielsen, K. Diagnosis of brucellosis by serology. *Vet. Microbiol.* **2002**, *90*, 447–459. [CrossRef]
5. Miles, A.A.; Pirie, N.W. The properties of antigenic preparations from Brucella melitensis: The hydrolysis of the formamino linkage. *Biochem. J.* **1939**, *33*, 1709–1715. [CrossRef] [PubMed]

6. Miles, A.A.; Pirie, N.W. The properties of antigenic preparations from Brucella melitensis: Hydrolysis and acetylation of the amino-polyhydroxy compound derived from the antigen. With an Addendum by J. St L. Philpot. *Biochem. J.* **1939**, *33*, 1716–1726. [CrossRef] [PubMed]
7. Caroff, M.; Bundle, D.; Perry, M.; Cherwonogrodzky, J.; Duncan, J. Antigenic S-type lipopolysaccharide of Brucella abortus 1119-3. *Infect. Immun.* **1984**, *46*, 384–388. [CrossRef]
8. Bundle, D.R.; Cherwonogrodzky, J.; Perry, M.B. Structural elucidation of the Brucella melitensis M antigen by high-resolution NMR at 500 MHz. *Biochemistry* **1987**, *26*, 8717–8726. [CrossRef]
9. Meikle, P.J.; Perry, M.B.; Cherwonogrodzky, J.; Bundle, D. Fine structure of A and M antigens from Brucella biovars. *Infect. Immun.* **1989**, *57*, 2820–2828. [CrossRef] [PubMed]
10. Kubler-Kielb, J.; Vinogradov, E. Reinvestigation of the structure of Brucella O-antigens. *Carbohydr. Res.* **2013**, *378*, 144–147. [CrossRef]
11. Zaccheus, M.V.; Ali, T.; Cloeckaert, A.; Zygmunt, M.S.; Weintraub, A.; Iriarte, M.; Moriyón, I.; Widmalm, G. The epitopic and structural characterization of Brucella suis biovar 2 O-polysaccharide demonstrates the existence of a new M-negative C-negative smooth Brucella serovar. *PLoS ONE* **2013**, *8*, e53941. [CrossRef]
12. Zygmunt, M.S.; Bundle, D.R.; Ganesh, N.V.; Guiard, J.; Cloeckaert, A. Monoclonal antibody-defined specific C epitope of Brucella O-polysaccharide revisited. *Clin. Vaccine Immunol.* **2015**, *22*, 979–982. [CrossRef]
13. Bundle, D.R.; Gidney, M.; Perry, M.; Duncan, J.; Cherwonogrodzky, J. Serological confirmation of Brucella abortus and Yersinia enterocolitica O: 9 O-antigens by monoclonal antibodies. *Infect. Immun.* **1984**, *46*, 389–393. [CrossRef]
14. Ganesh, N.V.; Sadowska, J.M.; Sarkar, S.; Howells, L.; McGiven, J.; Bundle, D.R. Molecular recognition of Brucella A and M antigens dissected by synthetic oligosaccharide glycoconjugates leads to a disaccharide diagnostic for brucellosis. *J. Am. Chem. Soc.* **2014**, *136*, 16260–16269. [CrossRef]
15. Bundle, D.; Cherwonogrodzky, J.; Gidney, M.; Meikle, P.; Perry, M.; Peters, T. Definition of Brucella A and M epitopes by monoclonal typing reagents and synthetic oligosaccharides. *Infect. Immun.* **1989**, *57*, 2829–2836. [CrossRef] [PubMed]
16. Guiard, J.; Paszkiewicz, E.; Sadowska, J.; Bundle, D.R. Design and synthesis of a universal antigen to detect brucellosis. *Angew. Chem. Int. Ed.* **2013**, *52*, 7181–7185. [CrossRef] [PubMed]
17. Mandal, S.S.; Ganesh, N.V.; Sadowska, J.M.; Bundle, D.R. Synthetic glycoconjugates characterize the fine specificity of Brucella A and M monoclonal antibodies. *Org. Biomol. Chem.* **2017**, *15*, 3874–3883. [CrossRef]
18. Rose, D.R.; Przybylska, M.; To, R.J.; Kayden, C.S.; Vorberg, E.; Young, N.M.; Bundle, D.R.; Oomen, R.P. Crystal structure to 2.45 Å resolution of a monoclonal Fab specific for the Brucella A cell wall polysaccharide antigen. *Protein Sci.* **1993**, *2*, 1106–1113. [CrossRef]
19. Oomen, R.P.; Young, N.M.; Bundle, D.R. Molecular modeling of antibody–antigen complexes between the Brucella abortus O-chain polysaccharide and a specific monoclonal antibody. *Protein Eng. Des. Sel.* **1991**, *4*, 427–433. [CrossRef] [PubMed]
20. Amon, R.; Grant, O.C.; Leviatan Ben-Arye, S.; Makeneni, S.; Nivedha, A.K.; Marshanski, T.; Norn, C.; Yu, H.; Glushka, J.N.; Fleishman, S.J.; et al. A combined computational-experimental approach to define the structural origin of antibody recognition of sialyl-Tn, a tumor-associated carbohydrate antigen. *Sci. Rep.* **2018**, *8*, 10786. [CrossRef]
21. Vyas, N.K.; Vyas, M.N.; Chervenak, M.C.; Johnson, M.A.; Pinto, B.M.; Bundle, D.R.; Quiocho, F.A. Molecular Recognition of Oligosaccharide Epitopes by a Monoclonal Fab Specific for Shigella flexneri Y Lipopolysaccharide: X-ray Structures and Thermodynamics. *Biochemistry* **2002**, *41*, 13575–13586. [CrossRef]
22. Li, X.; Grant, O.C.; Ito, K.; Wallace, A.; Wang, S.; Zhao, P.; Wells, L.; Lu, S.; Woods, R.J.; Sharp, J.S. Structural Analysis of the Glycosylated Intact HIV-1 gp120–b12 Antibody Complex Using Hydroxyl Radical Protein Footprinting. *Biochemistry* **2017**, *56*, 957–970. [CrossRef]
23. Kenne, L.; Unger, P.; Wehler, T. Synthesis and nuclear magnetic resonance studies of some N-acylated methyl 4-amino-4,6-dideoxy-α-D-mannopyranosides. *J. Chem. Soc. Perkin Trans. 1* **1988**, *5*, 1183–1186. [CrossRef]
24. Sood, A.; Gerlits, O.O.; Ji, Y.; Bovin, N.V.; Coates, L.; Woods, R.J. Defining the Specificity of Carbohydrate-Protein Interactions by Quantifying Functional Group Contributions. *J. Chem. Inf. Model.* **2018**, *58*, 1889–1901. [CrossRef] [PubMed]
25. Hudson, K.L.; Bartlett, G.J.; Diehl, R.C.; Agirre, J.; Gallagher, T.; Kiessling, L.L.; Woolfson, D.N. Carbohydrate–Aromatic Interactions in Proteins. *J. Am. Chem. Soc.* **2015**, *137*, 15152–15160. [CrossRef]
26. Cygler, M.; Rose, D.R.; Bundle, D.R. Recognition of a cell-surface oligosaccharide of pathogenic Salmonella by an antibody Fab fragment. *Science* **1991**, *253*, 442–445. [CrossRef] [PubMed]
27. Jeffrey, P.D.; Bajorath, J.; Chang, C.Y.; Yelton, D.; Hellström, I.; Hellström, K.E.; Sheriff, S. The x-ray structure of an anti-tumour antibody in complex with antigen. *Nat. Struct. Biol.* **1995**, *2*, 466–471. [CrossRef] [PubMed]
28. Vyas, N.K. Atomic features of protein-carbohydrate interactions. *Curr. Opin. Struct. Biol.* **1991**, *1*, 732–740. [CrossRef]

29. Caroff, M.; Bundle, D.R.; Perry, M.B. Structure of the O-chain of the phenol-phase soluble cellular lipopolysaccharide of Yersinia enterocolitica serotype O: 9. *Eur. J. Biochem.* **1984**, *139*, 195–200. [CrossRef]
30. Waterhouse, A.; Bertoni, M.; Bienert, S.; Studer, G.; Tauriello, G.; Gumienny, R.; Heer, F.T.; de Beer, T.A.P.; Rempfer, C.; Bordoli, L.; et al. SWISS-MODEL: Homology modelling of protein structures and complexes. *Nucleic Acids Res.* **2018**, *46*, W296–W303. [CrossRef]
31. Grant, O.C.; Wentworth, D.; Holmes, S.G.; Kandel, R.; Sehnal, D.; Wang, X.; Xiao, Y.; Sheppard, P.; Grelsson, T.; Coulter, A.; et al. Generating 3D Models of Carbohydrates with GLYCAM-Web. *BioRxiv* **2025**. [CrossRef]
32. Pettersen, E.F.; Goddard, T.D.; Huang, C.C.; Couch, G.S.; Greenblatt, D.M.; Meng, E.C.; Ferrin, T.E. UCSF Chimera—A visualization system for exploratory research and analysis. *J. Comput. Chem.* **2004**, *25*, 1605–1612. [CrossRef]
33. Morris, G.M.; Huey, R.; Lindstrom, W.; Sanner, M.F.; Belew, R.K.; Goodsell, D.S.; Olson, A.J. AutoDock4 and AutoDockTools4: Automated Docking with Selective Receptor Flexibility. *J. Comput. Chem.* **2009**, *30*, 2785–2791. [CrossRef]
34. Nivedha, A.K.; Thieker, D.F.; Hu, H.; Woods, R.J. Vina-Carb: Improving Glycosidic Angles during Carbohydrate Docking. *J. Chem. Theory Comput.* **2016**, *12*, 892–901. [CrossRef] [PubMed]
35. Nivedha, A.K.; Makeneni, S.; Foley, B.L.; Tessier, M.B.; Woods, R.J. Importance of ligand conformational energies in carbohydrate docking: Sorting the wheat from the chaff. *J. Comput. Chem.* **2014**, *35*, 526–539. [CrossRef]
36. Feig, M.; Karanicolas, J.; Brooks, C.L. MMTSB Tool Set: Enhanced sampling and multiscale modeling methods for applications in structural biology. *J. Mol. Graph. Model.* **2004**, *22*, 377–395. [CrossRef]
37. Wang, J.; Wolf, R.M.; Caldwell, J.W.; Kollman, P.A.; Case, D.A. Development and testing of a general amber force field. *J. Comput. Chem.* **2004**, *25*, 1157–1174. [CrossRef]
38. Maier, J.A.; Martinez, C.; Kasavajhala, K.; Wickstrom, L.; Hauser, K.E.; Simmerling, C. ff14SB: Improving the accuracy of protein side chain and backbone parameters from ff99SB. *J. Chem. Theory Comput.* **2015**, *11*, 3696–3713. [CrossRef] [PubMed]
39. Kirschner, K.N.; Yongye, A.B.; Tschampel, S.M.; González-Outeiriño, J.; Daniels, C.R.; Foley, B.L.; Woods, R.J. GLYCAM06: A generalizable biomolecular force field. Carbohydrates. *J. Comput. Chem.* **2008**, *29*, 622–655. [CrossRef] [PubMed]
40. Mahoney, M.W.; Jorgensen, W.L. A five-site model for liquid water and the reproduction of the density anomaly by rigid, nonpolarizable potential functions. *J. Chem. Phys.* **2000**, *112*, 8910–8922. [CrossRef]
41. Fadda, E.; Woods, R.J. On the Role of Water Models in Quantifying the Binding Free Energy of Highly Conserved Water Molecules in Proteins: The Case of Concanavalin A. *J. Chem. Theory Comput.* **2011**, *7*, 3391–3398. [CrossRef]
42. Case, D.A.; Babin, V.; Berryman, J.T.; Betz, R.M.; Cai, Q.; Cerutti, D.S.; Cheatham, T.E., III; Darden, T.A.; Duke, R.E.; Gohlke, H.; et al. *AMBER14*; University of California: San Francisco, CA, USA, 2014.
43. Roe, D.R.; Brooks, B.R. A protocol for preparing explicitly solvated systems for stable molecular dynamics simulations. *J. Chem. Phys.* **2020**, *153*, 054123. [CrossRef]
44. Roe, D.R.; Cheatham, T.E., III. PTRAJ and CPPTRAJ: Software for Processing and Analysis of Molecular Dynamics Trajectory Data. *J. Chem. Theory Comput.* **2013**, *9*, 3084–3095. [CrossRef] [PubMed]
45. Michaud-Agrawal, N.; Denning, E.J.; Woolf, T.B.; Beckstein, O. MDAnalysis: A toolkit for the analysis of molecular dynamics simulations. *J. Comput. Chem.* **2011**, *32*, 2319–2327. [CrossRef] [PubMed]
46. Miller, B.R., III; McGee, T.D., Jr.; Swails, J.M.; Homeyer, N.; Gohlke, H.; Roitberg, A.E. MMPBSA.py: An Efficient Program for End-State Free Energy Calculations. *J. Chem. Theory Comput.* **2012**, *8*, 3314–3321. [CrossRef]
47. Onufriev, A.; Bashford, D.; Case, D.A. Exploring protein native states and large-scale conformational changes with a modified generalized born model. *Proteins* **2004**, *55*, 383–394. [CrossRef]

*molecules*

MDPI

*Article*

# Cooperatively Catalyzed Activation of Thioglycosides with Iodine and Iron(III) Trifluoromethanesulfonate

**Ashley R. Dent, Aidan M. DeSpain and Alexei V. Demchenko ***

Department of Chemistry, Saint Louis University, 3501 Laclede Ave, St. Louis, MO 63103, USA; ashley.dent@slu.edu (A.R.D.)
* Correspondence: alexei.demchenko@slu.edu

**Abstract**

Reported herein is a further expansion of the cooperatively catalyzed Koenigs–Knorr glycosylation reaction, known as "the 4K reaction". It has been discovered that molecular iodine, along with a metal salt and an acid additive, can activate thioglycosides. Previous mechanistic studies showed the interaction of the anomeric sulfur with thiophilic iodine; this complex is stable until the halophilic metal salt and the acid additive are added. This new avenue has allowed for the investigation of halophilic promoters that would not activate thioglycosides without iodine. Presented herein is the recent discovery of iron(III) triflate as an efficient activator of thioglycosides via the 4K reaction pathway.

**Keywords:** glycosylation; ferric triflate; green chemistry; carbohydrates; thioglycosides

## 1. Introduction

Carbohydrates are a vital class of molecules that are found profusely throughout nature. Due to their fundamental significance, it is of the utmost importance that carbohydrates be synthesized efficiently. The most central aspect to carbohydrate synthesis is the glycosylation reaction; however, it continues to be the most difficult and demanding aspect as well [1]. While nature effortlessly achieves glycosylation reactions via an enzymatic mechanism, chemical glycosylation itself remains onerous [2–8]. Paulsen's statement written in 1982, "Although we have now learned to synthesize oligosaccharides, it should be emphasized that each oligosaccharide synthesis remains an independent problem, whose resolution requires considerable systematic research and a good deal of know-how. There are no universal reaction conditions for oligosaccharide syntheses" [2], is still current and very topical even now, many decades later – especially in relation to the synthesis of challenging linkages and targets. This article is dedicated to the late Professor Hans Paulsen (1922–2024), whose seminal publications on glycosylation, oligosaccharide synthesis, and building block reactivity have given enormous inspiration to the authors.

Early glycosylation reactions were performed by Michael [9], Fischer [10], and Koenigs/Knorr [11], making use of simple glycosyl halides or hemiacetals (Fischer). Although these simple glycosyl donors were appropriate for simple glycoside synthesis, the overall efficiency remained limited in scope and needed improvement. With classical Koenigs–Knorr reaction conditions [11–13], a glycosyl bromide donor is reacted with a glycosyl acceptor in the presence of silver(I) oxide or carbonate. Unfortunately, this reaction can be slow—particularly with unreactive bromides, such as those equipped with benzoyl protecting groups. Recently, we determined that a catalytic amount of a Lewis acid added to the Ag(I)-promoted glycosylation dramatically increases reaction rates and yields [14]. This has since been named "the 4K reaction" [15].

Thioglycosides are among the most common building blocks used in carbohydrate chemistry. For glycosylation, thioglycosides can be activated by the use of electrophilic or thiophilic promotors [16]. Among these, organosulfur compounds [17–22], halogen-based reagents [23–28], and photo-activators [29–32] are among the most popular. Activation with metal salts is also known [33]. Early work by Ferrier, who used mercury(II) salts for the activation of phenylthio glycosides [34], was complemented by Pohl, who used $Ph_3Bi(OTf)_2$ [35,36], and Sureshan [37] and Zhu [38], who used Au(III) salts. Our group previously reported the use of palladium(II) bromide [39], copper(II) bromide [40], and $FeCl_3$ [41].

Our lab also recently demonstrated that thioglycosides can be activated using molecular iodine along with a metal salt and an acid additive under the 4K reaction conditions [42]. While initial studies were aimed at silver sulfate, a recent expansion explored bismuth(III) triflate as an activator while listing other viable metal salts [43]. Presented herein is the discovery that iron(III) triflate is also a viable reagent for the 4K reaction with thioglycosides.

## 2. Results and Discussion

Ferric or iron(III) salts are formed from iron, the second most abundant metal on earth. Iron salts tend to be naturally abundant, inexpensive, and relatively benign. After preliminary optimization of the reaction conditions, we identified $I_2$ (1.5 equiv), $Fe(OTf)_3$ (1.5 equiv), and TfOH (0.2 equiv) in the presence of molecular sieves (3 Å) in 1,2-dichloroethane (DCE) as the most promising combination of reagents. $I_2$ and $Fe(OTf)_3$ without TfOH, or $Fe(OTf)_3$ and TfOH without $I_2$, will form the product too, but at much slower rates and lower yields. Reactions in the absence of $Fe(OTf)_3$ did not proceed for unreactive thioglycosides. Scaling back the amounts of $I_2$ and $Fe(OTf)_3$ to 1.0 equiv led to a loss in yield and decrease in reaction rates. All glycosylations with less-reactive per-*O*-benzoylated (disarmed) glycosyl donors were performed at rt, whereas reactions with all other, more reactive glycosyl donors were performed at −30 °C. Standard primary glycosyl acceptor **1** and secondary glycosyl acceptor **2** (Figure 1) were chosen to investigate the scope of this reaction.

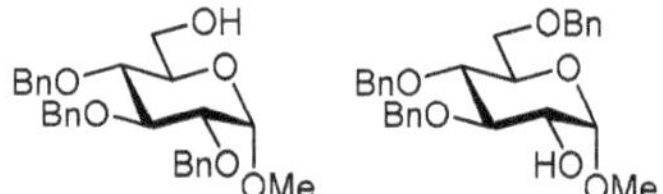


**1** **2**

**Figure 1.** Structures of glycosyl acceptors **1** and **2** used in this study.

The results of glycosylation reactions are shown in Table 1. The glycosidation of per-*O*-benzoylated (disarmed) glucosyl donor **3** with 6-OH acceptor **1** produced disaccharide **4** in 16 h in 96% yield (entry 1). Glycosidation of **3** with 2-OH acceptor **2** produced disaccharide **5** in 24 h in 87% yield (entry 2). Other classes of thioglucosides were explored to further expand the scope. Thus, the glycosidation of α-ethylthio glycoside **6** with 6-OH acceptor **1** gave disaccharide **4** in 30 h in 92% yield (entry 3). The glycosidation of phenylthio or tolylthio glycosyl donors **7** or **8** with 6-OH acceptor **1** gave disaccharide **4** in 30 h in 90–95% yield (entries 4–5). These disaccharides all exhibit complete β-selectivity due to the participation of the neighboring benzoyl protecting group at C-2. The glycosidation of per-*O*-benzylated glucosyl donor **9** with 6-OH acceptor **1** was faster as these donors are known to be more reactive (armed) [44]. Disaccharide **10** was formed in 16 h in 86% yield

($\alpha$:$\beta$ = 1:1.5, entry 6). The glycosidation of **9** with 2-OH acceptor **2** formed disaccharide **11** in 16 h in 73% yield ($\alpha$:$\beta$ = 1.0:1, entry 7). Poor stereoselectivity was due to lack of the stereocontrolling factors.

**Table 1.** Investigation of the scope of the iron(III) triflate-promoted 4K reaction.

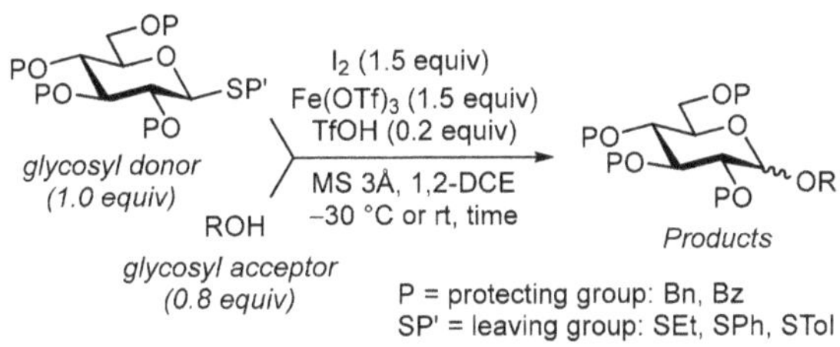


| Entry | Donor | Acceptor | Product, Time, Yield, Ratio $\alpha/\beta$ |
|---|---|---|---|
| 1 | 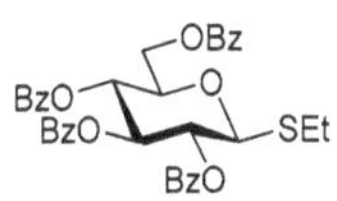 **3** | **1** | 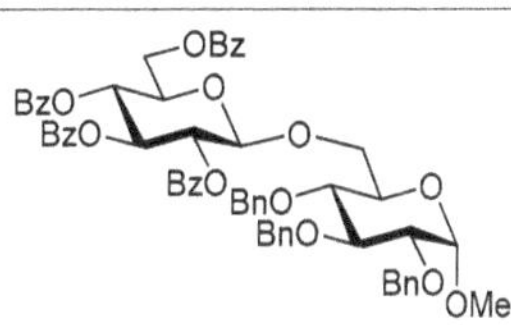 **4**, 16 h, 96%, $\beta$-only |
| 2 | **3** | **2** | 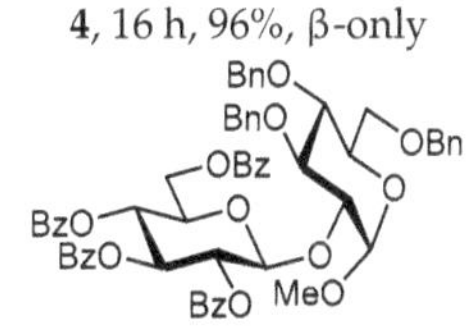 **5**, 24 h, 87%, $\beta$-only |
| 3 | 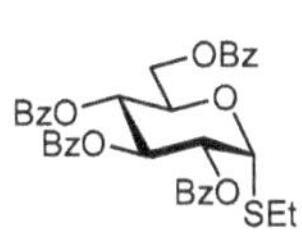 **6** | **1** | 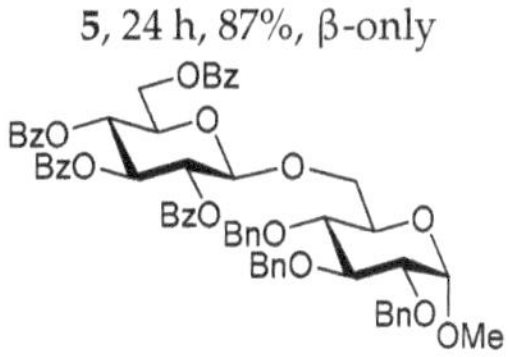 **4**, 30 h, 92%, $\beta$-only |
| 4 | 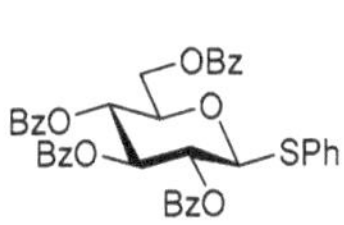 **7** | **1** | 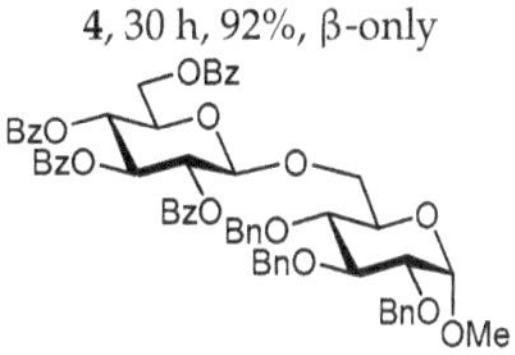 **4**, 30 h, 90%, $\beta$-only |
| 5 | 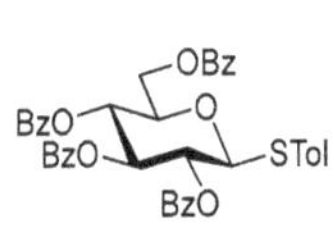 **8** | **1** | 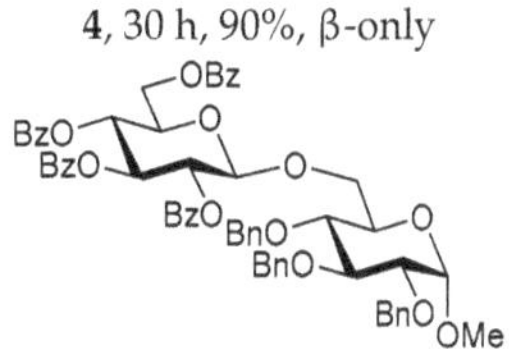 **4**, 30 h, 95%, $\beta$-only |

**Table 1.** *Cont.*

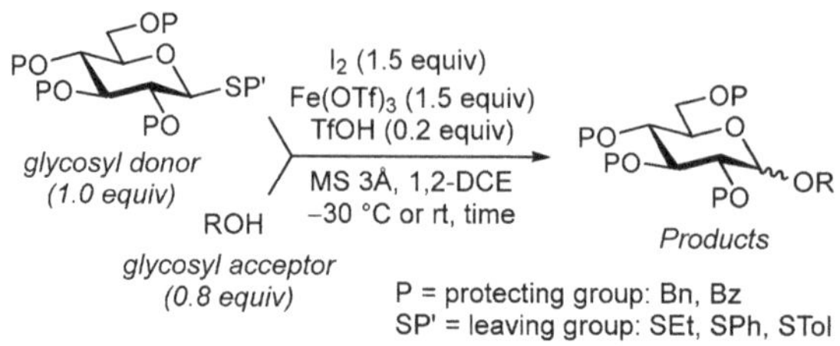


| Entry | Donor | Acceptor | Product, Time, Yield, Ratio α/β |
|---|---|---|---|
| 6 | 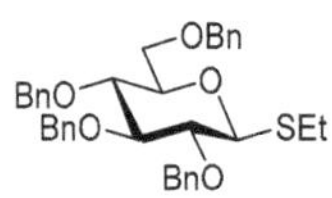 **9** | **1** | 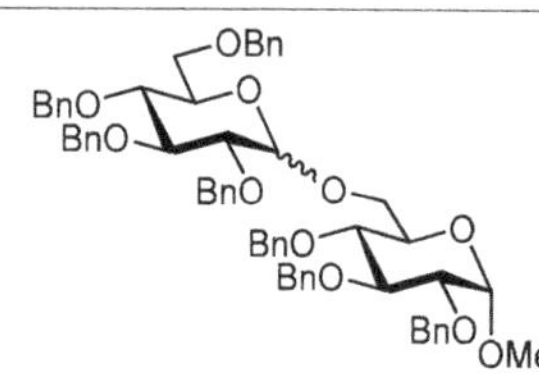 **10**, 16 h, 86%, α:β = 1:1.5 |
| 7 | **9** | **2** | 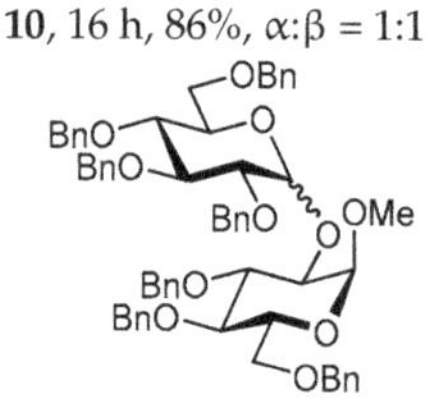 **11**, 16 h, 73%, α:β = 1:1 |
| 8 | 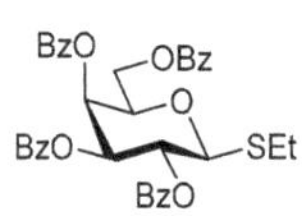 **12** | **1** | 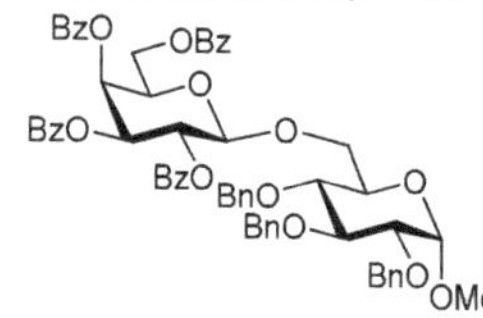 **13**, 16 h, 96%, β-only |
| 9 | **12** | **2** | 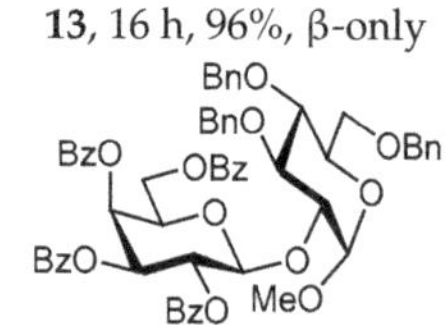 **14**, 24 h, 89%, β-only |
| 10 | 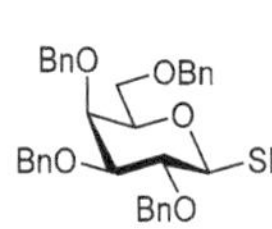 **15** | **1** | 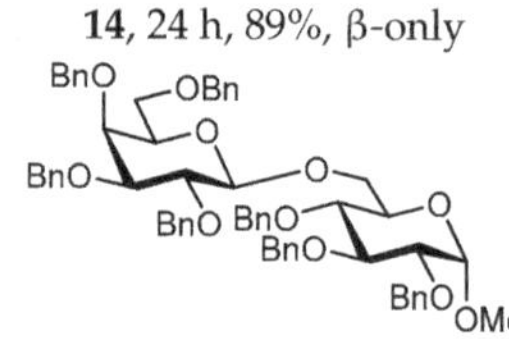 **16**, 16 h, 92%, β-only |
| 11 | **15** | **2** | 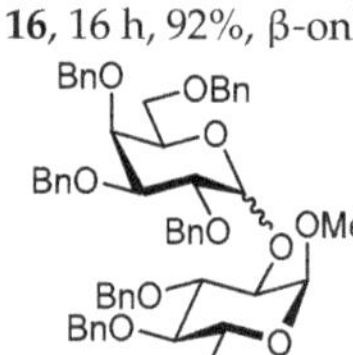 **17**, 16 h, 79%, α:β = 1.6:1 |

**Table 1.** *Cont.*

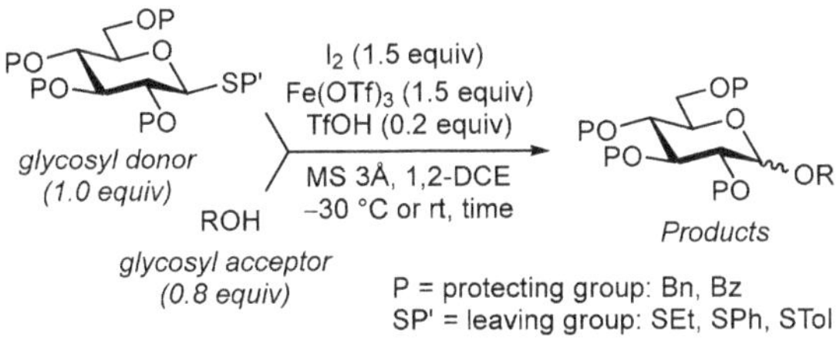


| Entry | Donor | Acceptor | Product, Time, Yield, Ratio α/β |
|---|---|---|---|
| 12 | 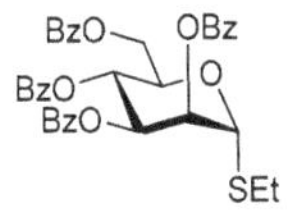 **18** | **1** | 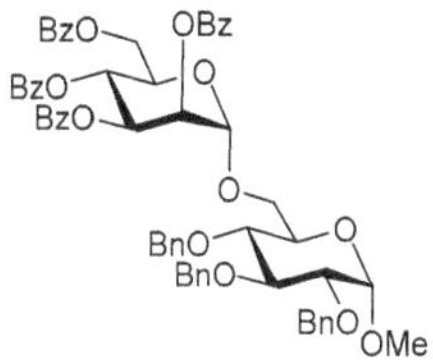 **19**, 24 h, 84%, α-only |
| 13 | **18** | **2** | 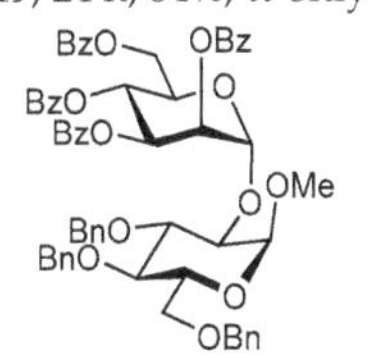 **20**, 24 h, 85%, α-only |
| 14 | 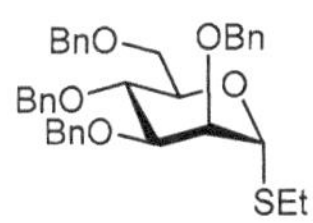 **21** | **1** | 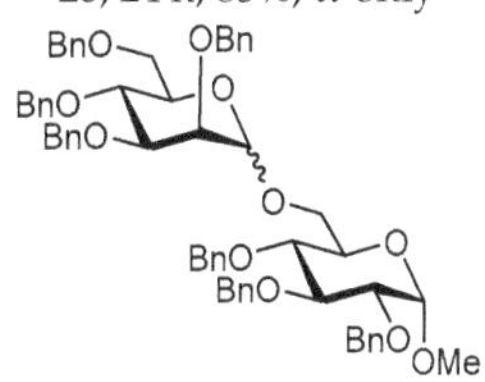 **22**, 16 h, 81%, α:β = 1:4 |
| 15 | **21** | **2** | 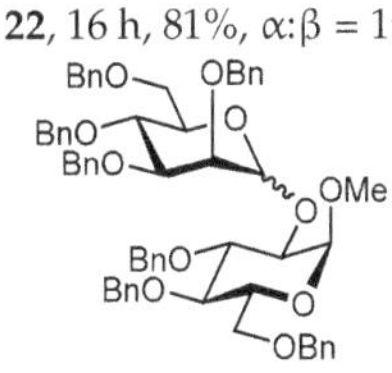 **23**, 16 h, 70%, α:β = 1:1.2 |
| 16 | 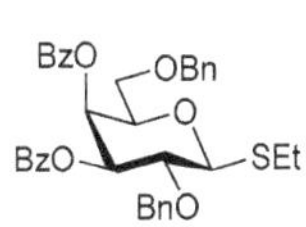 **24** | **1** | 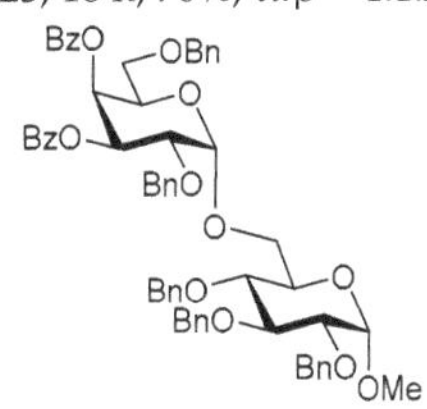 **25**, 16 h, 78%, α-only |

**Table 1.** *Cont.*

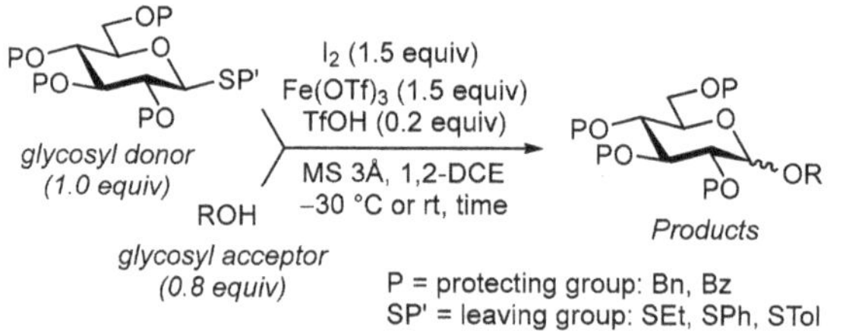


| Entry | Donor | Acceptor | Product, Time, Yield, Ratio α/β |
|---|---|---|---|
| 17 | 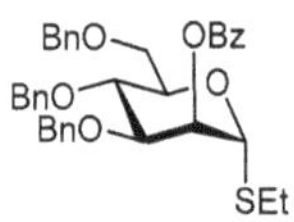 **26** | **1** | 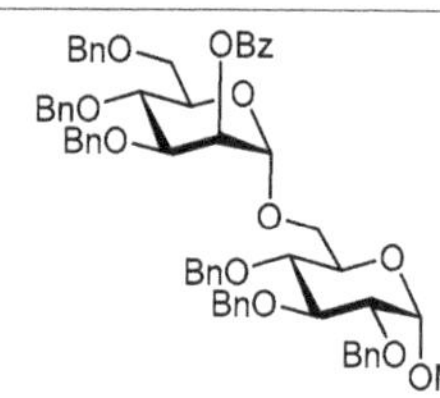 **27**, 16 h, 79%, α-only |

We then looked into reactions of galactosyl donors. The glycosidation of per-*O*-benzoylated galactosyl donor **12** with 6-OH acceptor **1** produced disaccharide **13** in 16 h in 96% yield (entry 8). The glycosidation of **12** with 2-OH acceptor **2** produced disaccharide **14** in 24 h in 89% yield (entry 9). These disaccharides all exhibit complete β-selectivity due to the participation of the neighboring benzoyl protecting group at C-2. The glycosidation of per-*O*-benzylated galactosyl donor **15** with 6-OH acceptor **1** formed disaccharide **16** in 16 h in 92% yield (β-only, entry 10). The glycosidation of **15** with 2-OH acceptor **2** formed disaccharide **17** in 16 h in 79% yield (α:β = 1.6:1, entry 11). The poor stereoselectivity of **17** was due to the lack of stereocontrolling factors.

The glycosidation of per-*O*-benzoylated mannosyl donor **18** with 6-OH acceptor **1** produced disaccharide **19** in 84% yield in 24 h (entry 12). The glycosidation of **18** with 2-OH acceptor **2** produced disaccharide **20** in 24 h in 85% yield (entry 13). These disaccharides all exhibit complete α-selectivity due to the participation of the neighboring benzoyl protecting group at C-2. The glycosidation of per-*O*-benzylated mannosyl donor **21** with 6-OH acceptor **1** formed disaccharide **22** in 16 h in 81% yield (α:β = 1:4, entry 14). The glycosidation of **21** with 2-OH acceptor **2** formed disaccharide **23** in 16 h in 70% yield (α:β = 1:1.2, entry 15).

We then investigated whether these new reaction conditions would work well in combination with the remote benzoyl groups for galactosyl donors to achieve α-selectivity [45,46]. The glycosidation of galactosyl donor **24** with 6-OH acceptor **1** produced disaccharide **25** in 16 h in 78% yield with complete α-selectivity (entry 16). Finally, we investigated the application of this methodology to a mannosyl donor equipped with the superarming protecting group pattern [47]. The glycosidation of this mannosyl donor **26** with 6-OH acceptor **1** formed disaccharide **27** in 16 h in 79% yield with complete α-selectivity (entry 17).

Undoubtedly, our studies have demonstrated that $I_2$/$Fe(OTf)_3$/TfOH-catalyzed 4K reactions are swift and high yielding. Poor stereocontrol in the case of 2-*O*-benzylated glycosyl donors is not uncommon, and while α-galactosylation could be effectively achieved by using remote benzoyl groups (vide infra), we have not yet determined how to improve the stereoselectivity of glucosylation. To gain a stereocontrolling mode with glucosyl donors, in the past we demonstrated that bismuth(III)-promoted 4K reactions are compatible with the H-bond-mediated Aglycone Delivery (HAD) pathway [43]. The HAD reaction is based on glycosyl donors' *O*-picoloyl (Pico) protecting group. These donors provide high or even complete *syn*-selectivities in respect to Pico [48]. The 4K reaction conditions developed herein, however, were found to be incompatible with 4-*O*-Pico-substituted building blocks. This is because ferric salts are known to be effective reagents for the removal of Pico groups [49].

## 3. Materials and Methods

### 3.1. General Methods

Column chromatography was performed on silica gel 60 (70–230 mesh); reactions were monitored by TLC on Kieselgel 60 F254. The compounds were detected by examination under UV light and by charring with 5% sulfuric acid in methanol. Solvents were removed under reduced pressure at <40 °C. $ClCH_2CH_2Cl$ (1,2-DCE) was distilled from $CaH_2$ directly prior to application. Molecular sieves (3 Å) used for reactions were first crushed and then activated in *vacuo* at 390 °C directly prior to application. Optical rotations were measured on a "Jasco P-2000" polarimeter (Jasco Corporation, Tokyo, Japan). $^1H$ NMR spectra were recorded in $CDCl_3$ at 400 or 700 MHz. $^{13}C\{^1H\}$ NMR spectra were recorded in $CDCl_3$ at 101 or 175 MHz. The $^1H$ NMR chemical shifts and the $^{13}C\{^1H\}$ NMR chemical shifts were referenced to $CDCl_3$ ($\delta_H$ = 7.26, $\delta_C$ = 77.00 ppm). Structural assignments were made with additional information from COSY experiments. Compound ratios were determined by comparing the integration intensities of the relevant signals in their $^1H$ NMR spectra. See Supplementary Materials for NMR spectra of all compounds. Accurate mass spectrometry determinations were performed using an Agilent 6230 ESI TOF LCMS mass spectrometer (Agilent, Santa Clara, CA, USA).

### 3.2. Synthesis of Building Blocks

**Methyl 2,3,4-tri-*O*-benzyl-α-D-glucopyranoside (1)** was synthesized as reported previously, and its analytical data was in accordance with that previously described [50].

**Methyl 3,4,6-tri-*O*-benzyl-α-D-glucopyranoside (2)** was synthesized as reported previously, and its analytical data was in accordance with that previously described [50].

**Ethyl 2,3,4,6-tetra-*O*-benzoyl-1-thio-β-D-glucopyranoside (3)** was synthesized as reported previously, and its analytical data was in accordance with that previously described [51].

**Ethyl 2,3,4,6-tetra-*O*-benzoyl-1-thio-α-D-glucopyranoside (6)** was synthesized as reported previously, and its analytical data was in accordance with that previously described [51].

**Phenyl 2,3,4,6-tetra-*O*-benzoyl-1-thio-β-D-glucopyranoside (7)** was synthesized as reported previously, and its analytical data was in accordance with that previously described [52].

**Tolyl 2,3,4,6-tetra-*O*-benzoyl-1-thio-β-D-glucopyranoside (8)** was synthesized as reported previously, and its analytical data was in accordance with that previously described [53].

**Ethyl 2,3,4,6-tetra-*O*-benzyl-1-thio-β-D-glucopyranoside (9)** was synthesized as reported previously, and its analytical data was in accordance with that previously described [54].

**Ethyl 2,3,4,6-tetra-*O*-benzoyl-1-thio-β-D-galactopyranoside (12)** was synthesized as reported previously, and its analytical data was in accordance with that previously described [51].

**Ethyl 2,3,4,6-tetra-*O*-benzyl-1-thio-β-D-galactopyranoside (15)** was synthesized as reported previously, and its analytical data was in accordance with that previously described [55].

**Ethyl 2,3,4,6-tetra-*O*-benzoyl-1-thio-β-D-mannopyranoside (18)** was synthesized as reported previously, and its analytical data was in accordance with that previously described [51].

**Ethyl 2,3,4,6-tetra-*O*-benzyl-1-thio-β-D-mannopyranoside (21)** was synthesized as reported previously, and its analytical data was in accordance with that previously described [56].

**Ethyl 3,4,-di-*O*-benzoyl-2,6-di-*O*-benzyl-1-thio-β-D-galactopyranoside (24)** was synthesized as reported previously, and its analytical data was in accordance with that previously described [45].

**Ethyl 2-*O*-benzoyl-3,4,6-tri-O-benzyl-1-thio-β-D-mannopyranoside (26)** was synthesized as reported previously, and its analytical data was in accordance with that previously described [57].

### 3.3. Synthesis of Disaccharides

**General procedure for glycosidation.** A mixture of glycosyl donor (0.05 mmol, 1.1 equiv), glycosyl acceptor (0.045 mmol, 1.0 equiv), and freshly activated molecular sieves (3 Å, 150 mg) in 1,2-dichloroethane (1.0 mL, 0.45 mM) was stirred under argon for 1 h at rt. $I_2$ (0.0675 mmol, 1.5 equiv), $Fe(OTf)_3$ (0.0675 mmol, 1.5 equiv), and TfOH (0.009 mmol, 0.2 equiv) were added, and the resulting mixture was stirred under argon at rt for the time specified in the tables and below. After that, the solids were filtered off through a pad of Celite and washed successively with $CH_2Cl_2$. The combined filtrate (~40 mL) was washed with 10% aq. $Na_2S_2O_3$ (10 mL). The organic phase was separated, dried with sodium sulfate, and concentrated under reduced pressure. The residue was purified by column chromatography on silica gel (ethyl acetate—hexane or toluene gradient elution) to afford the respective disaccharides in yields and stereoselectivities listed in the tables and below. Anomeric ratios (or anomeric purity) were determined by comparison of the integral intensities of relevant signals in $^1$H NMR spectra.

**Methyl 6-*O*-(2,3,4,6-tetra-*O*-benzoyl-β-D-glucopyranosyl)-2,3,4-tri-*O*-benzyl-α- D-glucopyranoside (4)** was obtained from donor 3 [51] (39.0 mg, 0.06 mmol) and acceptor 1 [50] (25.7 mg, 0.055 mmol) under the general glycosidation method as a colorless foam in 16 h in 96% yield (55.4 mg, 0.053 mmol). The title compound was also obtained from donor 6 [51] (24.3 mg, 0.038 mmol) and acceptor 1 (16.0 mg, 0.034 mmol) under the general glycosidation method as a colorless foam in 30 h in 92% yield (32.8 mg, 0.031 mmol). The title compound was also obtained from donor 7 [52] (35.4 mg, 0.051 mmol) and acceptor 1 (21.7 mg, 0.047 mmol) under the general glycosidation method as a colorless foam in 30 h in 90% yield (43.9 mg, 0.042 mmol). The title compound was also obtained from donor 8 [53] (30.6 mg, 0.044 mmol) and acceptor 1 (18.4 mg, 0.04 mmol) under the general glycosidation method as a colorless foam in 30 h in 95% yield (39.1 mg, 0.037 mmol). Analytical data for 4 was in accordance with that previously reported [58].

**Methyl 2-*O*-(2,3,4,6-tetra-*O*-benzoyl-β-D-glucopyranosyl)-3,4,6-tri-*O*-benzyl-α- D-glucopyranoside (5)** was obtained from donor 3 [51] (34.0 mg, 0.053 mmol) and acceptor 2 [50] (22.4 mg, 0.048 mmol) under the general glycosidation method as a colorless form in 24 h in 87% yield (43.9 mg, 0.042 mmol). Analytical data for 5 was in accordance with that previously reported [50,58]

**Methyl 2,3,4-tri-*O*-benzyl-6-*O*-(2,3,4,6-tetra-*O*-benzyl-α/β-D-glucopyranosyl)-α-D-glucopyranoside (10)** was obtained from donor **9** [54] (24.4 mg, 0.042 mmol) and acceptor

**1** [50] (17.6 mg, 0.038 mmol) under the general glycosidation method as a colorless form in 16 h in 86% yield (32.0 mg, 0.032 mmol). Analytical data for **10** was in accordance with that previously reported [59].

**Methyl 3,4,6-tri-*O*-benzyl-2-*O*-(2,3,4,6-tetra-*O*-benzyl-α/β-D-glucopyranosyl)-α-D-glucopyranoside (11)** was obtained from donor **9** [54] (24.8 mg, 0.042 mmol) and acceptor **2** [50] (17.9 mg, 0.039 mmol) under the general glycosidation method as a colorless form in 16 h in 73% yield (27.9 mg, 0.028 mmol). Analytical data for **11** was in accordance with that previously reported [59].

**Methyl 6-*O*-(2,3,4,6-tetra-*O*-benzoyl-β-D-galactopyranosyl)-2,3,4-tri-*O*-benzyl-α-D-glucopyranoside (13)** was obtained from donor 12 [51] (29.1 mg, 0.045 mmol) and acceptor 1 [50] (19.2 mg, 0.041 mmol) under the general glycosidation method as a colorless form in 16 h in 96% yield (41.4 mg, 0.040 mmol). Analytical data for 13 was in accordance with that previously reported [60].

**Methyl 2-*O*-(2,3,4,6-tetra-*O*-benzoyl-β-D-galactopyranosyl)-3,4,6-tri-*O*-benzyl-α-D-glucopyranoside (14)** was obtained from donor **12** [51] (24.4 mg, 0.038 mmol) and acceptor **2** [50] (16.1 mg, 0.035 mmol) under the general glycosidation method as a colorless form in 24 h in 89% yield (32.1 mg, 0.031 mmol). Analytical data for **14** was in accordance with that previously reported [61].

**Methyl 2,3,4-tri-*O*-benzyl-6-*O*-(2,3,4,6-tetra-*O*-benzyl-α/β-D-galactopyranosyl)-α-D-glucopyranoside (16)** was obtained from donor **15** [55] (22.3 mg, 0.038 mmol) and acceptor **1** [50] (16.1 mg, 0.035 mmol) under the general glycosidation method as a colorless foam in 16 h in 92% yield (31.5 mg, 0.032 mmol). Analytical data for **16** was in accordance with that previously reported [62].

**Methyl 3,4,6-tri-*O*-benzyl-2-*O*-(2,3,4,6-tetra-*O*-benzyl-α/β-D-galactopyranosyl)-α-D-glucopyranoside (17)** was obtained from donor **15** [55] (22.3 mg, 0.038 mmol) and acceptor **2** [50] (16.1 mg, 0.035 mmol) under the general glycosidation method as a colorless form in 16 h in 79% yield (27.1 mg, 0.027 mmol). Analytical data for **17** was in accordance with that previously reported [63].

**Methyl 6-*O*-(2,3,4,6-tetra-*O*-benzoyl-α-D-mannopyranosyl)-2,3,4-tri-*O*-benzyl-α-D-glucopyranoside (19)** was obtained from donor **18** [51] (24.4 mg, 0.038 mmol) and acceptor **1** [50] (16.1 mg, 0.035 mmol) under the general glycosidation method as a colorless foam in 24 h in 84% yield (30.1 mg, 0.029 mmol). Analytical data for **19** was in accordance with that previously reported [64].

**Methyl 2-*O*-(2,3,4,6-tetra-*O*-benzoyl-α-D-mannopyranosyl)-3,4,6-tri-*O*-benzyl-α-D-glucopyranoside (20)** was obtained from donor **18** [51] (21.6 mg, 0.034 mmol) and acceptor **2** [50] (14.2 mg, 0.031 mmol) under the general glycosidation method as a colorless form in 24 h in 85% yield (26.9 mg, 0.026 mmol). Analytical data for **20** was in accordance with that previously reported [65].

**Methyl 2,3,4-tri-*O*-benzyl-6-*O*-(2,3,4,6-tetra-*O*-benzyl-α/β-D-mannopyranosyl)-α-D-glucopyranoside (22)** was obtained from donor **21** [56] (30.4 mg, 0.052 mmol) and acceptor **1** [50] (21.9 mg, 0.047 mmol) under the general glycosidation method as a colorless foam in 16 h in 81% yield (37.5 mg, 0.038 mmol). Analytical data for **22** was in accordance with that previously reported [66].

**Methyl 3,4,6-tri-*O*-benzyl-2-*O*-(2,3,4,6-tetra-*O*-benzyl-α/β-D-mannopyranosyl)-α-D-glucopyranoside (23)** was obtained from donor **21** [56] (30.5 mg, 0.052 mmol) and acceptor **2** [50] (22.0 mg, 0.047 mmol) under the general glycosidation method as a colorless form in 16 h in 70% yield (32.9 mg, 0.033 mmol). Analytical data for **23** was in accordance with that previously reported [67].

**Methyl 6-O-(3,4-di-O-benzoyl-2,6-di-O-benzyl-α-D-galactopyranosyl)-2,3,4-tri-O-benzyl-α-D-glucopyranoside (25)** was obtained from donor 24 [45] (27.9 mg, 0.046 mmol) and acceptor 1 [50] (19.2 mg, 0.041 mmol) under the general glycosidation method as a colorless foam in 16 h in 78% yield (32.5 mg, 0.032 mmol). Analytical data for 25 was in accordance with that previously reported [45].

**Methyl 6-O-(2-O-benzoyl-3,4,6-tri-O-benzyl-α-D-mannopyranosyl)-2,3,4-tri-O-benzyl-α-D-glucopyranoside (27)** was obtained from donor 26 [57] (30.1 mg, 0.05 mmol) and acceptor 1 [50] (21.2 mg, 0.046 mmol) under the general glycosidation method as a colorless foam in 16 h in 79% yield (36.2 mg, 0.036 mmol). Analytical data for 27 was in accordance with that previously reported. [68]

## 4. Conclusions

Our previous studies of the 4K reaction with thioglycosides opened a new exciting avenue for discovery of new classes of thioglycoside activators. Previously, we demonstrated that $I_2$/Bi(OTf)$_3$/TfOH cooperatively catalyzed 4K reactions can be swift and efficient. Developed herein is an $I_2$/Fe(OTf)$_3$/TfOH cooperatively catalyzed 4K reaction for the direct activation of conventional thioglycosides. This methodology presents a promising expansion of the 4K reaction, which overall continues to be an exciting avenue for exploration. As demonstrated by several substrates and targets, this method offers new synthetic capabilities. The reaction conditions are relatively mild, and because of that, the glycosylation reactions are much slower than those in the presence of bismuth(III), albeit the product yields were similar.

Regardless of whether bismuth(III) or iron(III) were employed as co-catalysis, the reaction worked with a broad range of substrates, with both armed and disarmed glycosyl donors. In contrast, previously described activations in the presence of $FeCl_3$ alone worked well only for highly reactive, armed and superarmed thioglycoside donors, even in the presence of a large excess. Attempts to substitute Fe(OTf)$_3$ with $FeCl_3$ in the 4K reactions lead to decreased yields due to the accumulation of multiple side products.

It also became apparent that the 4K reactions with ferric triflate are not compatible with the HAD method because ferric salts are known to cleave the picoloyl group that is needed as a stereodirecting handle. Further exploration of the 4K reaction, the search for other effective promoters and catalysts, and application to the manual and automated synthesis of various linkages and glycan targets are currently underway.

**Supplementary Materials:** The following supporting information can be downloaded at: https://www.mdpi.com/article/10.3390/molecules30153058/s1, Figure S1. $^1$H NMR Spectrum ($CDCl_3$, 400 MHz) of Compound **4**; Figure S2. COSY NMR Spectrum ($CDCl_3$, 400 MHz) of Compound **4**; Figure S3. $^{13}$C NMR Spectrum ($CDCl_3$, 101 MHz) of Compound **4**; Figure S4. $^1$H NMR Spectrum ($CDCl_3$, 400 MHz) of Compound **5**; Figure S5. COSY NMR Spectrum ($CDCl_3$, 400 MHz) of Compound **5**; Figure S6. $^{13}$C NMR Spectrum ($CDCl_3$, 101 MHz) of Compound **5**; Figure S7. $^1$H NMR Spectrum ($CDCl_3$, 400 MHz) of Compound **10**; Figure S8. COSY NMR Spectrum ($CDCl_3$, 400 MHz) of Compound **10**; Figure S9. $^{13}$C NMR Spectrum ($CDCl_3$, 101 MHz) of Compound **10**; Figure S10. $^1$H NMR Spectrum ($CDCl_3$, 400 MHz) of Compound **11**; Figure S11. COSY NMR Spectrum ($CDCl_3$, 400 MHz) of Compound **11**; Figure S12. $^{13}$C NMR Spectrum ($CDCl_3$, 101 MHz) of Compound **11**; Figure S13. $^1$H NMR Spectrum ($CDCl_3$, 400 MHz) of Compound **13**; Figure S14. COSY NMR Spectrum ($CDCl_3$, 400 MHz) of Compound **13**; Figure S15. $^{13}$C NMR Spectrum ($CDCl_3$, 101 MHz) of Compound **13**; Figure S16. $^1$H NMR Spectrum ($CDCl_3$, 400 MHz) of Compound **14**; Figure S17. COSY NMR Spectrum ($CDCl_3$, 400 MHz) of Compound **14**; Figure S18. $^{13}$C NMR Spectrum ($CDCl_3$, 101 MHz) of Compound **14**; Figure S19. $^1$H NMR Spectrum ($CDCl_3$, 400 MHz) of Compound **16**; Figure S20. COSY NMR Spectrum ($CDCl_3$, 400 MHz) of Compound **16**;

Figure S21. $^{13}$C NMR Spectrum ($CDCl_3$, 101 MHz) of Compound **16**; Figure S22. $^{1}$H NMR Spectrum ($CDCl_3$, 400 MHz) of Compound **17**; Figure S23. COSY NMR Spectrum ($CDCl_3$, 400 MHz) of Compound **17**; Figure S24. $^{13}$C NMR Spectrum ($CDCl_3$, 101 MHz) of Compound **17**; Figure S25. $^{1}$H NMR Spectrum ($CDCl_3$, 400 MHz) of Compound **19**; Figure S26. COSY NMR Spectrum ($CDCl_3$, 400 MHz) of Compound **19**; Figure S27. $^{13}$C NMR Spectrum ($CDCl_3$, 101 MHz) of Compound **19**; Figure S28. $^{1}$H NMR Spectrum ($CDCl_3$, 400 MHz) of Compound **20**; Figure S29. COSY NMR Spectrum ($CDCl_3$, 400 MHz) of Compound **20**; Figure S30. $^{13}$C NMR Spectrum ($CDCl_3$, 101 MHz) of Compound **20**; Figure S31. $^{1}$H NMR Spectrum ($CDCl_3$, 400 MHz) of Compound **22**; Figure S32. COSY NMR Spectrum ($CDCl_3$, 400 MHz) of Compound **22**; Figure S33. $^{13}$C NMR Spectrum ($CDCl_3$, 101 MHz) of Compound **22**; Figure S34. $^{1}$H NMR Spectrum ($CDCl_3$, 400 MHz) of Compound **23**; Figure S35. COSY NMR Spectrum ($CDCl_3$, 400 MHz) of Compound **23**; Figure S36. $^{13}$C NMR Spectrum ($CDCl_3$, 101 MHz) of Compound **23**; Figure S37. $^{1}$H NMR Spectrum ($CDCl_3$, 400 MHz) of Compound **25**; Figure S38. COSY NMR Spectrum ($CDCl_3$, 400 MHz) of Compound **25**; Figure S39. $^{13}$C NMR Spectrum ($CDCl_3$, 101 MHz) of Compound **25**; Figure S40. $^{1}$H NMR Spectrum ($CDCl_3$, 400 MHz) of Compound **27**; Figure S41. COSY NMR Spectrum ($CDCl_3$, 400 MHz) of Compound **27**; Figure S42. $^{13}$C NMR Spectrum ($CDCl_3$, 101 MHz) of Compound **27**.

**Author Contributions:** Conceptualization, A.R.D. and A.V.D.; methodology, A.R.D.; formal analysis, A.R.D.; investigation, A.R.D. and A.M.D.; data curation, A.R.D. and A.M.D.; writing—original draft preparation, A.R.D.; writing—review and editing, A.R.D. and A.V.D.; supervision, A.V.D.; project administration, A.V.D.; funding acquisition, A.V.D. All authors have read and agreed to the published version of the manuscript.

**Funding:** The authors are indebted to the NSF (CHE-2350461) for support of this work.

**Institutional Review Board Statement:** Not applicable.

**Informed Consent Statement:** Not applicable.

**Data Availability Statement:** All data is available from the authors.

**Conflicts of Interest:** The authors declare no conflicts of interest.

## References

1. Demchenko, A.V. *Handbook of Chemical Glycosylation: Advances in Stereoselectivity and Therapeutic Relevance*; Wiley-VCH: Weinheim, Germany, 2008.
2. Paulsen, H. Advances in selective chemical syntheses of complex oligosaccharides. *Angew. Chem. Int. Ed. Engl.* **1982**, *21*, 155–173. [CrossRef]
3. Paulsen, H. Synthesis of complex oligosaccharide chains of glycoproteins. *Chem. Soc. Rev.* **1984**, *13*, 15–45. [CrossRef]
4. Toshima, K.; Tatsuta, K. Recent progress in O-glycosylation methods and its application to natural-products synthesis. *Chem. Rev.* **1993**, *93*, 1503–1531. [CrossRef]
5. Adero, P.O.; Amarasekara, H.; Wen, P.; Bohe, L.; Crich, D. The experimental evidence in support of glycosylation mechanisms at the SN1-SN2 Interface. *Chem. Rev.* **2018**, *118*, 8242–8284. [CrossRef] [PubMed]
6. Kulkarni, S.S.; Wang, C.C.; Sabbavarapu, N.M.; Podilapu, A.R.; Liao, P.H.; Hung, S.C. "One-Pot" Protection, Glycosylation, and Protection-Glycosylation Strategies of Carbohydrates. *Chem. Rev.* **2018**, *118*, 8025–8104. [CrossRef] [PubMed]
7. Nielsen, M.M.; Pedersen, C.M. Catalytic Glycosylations in Oligosaccharide Synthesis. *Chem. Rev.* **2018**, *118*, 8285–8358. [CrossRef] [PubMed]
8. Panza, M.; Pistorio, S.G.; Stine, K.J.; Demchenko, A.V. Automated chemical oligosaccharide synthesis: Novel approach to traditional challenges. *Chem. Rev.* **2018**, *118*, 8105–8150. [CrossRef] [PubMed]
9. Michael, A. On the synthesis of helicin and phenolglucoside. *Am. Chem. J.* **1879**, *1*, 305–312.
10. Fischer, E. Über die glucoside der alkohole. *Ber. Dtsch. Chem. Ges.* **1893**, *26*, 2400–2412. [CrossRef]
11. Koenigs, W.; Knorr, E. Über einige derivate des traubenzuckers und der galactose. *Ber. Dtsch. Chem. Ges.* **1901**, *34*, 957–981. [CrossRef]
12. Wulff, G.; Rohle, G. Results and problems of O-glycoside synthesis. *Angew. Chem. Int. Ed. Engl.* **1974**, *13*, 157–170. [CrossRef] [PubMed]
13. Igarashi, K. The Koenigs-Knorr reaction. *Adv. Carbohydr. Chem. Biochem.* **1977**, *34*, 243–283.

14. Singh, Y.; Demchenko, A.V. Koenigs-Knorr glycosylation reaction catalyzed by trimethylsilyl trifluoromethanesulfonate. *Chem. Eur. J.* **2019**, *25*, 1461–1465. [CrossRef] [PubMed]
15. Demchenko, A.V.; De Meo, C. The 4K reaction. *Carbohydr. Res.* **2024**, *538*, 109102. [CrossRef] [PubMed]
16. Lian, G.; Zhang, X.; Yu, B. Thioglycosides in Carbohydrate research. *Carbohydr. Res.* **2015**, *403*, 13–22. [CrossRef] [PubMed]
17. Fugedi, P.; Garegg, P.J. A novel promoter for the efficient construction of 1,2-trans linkages in glycoside synthesis, using thioglycosides as glycosyl donors. *Carbohydr. Res.* **1986**, *149*, c9–c12. [CrossRef]
18. Dasgupta, F.; Garegg, P.J. Use of sulfenyl halides in carbohydrate reactions. Part I. Alkyl sulfenyl triflate as activator in the thioglycoside-mediated formation of beta-glycosidic linkages during oligosaccharide synthesis. *Carbohydr. Res.* **1988**, *177*, C13–C17. [CrossRef]
19. Crich, D.; Smith, M. S-(4-Methoxyphenyl) benzenethiosulfinate(MPBT)/trifluoromethanesulfonic anhydride (Tf2O): A convenient system for the generation of glycosyl triflates from thioglycosides. *Org. Lett.* **2000**, *2*, 4067–4069. [CrossRef] [PubMed]
20. Crich, D.; Smith, M. 1-Benzenesulfinyl piperidine/trifluoromethanesulfonic anhydride: A potent combination of shelf-stable reagents for the low-temperature conversion of thioglycosides to glycosyl triflates and for the formation of diverse glycosidic linkages. *J. Am. Chem. Soc.* **2001**, *123*, 9015–9020. [CrossRef] [PubMed]
21. Codee, J.D.C.; Litjens, R.E.J.N.; Heeten, R.; Overkleeft, H.S.; van Boom, J.H.; van der Marel, G.A. $Ph_2SO/Tf_2O$: A powerful promotor system in chemoselective glycosylations using thioglycosides. *Org. Lett.* **2003**, *5*, 1519–1522. [CrossRef] [PubMed]
22. Duron, S.G.; Polat, T.; Wong, C.H. N-(Phenylthio)-ε-caprolactam: A new promoter for the activation of thioglycosides. *Org. Lett.* **2004**, *6*, 839–841. [CrossRef] [PubMed]
23. Veeneman, G.H.; van Leeuwen, S.H.; van Boom, J.H. Iodonium ion promoted reactions at the anomeric centre. II. An efficient thioglycoside mediated approach toward the formation of 1,2-trans linked glycosides and glycosidic esters. *Tetrahedron Lett.* **1990**, *31*, 1331–1334. [CrossRef]
24. Nicolaou, K.C.; Seitz, S.P.; Papahatjis, D.P. A mild and general method for the synthesis of O-glycosides. *J. Am. Chem. Soc.* **1983**, *105*, 2430–2434. [CrossRef]
25. Kihlberg, J.O.; Leigh, D.A.; Bundle, D.R. The in situ activation of thioglycosides with bromine: An improved glycosylation method. *J. Org. Chem.* **1990**, *55*, 2860–2863. [CrossRef]
26. Kartha, K.P.M.; Aloui, M.; Field, R.A. Iodine: A versatile reagent in carbohydrate chemistry II. Efficient chemospecific activation of thiomethylglycosides. *Tetrahedron Lett.* **1996**, *37*, 5175–5178. [CrossRef]
27. Burkart, M.D.; Zhang, Z.; Hung, S.-C.; Wong, C.-H. A new method for the synthesis of fluoro-carbohydrates and glycosides using selectfluor. *J. Am. Chem. Soc.* **1997**, *119*, 11743–11746. [CrossRef]
28. Ercegovic, T.; Meijer, A.; Magnusson, G.; Ellervik, U. Iodine monochloride/silver trifluoromethanesulfonate (ICl/AgOTf) as a convenient promoter system for O-glycoside synthesis. *Org. Lett.* **2001**, *3*, 913–915. [CrossRef] [PubMed]
29. Mao, R.-Z.; Guo, F.; Xiong, D.-C.; Li, Q.; Duan, J.; Ye, X.-S. Photoinduced C-S Bond Cleavage of Thioglycosides and Glycosylation. *Org. Lett.* **2015**, *17*, 5606–5609. [CrossRef] [PubMed]
30. Wever, W.J.; Cinelli, M.A.; Bowers, A.A. Visible Light Mediated Activation and O-Glycosylation of Thioglycosides. *Org. Lett.* **2013**, *15*, 30–33. [CrossRef] [PubMed]
31. Spell, M.L.; Deveaux, K.; Bresnahan, C.G.; Bernard, B.L.; Sheffield, W.; Kumar, R.; Ragains, J.R. A Visible-Light-Promoted O-Glycosylation with a Thioglycoside Donor. *Angew. Chem. Int. Ed.* **2016**, *55*, 6515–6519. [CrossRef] [PubMed]
32. Mao, R.-Z.; Xiong, D.-C.; Guo, F.; Li, Q.; Duan, J.; Ye, X.-S. Light-driven highly efficient glycosylation reactions. *Org. Chem. Front.* **2016**, *3*, 737–743. [CrossRef]
33. Escopy, S.; Demchenko, A.V. Transition-Metal-Mediated Glycosylation with Thioglycosides. *Chem. Eur. J.* **2022**, *28*, e202103747. [CrossRef] [PubMed]
34. Ferrier, R.J.; Hay, R.W.; Vethaviyasar, N. A potentially versatile synthesis of glycosides. *Carbohydr. Res.* **1973**, *27*, 55–61. [CrossRef]
35. Goswami, M.; Ellern, A.; Pohl, N.L. Bismuth(V)-mediated thioglycoside activation. *Angew. Chem. Int. Ed. Engl.* **2013**, *52*, 8441–8445. [CrossRef] [PubMed]
36. Goswami, M.; Ashley, D.C.; Baik, M.H.; Pohl, N.L. Mechanistic Studies of Bismuth(V)-Mediated Thioglycoside Activation Reveal Differential Reactivity of Anomers. *J. Org. Chem.* **2016**, *81*, 5949–5962. [CrossRef] [PubMed]
37. Vibhute, A.M.; Dhaka, A.; Athiyarath, V.; Sureshan, K.M. A versatile glycosylation strategy via Au (III) catalyzed activation of thioglycoside donors. *Chem. Sci.* **2016**, *7*, 4259–4263. [CrossRef] [PubMed]
38. Zhu, X.; Schmidt, R.R. New principles for glycoside-bond formation. *Angew. Chem. Int. Ed.* **2009**, *48*, 1900–1934. [CrossRef] [PubMed]
39. Escopy, S.; Singh, Y.; Demchenko, A.V. Palladium(II)-assisted activation of thioglycosides. *Org. Biomol. Chem.* **2021**, *19*, 2044–2054. [CrossRef] [PubMed]
40. Pooladian, F.; Escopy, S.; Demchenko, A.V. Activation of thioglycosides with copper(II) bromide. *Molecules* **2022**, *27*, 7354. [CrossRef] [PubMed]

41. Ridgway, L.M.; Das, A.; Shadrick, M.L.; Demchenko, A.V. Ferric Chloride Promoted Glycosidation of Alkyl Thioglycosides. *Molecules* **2024**, *29*, 4845. [CrossRef] [PubMed]
42. Dent, A.; Escopy, S.; Demchenko, A.V. Cooperatively Catalyzed Activation of Thioglycosides That Bypasses Intermediacy of Glycosyl Halides. *Chem. Eur. J.* **2023**, *29*, e202300873. [CrossRef] [PubMed]
43. Dent, A.R.; Demchenko, A.V. Halophilic Metal Salts for the Cooperatively Catalyzed Activation of Thioglycosides. *J. Org. Chem.* **2025**, *90*, 6478–6490. [CrossRef] [PubMed]
44. Bandara, M.D.; Yasomanee, J.P.; Demchenko, A.V. Application of armed, disarmed, superarmed and superdisarmed building blocks in stereocontrolled glycosylation and expeditious oligosaccharide synthesis. In *Selective Glycosylations: Synthetic Methods and Catalysts*; Bennett, C.S., Ed.; Wiley: Weinheim, Germany, 2017; pp. 29–58.
45. Shadrick, M.; Singh, Y.; Demchenko, A.V. Stereocontrolled α-galactosylation under cooperative catalysis. *J. Org. Chem.* **2020**, *85*, 15936–15944. [CrossRef] [PubMed]
46. Shadrick, M.; Stine, K.J.; Demchenko, A.V. Expanding the scope of stereoselective α-galactosylation using glycosyl chlorides. *Bioorg. Med. Chem.* **2022**, *73*, 117031. [CrossRef] [PubMed]
47. Premathilake, H.D.; Mydock, L.K.; Demchenko, A.V. Superarming common glycosyl donors by simple 2-O-benzoyl-3,4,6-tri-O-benzyl protection. *J. Org. Chem.* **2010**, *75*, 1095–1100. [CrossRef] [PubMed]
48. Yasomanee, J.P.; Demchenko, A.V. The effect of remote picolinyl and picoloyl substituents on the stereoselectivity of chemical glycosylation. *J. Am. Chem. Soc.* **2012**, *134*, 20097–20102. [CrossRef] [PubMed]
49. Geringer, S.A.; Mannino, M.P.; Bandara, M.D.; Demchenko, A.V. Picoloyl protecting group in synthesis: Focus on a highly chemoselective catalytic removal. *Org. Biomol. Chem.* **2020**, *18*, 4863–4871. [CrossRef] [PubMed]
50. Ranade, S.C.; Kaeothip, S.; Demchenko, A.V. Glycosyl alkoxythioimidates as complementary building blocks for chemical glycosylation. *Org. Lett.* **2010**, *12*, 5628–5631. [CrossRef] [PubMed]
51. Sail, D.; Kovac, P. Benzoylated ethyl 1-thioglycosides: Direct preparation from per-O-benzoylated sugars. *Carbohydr. Res.* **2012**, *357*, 47–52. [CrossRef] [PubMed]
52. Ferrier, R.J.; Furneaux, R.H. 1,2-trans-1-Thioglycosides. In *Methods in Carbohydrate Chemistry*; Whistler, R.L., BeMiller, J.N., Eds.; Academic Press: Cambridge, MA, USA, 1980; Volume 8, pp. 251–253.
53. France, R.R.; Rees, N.V.; Wadhawman, J.D.; Fairbanks, A.J.; Compton, R.G. Selective activation of glycosyl donors using electrochemical techniques: A study of the thermodynamic oxidation potentials of a range of chalcoglycosides. *Org. Biomol. Chem.* **2004**, *2*, 2188–2194. [CrossRef] [PubMed]
54. Andersson, F.; Fugedi, P.; Garegg, P.J.; Nashed, M. Synthesis of 1,2-cis-linked glycosides using dimethyl(methylthio) sulfonium triflate as promoter and thioglycosides as glycosyl donors. *Tetrahedron Lett.* **1986**, *27*, 3919–3922. [CrossRef]
55. Weïwer, M.; Sherwood, T.; Linhardt, R.J. Synthesis of Floridoside. *J. Carbohydr. Chem.* **2008**, *27*, 420–427. [CrossRef]
56. Wang, C.; Sanders, B.; Baker, D.C. Synthesis of a glycodendrimer incorporating multiple mannosides on a glucoside core. *Can. J. Chem.* **2011**, *89*, 959–963. [CrossRef]
57. Elie, C.; Verduyn, R.; Dreef, C.; Brounts, D.; van der Marel, G.; van Boom, J. Synthesis of 6-0-(α-D-mannopyranosyl)-D-myo-inositol: A fragment from mycobacteria phospholipids. *Tetrahedron* **1990**, *46*, 8243–8254. [CrossRef]
58. Garcia, B.A.; Gin, D.Y. Dehydrative glycosylation with activated diphenyl sulfonium reagents. Scope, mode of C(1)-hemiacetal activation, and detection of reactive glycosyl intermediates. *J. Am. Chem. Soc.* **2000**, *122*, 4269–4279. [CrossRef]
59. Nigudkar, S.S.; Parameswar, A.R.; Pornsuriyasak, P.; Stine, K.J.; Demchenko, A.V. O-Benzoxazolyl imidates as versatile glycosyl donors for chemical glycosylation. *Org. Biomol. Chem.* **2013**, *11*, 4068–4076. [CrossRef] [PubMed]
60. Codee, J.D.C.; Van den Bos, L.J.; Litjens, R.E.J.N.; Overkleeft, H.S.; Van Boeckel, C.A.A.; Van Boom, J.H.; Van der Marel, G.A. Chemoselective glycosylations using sulfonium triflate activator systems. *Tetrahedron* **2004**, *60*, 1057–1064. [CrossRef]
61. Hasty, S.J.; Kleine, M.A.; Demchenko, A.V. S-Benzimidazolyl glycosides as a platform for oligosaccharide synthesis by an active-latent strategy. *Angew. Chem. Int. Ed.* **2011**, *50*, 4197–4201. [CrossRef] [PubMed]
62. Vankar, Y.D.; Vankar, P.S.; Behrendt, M.; Schmidt, R.R. Synthesis of β-O-glycosides using enol ether and imidate derived leaving groups. Emphasis on the use of nitriles as a solvent. *Tetrahedron* **1991**, *47*, 9985–9992. [CrossRef]
63. Premathilake, H.D.; Demchenko, A.V. 2-Allylphenyl glycosides as complementary building blocks for oligosaccharide and glycoconjugate synthesis. *Beilstein J. Org. Chem.* **2012**, *8*, 597–605. [CrossRef] [PubMed]
64. Pornsuriyasak, P.; Demchenko, A.V. S-Thiazolinyl (STaz) glycosides as versatile building blocks for convergent selective, chemoselective, and orthogonal oligosaccharide synthesis. *Chem. Eur. J.* **2006**, *12*, 6630–6646. [CrossRef] [PubMed]
65. Singh, Y.; Wang, T.; Geringer, S.A.; Stine, K.J.; Demchenko, A.V. Regenerative glycosylation. *J. Org. Chem.* **2018**, *83*, 374–381. [CrossRef] [PubMed]

66. Mishra, B.; Neralkar, M.; Hotha, S. Stable alkynyl glycosyl carbonates: Catalytic anomeric activation and synthesis of a tridecasaccharide reminiscent of Mycobacterium tuberculosis cell wall lipoarabinomannan. *Angew. Chem. Int. Ed.* **2016**, *128*, 7917–7922. [CrossRef]
67. Geringer, S.A.; Demchenko, A.V. Iron(III) chloride-catalyzed activation of glycosyl chlorides. *Org. Biomol. Chem.* **2018**, *16*, 9133–9137. [CrossRef] [PubMed]
68. Ravidà, A.; Liu, X.; Kovacs, L.; Seeberger, P.H. Synthesis of glycosyl phosphates from 1,2-orthoesters and application to in situ glycosylation reactions. *Org. Lett.* **2006**, *8*, 1815–1818. [CrossRef] [PubMed]

molecules

MDPI

Article

# Bivalent Inhibitors of Mannose-Specific Bacterial Adhesion: A Xylose-Based Conformational Switch to Control Glycoligand Distance †

**Sven Ole Jaeschke, Ingo vom Sondern and Thisbe K. Lindhorst ***

Otto Diels Institute of Organic Chemistry, Christiana Albertina University of Kiel, 24098 Kiel, Germany
* Correspondence: tklind@oc.uni-kiel.de
† In fond remembrance dedicated to the memory of Professor Dr. Hans Paulsen.

**Abstract**

Functional glycomimetics is suited to study the parameters of carbohydrate recognition that forms the basis of glycobiology. It is particularly attractive when a glycoligand allows for the investigation of two different states, such as varying distance between multiple glycoligands. Here, a xylopyranoside was employed as a scaffold for the presentation of two mannoside units which are ligands of the bacterial lectin FimH. The chair conformation of the central xyloside can be switched between a $^4C_1$ and a $^1C_4$ conformation whereby the two conjugated mannoside ligands are flipped from a di-equatorial into a di-axial position. Concomitantly, the distance between the two glycoligands changes and, as a consequence, so does the biological activity of the respective bivalent glycocluster, as shown in adhesion–inhibition assays with live bacteria. Molecular modeling was employed to correlate the inter-ligand distance with the structure of the formed glycocluster–FimH complex. Our study suggests that conformational switches can be employed and further advanced as smart molecular tools to study structural boundary conditions of carbohydrate recognition in a bottom-up approach.

**Keywords:** carbohydrate chemistry; carbohydrate recognition; conformational switch; bacterial adhesion; FimH; lectin inhibitors; inter-ligand distance; molecular modeling

## 1. Introduction

The carbohydrate-specific adhesion of *E. coli* bacteria to glycosylated surfaces is an important topic in glycoscience that arouses interest in two respects. First, bacterial adhesion is a prerequisite for infectious diseases and for biofilm formation. Knowing the involved mechanisms and being able to prevent the adhesion of bacteria, for example with carbohydrate-based inhibitors, can contribute to progress in diagnostics and therapy [1–3]. Second, bacterial adhesion can serve as a relevant biosystem with which to probe the parameters of carbohydrate recognition in a bottom-up approach using bespoke glycomimetics [4–7].

To accomplish adhesion to the glycocalyx of their host cells, bacteria use proteinaceous adhesive organelles, so-called fimbriae that project from the bacterial cell surface in multiple copies. Fimbriae have various carbohydrate specificities. Type 1 fimbriae are the best-known and most often studied fimbriae displaying a specificity for α-D-mannosides and are mediated by a lectin domain at the fimbrial tips called FimH [8]. Many different glycomimetics have been designed as FimH antagonists aiming at the inhibition of type 1 fimbriae-mediated bacterial adhesion [9–12]. It has become clear that various structural aspects govern the inhibitory power of carbohydrate-based inhibitors. These include the nature of the aglycone portion of a mannoside, which can exert favorable interactions at the entrance of the FimH carbohydrate recognition domain (CRD), specifically with the so-called tyrosine gate formed by amino acid residues Tyr48 and Tyr137 [13]. Hence, mannosides with aromatic aglycon moieties are especially powerful inhibitors of FimH as they can exert $\pi$-$\pi$ interactions with the tyrosine gate [14]. Additionally, multivalency effects have often been observed with bivalent and multivalent glycoclusters [15,16]. Also, the distance between multiple mannoside ligands has been recognized as an important parameter in carbohydrate recognition, for example when glycoligands were studied, which are based on a rigid biphenyl scaffold [17–19].

Here, we introduce a new type of bivalent inhibitor of type 1 fimbriae-mediated bacterial adhesion. The special feature of this type of glycocluster is that the distance between the two glycoligands can be varied through a conformational switch that is represented by a central xylopyranoside scaffold. Two glycoligand moieties which are attached to positions 1 and 3 of a xyloside scaffold differ greatly in their distance from each other depending on the conformation of the central sugar ring. This is so because the ring flip between a $^4C_1$ and $^1C_4$ xylose chair conformation forces a maximal orientational change of the ring substituents from all-equatorial to all-axial. Hence, while the two equatorially positioned glycoligand units at C-1 and C-2 point away from each other in the $^4C_1$ conformation of the xyloside, they come much closer together in the $^1C_4$ chair (Figure 1a).

The design of this "switchable" glycoligand is based on a 2,4-diamino xylopyranoside which has been employed previously to realize flipping between the two complementary chair conformations through metal complexation [21–23]. We have recently demonstrated that the iodosulfonamidation of a suitable glycal is the method of choice to achieve substituted 2,4-diamino xyloside stereo- and regioselectively [24]. Following this synthetic route, the substituted *para*-bromophenyl xyloside **1** was produced [24] (Figure 1b). It offers two features which are essential for the herein described approach of controlling glycoligand distance through conformational switching. First, the Boc-protected amino functions at positions 2 and 4 of the xyloside ring allow the switching of the pyranose $^4C_1$ conformation into the flipped $^1C_4$ conformation after removal of the Boc-protecting groups. This can be accomplished, e.g., through metal complexation of the liberated 2,4-diamine or through the covalent locking of the $^1C_4$ conformation by formation of a cyclic urea derivative (vide infra, Scheme 2). Second, the *para*-bromophenyloxy residues at positions 1 and 3 of the sugar ring provide the option to ligate an arbitrary glycoligand that is functionalized with a phenylboronic acids moiety in a Pd-catalyzed cross-coupling reaction.

This approach enables the synthesis of the targeted bivalent glycoligands carrying two biphenyl mannoside moieties as ligands of the bacterial lectin FimH where the inter-ligand distance will vary depending on the chair conformation of the central xyloside scaffold, $^4C_1$ or $^1C_4$, respectively. The biphenyl linker units were chosen on purpose for several reasons. First, biphenyl α-D-mannopyranosides are known as high-affinity ligands of FimH [25,26]; second, their rigidity facilitates direct transfer of the spatial effect of the ring flip of the central xyloside to the relative orientation of the ligated mannoside ligands. Finally, in the glycocluster based on the xyloside in $^1C_4$ conformation, the glycoligands can undergo intramolecular $\pi$-$\pi$ interactions stabilizing a close inter-ligand distance [22].

Using a bivalent glycoligand based on such a conformational switch allows, firstly, the determination of how differential distances between two ligands affect the inhibitory potency of the corresponding compound. Second, the principle of a conformational switch can be explored which is attractive since such glycomimetics have not been used in carbohydrate recognition studies so far.

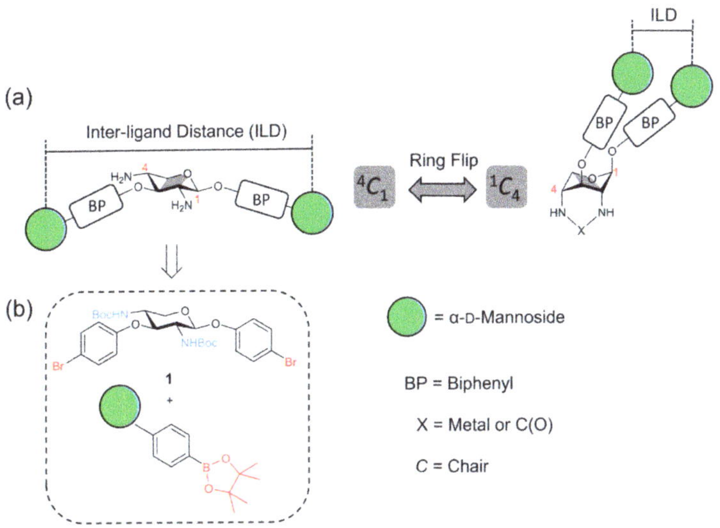


**Figure 1.** (**a**) The idea behind this account consists of switching the inter-ligand distance (ILD) within a bivalent glycocluster through the ring flip of a central xyloside scaffold. When the xyloside scaffold exists in the $^4C_1$ conformation, the ILD is maximal whereas it is reduced when the xylopyranoside ring is forced into the $^1C_4$ conformation. Additionally, in the $^1C_4$ conformer the two biphenyl moieties can exert intramolecular $\pi$-$\pi$ interactions which further contribute to the tight spatial organization of ligand presentation. (**b**) The retrosynthetic analysis of such a conformational switch leads back to the xyloside **1** which can be cross-coupled with a carbohydrate phenylboronic acid. Mannoside moieties are depicted as green circles according to the symbol nomenclature for glycans (SNFG) [20].

Hence, in this account we describe the synthesis of two bivalent glycoligands based on a switchable xylopyranoside scaffold in the $^4C_1$ and in $^1C_4$ conformation, respectively, and their investigation as inhibitors of type 1 fimbriae-mediated adhesion of *E. coli* on mannan-coated microtiter plates. Moreover, molecular dynamics and molecular modeling simulations were performed to provide a basis for the interpretation of the observed structure–function relationships.

## 2. Results

### 2.1. Synthesis

Synthesis started with the *N*-Boc-protected 2,4-diamino xyloside **1** [24] in which the two *para*-bromophenyloxy residues at positions 1 and 3 of the sugar ring were intended for ligation with a suitable phenylboronic acids moiety in a Suzuki–Miyaura cross-coupling reaction [27]. Here, we employed the literature-known 4-mannopyranosyloxyphenyl boronate **2** [28,29], which can be obtained from the respective bromophenyl mannoside by standard procedures. In order to achieve the divalent biphenylmannoside glycocluster **3**, the Suzuki–Miyaura cross-coupling reaction was first attempted with $Pd(PPh_3)_4$ as the catalyst and $Cs_2CO_3$ or $K_3PO_4$ as the base (Scheme 1) [5]. However, no product was formed under these conditions. Also, variation of the temperature or the solvent did not lead to success. However, it was recognized that the xyloside **1** reacted with the Pd catalyst in an oxidative addition reaction, whereas the transmetallation and reductive elimination reaction steps did not occur. Because for these latter steps of the cross-coupling reaction the base is important, we next focused on the addition of fluorides. Fluorides are potent bases for Suzuki–Miyaura reactions due to the fluorophilicity of the organoboranes [30]. Indeed, when cesium fluoride (CsF) was employed as the base [31], cross-coupling of **1** and **2** in THF for 8 h at 80 °C using $Pd(dppf)Cl_2{\cdot}CH_2Cl_2$ as the catalyst led to the desired product **3**, however only in a moderate yield of 40%. By increasing the reaction time to 16 h, the yield was almost doubled to 79% (Scheme 1).

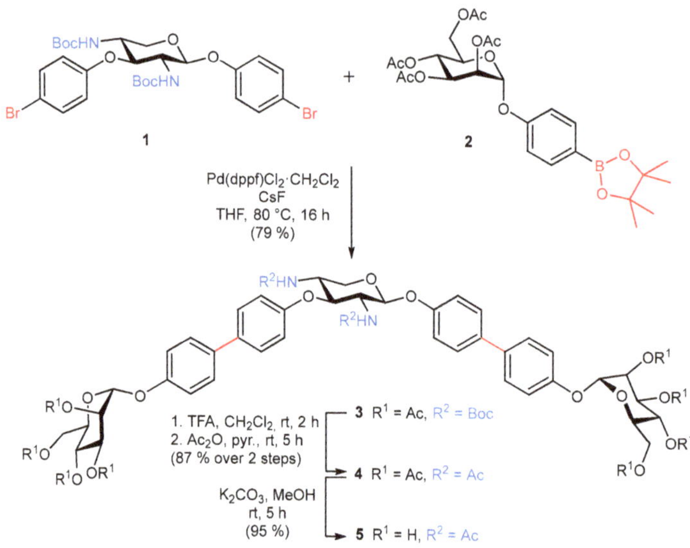


**Scheme 1.** Synthesis of the divalent biphenylmannoside glycoclusters **3**, **4**, and **5**. The central xyloside scaffold unit in **4** and **5** adopts the $^4C_1$ conformation (see text). The key step of this synthesis is a double Suzuki–Miyaura cross-coupling of the bis-arylhalide **1** and the phenylboronic acid **2**. TFA: trifluoroacetic acid.

The investigation of the target glycocluster **3** by $^1$H NMR spectroscopy indicated a conformational equilibrium between various conformations of the xylopyranoside ring owing to line broadening of the signals. An analogous observation has been reported earlier for the xyloside **1** [24] and the occurring conformational dynamic has been further analyzed in a study combining spectroscopic and computational data [32]. Here however, two defined conformers, one in a $^4C_1$ and the other in a $^1C_4$ conformation, are needed in order to alter the distance between the scaffolded biphenyl mannoside ligands in a controlled way. Therefore, the *N*-protecting groups were exchanged in an attempt to omit the conformational dynamic of the xyloside scaffold. Removal of the Boc-protecting groups with TFA furnished the unprotected diamine, which was subsequently acetylated in the same pot to give the *N*-acetylated mannosyloxybiphenyl xyloside **4** in a yield of 87% over two steps. Note that free amino groups had to be avoided as they can lead to false results in biological testing. The $^1$H NMR-spectroscopic analysis of **4** then indicated a clear $^4C_1$ conformation for the xyloside scaffold (as well as for the mannoside ligands) due to sharp and well-resolved peaks in the spectrum (see Supplementary Materials). The same was true for the OH-free xyloside-based glycocluster **5** which was obtained after removal of the *O*-acetyl groups under Zemplén conditions [33] in 95% yield (Scheme 1).

The biphenylmannoside glycocluster **5** is suitable as a divalent FimH antagonist in biological testing (vide infra). Here, the two biphenyl mannoside ligands are scaffolded on a xylopyranoside ring in $^4C_1$ conformation. In order to achieve the complementary FimH antagonist in which the two biphenyl mannoside ligands are differently oriented, that is arranged on a xylopyranoside ring in $^1C_4$ conformation, the ring flip of the central carbohydrate had to be achieved. To this end, the 2- and 4-amino groups were incorporated into a cyclic urea with an electrophilic C1 reagent. This reaction forces the xylopyranoside scaffold to adopt the $^1C_4$ conformation. Starting from the bis-Boc-protected glycocluster **3**, first the Boc-protecting groups were cleaved with TFA and then the unprotected intermediate was subjected to an intramolecular cyclization reaction with carbonyldiimidazole (CDI) in the presence of $Et_3N$ in DMF [34]. Under these conditions, the bicyclic urea derivative **6** was obtained in 74% yield over two steps (Scheme 2). Its global deprotection under Zemplén conditions gave the target glycocluster **7** in which the two biphenyl mannoside ligands are differently oriented compared to **5** owing to the ring flip of the central xyloside hinge.

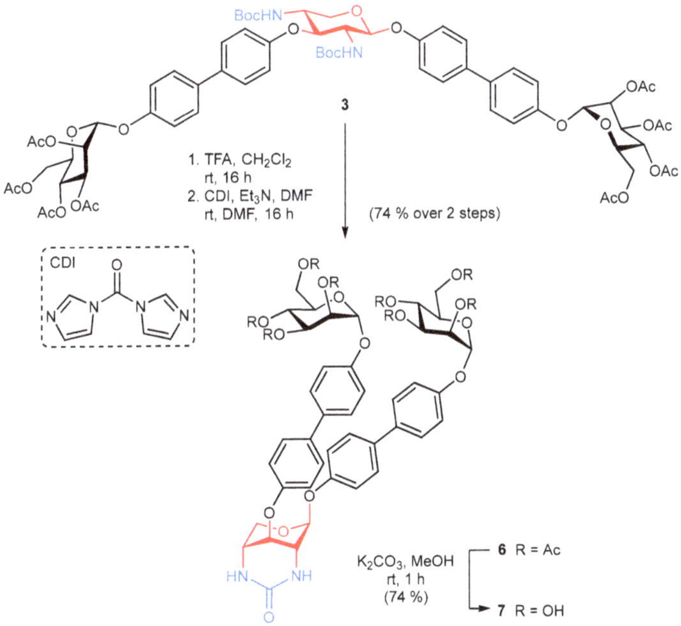


**Scheme 2.** Synthesis of the divalent biphenylmannoside glycocluster **7** in which the central xyloside scaffold is locked in the $^1C_4$ conformation. CDI: carbonyldiimidazole.

To obtain a feeling for the distance distribution of the two glycoligands conjugated to the central scaffold xyloside in **5** and **7**, molecular dynamics studies were carried out.

### 2.2. Molecular Dynamics

At first glance, the two biphenyl mannoside units in **5** and **7** seem to project from the xyloside scaffold very differently. However, there is quite some conformational freedom due to free rotation around the connecting single bonds which might lead to a different picture in reality. To assess the actual distance between the two mannose ligands in both glycoclusters, they were subjected to a molecular dynamics (MD) simulation. For this, the program Desmond [35,36] as implemented in the Schrödinger Maestro software package 2021-1 [37] was employed. The MD simulations were carried out in explicit water (SPC model [38] at 310 K for 200 ns; see Supplementary Materials). For both glycoclusters, the respective chair conformations ($^4C_1$ for **5** and $^1C_4$ for **7**) of the central xyloside ring were assumed to remain stable over the simulation period. On the other hand, the spatial orientation of the biphenyl linker units is flexible during the simulation, influencing the distance between the glycosidically bound mannosyl residues.

The obtained MD trajectories were analyzed to evaluate the distance between the centres of the two mannopyranoside rings projecting from the xyloside scaffold in **5** and **7** (Supplementary Material Figure S2). In compound **5**, the inter-ring distance fluctuated between 7 and 27 Å, while in compound **7**, it ranged from 5 to 25 Å. To characterize the most frequently occurring distances, the probability distribution of the distances between the two mannoside ligands were plotted for **5** and **7** (Figure 2). It can be seen that the change of the conformation of the central xyloside ring from $^4C_1$ to $^1C_4$ clearly impacts the amplitude of the inter-ligand distances as well as their distribution. However, the differences are less pronounced than intuitively expected. In glycocluster **5**, the distance between the two mannoside rings reaches a maximum in comparison to **7** owing to the $^4C_1$ conformation of the xyloside scaffold and the conformational flexibility of the biphenyl units (Figure 2, structure A). Here, the most frequently occurring distance is ~25 Å. On the contrary, the inter-ligand distances in **7** are smaller due to the $^1C_4$ conformation of the xyloside scaffold. Note that two main populations are seen in **7**. One population shows an inter-ligand distance amplitude at ~20 Å (Figure 2, structure B), whereas in the other population (Figure 2, structure C), the two mannoside ligands are very close to each other (~6 Å) due to $\pi$-$\pi$ interactions between the biphenyl units. This is a phenomenon which has been seen before in arylether-substituted 2,4-diamino xyloside switches where $\pi$-$\pi$ stacking can lead to stabilization of the $^1C_4$ conformation of the xyloside [22].

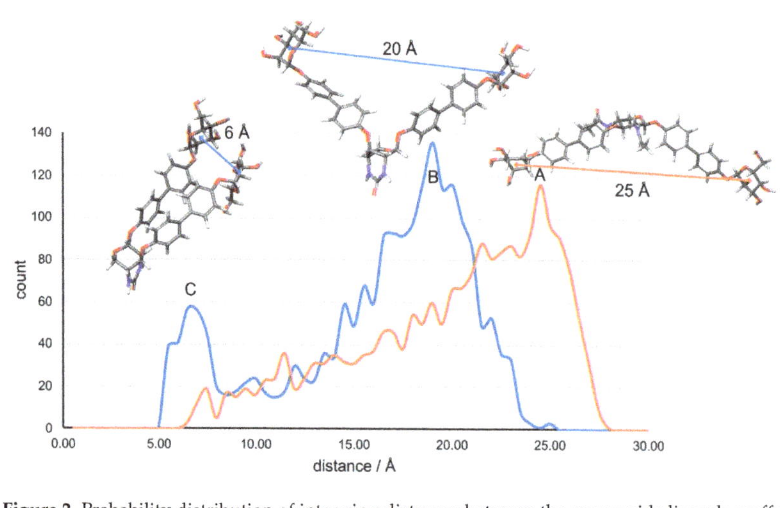


**Figure 2.** Probability distribution of inter-ring distances between the mannoside ligands scaffolded in **5** (bronze) and **7** (blue), respectively, and their occurrence as determined by molecular dynamics. Distances are measured between the centers of the mannopyranoside rings. Structures A, B, and C are representative snapshots of the MD simulation, illustrating the consequences of the conformational change of the central xylopyranoside chair from $^4C_1$ (in **5**, bronze) to $^1C_4$ (in **7**, blue). $\pi$-$\pi$ interactions between the biphenyl moieties, which are axially aligned in **7**, result in structures like C. Structures rendered with Schrödinger. Color code for the structures: carbons (grey), nitrogens (blue), oxygen (red).

### 2.3. Biological Testing

Although the inter-ligand distances in glycoclusters **5** and **7** cover a relatively wide range (cf. Section 2.2), in the most frequently occurring conformers the distance between the biphenyl mannoside antennas is different (~25 Å for **5** and ~20 and ~6 Å for **7**). In order to test if this difference, which is a result of the inverse ring conformation of the central xyloside scaffold, effects the properties of **5** and **7** as bivalent FimH ligands, an adhesion–inhibition assay with type 1 fimbriated *E. coli* bacteria (PKL1162) [39] was performed. In the literature-known assay [40], adhesion of GFP (green fluorescent protein)-transfected *E. coli* to a mannan-coated surface is determined (for details see Supplementary Materials). Remaining adhered bacteria after inhibition and washing can be determined by fluorescence read-out. The glycoclusters **5** and **7** were employed in serial dilutions on microtiter plates where they compete with the mannan coating for FimH binding. As a result, dose-response inhibition curves are obtained (Supplementary Material Figure S1) of which $IC_{50}$ values for each inhibitor can be deduced (see Supplementary Material Table S1). Since biological experiments and $IC_{50}$ values can vary significantly between independent experiments with live bacteria, methyl α-D-mannopyranoside (MeMan) was tested on the same plate serving as internal reference inhibitor. The inhibitory potencies of the tested glycoclusters were then referenced to MeMan leading to so-called relative inhibitory potencies (RIP values). Since the solubility of the glycoclusters **5** and **7** in aqueous buffer is limited due to the included hydrophobic biphenyl moieties, 5% DMSO were added to the glycocluster solutions (0.4 mm). This does, however, not compromise the assay.

The results of the adhesion–inhibition assay are summarized in Table 1. As the biphenyl moieties glycosidically linked to the mannoside ligands can exert favorable interactions with the tyrosine gate at the entrance of the FimH CRD, their FimH affinity is considerably higher than that of MeMan [41]. The resulting RIP values of **5** and **7** were in the range of literature-known aryl mannoside ligands (cf. Supplementary Materials) [42]. However strikingly, the inhibitory potencies of the glycoclusters **5** and **7** differ from each other by a factor of close to 3 in spite of the fact that both inhibitors display two copies of exactly the same biphenyl α-D-mannopyranoside ligand.

**Table 1.** Inhibitory potencies of the xylose-scaffolded glycoclusters **5** and **7** as determined in adhesion–inhibition assays with type 1 fimbriated *E. coli* bacteria (PKL1162).

| Glycocluster | $IC_{50}$ [a] [μmol] | $IC_{50}$ MeMan [b] [mmol] | RIP [c] | Average RIP [d] |
|---|---|---|---|---|
| **5** | 15.2 (±1.5) | 7.08 (±0.44) | 466 (±76) | |
| $^4C_1$ conformation of the xyloside scaffold | | | | 410 (±68) |
| | 15.7 (±1.3) | 5.34 (±0.48) | 339 (±60) | |
| **7** | 63.9 (±14.1) | 7.08 (±0.44) | 111 (±31) | |
| $^1C_4$ conformation of the xyloside scaffold | | | | 157 (±45) |
| | 36.0 (±11.1) | 5.34 (±0.48) | 148 (±59) | |

[a] $IC_{50}$ values are average values of duplicate or triplicate results on one plate. The fitting error from non-linear regression is given in brackets. [b] $IC_{50}$ (MeMan) values are average values of triplicate results on the same plate. The fitting error from non-linear regression is given in brackets. [c] RIP values are based on the inhibitory potency of MeMan tested on the same microplate IP(MeMan) ≡ 1); RIP(glycocluster) = $IC_{50}$(MeMan)/$IC_{50}$(glycocluster). The fitting error from error propagation is given in brackets (see Supplementary Materials for details). [d] Average RIPs are mean values of two independent experiments (I and II) with error propagation in brackets (see Supplementary Materials for details).

As the differences seen in the inhibitory potencies between **5** and **7** cannot be attributed to the nature of the ligand or the aglycon moiety of the mannoside, nor to the valency of the tested glycocluster (both are bivalent), it seems likely that the conformational dynamics of the glycoligand antennas and, more specifically, the various inter-ligand distances displayed in **5** and **7** are decisive for the different ligand properties. In **5**, the inter-ligand distance is maximal and this seems to be favorable for the inhibitory potency of the respective glycocluster. Even binding of two FimH proteins by a bivalent glycoligand has been described in the literature [43], however, it is not sure, that the same binding mode is happening in an assay with live type 1 fimbriated bacteria. Nevertheless, a greater conformational availability of glycoligands such as in **5** apparently facilitates lectin binding leading to higher FimH affinity (lower $IC_{50}$ values). In glycocluster **7**, the mannoside ligands span shorter inter-ligand distances resulting in lower lectin affinity. Especially those conformers occurring in **7** where the two glycoligands stack to each other through $\pi$-$\pi$ interactions (Figure 2, structure C) can be assumed to show a relatively weak FimH affinity as any of the two the mannoside ligands is poorly available for complexation within the CRD. This interpretation is supported by an analogous observation described in the literature [44].

The biological assays show that the inter-ligand distance in a bivalent glycocluster together with the conformational availability of the glycoligands displayed in the respective inhibitor impact the affinity for a specific lectin. Note, that in the herein described approach, this difference is caused by the flip between the $^4C_1$ and $^1C_4$ chair conformation of a central carbohydrate scaffold. The idea behind this approach bears the potential to modulate and control, respectively, ligand affinity by switching carbohydrate conformation.

In order to rationalize the measured effects, the interactions of the glycoclusters with the bacterial lectin FimH were further studied by molecular modeling.

### 2.4. Molecular Modeling

To investigate the structure of the glycocluster–FimH complexes, docking studies were performed with the tested glycoclusters **5** and **7** and the lectin FimH using the software Glide 2024-2 [45] as implemented in the Schrödinger software package 2024-2 [46] (see Supplementary Materials for details). The tyrosine residue Tyr48 as part of the FimH tyrosine gate shows a pronounced flexibility [25,47–49]. Consequently, depending on the complexed ligand, varying conformations of the tyrosine gate were observed. As the open gate conformation can favorably stabilize complexes with Man ligands carrying an aromatic aglycon moiety, the respective FimH structure (PDB code 6G2S) [50] was employed in the modeling. The top scoring results obtained in the docking experiments for both glycoclusters were subjected to a MM-GBSA calculation [51] (molecular mechanics energies combined with generalized born and surface area continuum solvation) to yield binding energies for both FimH ligands, **5** and **7**. The docking scores and corresponding binding energies are collected in Table 2. While the docking scores are similar for both glycoclusters, the binding energies differ which is in consistency with the different RIP values measured in the bioassays.

**Table 2.** Computed results from molecular modeling studies with the synthetic ligands **5** and **7** and the open gate conformer of FimH (pdb: 6G2S) [50]. The best docking scores and corresponding binding energies (MM-GBSA) using Glide and Prime are listed. Lower values suggest higher FimH affinity. RIP values (higher values relating to stronger FimH binding) are shown for comparison.

| Glycocluster (RIP) | Glide Score | Binding Energy [kcal mol$^{-1}$] |
|---|---|---|
| $^4C_1$ glycocluster **5** (410 ± 68) | −10.468 | −83.02 |
| $^1C_4$ glycocluster **7** (157 ± 45) | −10.581 | −67.85 |

Docking showed that in both inspected cases, an α-D-mannoside moiety is buried within the FimH CRD interacting through the typical hydrogen bond network [1]. Both glycoclusters show explicit π-π interactions between the biphenyl aglycone of the mannoside ligand and Tyr48 (light blue dotted lines in Figure 3), which are responsible for a high inhibitory potency in comparison to MeMan (cf. Table 1). The docking studies suggest, that glycocluster **5** with the central xyloside in $^4C_1$ conformation binds to FimH such that the anomeric mannoside unit occupies the FimH CRD, while the second mannoside ligand conjugated to the 3-position of the xyloside scaffold protrudes outwards, being poised for potential multivalent interactions with another FimH domain in close proximity (Figure 3) [43]. In addition, the FimH-**5** complex is further stabilized by a hydrogen bond between the amido group at position 2 of the xyloside scaffold and the Tyr48 residue of FimH. The inter-ligand distance between the two mannoside glycoligands in the complex is 25 Å, paralleling with the results from the MD simulation (Section 2.2).

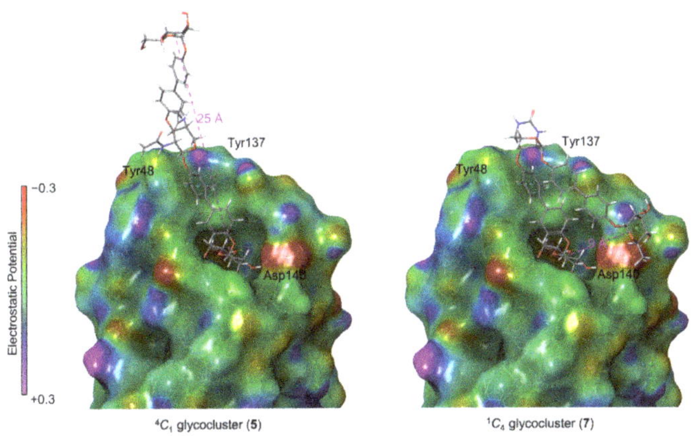


**Figure 3.** Connolly surface representations of FimH–glycocluster complexes as derived from molecular docking using the open gate conformation of FimH (pdb: 6G2S) [50]. The xylose-scaffolded glycoclusters **5** and **7** are depicted as stick models. Protein coloring represents the electrostatic potential of the surface (positive charges in purple, neutral in green, negative charges in red, cf. depicted color bar). Hydrogen bonds, π-π interactions, and inter-ring distance are indicated as yellow, blue, and purple dashed lines, respectively.

In contrast, in the bivalent glycocluster **7**, where the xyloside scaffold is locked in a $^1C_4$ conformation, the mannoside ligand that is conjugated with the 3-position of the scaffold is complexed within the FimH CRD. Here, the glycocluster adopts an angled conformation owing to the $^1C_4$ conformation of the xyloside ring which is flipped in comparison to **5**. The $^1C_4$ conformation of the central sugar also results in a shorter inter-ligand distance of about 9 Å in the ligand–FimH complex (cf. Section 2.2). Additional secondary interactions are observed for the anomeric mannoside unit forming hydrogen bonds with the polar amino acid residue Asp140. Here, the more restricted structure of the molecule most likely precludes the option for an effective multivalent interaction of the glycocluster. Hence, the simulated FimH–glycocluster complexes show differences between glycoclusters **5** and **7** which can be related to the reduced inhibitory potency (lower RIP value) of glycocluster **7** in comparison to glycocluster **5**.

## 3. Materials and Methods

### 3.1. General Information Regarding Synthesis and Spectroscopy

Moisture-sensitive reactions were carried out in flame-dried glassware and under a positive pressure of nitrogen. Analytical thin layer chromatography (TLC) was performed on silica gel plates (GF 254, Merck, Darmstadt, Germany). Visualization was achieved by UV light and/or with 10 % sulfuric acid in ethanol, vanillin (3.0 g vanillin and 0.5 mL $H_2SO_4$ in 100 mL EtOH) or ninhydrin, followed by heat treatment at approx. 200 °C. The products were purified by flash chromatography on silica gel columns (Merck, 230–400 mesh, particle size 0.040–0.063 mm, Darmstadt, Germany) or by automated flash chromatography using a puriFlash 450 device from the Interchim® company (Montluçon, France). MeOH was dried over magnesium under a nitrogen atmosphere. Optical rotations were measured with a PerkinElmer 241 polarimeter (PerkinElmer, Waltham, MA, USA) with a sodium D-line (589 nm) and a cuvette of 10 cm path length, in the solvents indicated. Proton ($^1$H) nuclear magnetic resonance spectra and carbon ($^{13}$C) nuclear magnetic resonance spectra were recorded on a Bruker DRX-500 and AV-600 spectrometer (Bruker, Billerica, MA, USA) at 300 K. Chemical shifts are referenced to the internal standard tetramethylsilane (TMS) or to the residual proton of the NMR solvent. Multiplets (multiplicity s = singlet, d = doublet, t = triplet, m = multiplet) are listed according to chemical shift, coupling constants are given in Hertz (Hz). Full assignment of the signals was achieved by using 2D NMR techniques ($^1$H-$^1$H COSY, $^1$H-$^{13}$C HMBC and $^1$H-$^{13}$C HSQC). Infrared (IR) spectra were measured with a PerkinElmer FT-IR Paragon 1000 (ATR) spectrometer (PerkinElmer, Waltham, MA, USA) and are reported in cm$^{-1}$. ESI mass spectra were recorded on a LCQ Classic from Thermo Finnigan (Somerset, NJ, USA).

### 3.2. Synthesis

*4′-[(1,1′-Biphenyl)-2,3,4,6-tetra-O-acetyl-α-D-mannopyranoside]-4-yl 2,4-dideoxy-2,4-N-Boc-3-O-{4′-[(1,1′-biphenyl)-2,3,4,6-tetra-O-acetyl-α-D-mannopyranoside]-4-yl}-β-D-xylopyranoside* (**3**). The xyloside **1** [26] (60.0 mg, 87.2 µmol), the pinacoyl ester **2** [28] (144 mg, 261 µmol), CsF (79.4 mg, 523 µmol) and Pd(dppf)$Cl_2$·$CH_2Cl_2$ (14.2 mg, 17.4 µmol) were dissolved in dry THF (6.00 mL) and the reaction mixture was heated to 80 °C and stirred for 16 h. After cooling to room temperature, it was diluted with ethyl acetate and the organic layer was washed with $H_2O$. The combined organic layers were dried over $MgSO_4$, it was filtered and concentrated. Purification by column chromatography on silica gel (cyclohexane/ethyl acetate, 1:1) yielded **3** (93.0 mg, 79%) as a colorless solid; $R_f$ = 0.22 (cyclohexane/ethyl acetate, 1:1); $[\alpha]_D^{23}$ = +31.2 (c 0.02, acetone). $^1$H NMR (600 MHz, acetone-d$_6$, 298 K, TMS): δ = 7.63–7.57 (m, 6H, 6 $H_{arom}$), 7.52 (d, $^3J$ = 8.7 Hz, 2H, $H_{arom}$), 7.27–7.21 (m, 4H, 4 $H_{arom}$), 7.20–7.11 (m, 4H, 4 $H_{arom}$), 6.50 (d, $^3J_{NH,2}$ = 8.6 Hz, 1H, NH), 6.33 (d, $^3J_{NH,4}$ = 7.9 Hz, 1H, NH), 5.71 (d, $^3J_{1,2}$ = 1.1 Hz, 1H, H-1$_{Man}$), 5.70 (d, $^3J_{1,2}$ = 1.1 Hz, 1H, H-1$_{Man}$), 5.51–5.46 (m, 4H, 2 H-2$_{Man}$, 2 H-3$_{Man}$), 5.43 (d, $^3J_{1,2}$ = 7.2 Hz, 1H, H-1$_{Xyl}$), 5.34 (dd~t, $^3J_{3,4}$ = 9.9 Hz, $^3J_{4,5}$ = 9.9 Hz, 2H, 2 H-4$_{Man}$), 4.93 (t, $^3J_{3,4}$ = 9.2 Hz, $^3J_{2,3}$ = 9.2 Hz, 1H, H-3$_{Xyl}$), 4.25 (dd, $^2J_{6a,6b}$ = 12.0 Hz, $^3J_{5,6a}$ = 5.9 Hz, 1H, H-6a$_{Man}$), 4.24 (dd, $^2J_{6a,6b}$ = 12.0 Hz, $^3J_{5,6a}$ = 5.9 Hz, 1H, H-6a$_{Man}$), 4.20–4.15 (m, 2H, 2 H-5$_{Man}$), 4.10–4.07 (m, 2H, 2 H-6b$_{Man}$), 4.03 (dd, $^2J_{5a,5b}$ = 11.1 Hz, $^3J_{5a,4}$ = 4.7 Hz, 1H, H-5a$_{Xyl}$), 3.96–3.90 (m, 1H, H 4$_{Xyl}$), 3.84–3.78 (m, 1H, H-2$_{Xyl}$), 3.73 (t, $^2J_{5a,5b}$ = 10.9 Hz, $^3J_{5a,4}$ = 10.9 Hz, 1H, H-5b$_{Xyl}$), 2.16, 2.06, 1.99, 1.95, 1.94 (each s, 24H, 8 C(O)$CH_3$), 1.33, 1.31 (each s, 18H, 2 C($CH_3$)$_3$) ppm. $^{13}$C NMR (125 MHz, acetone-d$_6$, 298 K, TMS): δ = 170.62, 170.40, 170.37, 170.28 (8C, C(O)$CH_3$), 160.33 (NH-C(O)), 158.03 (NH-C(O)), 155.91 ($C_{quart}$), 155.75 ($C_{quart}$), 138.50 (2 $C_{quart}$), 136.50 ($C_{quart}$), 136.54 ($C_{quart}$), 136.27 ($C_{quart}$), 135.53 ($C_{quart}$), 128.61 (2 $CH_{arom}$), 128.57 (2 $CH_{arom}$), 128.47 (2 $CH_{arom}$), 128.24 (2 $CH_{arom}$), 118.24 (2 $CH_{arom}$), 118.23 (2 $CH_{arom}$), 118.15 (2 $CH_{arom}$), 118.12 (2 $CH_{arom}$),

100.54 (C-$1_{Xyl}$), 96.97 (2 C-$1_{Man}$), 79.06 (C-$3_{Xyl}$), 70.28 (2 C-$5_{Man}$), 69.88 (2 C-$2_{Man}$), 69.80 (2 C-$3_{Man}$), 66.61 (2 C-$4_{Man}$), 64.64 (C-$5_{Xyl}$), 62.94 (2 C-$6_{Man}$), 57.89 (C-$2_{Xyl}$), 53.26 (C-$4_{Xyl}$), 28.55 (3C, $C(CH_3)_3$), 28.53 (3C, $C(CH_3)_3$), 21.71, 20.65, 20.61, 20.59 (8C, $C(O)CH_3$) ppm. IR (ATR) $\nu_{max}/cm^{-1}$ = 3309, 1750, 1683, 1497, 1367, 1220, 1039, 824, 601; ESI-HRMS $m/z$ = 1362.5278 $[M + NH_4]^+$, calcd for $[M + NH_4]^+$ 1362.5292.

*4′-[(1,1′-Biphenyl)-2,3,4,6-tetra-O-acetyl-α-D-mannopyranoside]-4-yl 2,4-dideoxy-2,4-N-acetyl-3-O-{4′-[(1,1′-biphenyl)-2,3,4,6-tetra-O-acetyl-α-D-mannopyranoside]-4-yl}-β-D-xylopyranoside* (**4**). The N-Boc-protected glycocluster **3** (20.0 mg, 14.9 µmol) was dissolved in $CH_2Cl_2$ (1.00 mL), TFA (200 µL) was added and the reaction mixture stirred for 2 h at room temperature. Then all volatiles were removed in vacuo and the residue co-evaporated with toluene (3×). The crude product was dissolved in dry pyridine (2.00 mL), $Ac_2O$ (500 µL) was added and the reaction mixture stirred at room temperature for 5 h. Then it was diluted with ethyl acetate and the organic layer was subsequently washed with 1N HCl, sodium bicarbonate and brine. The organic layer was dried over $MgSO_4$, its was filtered and concentrated. Purification by column chromatography on silica (toluene/acetone, 1:1) gave the title compound **4** (16.0 mg, 87%) as a colorless solid; $R_f$ = 0.32 (toluene/acetone, 1:1); $[\alpha]_D^{23}$ = +16.6 (c 0.02, acetone). $^1H$ NMR (500 MHz, acetone-$d_6$, 298 K, TMS): δ = 7.63–7.59 (m, 4H, 4 $H_{arom}$), 7.58–7.54 (m, 4H, 4 $H_{arom}$), 7.51 (d, $^3J_{NH,2}$ = 8.6 Hz, 1H, NH), 7.36 (d, $^3J_{NH,4}$ = 8.4 Hz, 1H, NH), 7.27–7.22 (m, 4H, 4 $H_{arom}$), 7.19 (d, $^3J$ = 8.9 Hz, 2H, $H_{arom}$), 7.11 (d, $^3J$ = 8.8 Hz, 2H, $H_{arom}$), 5.71–5.69 (m, 2H, 2 H-$1_{Man}$), 5.57 (d, $^3J_{1,2}$ = 6.5 Hz, 1H, H-$1_{Xyl}$), 5.50-5.46 (m, 4H, 2 H-$2_{Man}$, 2 H-$3_{Man}$), 5.33 (dd~t, $^3J_{3,4}$ = 9.9 Hz, $^3J_{4,5}$ = 9.9 Hz, 2H, 2 H-$4_{Man}$), 4.99 (t, $^3J_{3,4}$ = 7.9 Hz, $^3J_{2,3}$ = 7.9 Hz, 1H, H-$3_{Xyl}$), 4.23 (dd, $^2J_{6a,6b}$ = 12.0 Hz, $^3J_{5,6a}$ = 5.9 Hz, 2H, 2 H-$6a_{Man}$), 4.21–4.10 (m, 4H, 2 H-$5_{Man}$, H-$5a_{Xyl}$, H-$4_{Xyl}$), 4.08 (dd, $^2J_{6a,6b}$ = 12.0 Hz, $^3J_{5,6b}$ = 2.3 Hz, 2H, 2 H-$6b_{Man}$), 4.05–3.99 (m, 1H, H-$2_{Xyl}$), 3.73 (dd, $^2J_{5a,5b}$ = 11.6 Hz, $^3J_{5a,4}$ = 8.4 Hz, 1H, H-$5b_{Xyl}$), 2.16, 2.05, 1.99, 1.94 (each s, 24H, 8 $C(O)CH_3$), 1.80, 1.75 (each s, 6H, $NHC(O)\underline{CH_3}$) ppm. $^{13}C$ NMR (125 MHz, acetone-$d_6$, 298 K, TMS): δ = 170.65, 170.43, 170.40, 170.31 (8C, $\underline{C}(O)CH_3$), 170.54, 170.34 (2C, $NH\underline{C}(O)CH_3$), 159.70 ($C_{quart}$), 159.62 ($C_{quart}$), 155.94 ($C_{quart}$), 155.80 ($C_{quart}$), 136.45 ($C_{quart}$), 136.31 ($C_{quart}$), 134.33 ($C_{quart}$), 128.63 (2 $CH_{arom}$), 128.59 (2 $CH_{arom}$), 128.49 (2 $CH_{arom}$), 128.23 (2 $CH_{arom}$), 118.28 (4C, $CH_{arom}$), 118.10 (2 $CH_{arom}$), 117.97 (2 $CH_{arom}$), 99.43 (C-$1_{Xyl}$), 97.01 (2 C-$1_{Man}$), 77.09 (C-$3_{Xyl}$), 70.31 (2 C-$5_{Man}$), 69.9 2 (2 C-$2_{Man}$), 69.83 (2 C-$3_{Man}$), 66.64 (2 C-$4_{Man}$), 63.38 (C-$5_{Xyl}$), 62.97 (2 C-$6_{Man}$), 55.57 (C-$2_{Xyl}$), 50.93 (C-$4_{Xyl}$), 23.09, 23.00 (2C, $NHC(O)\underline{C}H_3$), 21.74, 20.68, 20.64, 20.61 (8C, $C(O)\underline{C}H_3$) ppm. IR (ATR) $\nu_{max}/cm^{-1}$ = 2298, 1750, 1662, 1496, 1370, 1221, 1039, 825, 600. ESI-HRMS $m/z$ = 1246.4428 $[M + NH_4]^+$, calcd for $[M + NH_4]^+$ = 1246.4449.

*4′-[(1,1′-Biphenyl)-α-D-mannopyranosyloxy]-4-yl 2,4-dideoxy-2,4-N-acetyl-3-O-{4′-[(1,1′-biphenyl)-α-D-mannopyranosyloxy]-4-yl}-β-D-xylopyranoside* (**5**). The protected glycocluster **4** (13.0 mg, 10.6 µmol) was dissolved in MeOH (1.00 mL), $K_2CO_3$ (414 µg, 3.00 µmol) was added and it was stirred for 5 h at room temperature. It was neutralized with Amberlite® IR-120, filtered and concentrated to dryness. Co-evaporation with toluene and $CH_2Cl_2$ yielded **5** (9.00 mg, 95%) as a colorless solid; $[\alpha]_D^{23}$ = +26.3 (c 0.02, MeOH). $^1H$ NMR (600 MHz, MeOD-$d_3$, 298 K, TMS): δ = 7.53–7.47 (m, 8H, 8 $H_{arom}$), 7.24–7.05 (m, 8H, 8 $H_{arom}$), 5.51 (s, 2H, 2 H-$1_{Man}$), 5.35 (d, $^3J_{1,2}$ = 7.7 Hz, 1H, H-$1_{Xyl}$), 4.85–4.82 (m, 1H, H-$3_{Xyl}$), 4.24–4.18 (m, 1H, H-$4_{Xyl}$), 4.06–4.00 (m, 4H, H-$2_{Xyl}$, H-$5a_{Xyl}$, 2 H-$2_{Man}$), 3.92 (dd, $^3J_{3,4}$ = 8.3 Hz, $^3J_{2,3}$ = 4.0 Hz, 2H, 2 H-$3_{Man}$), 3.80–3.70 (m, 6H, 2 H-$4_{Man}$, 2 H-$6a_{Man}$, 2 H-$6b_{Man}$), 3.66–3.59 (m, 3H, 2 H-$5_{Man}$, H-$5b_{Xyl}$), 1.81, 1.76 (each s, 6H, $NHC(O)\underline{CH_3}$) ppm. $^{13}C$ NMR (125 MHz, MeOD-$d_3$, 298 K, TMS): δ = 173.68, 173.53 (2C, $NH\underline{C}(O)CH_3$), 159.96 ($C_{quart}$), 157.96 ($C_{quart}$), 157.23 ($C_{quart}$), 157.14 ($C_{quart}$), 136.70 ($C_{quart}$), 136.11 ($C_{quart}$), 136.06 ($C_{quart}$), 135.69 ($C_{quart}$), 129.22 (2 $CH_{arom}$), 128.77 (2 $CH_{arom}$), 128.64 (2 $CH_{arom}$), 128.61 (2 $CH_{arom}$), 118.21 (4C, $CH_{arom}$),

118.11 (4C, $CH_{arom}$), 100.47 (C-$1_{Xyl}$), 100.23 (2 C-$1_{Man}$), 78.89 (C-$3_{Xyl}$), 75.40 (2 C-$5_{Man}$), 72.43 (2 C-$2_{Man}$), 72.03 (2 C-$3_{Man}$), 68.36 (2 C-$4_{Man}$), 64.40 (C-$5_{Xyl}$), 62.70 (2 C-$6_{Man}$), 57.20 (C-$2_{Xyl}$), 52.18 (C-$4_{Xyl}$), 22.74, 22.66 (2C, NHC(O)$\underline{C}H_3$) ppm. IR (ATR) $\nu_{max}/cm^{-1}$ = 3364, 2467, 1494, 1220, 999, 824, 677, 575. ESI-HRMS *m*/*z* = 910.3593 $[M + NH_4]^+$, calcd for $[M + NH_4]^+$ 910.3604.

*4′-[(1,1′-Biphenyl)-2,3,4,6-tetra-O-acetyl-α-D-mannopyranoside]-4-yl 2,4-N-carbonyl-2,4-dideoxy-3-O-{4′-[(1,1′-biphenyl)-2,3,4,6-tetra-O-acetyl-α-D-mannopyranoside]-4-yl}-β-D-xylopyranoside* (**6**). The glycocluster **3** (20.0 mg, 14.9 μmol) was dissolved in $CH_2Cl_2$ (1.00 mL), TFA (200 μL) was added and it was stirred for 2 h at room temperature. Then all volatiles were removed under reduced pressure and the residue was co-evaporated with toluene (3×). The crude product was dissolved in DMF (150 μL) and N,N-carbonyldiimidazole (2.90 mg, 17.9 μmol) was added, followed by dropwise addition of triethylamine (10.4 μL, 74.5 μmol). The reaction mixture was stirred for 16 h at room temperature, then it was diluted with ethyl acetate and washed with 1N HCl, sodium bicarbonate and brine. The combined organic layer were dried over $MgSO_4$, it was filtered and concentrated. Purification by column chromatography on silica gel (toluene/acetone, 4:6) gave the title compound **6** (13.0 mg, 74%) as a colorless solid; $R_f$ 0.15 (toluene/acetone, 4:6); $[\alpha]_D^{23}$ = −27.3 (c 0.02, acetone). $^1$H NMR (600 MHz, acetone-$d_6$, 298 K, TMS): δ = 7.68 (d, $^3J$ = 8.7 Hz, 2H, 2 $CH_{arom}$), 7.65 (d, $^3J$ = 8.7 Hz, 2H, 2 $CH_{arom}$), 7.59 (d, $^3J$ = 8.7 Hz, 2H, 2 $CH_{arom}$), 7.55 (d, $^3J$ = 8.7 Hz, 2H, 2 $CH_{arom}$), 7.29 (d, $^3J$ = 8.7 Hz, 2H, 2 $CH_{arom}$), 7.26 (d, $^3J$ = 8.7 Hz, 2H, 2 $CH_{arom}$), 7.23 (d, $^3J$ = 8.7 Hz, 2H, 2 $CH_{arom}$), 7.09 (d, $^3J$ = 8.7 Hz, 2H, 2 $CH_{arom}$), 6.17 (d, $^3J_{NH,2}$ = 4.2 Hz, 1H, NH), 6.07 (d, $^3J_{NH,4}$ = 4.6 Hz, 1H, NH), 5.72–5.68 (m, 2H, 2 H-$1_{Man}$), 5.52 (s, 1H, H-$1_{Xyl}$), 5.51–5.45 (m, 4H, 2 H-$2_{Man}$, 2 H-$3_{Man}$), 5.33 (dd~t, $^3J_{3,4}$ = 10.0 Hz, $^3J_{4,5}$ = 10.0 Hz, 2H, 2 H-$4_{Man}$), 5.06 (t, $^3J_{3,4}$ = 3.1 Hz, $^3J_{2,3}$ = 3.1 Hz, 1H, H-$3_{Xyl}$), 4.45 (d, $^2J_{5a,5b}$ = 11.4 Hz, 1H, H-$5a_{Xyl}$), 4.23 (dd, $^2J_{6a,6b}$ = 11.9 Hz, $^3J_{5,6a}$ = 5.9 Hz, 2H, 2 H-$6a_{Man}$), 4.20–4.14 (m, 2H, 2 H-$5_{Man}$), 4.08 (dd, $^2J_{6a,6b}$ = 12.0 Hz, $^3J_{5,6b}$ = 2.2 Hz, 2H, 2 H-$6b_{Man}$), 3.95 (s, 1H, H-$4_{Xyl}$), 3.69 (s, 1H, H-$2_{Xyl}$), 3.60 (d, $^2J_{5a,5b}$ = 12.0 Hz, 1H, H-$5b_{Xyl}$), 2.16, 2.16, 2.06, 2.04, 1.99, 1.99, 1.95, 1.93 (each s, 24H, 8 C(O)$CH_3$) ppm. $^{13}$C NMR (125 MHz, acetone-$d_6$, 298 K, TMS): δ = 170.66, 170.43, 170.40, 170.31 (8 $\underline{C}$(O)$CH_3$), 157.37, 157.22, 155.86, 155.82, 136.35, 136.31, 134.81, 134.59 (8 $C_{quart}$), 128.71, 128.53, 128.52, 128.49, 118.27, 118.23, 117.57, 117.32 (16 $CH_{arom}$), 99.31 (C-$1_{Xyl}$), 96.98, 96.96 (2 C-$1_{Man}$), 70.26 (2 C-$5_{Man}$), 69.88 (2 C-$2_{Man}$), 69.80 (2 C-$3_{Man}$), 68.88 (C-$3_{Xyl}$), 66.60 (2 C-$4_{Man}$), 62.94 (2 C-$6_{Man}$), 61.98 (C-$5_{Xyl}$), 48.57 (C-$2_{Xyl}$), 47.85 (C-$4_{Xyl}$), 21.71, 20.65, 20.61, 20.59 (8 C(O)$\underline{C}H_3$) ppm. IR (ATR) $\nu_{max}/cm^{-1}$ = 2925, 1749, 1607, 1496, 1369, 1217, 1039, 825, 598. ESI-HRMS *m*/*z* = 1188.4013 $[M + NH_4]^+$, calcd for $[M + NH_4]^+$ 1188.4031.

*4′-[(1,1′-Biphenyl)-α-D-mannopyranoside]-4-yl 2,4-N-carbonyl-2,4-dideoxy-3-O-{4′-[(1,1′-biphenyl)-α-D-mannopyranoside]-4-yl}-β-D-xylopyranoside* (**7**). The protected glycocluster **6** (10.0 mg, 8.54 μmol) was dissolved in MeOH (1.00 mL), $K_2CO_3$ (354 μg, 2.56 μmol) was added and the reaction mixture was stirred for 4 h at room temperature. It was neutralized with Amberlite® IR-120 (Sigma-Aldrich, St. Louis, MO, USA), filtered and concentrated to dryness. Co-evaporation with toluene and $CH_2Cl_2$ gave **7** (5.30 mg, 74%) as a colorless solid; $[\alpha]_D^{23}$ = +36.2 (c 0.02, MeOH). $^1$H NMR (600 MHz, MeOD-$d_3$, 298 K, TMS): δ = 7.63–7.46 (m, 8H, 8 $H_{arom}$), 7.25–7.01 (m, 8H, 8 $H_{arom}$), 5.52 (s, 1H, H-$1_{Man}$), 5.50 (s, 1H, H-$1_{Man}$), 5.43 (s, 1H, H-$1_{Xyl}$), 5.01 (t, $^3J_{3,4}$ = 3.7 Hz, $^3J_{2,3}$ = 3.7 Hz, 1H, H-$3_{Xyl}$), 4.50 (dd, $^2J_{5a,5b}$ = 11.5 Hz, $^3J_{4,5a}$ = 11.5 Hz, H, H-$5a_{Xyl}$), 4.04-4.01 (m, 2H, 2 H-$2_{Man}$), 3.94-3.90 (m, 2H, 2 H-$3_{Man}$), 3.89 (s, 1H, H-$2_{Xyl}$), 3.80-3.57 (m, 10H, 2 H-$4_{Man}$, 2 H-$5_{Man}$, 2 H-$6a_{Man}$, 2 H-$6b_{Man}$, H-$4_{Xyl}$, H-$5b_{Xyl}$), ppm. $^{13}$C NMR (125 MHz, MeOD-$d_3$, 298 K, TMS): δ = 160.15 (NHC(O)NH), 157.41 ($C_{quart}$), 157.36 ($C_{quart}$), 157.19 ($C_{quart}$), 157.14 ($C_{quart}$), 136.18 ($C_{quart}$), 136.15 ($C_{quart}$), 136.00 ($C_{quart}$), 135.85 ($C_{quart}$), 128.96 (2 $CH_{arom}$), 128.75 (2 $CH_{arom}$), 128.72 (2 $CH_{arom}$), 128.68 (2 $CH_{arom}$), 118.14 (2C, $CH_{arom}$), 118.08 (2C, $CH_{arom}$), 117.79 (2C, $CH_{arom}$), 117.46 (2C, $CH_{arom}$), 99.64

(C-$1_{Xyl}$), 100.24 (2 C-$1_{Man}$), 75.39(2 C-$5_{Man}$), 72.43 (2 C-$3_{Man}$), 72.03 (2 C-$2_{Man}$), 68.68 (C-$3_{Xyl}$), 68.36 (2 C-$4_{Man}$), 62.69 (2 C-$6_{Man}$), 61.94 (C-$5_{Xyl}$), 52.79 (C-$4_{Xyl}$), 47.85 (C-$2_{Xyl}$) ppm. IR (ATR) $\nu_{max}/cm^{-1}$ = 3251, 2013, 1495, 1220, 999, 825, 674, 584. ESI-HRMS $m/z$ = 852.3178 $[M + NH_4]^+$, calcd for $[M + NH_4]^+$ 852.3185.

### 3.3. Cultivation of Bacteria

GFP-expressing *E. coli* bacteria (strain PKL1162, see Supplementary Materials) were cultured in 5.00 mL LB medium and incubated overnight at 37 °C and 100 rpm. Afterwards the mixture was centrifuged at 4 °C and 5000 rpm for 15 min. The bacteria pellet was washed twice with PBS buffer (2.00 mL) and then resuspended in PBS buffer. Finally, the suspension was adjusted to OD600 = 0.4.

### 3.4. Adhesion–Inhibition Assay with GFP-PKL1162 E. coli Bacteria

Inhibitor solutions of the respective glycosides **5** and **7** (0.4 mM in PBS buffer) as well as methyl α-D-mannopyranoside (MeMan, 200 mM in PBS buffer) were prepared and serial dilutions (1:2, 10 steps) of each solution added to the mannan-coated microtiter plates (50 μL/well, see Supplementary Materials). Next, the prepared bacterial suspension ($OD_{600}$ = 0.4, 50 μL/well) was added and the microtiter plates were incubated at 37 °C and 100 rpm for 45 min. The plates were washed with PBS buffer (3 × 150 μL/well) and then the wells were filled with PBS (100 μL/well) and the fluorescence intensity (485 nm/535 nm) was determined. On each individual plate the standard inhibitor MeMan was tested in parallel. Each compound was tested in duplicates or triplicates, respectively.

### 3.5. Molecular Dynamics

The structures of **5** and **7** were built by using Maestro and minimized by use of MacroModel with the OPLS3 force field in implicit water (GB/SA continuum solvation model); for details, see the Supplementary Materials.

### 3.6. Molecular Modeling

For molecular modeling, the Schrödinger software package 2024-2 implementing the Maestro interface was used [46]; for details see the Supplementary Materials.

## 4. Conclusions

Carbohydrate recognition is governed by a multitude of parameters, among which the inter-ligand distance in a bivalent glycocluster is an important aspect. The idea behind this account has been to adjust the distance between two glycoligands by flipping the chair conformation of a central carbohydrate scaffold. This approach supplements a related concept that uses photoswitchable glycomimetics to alter the spatial orientation of glycoligands [7,52–54]. The herein introduced conformational switch is based on a known 2,4-diamino-xyloside, which can be flipped from a $^4C_1$ into a $^1C_4$ conformation through the chemical fixation of the two amino functions. Concomitantly, the distance between two biphenyl mannoside units conjugated to positions 1 and 3 of the central xyloside ring changes. In the $^4C_1$ glycocluster (**5**) the most populated conformers show inter-ligand distances of ~25 Å, whereas the $^1C_4$ glycocluster (**7**) the inter-ligand distances are shorter. Two populations can be seen in the case of **7** with most common inter-ligand distances of ~20 Å and ~6 Å, respectively.

As biphenyl mannosides are excellent ligands of the bacterial lectin FimH that mediates bacterial adhesion, both bivalent glycoclusters were tested as inhibitors of FimH-mediated adhesion of live *E. coli*. It turned out that the inhibitory potency of **5** is almost 3 times higher than that of **7**. This finding can be attributed to a difference in the distance between the glycoligands, since the glycoclusters **5** and **7** do not differ in any other structural aspect. Nevertheless, it is difficult to conclusively interpret the results of the bioassays, why molecular modeling studies were performed. Modeling revealed binding energies for the two carbohydrate-FimH complexes which parallel with the results from the adhesion–inhibition tests. In other words, the comparably higher FimH affinity of **5** can be correlated with its higher potency as inhibitor of mannose-specific bacterial adhesion. Furthermore, molecular modeling suggests that **5** is complexed within the FimH CRD such, that multivalent interactions with an adjacent lectin domain are possible, whereas this is excluded in the more densely packed structure of **7**.

In conclusion, modeling results rationalize the different RIP values measured for the bivalent glycoclusters **5** and **7** by highlighting how inter-ligand distance and the spatial orientation of ligands modulate both binding affinity and the potential for multivalent interactions. Furthermore, the idea of exploring differential carbohydrate binding by utilizing a conformational glycoswitch is a valuable new approach in glycoscience. We will further advance this concept also in order to shed more light on the so far underexplored dynamic processes in carbohydrate recognition.

**Supplementary Materials:** The following supporting information can be downloaded at: https://www.mdpi.com/article/10.3390/molecules30153074/s1, Figure S1: Inhibition curves obtained with glycoclusters **5** and **7** as inhibitors of type 1 fimbriae-mediated bacterial adhesion to mannan; Table S1: $IC_{50}$ values as deduced from the inhibition curves obtained with MeMan, $^4C_1$ cluster **5** and $^1C_4$ cluster **7** and corresponding RIP values; Figure S2: Distances between the Man-ligand residues as a function of the simulation time; Table S2: Scoring values for docking of glycoclusters **5** and **7** into the open gate (pdb: 6G2S) conformation of FimH using Glide; Table S3: Values of computed binding energies $\Delta G_{Bind}$ (in kcal $mol^{-1}$) obtained from MM-GBSA calculations for **5** and **7** into the open gate (pdb: 6G2S) conformation of FimH; Figures S3–S12: $^1$H NMR and $^{13}$C NMR spectra of synthesized compounds. Reference [55] is cited in the supplementary materials.

**Author Contributions:** Conceptualization, S.O.J. and T.K.L.; methodology, S.O.J. and I.v.S.; modeling, S.O.J. and I.v.S.; validation, S.O.J., I.v.S. and T.K.L.; formal analysis, S.O.J. and I.v.S.; investigation, S.O.J., I.v.S. and T.K.L.; resources, T.K.L.; data curation, S.O.J., I.v.S. and T.K.L.; writing—original draft preparation, S.O.J.; writing—review and editing, S.O.J., I.v.S. and T.K.L.; visualization, S.O.J., I.v.S. and T.K.L.; supervision, T.K.L.; project administration, T.K.L.; funding acquisition, T.K.L. All authors have read and agreed to the published version of the manuscript.

**Funding:** This research received no external funding.

**Institutional Review Board Statement:** Not applicable.

**Informed Consent Statement:** Not applicable.

**Data Availability Statement:** The data from this study are included in the article and Supplementary Materials.

**Acknowledgments:** Support from Christiana Albertina University is gratefully acknowledged.

**Conflicts of Interest:** The authors declare no conflicts of interest.

## References

1. Mydock-McGrane, L.K.; Hannan, T.J.; Janetka, J.W. Rational Design Strategies for FimH Antagonists: New Drugs on the Horizon for Urinary Tract Infection and Crohn's Disease. *Expert Opin. Drug Discov.* **2017**, *12*, 711–731. [CrossRef] [PubMed]
2. Sarshar, M.; Behzadi, P.; Ambrosi, C.; Zagaglia, C.; Palamara, A.T.; Scribano, D. FimH and Anti-Adhesive Therapeutics: A Disarming Strategy Against Uropathogens. *Antibiotics* **2020**, *9*, 397. [CrossRef] [PubMed]

3. Fares, M.; Imberty, A.; Titz, A. Bacterial Lectins: Multifunctional Tools in Pathogenesis and Possible Drug Targets. *Trends Microbiol.* **2025**. [CrossRef] [PubMed]
4. Müller, C.; Despras, G.; Lindhorst, T.K. Organizing Multivalency in Carbohydrate Recognition. *Chem. Soc. Rev.* **2016**, *45*, 3275–3302. [CrossRef] [PubMed]
5. Cutolo, G.; Reise, F.; Schuler, M.; Nehmé, R.; Despras, G.; Brekalo, J.; Morin, P.; Renard, P.-Y.; Lindhorst, T.K.; Tatibouët, A. Bifunctional Mannoside–Glucosinolate Glycoconjugates as Enzymatically Triggered Isothiocyanates and FimH Ligands. *Org. Biomol. Chem.* **2018**, *16*, 4900–4913. [CrossRef] [PubMed]
6. Tamburrini, A.; Colombo, C.; Bernardi, A. Design and Synthesis of Glycomimetics: Recent Advances. *Med. Res. Rev.* **2020**, *40*, 495–531. [CrossRef] [PubMed]
7. Friedrich, L.M.; Lindhorst, T.K. Orthogonal Photoswitching of Heterobivalent Azobenzene Glycoclusters: The Effect of Glycoligand Orientation in Bacterial Adhesion. *Beilstein J. Org. Chem.* **2025**, *21*, 736–748. [CrossRef] [PubMed]
8. Sauer, M.M.; Jakob, R.P.; Eras, J.; Baday, S.; Eriş, D.; Navarra, G.; Bernèche, S.; Ernst, B.; Maier, T.; Glockshuber, R. Catch-Bond Mechanism of the Bacterial Adhesin FimH. *Nat. Commun.* **2016**, *7*, 10738. [CrossRef] [PubMed]
9. Sharon, N. Carbohydrates as Future Anti-Adhesion Drugs for Infectious Diseases. *Biochim. Biophys. Acta Gen. Subj.* **2006**, *1760*, 527–537. [CrossRef] [PubMed]
10. Ernst, B.; Magnani, J.L. From Carbohydrate Leads to Glycomimetic Drugs. *Nat. Rev. Drug Discov.* **2009**, *8*, 661–677. [CrossRef] [PubMed]
11. Hartmann, M.; Lindhorst, T.K. The Bacterial Lectin FimH, a Target for Drug Discovery—Carbohydrate Inhibitors of Type 1 Fimbriae-Mediated Bacterial Adhesion. *Eur. J. Org. Chem.* **2011**, *2011*, 3583–3609. [CrossRef]
12. Leusmann, S.; Ménová, P.; Shanin, E.; Titz, A.; Rademacher, C. Glycomimetics for the Inhibition and Modulation of Lectins. *Chem. Soc. Rev.* **2023**, *52*, 3663–3740. [CrossRef] [PubMed]
13. Fiege, B.; Rabbani, S.; Preston, R.C.; Jakob, R.P.; Zihlmann, P.; Schwardt, O.; Jiang, X.; Maier, T.; Ernst, B. The Tyrosine Gate of the Bacterial Lectin FimH: A Conformational Analysis by NMR Spectroscopy and X-Ray Crystallography. *ChemBioChem* **2015**, *16*, 1235–1246. [CrossRef] [PubMed]
14. Firon, N.; Ashkenazi, S.; Mirelman, D.; Ofek, I.; Sharon, N. Aromatic Alpha-Glycosides of Mannose Are Powerful Inhibitors of the Adherence of Type 1 Fimbriated *Escherichia coli* to Yeast and Intestinal Epithelial Cells. *Infect. Immun.* **1987**, *55*, 472–476. [CrossRef] [PubMed]
15. Chalopin, T.; Brissonnet, Y.; Sivignon, A.; Deniaud, D.; Cremet, L.; Barnich, N.; Bouckaert, J.; Gouin, S.G. Inhibition Profiles of Mono- and Polyvalent FimH Antagonists against 10 Different *Escherichia coli* Strains. *Org. Biomol. Chem.* **2015**, *13*, 11369–11375. [CrossRef] [PubMed]
16. Hatton, N.E.; Nabarro, J.; Yates, N.D.J.; Parkin, A.; Wilson, L.G.; Baumann, C.G.; Fascione, M.A. Mannose-Presenting "Glyco-Colicins" Convert the Bacterial Cell Surface into a Multivalent Adsorption Site for Adherent Bacteria. *JACS Au* **2024**, *4*, 2122–2129. [CrossRef] [PubMed]
17. Roy, R.; Das, S.K.; Santoyo-González, F.; Hernández-Mateo, F.; Dam, T.K.; Brewer, C.F. Synthesis of "Sugar-Rods" with Phytohemagglutinin Cross-Linking Properties by Using the Palladium-Catalyzed Sonogashira Reaction. *Chem. Eur. J.* **2000**, *6*, 1757–1762. [CrossRef]
18. Roy, R.; Trono, M.C.; Giguère, D. Effects of Linker Rigidity and Orientation of Mannoside Cluster for Multivalent Interactions with Proteins. In *Glycomimetics: Modern Synthetic Methodologies*; ACS Symposium Series; American Chemical Society: Washington, DC, USA, 2005; Volume 896, pp. 137–150, ISBN 978-0-8412-3880-0.
19. Bergeron-Brlek, M.; Shiao, T.C.; Trono, M.C.; Roy, R. Synthesis of a Small Library of Bivalent α-D-Mannopyranosides for Lectin Cross-Linking. *Carbohydr. Res.* **2011**, *346*, 1479–1489. [CrossRef] [PubMed]
20. Varki, A.; Cummings, R.D.; Aebi, M.; Packer, N.H.; Seeberger, P.H.; Esko, J.D.; Stanley, P.; Hart, G.; Darvill, A.; Kinoshita, T.; et al. Symbol Nomenclature for Graphical Representations of Glycans. *Glycobiology* **2015**, *25*, 1323–1324. [CrossRef] [PubMed]
21. Yuasa, H.; Hashimoto, H. Bending Trisaccharides by a Chelation-Induced Ring Flip of a Hinge-Like Monosaccharide Unit. *J. Am. Chem. Soc.* **1999**, *121*, 5089–5090. [CrossRef]
22. Yuasa, H.; Miyagawa, N.; Izumi, T.; Nakatani, M.; Izumi, M.; Hashimoto, H. Hinge Sugar as a Movable Component of an Excimer Fluorescence Sensor. *Org. Lett.* **2004**, *6*, 1489–1492. [CrossRef] [PubMed]
23. Takeuchi, J.; Ohkubo, A.; Yuasa, H. A Ring-Flippable Sugar as a Stimuli-Responsive Component of Liposomes. *Chem. Asian J.* **2015**, *10*, 586–594. [CrossRef] [PubMed]
24. Jaeschke, S.O.; Lindhorst, T.K. Versatile Synthesis of Diaminoxylosides via Iodosulfonamidation of Xylal Derivatives. *Eur. J. Org. Chem.* **2021**, *2021*, 6312–6318. [CrossRef]

25. Hung, C.-S.; Bouckaert, J.; Hung, D.; Pinkner, J.; Widberg, C.; DeFusco, A.; Auguste, C.G.; Strouse, R.; Langermann, S.; Waksman, G.; et al. Structural Basis of Tropism of *Escherichia coli* to the Bladder during Urinary Tract Infection. *Mol. Microbiol.* **2002**, *44*, 903–915. [CrossRef] [PubMed]
26. Mayer, K.; Eris, D.; Schwardt, O.; Sager, C.P.; Rabbani, S.; Kleeb, S.; Ernst, B. Urinary Tract Infection: Which Conformation of the Bacterial Lectin FimH Is Therapeutically Relevant? *J. Med. Chem.* **2017**, *60*, 5646–5662. [CrossRef] [PubMed]
27. Miyaura, N.; Yanagi, T.; Suzuki, A. The Palladium-Catalyzed Cross-Coupling Reaction of Phenylboronic Acid with Haloarenes in the Presence of Bases. *Synth. Commun.* **1981**, *11*, 513–519. [CrossRef]
28. Schwizer, D.; Gäthje, H.; Kelm, S.; Porro, M.; Schwardt, O.; Ernst, B. Antagonists of the Myelin-Associated Glycoprotein: A New Class of Tetrasaccharide Mimics. *Bioorg. Med. Chem.* **2006**, *14*, 4944–4957. [CrossRef] [PubMed]
29. Han, Z.; Pinkner, J.S.; Ford, B.; Chorell, E.; Crowley, J.M.; Cusumano, C.K.; Campbell, S.; Henderson, J.P.; Hultgren, S.J.; Janetka, J.W. Lead Optimization Studies on FimH Antagonists: Discovery of Potent and Orally Bioavailable Ortho-Substituted Biphenyl Mannosides. *J. Med. Chem.* **2012**, *55*, 3945–3959. [CrossRef] [PubMed]
30. Amatore, C.; Jutand, A.; Le Duc, G. The Triple Role of Fluoride Ions in Palladium-Catalyzed Suzuki–Miyaura Reactions: Unprecedented Transmetalation from [$ArPdFL_2$] Complexes. *Angew. Chem. Int. Ed.* **2012**, *51*, 1379–1382. [CrossRef] [PubMed]
31. Hirano, T.; Hiromoto, K.; Kagechika, H. Development of a Library of 6-Arylcoumarins as Candidate Fluorescent Sensors. *Org. Lett.* **2007**, *9*, 1315–1318. [CrossRef] [PubMed]
32. Jaeschke, S.O.; Lindhorst, T.K.; Auer, A. Between Two Chairs: Combination of Theory and Experiment for the Determination of the Conformational Dynamics of Xylosides. *Chem. Eur. J.* **2022**, *28*, e202201544. [CrossRef] [PubMed]
33. Zemplén, G.; Pacsu, E. Über Die Verseifung Acetylierter Zucker Und Verwandter Substanzen. *Berichte Dtsch. Chem. Ges. B Ser.* **1929**, *62*, 1613–1614. [CrossRef]
34. Chen, Y.-L.; Pyplo-Schnieders, J.; Redlich, H.; Luftmann, H.; Fröhlich, R. Preparation of 2-Deoxystreptamine Derivatives with All-Axial Substituents for Desymmetrization. *Tetrahedron Lett.* **2007**, *48*, 8145–8148. [CrossRef]
35. *Schrödinger Release 2021-1: Desmond Molecular Dynamics System*; D. E. Shaw Research: New York, NY, USA, 2021.
36. *Schrödinger Release 2021-1: Maestro-Desmond Interoperability Tools*; Schrödinger: New York, NY, USA, 2021.
37. *Schrödinger Release 2021-1: Maestro*; Schrödinger LLC: New York, NY, USA, 2021.
38. Mark, P.; Nilsson, L. Structure and Dynamics of the TIP3P, SPC, and SPC/E Water Models at 298 K. *J. Phys. Chem. A* **2001**, *105*, 9954–9960. [CrossRef]
39. Reisner, A.; Haagensen, J.A.J.; Schembri, M.A.; Zechner, E.L.; Molin, S. Development and Maturation of *Escherichia coli* K-12 Biofilms. *Mol. Microbiol.* **2003**, *48*, 933–946. [CrossRef] [PubMed]
40. Hartmann, M.; Horst, A.K.; Klemm, P.; Lindhorst, T.K. A Kit for the Investigation of Live *Escherichia coli* Cell Adhesion to Glycosylated Surfaces. *Chem. Commun.* **2010**, *46*, 330–332. [CrossRef] [PubMed]
41. Klein, T.; Abgottspon, D.; Wittwer, M.; Rabbani, S.; Herold, J.; Jiang, X.; Kleeb, S.; Lüthi, C.; Scharenberg, M.; Bezençon, J.; et al. FimH Antagonists for the Oral Treatment of Urinary Tract Infections: From Design and Synthesis to in Vitro and in Vivo Evaluation. *J. Med. Chem.* **2010**, *53*, 8627–8641. [CrossRef] [PubMed]
42. Sperling, O.; Fuchs, A.; Lindhorst, T.K. Evaluation of the Carbohydrate Recognition Domain of the Bacterial Adhesin FimH: Design, Synthesis and Binding Properties of Mannoside Ligands. *Org. Biomol. Chem.* **2006**, *4*, 3913–3922. [CrossRef] [PubMed]
43. Sauer, M.M.; Jakob, R.P.; Luber, T.; Canonica, F.; Navarra, G.; Ernst, B.; Unverzagt, C.; Maier, T.; Glockshuber, R. Binding of the Bacterial Adhesin FimH to Its Natural, Multivalent High-Mannose Type Glycan Targets. *J. Am. Chem. Soc.* **2019**, *141*, 936–944. [CrossRef] [PubMed]
44. von der Lieth, C.-W.; Frank, M.; Lindhorst, T.K. Molecular Dynamics Simulations of Glycoclusters and Glycodendrimers. *Rev. Mol. Biotechnol.* **2002**, *90*, 311–337. [CrossRef] [PubMed]
45. *Schrödinger Release 2024-2: Glide*; Schrödinger LLC: New York, NY, USA, 2024.
46. *Schrödinger Release 2024-2: Maestro*; Schrödinger LLC: New York, NY, USA, 2024.
47. Bouckaert, J.; Berglund, J.; Schembri, M.; Genst, E.D.; Cools, L.; Wuhrer, M.; Hung, C.-S.; Pinkner, J.; Slättegård, R.; Zavialov, A.; et al. Receptor Binding Studies Disclose a Novel Class of High-Affinity Inhibitors of the *Escherichia coli* FimH Adhesin. *Mol. Microbiol.* **2005**, *55*, 441–455. [CrossRef] [PubMed]
48. Wellens, A.; Garofalo, C.; Nguyen, H.; Gerven, N.V.; Slättegård, R.; Hernalsteens, J.-P.; Wyns, L.; Oscarson, S.; Greve, H.D.; Hultgren, S.; et al. Intervening with Urinary Tract Infections Using Anti-Adhesives Based on the Crystal Structure of the FimH–Oligomannose-3 Complex. *PLoS ONE* **2008**, *3*, e2040. [CrossRef]
49. Wellens, A.; Lahmann, M.; Touaibia, M.; Vaucher, J.; Oscarson, S.; Roy, R.; Remaut, H.; Bouckaert, J. The Tyrosine Gate as a Potential Entropic Lever in the Receptor-Binding Site of the Bacterial Adhesin FimH. *Biochemistry* **2012**, *51*, 4790–4799. [CrossRef] [PubMed]

50. Schönemann, W.; Cramer, J.; Mühlethaler, T.; Fiege, B.; Silbermann, M.; Rabbani, S.; Dätwyler, P.; Zihlmann, P.; Jakob, R.P.; Sager, C.P.; et al. Improvement of Aglycone π-Stacking Yields Nanomolar to Sub-Nanomolar FimH Antagonists. *ChemMedChem* **2019**, *14*, 749–757. [CrossRef] [PubMed]
51. *Schrödinger Release 2024-2: Prime;* Schrödinger LLC: New York, NY, USA, 2024.
52. Despras, G.; Poonthiyil, V.; Lindhorst, T.K. Photochromic Carbohydrate Conjugates. In *Molecular Photoswitches;* John Wiley & Sons, Ltd.: Hoboken, NJ, USA, 2022; pp. 1015–1045, ISBN 978-3-527-82762-6.
53. Friedrich, L.M.; Hartke, B.; Lindhorst, T.K. Advancing Optoglycomics: Two Orthogonal Azobenzene Glycoside Antennas in One Glycocluster—Synthesis, Switching Cycles, Kinetics and Molecular Dynamics. *Chem. Eur. J.* **2024**, *30*, e202402125. [CrossRef] [PubMed]
54. Rivero-Barbarroja, G.; Maisonneuve, S.; Xie, J.; García Fernández, J.M.; Ortiz Mellet, C. Light-Responsive Glycosidase Inhibitors: Tuning Enzyme Selectivity and Switching Factors through Integrated Chemical and Optoglycomic Strategies. *Bioorg. Chem.* **2025**, *162*, 108575. [CrossRef] [PubMed]
55. Lu, C.; Wu, C.; Ghoreishi, D.; Chen, W.; Wang, L.; Damm, W.; Ross, G.A.; Dahlgren, M.K.; Russell, E.; Von Bargen, C.D.; et al. OPLS4: Improving Force Field Accuracy on Challenging Regimes of Chemical Space. *J. Chem. Theory Comput.* **2021**, *17*, 4291–4300. [CrossRef] [PubMed]

*molecules*

MDPI

*Article*

# Characterization of Human Recombinant β1,4-GalNAc-Transferase B4GALNT1 and Inhibition by Selected Compounds

**Iram Abidi [1], Alexander N. Kocev [1], Jonathan L. Babulic [1], Chantelle J. Capicciotti [1,2], Jagdeep Walia [1,3] and Inka Brockhausen [1,*]**

1 Department of Biomedical and Molecular Sciences, Queen's University, Kingston, ON K7L 3N6, Canada
2 Department of Chemistry, Queen's University, Kingston, ON K7L 3N6, Canada
3 Department of Pathology, Queen's University, Kingston, ON K7L 3N6, Canada
* Correspondence: brockhau@queensu.ca

**Abstract**

Gangliosides are essential for membrane functions, cell recognition, and maintenance of the nervous system. GM2 gangliosidosis is a group of rare genetic lysosomal storage diseases that includes Tay-Sachs disease (TSD), Sandhoff disease (SD), and AB variant. TSD and SD are characterized by deficient β-N-acetyl-hexosaminidase activity. This leads to decreased catabolism of β-N-acetyl-hexosamine-containing ganglioside GM2 in the lysosomes, damage to cells and tissues, and severe neurological symptoms. GM2 is a major ganglioside accumulating in TSD and SD, and is synthesized from GM3 by β1,4-N-acetylgalactosaminyltransferase 1 (B4GALNT1, GM2 synthase). Therapies under development for GM2 gangliosidosis include adeno-associated virus gene therapy, enzyme replacement, and substrate reduction therapy (SRT). The goal of this work was to express and purify human B4GALNT1, characterize its activity, and explore its structural features by protein modeling and substrate docking. We used a panel of synthetic compounds to study their potential inhibition of B4GALNT1 activity. This work can serve to develop SRT for GM2 gangliosidosis.

**Keywords:** B4GALNT1; GalNAc-transferase; GM2 gangliosidosis; inhibition; bioinformatics

## 1. Introduction

Glycosphingolipids (GSLs) are integral components of the cell membrane and are especially rich in the brain as sialic acid (Sia)-containing gangliosides [1] that are based on a lactosyl-ceramide (Lac-Cer, Galβ1-4Glcβ-Cer) core (Figure 1). GSLs are concentrated in membrane microdomains (lipid rafts). They help to maintain membrane fluidity and integrity, they act as cell surface epitopes and markers, participate in the functions and folding of membrane proteins and receptors, and help in cell–cell recognition and communication and regulation of the immune system. Gangliosides play crucial roles in early childhood and brain development and are involved in the maintenance of the nervous system [2–5].

In addition, many GSLs are receptors for bacterial toxins. For example, GM2 (GalNAcβ1-4Galβ1-4Glcβ-Cer) was shown to bind to Delta-toxin from *Clostridium perfringens* [6]. The simplest ganglioside is GM3, Siaα2-3Galβ1-4Glcβ-Cer, which is converted in the Golgi by an α2,8-sialyltransferase (e.g., ST8SiaI) to GD3, Siaα2-8Siaα2-3Galβ1-4Glcβ-Cer. GM3 and GD3 are prominent gangliosides found in the central nervous system. GD3 is upregulated in many invasive ductal breast carcinoma cases [7]. GD3 is also the predominant ganglioside in neural stem cells and is important for neuronal functions [8]. GD3 associates with epidermal growth factor receptor in breast cancer cells, and both GD3 and GD2 (Siaα2-8Siaα2-3[GalNAcβ1-4] Galβ1-4Glcβ-Cer) were found to be involved in signal transduction [7]. GD2 is also considered a marker for mesenchymal stromal cells [9].

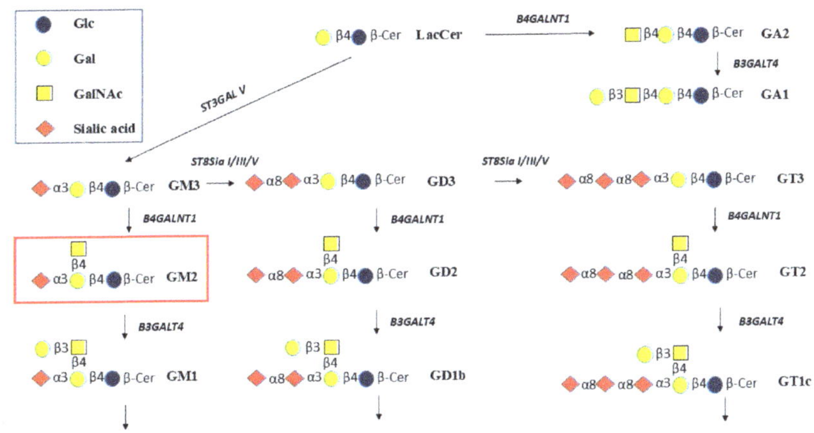


**Figure 1.** Biosynthetic pathways from lactosylceramide (Lac-Cer) to GA2 and GA1 and to gangliosides GM3, GM2, and GM1, and to gangliosides having more than 1 sialic acid residue. B4GALNT1 is required to synthesize GA2, GM2, GD2, and GT2. The GalNAc residue is then extended by Galβ1-3. Galβ1-4 and Galβ1-3 residues may be substituted by sialic acid residues (modified from [10]).

GM3 and GD3 are converted by β1,4-GalNAc-transferase B4GALNT1 to form GM2 and GD2, respectively (Figure 1). Thus, B4GALNT1 has been named GM2 synthase or GD2 synthase. The enzyme also transfers GalNAc to Lac-Cer to form GA2, GalNAcβ1-4Galβ1-4Glcβ-Cer, and is expected to synthesize GT2 from GT3.

In the lysosomal catabolic pathways, the GalNAc residues are cleaved by β-*N*-acetyl-hexosaminidase (HEX) (Figure 1) [10]. Substrates for HEX include GA2, GM2, GD2, and GT2. Lacto- and neolacto-series and Gal-Cer series of GSLs also include structures having terminal GlcNAcβ1-3 residues, while Gb4 of the globo-series has a terminal GalNAcβ1-3. These GSLs could also be cleaved by HEX [11]. However, in lysosomal storage diseases (LSD) such as GM2 gangliosidosis, Tay-Sachs disease (TSD), and Sandhoff disease (SD), the cleavage of GalNAc is decreased or blocked due to mutations of HEX or an essential activator protein, GM2A. As a consequence, primarily GM2 accumulates in the brain of children, leading to progressive neurodegeneration and early death. Lyso-GM2 is the product of *N*-deacylase that cleaves the fatty acid residue of GM2 and other GSLs. Thus, with a high concentration of GM2, lyso-GM2 also accumulates. GD2 positivity in bone marrow cells may represent a useful prognostic marker for patients with non-metastatic neuroblastoma [12], and B4GALNT1 has been suggested to be a pan-cancer biomarker [13]. An inhibition of GM2 synthesis may thus be useful to study the role of GM2/GD2 in human tumor cells (malignant melanoma lines, neuroblastoma lines, and some glioma lines) that express high levels of B4GALNT1 mRNA [14]. Blocking the pathway to GM2 synthesis causes a deficiency of the next metabolites in the pathway, such as GM1, which has been shown to be linked to Parkinsonian symptoms in *B4galnt1* knock-out mice. The GM1 oligosaccharide without the Cer moiety was used to successfully treat these symptoms [15]. Thus, the functions of membrane-bound gangliosides are based on both the glycan and lipid moieties and their interactions with proteins.

Therapy for GM2 gangliosidosis is currently being developed by introducing the correct HEXA and/or HEXB coding genes through adeno-associated virus (AAV)-based gene therapies [16,17]. In a mouse model of HEX-deficiency, a dose-dependent correction of GM2 accumulation in the brain was achieved after AAV9 transfer. In addition, mice lacking GM2A also responded to AAV-based gene transfer [18]. An alternative potential treatment for GSL accumulation is enzyme replacement. For example, in Gaucher disease (glucoceramidase deficiency), a functional, modified enzyme can be repeatedly injected into patients to minimize the effect of GSL accumulation. However, these enzymes do not cross the blood/brain barrier, and there may be side effects; the underlying, progressing pathology is treated peripherally but not in the brain, and it is not a cure. The delivery of Golgi- and lysosomal-resident enzymes is difficult, since proper post-translational modifications to avoid immunogenicity and protection from protease digestion, as well as for targeting, may be required.

Another possibility for LSD treatment is substrate reduction therapy (SRT), which can reduce biosynthetic precursors and can, thus, reduce pathogenicity due to accumulation of GSL [19,20]. The Glc derivative *N*-butyldeoxynojirimycin (Miglustat) [21] functions as a competitive and reversible inhibitor of the enzyme glucosylceramide synthase (GCS) that forms Glc-Cer and is used to reduce the synthesis of GSL and ganglioside metabolites based on Glc-Cer. Miglustat is a treatment for Gaucher disease, but also leads to serious side effects. Another potential treatment to correct genetic abnormalities is based on cell therapy with induced pluripotent stem cells (iPSCs). iPSCs offer disease models and can also be developed into established healthy cell populations to be administered to patients. Although there is promise in these therapies, currently, there is no established cure for TSD and SD.

B4GALNT1 is a type-II Golgi transmembrane homodimeric protein [22] and belongs to the CAZy GT12 family of glycosyltransferases (GTs), having a GT-A fold and an inverting mechanism [23]. In humans, B4GALNT1 mutations that lead to decreased enzyme activity are associated with autosomal recessive spastic paraplegia, termed SPG26, a slowly progressive neurodegenerative disorder with increasing muscle weakness (Table 1) [24–26].

In mice lacking B4GALNT1 and GM2, the symptoms include Parkinsonian neuropathology [27], suggesting that B4GALNT1 is an essential enzyme. Forced expression of B4GALNT1 in human melanoma cells increased GM2 levels as well as cleavage of amyloid precursor protein that is associated with the pathophysiology of Alzheimer's disease (AD) [28]. Levels of GM3 are increased in the brains of AD patients and in a mouse model [29]. Kaya et al. [5] determined by mass spectrometry that gangliosides GM3, GM2, and GM1 accumulate in amyloid plaques of AD mice. A GCS inhibitor that penetrates into the brain was shown to reduce GM3 levels as well as amyloid plaques in AD mice [30]. Inhibition of GlcCer synthesis would also reduce the levels of GM2 and other metabolites, and this could be beneficial in dementia. Thus, balanced levels of gangliosides play a critical role in cell membrane functions, and a complete inhibition of B4GALNT1 may lead to neuropathology. SRT should, therefore, aim to maintain an optimal level of enzyme activity and maintain a balance of gangliosides.

The primary objective of this study is to express, purify, and characterize recombinant human B4GALNT1 and to advance the development of appropriate inhibitors that reduce GM2 synthesis and thus may prevent pathological accumulation of GM2.

## 2. Results and Discussion

### 2.1. Protein Expression and Purification

The expression of recombinant $His_8$-tagged B4GALNT1 in HEK293 and HEK293T cells showed very low levels and low enzyme activity. SDS-Page and Western blots indicated that the enzyme was associated with the pellet in an insoluble form and showed 0.01 nmol/h/mg with GM3 acceptor in the lysed pellet. We, therefore, used Expi293 suspension cells that were expected to produce large amounts of soluble recombinant protein. Using the Expi293 cell expression system, Western blot of the medium revealed a protein band at 88 kDa on Day 2, which became strongest by Day 5 of transfection. The lysed cell pellet showed no protein band at 88.8 kDa (Figure S1), indicating that soluble $His_8$-B4GALNT1 was secreted into the culture medium with a maximal production on Day 5.

Purification of $His_8$-B4GALNT1 was performed using Ni-NTA chromatography. Most of the enzyme protein eluted with 500 mM imidazole (Figure S2). B4GALNT1 appeared as a strong band on a 12% SDS-PAGE gel in elution fractions, corresponding to the theoretical size of 88.8 kDa, with a protein concentration of 0.4 mg/mL. Purified enzyme was concentrated using a 10MWCO Centricon column.

### 2.2. Characterization of Purified B4GALNT1 Activity

The activity of B4GALNT1 produced in Expi293 cells was confirmed with 0.2 mM GSL acceptor substrates using crude enzyme from the medium or highly purified enzyme. GM3 produced 6.1 nmol/h/mg activity with purified enzyme, while GD3 produced 3.2 nmol/h/mg. Lac-Cer and GM1, Galβ1-3GalNAc1-4(Siaα2-3)Galβ1-4Glcβ-Cer (Figure 1), showed <0.1 nmol/h/mg activity (Figure 2). This indicates that the Siaα2-3Galβ1-4 linkage in the acceptor is important for enzyme recognition, but an additional Sia residue in GD3 is less preferred. Thus, B4GALNT1 activity with GT3 acceptor is expected to be very low, although this remains to be tested in vitro. The activity was maximal with purified enzyme using GM3 as its major acceptor substrate. The apparent $K_M$ for GM3 substrate with purified enzyme was 0.5 mM. This value is within the range reported for other B4GALNT1 activities from membrane fractions. The apparent $K_M$ values reported for GM3 acceptor varied from 0.69 mM in bovine thyroid extracts [31] to 0.1 mM [32] for the rat enzyme and 0.5 mM for the human enzyme [33]. The absolute activities and $K_M$ values, however, depend on the type of enzyme (membrane-bound or purified, type of construct), the expression system, and the conditions of the assay, including length of incubation and the type and concentration of detergent and cofactors. Thus, it has been difficult to determine an accurate apparent $V_{max}$ value. In a cell extract, different species of the same enzyme and potentially different variants may be present.

The human enzyme can be proteolytically released as a soluble form [22]. Previous studies with various forms of B4GALNT1 from human, mouse, and rat sources also showed that GM3 is the major acceptor substrate and GM2 is, therefore, the major GSL product of B4GALNT1 [4,22,31–36]. The rat enzyme from ascites hepatoma cells, for example, has 77.3% activity towards GD3 compared to GM3 and only 1.2% activity with LacCer [34]. A high B4GALNT1 activity was achieved by Welland et al. [37] using a liposome-based assay with the enzyme tethered to the membrane via a His-tag.

Several nucleotide sugars were tested as donor substrates for B4GALNT1 with GM3 acceptor, replacing UDP-GalNAc in standard assays. UDP-GalNAc was the only active donor substrate while UDP-Gal, UDP-Glc, and UDP-GlcNAc failed to serve as donor substrates. Thus, B4GALNT1 has a strict donor specificity.

Although recombinant B4GALNT1 appeared to be soluble when expressed in Expi293 cells, Kyte and Doolittle hydrophobicity analysis revealed several shorter hydrophobic sequences that may possibly be involved in binding the hydrophobic moiety of the acceptor substrates. These hydrophobic sequences may also be responsible for an accumulation of misfolded protein in the pellet when expressed in HEK cells. We, therefore, tested whether detergent that may aid the solubility of the substrate or affect the folding of purified B4GALNT1 could improve the activity. The activity increased slightly from 0.06 to 0.12% Triton X-100 in the assay (Figure S3). Although the enzyme has previously been assayed with various types of detergents up to 2% [22], we used 0.06% Triton X-100 in order to prevent problems with GSL product isolation after the incubation. CDP-choline was previously used in B4GALNT1 assays as a potential inhibitor of UDP-GalNAc degradation [32]. In our standard assays of highly purified B4GALNT1 that should lack hydrolytic enzymes, the presence of 10 or 30 mM CDP-choline did not inhibit or activate B4GALNT1 activity and was therefore omitted in further assays.

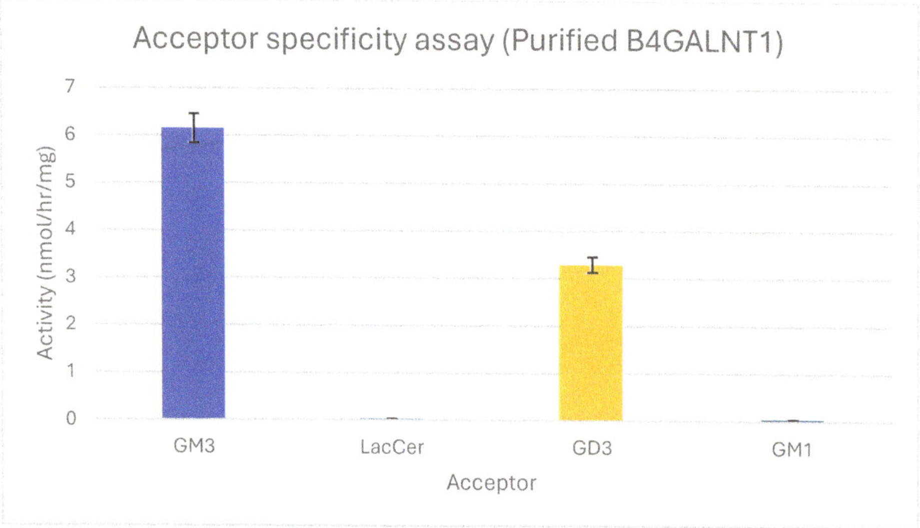


**Figure 2.** Acceptor substrate specificity of purified B4GALNT1. The activities of the enzyme are shown using 0.2 mM acceptor substrates GM3, Lac-Cer, GD3, or GM1 in standard assays. Error bars show variation between duplicate assays. All values were standardized against the negative control that lacks acceptors. Only GSLs having a Sia$\alpha$2-3Gal$\beta$1-4 sequence served as significant acceptors.

Generally, buffer pH values have been used in B4GALNT1 assays between 6.6 and 7.6 [22,32,33] while we used pH 7 due to a broad pH optimum in our assays. B4GALNT1, as an inverting GT12 enzyme, may require the presence of metal ions for donor binding and catalysis. The effect of metal ions was tested with 8 mM $MnCl_2$, $MgCl_2$, Zn-acetate, Co-acetate, or EDTA in standard assays of B4GALNT1 activity using GM3 as acceptor substrate. Compared to $MnCl_2$, $MgCl_2$, Zn-acetate and Co-acetate resulted in activities of 67%, 22%, and 15% respectively, while assays conducted with EDTA resulted in complete loss of activity (Figure 3). This showed the dependency of B4GALNT1 activity on divalent metal ions, with $Mn^{2+}$ being optimal. In contrast, the membrane-bound rat B4GALNT1, which has about 97% sequence identity to the human enzyme, has been reported to have higher activity in the presence of 10 mM $Co^{2+}$ compared to $Mn^{2+}$ [32].

*2.3. Bioinformatics*

The CAZy data bank classified B4GALNT1 into the GT12 family, characterized as inverting GTs having a GT-A fold and one Rossmann domain where the central beta-sheet is expected to contain the active site. Only four GT12 enzymes have been characterized, and these are all GalNAc-transferases from man, mouse, and rat with distant similarity to GT2, GT21, and GT27 families. The two subunits of B4GALNT1 homodimers have two catalytic domains in antiparallel orientation [36]. B4GALNT1 is disulfide bonded with C429-C476, forming an intra-chain bond, and C80-412 and C82-C529 forming inter-subunit bonds [36,37].

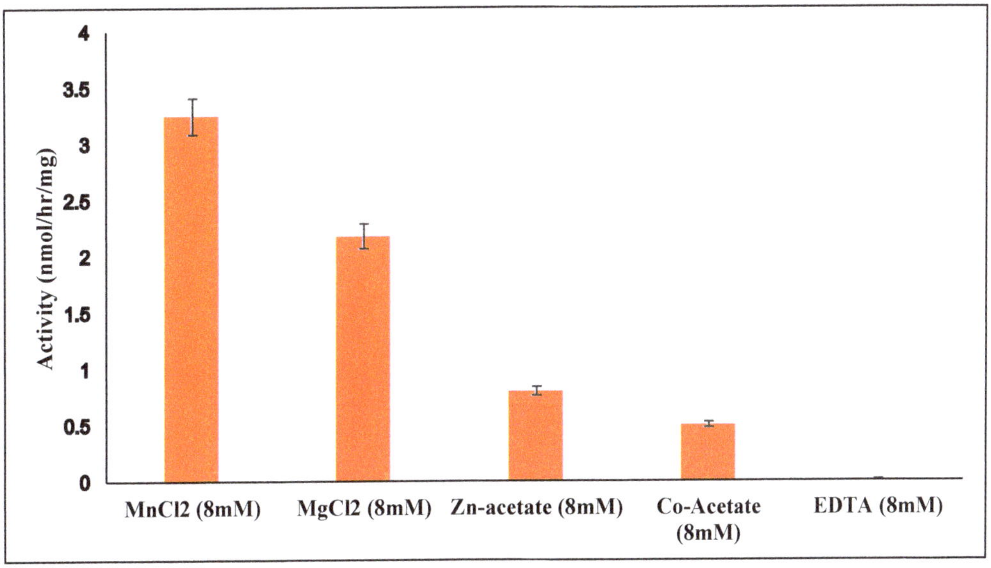


**Figure 3.** Dependence of purified B4GALNT1 activity on divalent metal ions. $Mn^{2+}$ ions in the standard assay with 0.2 mM GM3 acceptor were replaced by other divalent metal ions or EDTA. All values were standardized against the negative control. Error bars indicate the variation between duplicate assays.

All of these Cys residues are highly conserved, suggesting that dimer formation and three-dimensional protein structure are essential for activity (Figure 4). Human B4GALNT1 has a high degree of amino acid sequence identity with the enzyme from mouse (86.87%), rat (87.62%), bovine (90.99%), and Macaca fascicularis (98.87%). All of the Cys residues involved in disulfide bonding, as well as a putative catalytic DDD sequence, are present in these species.

Table 1 lists the conserved amino acids that may represent important residues for the biological function of B4GALNT1. The consensus sequence aligned by MUSCLE using 352 aligned protein sequences and generated with WebLogo is shown in Figure 5. Two centrally located DxD sequences are found in GM2 synthases. However, only D356, D357, and D358 are highly conserved sequences in GM2 synthases that line the active site, and D356 may be the critical Asp residue involved in catalysis [23]. DxD is a widely conserved motif in GT-A folded enzymes and is thought to coordinate with metal ions and play a role as a catalyst (Table 1) [38]. Modeling showed that the DxD motif (D356 to D358) is within the catalytic site, explaining that mutations of D356 and D358 abolished activity. The DD sequence is also highly conserved in B4GALNT1. Inamori et al. [39] identified an inactive SPG variant of D313, the second Asp of the DD sequence, that is thought to bind UDP-GalNAc. Other inactive SPG variants of highly conserved residues include N437K, A441E, P453H, and R519W. However, the F438L mutant exhibited low (<10%) activity.

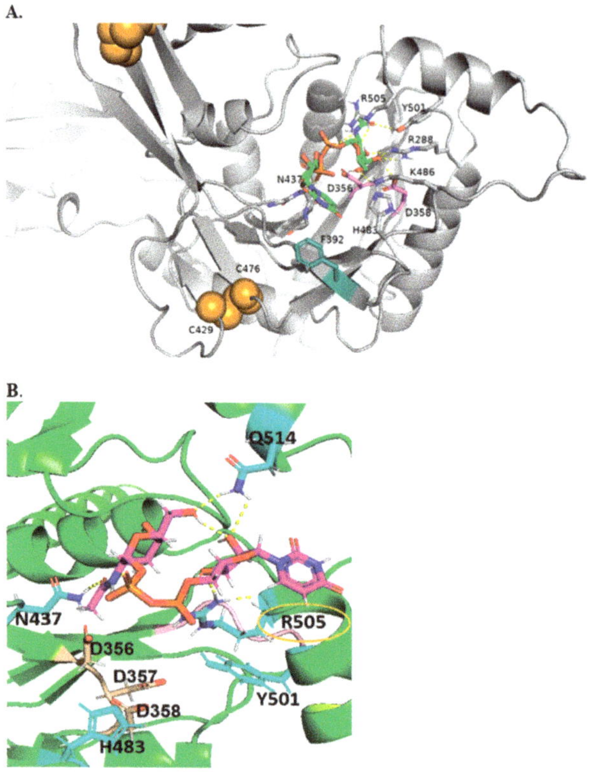


**Figure 4.** Alphafold2 models of B4GALNT1 with DXD motif in magenta and intrachain disulfide bonds in orange. (**A**). Global view of multi-domain B4GALNT1. (**B**). Model of B4GALNT1 docked with donor UDP-GalNAc making predicted interactions with ligand shown with labels, with yellow dashed lines representing H-bonds and potential pi–pi stacking partner shown in teal.

**Table 1.** Amino acid residues important for activity of human B4GALNT1.

| MSA | Variants | Activity of Mutants | References |
|---|---|---|---|
| K284N | SPG | | [25] |
| R288 | SPG | | [24] |
| R300C | SPG | | [24] |
| D313A | SPG | | [24,39] |
| V352 | | 167% | [38] |
| W354 | | 24% | [38] |
| D356 | | nd | [38] |
| D357 | | | |
| D358 | | nd | [38] |

**Table 1.** *Cont.*

| MSA | Variants | Activity of Mutants | References |
|---|---|---|---|
| (F392) | | | |
| C429 | | | |
| D433A | SPG | | [24] |
| N437K | SPG | nd | [39] |
| F438L | SPG | | [39] |
| F439 | SPG | | [24] |
| A441E | SPG | nd | [39] |
| P453H | SPG | nd | [39] |
| C476 | | | |
| H483 | | | |
| K486 | | | |
| Y501 | | | |
| R505H | SPG | nd | [25] |
| Q514 | | | |
| A516 | SPG | | [24] |
| R519P/W | SPG | nd | [26,39] |

**MSA**, multiple sequence alignments, showing conserved amino acids determined by BLAST. Numbers in brackets denote less conserved residues. SPG mutants are included. **Variants**, natural variants in patients with hereditary spastic paraplegia (SPG). Mutations in Variants affect highly conserved amino acids, with the exception of W354, A516, and R519, which are not highly conserved in human B4GALNT1. **Activity of mutants**, created by single amino acid replacements. nd, not detected.

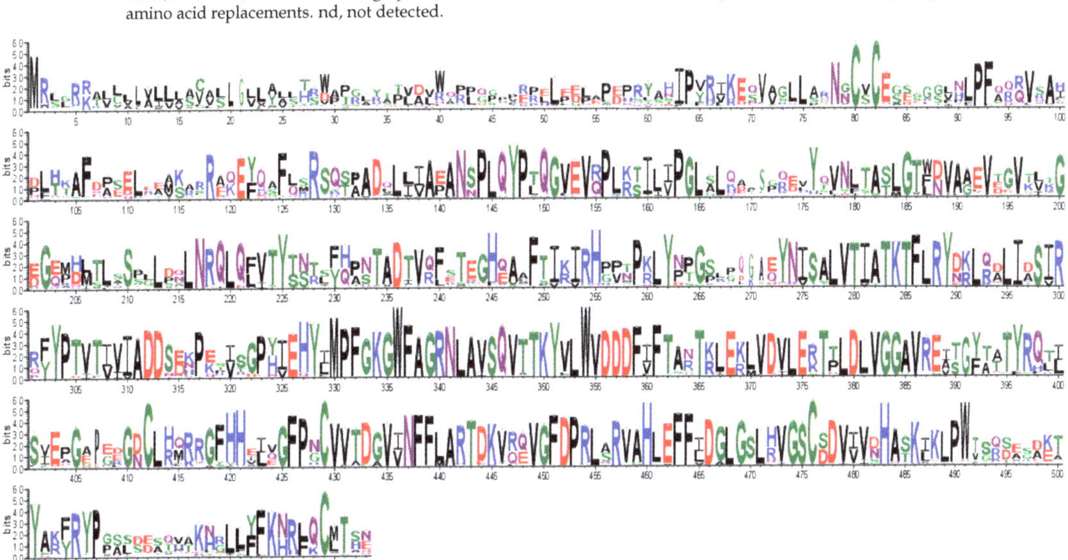

**Figure 5.** Consensus sequence logos generated by Weblogo. Sequence conservation in full-length B4GALNT1, showing conserved residues as large letters. Among highly conserved residues are D356 (DxD motif), Cys80, Cys82, Cys412,Cys 429, Cys476, Cys529 (disulfide bonds), R300 and F439 (SPG), and N179, N274 (N-glycosylation sites).

Interestingly, the V352A mutant of B4GALNT1 had 167% activity compared to the wild type and may facilitate D356-mediated catalysis, possibly by widening the substrate binding groove. In contrast, the W354A mutant was only 23.5% active [38].

Other conserved amino acids that appear to be within the active site are highly conserved N437, H483, Y501, and R505. Welland et al. [37] suggested that R505 is coordinated with UDP in acceptor and product, likely by binding to phosphate groups. Several variants of B4GALNT1 were found in SPG patients (Table 1). Conserved variant amino acids include R300, D433, and F439, but according to modeling, are not located within the active site. Less conserved variant amino acids include A516, R519 in SPG, and the roles of these residues are unknown. Interestingly, mutations in SPG26 [24] have been identified for residues R300 (to C) and D433 (to A). Since these may not be critical for catalysis or UDP-GalNAc binding, they may be involved in shaping the overall substrate binding site for UDP-GalNAc.

The amino acids near the proposed substrate binding and catalytic site in our study align well with those suggested by Welland et al. [37] based on the crystal structure and molecular dynamics simulations of the dimer of the soluble lumenal domain of B4GALNT1. The proposed catalytic domain encompasses amino acids 258 to 533. Even in the absence of the tm domain, amino acids flanking the active site were suggested to form surface loops that contribute to membrane insertion.

There are three potential N-Glycosylation sites at N79, N179, and N274 that are distant from the UDP-GalNAc binding site. Mutations in these amino acids led to decreased activity [40]. Only N179 and 274, but not N79, are highly conserved and may contribute to maintenance of protein structure. Some of these N-glycans are sialylated and could contribute to the overall folding, dimer formation, and stability of the enzyme in the late Golgi compartment [22].

It is not yet known if any of the Thr and Ser residues of B4GALNT1 carry O-glycans. For example, S314 and S401 are potential O-glycosylation sites [41]. HEK cells were previously shown to have the enzymes that form sialylated core 1 and 2 O-glycans as well as complex N-glycans [42] and could modify B4GALNT1.

### 2.4. Molecular Docking

Figure 4 shows a model of B4GALNT1 docked with the essential donor substrate UDP-GalNAc and suggests that conserved N437, R505, R288, and K486 are engaged in hydrogen bonding interactions with donor UDP-GalNAc and could be associated with the UDP reaction product. In addition, H483 and Y501 are near the UDP-GalNAc binding site and may be crucial residues for other interactions. The proximity of the F392 side chain to the uridine moiety suggests the possibility of coordination via pi–pi stacking. The DxD motif (356-358) that can coordinate $Mn^{2+}$ ion is in proximity to Y501, which may be necessary for coordinating the hydrophobic acceptor substrate. R505 is coordinated near C1 of the GalNAc moiety and is mutated in GM2 synthase deficiency [25] (Figure 4). It seems likely that R505 and H483 are also involved in binding sialylated acceptors such as GM3. As with other GTs, it was not possible to successfully dock the acceptor substrates into the protein, likely because of the complexity in binding a ceramide moiety, as well as a complex glycan chain.

### 2.5. Inhibitors

Currently, inhibitors for B4GALNT1 are not available. We tested several synthetic compounds as inhibitors of B4GALNT1 activity, but many of them did not inhibit the activity. These compounds are not directly related to structures of B4GALNT1 substrates but could potentially bind to the protein via hydrophobic or electrostatic interactions. A series of compounds named QT [43] showed various degrees of inhibition. QT compounds have two positively charged bis-imidazolium rings connected to an aliphatic spacer between rings (**n**) and symmetrical aliphatic wing chains (**m**) attached to the nitrogen of imidazolium groups. The highly flexible aliphatic chains have varying numbers of carbons, and the compounds are in dimesylate salt forms (Table 2, Figure 6). QT compounds with a higher number of carbons in the aliphatic chains previously showed inhibition for selected enzymes, and this was unrelated to the type of substrates used [43]. Another potential inhibitor was 2-naphthyl 2-butanamido-2-deoxy-1-thio-β-D-glucopyranoside (**612**). GlcNAc-naphthyl derivatives such as **612** were shown to inhibit bovine milk β1,4-Gal-transferase as acceptor substrate analogs [44,45].

**Table 2.** List of bis-imidazolium inhibitors tested for activity of purified B4GALNT1.

| Inhibitor | n | m | % Inhibition |
|---|---|---|---|
| **QT149** | 22 | 1 | 90 |
| **QT163** | 6 | 11 | 80 |
| **QT160** | 10 | 10 | 89 |
| **QT161** | 10 | 11 | 84 |
| **QT162** | 10 | 12 | 94 |
| **QT166** | 16 | 7 | 92 |
| **QT169** | 16 | 10 | 74 |
| **QT170** | 16 | 11 | 40 |
| **QT171** | 16 | 12 | 47 |
| **612** | - | - | 44 |

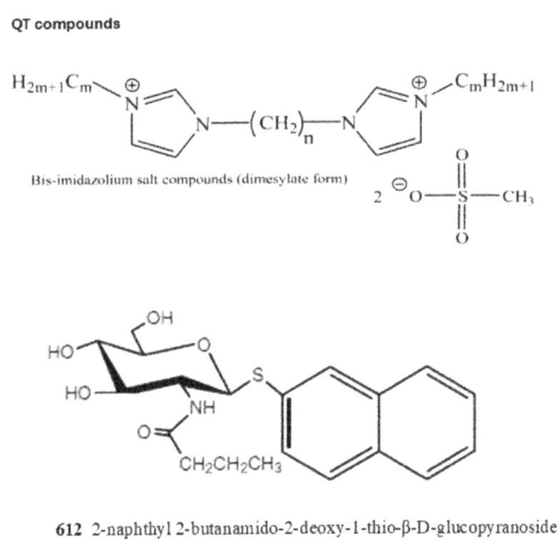


**Figure 6.** The structures of inhibitors used in this study are shown. The effects on activity are listed in Table 2. For QT compounds: **n**, number of aliphatic core carbons; **m**, number of symmetrical aliphatic carbons attached to the imidazolium rings.

To determine inhibition, purified B4GALNT1 was assayed under standard assay conditions utilizing 0.2 mM GM3 as the acceptor substrate, 0.39 mM UDP-GalNAc as donor substrate, and 1 mM **QT** compound or **612**. Table 2 summarizes the variable inhibitory effects of **QT** compounds on B4GALNT1 activity. Compound **612** showed 44% inhibition. **QT163** (6/11), having a 6-carbon spacer chain and 11-carbon wing chains, showed 80% inhibition, with an $IC_{50}$ value of 0.2 mM. Other **QT** compounds inhibited the activity between 40 and 94%. Our results suggest that it is not the length of either the core or the wing aliphatic chain that determines the inhibitory potential, but a combination of both and, likely, the overall properties of the compounds. In future studies, we will try to dock **QT163** to the B4GALNT1 model to determine which hydrophobic or negatively charged groups in the enzyme could be responsible for inhibitor binding. This binding is expected to affect the structure of the catalytic domain or block the access of substrates.

B4GALNT1 assays were carried out with 0.2 mM GM3 as the acceptor, 0.07 mM donor, and 1 mM inhibitor concentration. The values shown in the Table are averages of at least duplicate determinations. The structures of inhibitors are shown in Figure 6. Number of aliphatic core carbons (**n**); number of symmetrical aliphatic carbons attached to the imidazolium rings (**m**).

## 3. Materials and Methods

### 3.1. Materials

Materials were purchased from Sigma-Aldrich (Burlington, MA, USA), unless otherwise stated. UDP-[$^3$H]GalNAc, UDP-[$^3$H]Gal, UDP-[$^3$H]GlcNAc, and UDP-[$^{14}$C]Glc were obtained from American Radiolabeled Chemicals (St. Louis, MO, USA). GSLs were from Cayman: Ann Arbor, MI, USA. Potential inhibitors were synthesized as described [46]. FBS was from Wisent: Saint-Jean-Baptiste, QC, Canada.

### 3.2. Plasmid Isolation and Transformation

The plasmid pGEn2-DEST was developed and donated by Kelly Moremen, University of Georgia. This mammalian expression vector has a CMV promoter, N-terminal $His_8$ tag, AviTag, Super GFP tags, and ampicillin resistance genes. Sequences were confirmed by Sanger sequencing (The Centre for Applied Genomics, Toronto, ON, Canada). To enhance protein solubility, the short cytoplasmic and the transmembrane (tm) domain (amino acids 1–25) at the N-terminus were deleted in the recombinant B4GALNT1 construct.

Plasmid pGEn2-DEST was transformed into *E. coli* (DH5α) using the heat shock method. Positive transformants were selected from LB (Lennox)-agar plates containing 100 µg/mL ampicillin (pGEn2-DEST). An overnight small-scale culture from a single colony was prepared in 3 mL of LB Broth (BioShop) with 100 µg/mL ampicillin, maintained at 37 °C, 225 RPM for 16 h. An aliquot of the overnight culture (100 µL) was added to a 250 mL Erlenmeyer flask, containing 100 mL of LB broth and 100 µg/mL ampicillin. The flask was shaken at 37 °C, 225 RPM, to reach absorbance OD600 = 0.6.

For isolation of high-quality plasmid DNA from *E. coli* DH5α cultures, Gene JET Plasmid Midiprep Kit (Thermo Scientific, Waltham, MA USA) was used. The above prepared 100 mL culture was centrifuged for 10 min, and supernatant was discarded. The bacterial pellet was resuspended using 2 mL Resuspension Buffer, and the suspension was vortexed. The plasmid quality and quantity of the isolated DNA were examined using a NanoDrop spectrophotometer. The total yield was 145 µg DNA per 100 mL culture. Plasmid DNA was filtered through a 0.22 µm filter prior to transfection.

### 3.3. Expression of B4GALNT1 in Expi293 Cells

The mammalian cell-based expression system of human embryonic kidney (HEK) Expi293 suspension cell lines was used to express soluble protein in sufficient quantities for characterization, based on previously described protocols [47]. Expi293 cells were cultured in expression medium (Thermo-Fisher, Waltham, MA, USA). Cells were maintained in a humid 5% $CO_2$ atmosphere at 37 °C, shaking at 120 RPM. Cells were passaged for maintenance after reaching $4 \times 10^6$ viable cells/mL. Cells were cultured for at least 3 passages following thaw prior to transfection. Expi293 cells were transiently transfected with the pGEn2-DEST plasmid containing the human B4GALNT1 gene [48] using Expifectamine™ 293 Transfection Kit (Thermo-Fisher, Waltham, MA, USA).

On the day of transfection, the cells were diluted to 3 million cells/mL in 100 mL of culture medium. Plasmid DNA containing the B4GALNT1 gene (100 μg) was diluted with 5.6 mL of Opti-MEM GIBCO medium (Thermo-Fisher, Waltham, MA, USA). A total of 320 μL of Expifectamine lipid-based transfection reagent was diluted with 6 mL of Opti-MEM GIBCO medium. The diluted transfection agent was added to the DNA dropwise and mixed gently by inverting the tube 4 times, followed by incubation for 15 min at room temperature. The pre-complexed DNA and Expifectamine mixture was then added dropwise to the cells, swirling gently. Transfected cells were then maintained in a humid 5% CO2 atmosphere at 37 °C, shaking at 120 RPM. Expifectamine™ 293 Transfection Kit (Thermo-Fisher, Waltham, MA, USA) enhancer reagents were added 16 h following transfection. Cell aliquots were taken daily to monitor cell growth and viability with a hemocytometer using trypan blue staining. On day 5 following transfection, cells were harvested by centrifugation (25 min, 4000 RCF), and cells and supernatants were kept at 4 °C until purification. Enzyme expression was confirmed by SDS-PAGE and Western blot using anti-His antibody to reveal a protein of 88.8 kDa [49].

### 3.4. *Purification of B4GALNT1 Using Ni-NTA Column Chromatography*

Cells were disrupted by hand homogenization in 50 mM sucrose, followed by centrifugation. $His_8$-tagged B4GALNT1 produced by transient transfection in Expi293 cells was purified from the supernatant using a column of 5 mL HisPur Ni-NTA agarose resin slurry (ThermoFisher Scientific, Waltham, MA, USA). The medium was adjusted with 10× media adjustment buffer (200 mM imidazole, 2 M NaCl, and 300 mM sodium phosphate, pH 7.2) to make the final media. The column was equilibrated with 10 column volumes of Buffer A (20 mM HEPES, 300 mM NaCl, 20 mM imidazole, pH 7.2). The sample was adjusted with media adjustment buffer (200 mM imidazole, 2 M NaCl, and 300 mM sodium phosphate, pH 7.2) and was loaded on the equilibrated column, and the flowthrough was collected. The column was washed first with 50 mL of Wash buffer 1 (2 mM HEPES, 30 mM NaCl, 20 mM imidazole, pH 7.2), followed by Wash buffer 2 (2 mM HEPES, 30 mM NaCl, 50 mM imidazole, pH 7.2) and Wash buffer 3 (2 mM HEPES, 30 mM NaCl, 100 mM imidazole). Protein was eluted with 50 mL of Elution buffer (2 mM HEPES, 30 mM NaCl, 300 mM imidazole, pH 7.2). Lastly, the column was washed with 15 mL 2 mM HEPES, 30 mM NaCl, 500 mM imidazole. The eluted protein was concentrated with a 10 kDa MWCO filter (4000× *g*) at 6000 RPM for 10 min at 4 °C (for 10–12 cycles) and buffer-exchanged into 20 mM Tris HCl, pH 7.5 (3 to 4 cycles). The protein concentration was determined using the Bicinchoninic acid assay. Purified protein was stored in aliquots with 20% glycerol at 4 °C, −20 °C, and for long-term storage at −80 °C.

### 3.5. *Glycosyltransferase Standard Assays for Human B4GALNT1*

B4GALNT1 activity was measured by the transfer of radiolabeled GalNAc from the UDP-[$^3$H]GalNAc donor using GM3 as an acceptor substrate. GM3 was dissolved in chloroform/methanol (2:1), which was removed before the assay with a stream of nitrogen. Standard assays contained in a total volume of 50 μL: 20 μL (8 μg protein) purified B4GALNT1 in

20 mM Tris pH 7, 20% glycerol, 0.39 mM UDP-[$^3$H]GalNAc (1900 CPM/nmol), 0.2 mM dried acceptor GM3, 40 mM Na-cacodylate buffer pH 7, 0.06% Triton X-100, and 8 mM $MnCl_2$. Assay mixtures were incubated at 37 °C for 1 h, and reactions were quenched by freezing. Assays were carried out in duplicate determinations. The negative control assays did not contain the acceptor substrate. The reaction product was isolated via C18 Sep-Pak cartridges and was eluted with MeOH. Radioactivity was measured by scintillation counting.

### 3.6. Bioinformatics

Predicted B4GALNT1 3D structure was obtained from the AlphaFold database, which was used to understand structural features. A docked model of B4GALNT1 binding UDP-GalNAc donor substrate was obtained with MolSoft ICM 3.9. BLAST search was run in Uniprot against the UniRef50 database, with the IDs of the following families mapped against the UniprotKB database: UniRef50_A0A8K1LFD9, UniRef50_C0H8Y3, UniRef50_A0A8I3NPC1, UniRef50_H3DAS0, UniRef50_Q00973, UniRef50_A0AD0GQI1, and UniRef50_A0A974BWB6. Removal of outliers resulted in a final set of 352 sequences, which were aligned using MUSCLE. A consensus sequence, following gap excision, was generated with Weblogo.

### 3.7. Inhibition of B4GALNT1

Inhibition was measured under standard assay conditions for purified B4GALNT1. Compounds were dissolved in MeOH and dried before the addition of the enzyme and other assay components. Positive control assay mixtures contained 0.2 mM GM3 and lacked inhibitor. To determine the $IC_{50}$, compound **QT163** was used at 0.2, 0.4, 0.6, 0.8, and 1 mM concentrations in the assay.

## 4. Conclusions

The aim of this study was to characterize a soluble form of human B4GALNT1, the enzyme that transfers a GalNAc residue in β1-4 linkage to the Galβ1-4 residue of gangliosides GM3, GD3, and other GSLs. The inhibition of this enzyme would reduce the synthesis of GM2, which could be of therapeutic value for patients who accumulate GM2 due to the inability to hydrolyze the GalNAc residue. This inhibition would shift the balance of GSL and may impact cellular functions that rely on specific GSLs on membrane and lipid rafts. Thus, an application of these inhibitors in SRT has to be carefully studied and monitored.

Previous substrate specificity assays [4,33] revealed that ganglioside GM3 containing Neu5Ac was a major acceptor substrate for B4GALNT1 in microsomal membranes and for a soluble recombinant protein. GM3 having Neu5Gc was about 12.8% less active. Our specificity assays using highly purified soluble B4GALNT1 similarly showed that GM3 was the preferred acceptor and confirmed that GD3 exhibited half of that activity, while Lac-Cer and GM1 showed minimal activity. This indicates that the Siaα2-3Galβ1-4Glc-linkage in the acceptor is important for enzyme recognition and that one terminal Sia residue is preferred.

It is not clear yet if milk oligosaccharides having the Siaα2-3Galβ1-4Glc linkage and glycoproteins with Siaα2-3Galβ1-4GlcNAc linkage form acceptors for B4GALNT1 or if the enzyme requires the hydrophobic Glc-Cer moiety in the acceptor. Positively charged residues in the acceptor binding site (R505 and H483) may aid in recognition of Sia, and mutations at these residues may reveal a change in acceptor specificity. Y501 and other hydrophobic amino acids could aid in binding the Cer group of the acceptor.

A comparison of published properties of B4GALNT1 with our work shows that the GM2 synthase activity is dependent on many factors, including the protein sequence and glycosylation, presence of membranes, and assay conditions. These various results allow only an approximate estimate of the activity in vivo, which may differ according to the cell type expressing B4GALNT1.

The aim of this study was to identify inhibitors for B4GALNT1 that may be developed for treatment of GM2 gangliosidosis but not cause symptoms of SPG. The compounds we chose might be appropriate for this purpose since they inhibited B4GALNT1 activity between 40 and 94%. A complex of B4GALNT1 docked with its donor substrate UDP-GalNAc showed amino acids near the docking site that may play a role in UDP-GalNAc binding and catalysis. This information, together with protein structure analysis of B4GALNT1 to confirm the role of conserved amino acids, will aid in the design of additional, specific, non-covalent inhibitors suitable for SRT. Studies of drug development should also include the kinetics of inhibition, test dosage, and toxicity, and address drug delivery to neurons that accumulate GM2.

**Supplementary Materials:** The following supporting information can be downloaded at: https://www.mdpi.com/article/10.3390/molecules30173615/s1, Figure S1: Expression of His8-tagged B4GALNT1 in Expi293 cells.; Figure S2: SDS-PAGE after purification of His8-tagged B4GALNT1 expressed in Expi293 cells and secreted into the culture medium.; Figure S3. Detergent effect.

**Author Contributions:** Conceptualization, I.B.; Writing—Original Draft Preparation, I.A.; Data acquisition, I.A., A.N.K., J.L.B., C.J.C.; Writing—Review and Editing, I.B.; Supervision, I.B.; Funding Acquisition, I.B., J.W. All authors have read and agreed to the published version of the manuscript.

**Funding:** Funding for this project was provided by GlycoNet (to J.W., I.B.) and an NSERC Discovery grant 2022-05389 (to I.B.).

**Institutional Review Board Statement:** Not applicable.

**Informed Consent Statement:** Not applicable.

**Data Availability Statement:** The authors will make data available upon request.

**Acknowledgments:** The authors thank Zhenbo Zhang for the initial expression experiments of B4GALNT1 in HEK293 cells. We thank Xinran Wang for GT assays and Dylan Falconer for critical discussions.

**Conflicts of Interest:** The authors declare that they have no conflicts of interest.

## References


1. Ledeen, R.; Wu, G. Gangliosides of the nervous system. *Methods Mol. Biol.* **2018**, *1804*, 19–55. [PubMed]
2. Ohmi, Y.; Tajima, O.; Ohkawa, Y.; Yamauchi, Y.; Sugiura, Y.; Furukawa, K.; Furukawa, K. Gangliosides are essential in the protection of inflammation and neurodegeneration via maintenance of lipid rafts: Elucidation by a series of ganglioside-deficient mutant mice. *J. Neurochem.* **2011**, *116*, 926–935. [CrossRef] [PubMed]
3. Lingwood, D.; Simons, K. Lipid rafts as a membrane-organizing principle. *Science* **2010**, *327*, 46–50. [CrossRef]
4. Furukawa, K.; Takamiya, K.; Furukawa, K. Beta1,4-N-acetylgalactosaminyltransferase-GM2/GD2 synthase: A key enzyme to control the synthesis of brain-enriched complex gangliosides. *Biochim. Biophys. Acta* **2002**, *1573*, 356–362. [CrossRef]
5. Kaya, I.; Jennische, E.; Dunevall, J.; Lange, S.; Ewing, A.G.; Malmberg, P.; Baykal, A.T.; Fletcher, J.S. Spatial Lipidomics Reveals Region and Long Chain Base Specific Accumulations of Monosialogangliosides in Amyloid Plaques in Familial Alzheimer's Disease Mice (5xFAD) Brain. *ACS Chem. Neurosci.* **2020**, *11*, 14–24. [CrossRef]
6. Manich, M.; Knapp, O.; Gibert, M.; Maier, E.; Jolivet-Reynaud, C.; Geny, B.; Benz, R.; Popoff, M.R. Clostridium perfringens delta toxin is sequence related to beta toxin, NetB, and Staphylococcus pore-forming toxins, but shows functional differences. *PLoS ONE* **2008**, *3*, e3764. [CrossRef]

7. Liang, Y.J.; Wang, C.Y.; Wang, I.A.; Chen, Y.W.; Li, L.T.; Lin, C.Y.; Ho, M.Y.; Chou, T.L.; Wang, Y.H.; Chiou, S.P.; et al. Interaction of glycosphingolipids GD3 and GD2 with growth factor receptors maintains breast cancer stem cell phenotype. *Oncotarget* **2017**, *8*, 47454–47473. [CrossRef]
8. Itokazu, Y.; Ariga, T.; Fuchigami, T.; Li, D. Gangliosides in neural stem cell fate determination and nerve cell specification—preparation and administration. *bioRxiv* **2024**, 598109. [CrossRef] [PubMed]
9. Martinez, C.; Hofmann, T.J.; Marino, R.; Dominici, M.; Horwitz, E.M. Human bone marrow mesenchymal stromal cells express the neural ganglioside GD2: A novel surface marker for the identification of MSCs. *Blood* **2007**, *109*, 4245–4248. [CrossRef]
10. Ryckman, A.E.; Brockhausen, I.; Walia, J.S. Metabolism of Glycosphingolipids and Their Role in the Pathophysiology of Lysosomal Storage Disorders. *Int. J. Mol. Sci.* **2020**, *21*, 6881. [CrossRef]
11. Tropak, M.B.; Yonekawa, S.; Karumuthil-Melethil, S.; Thompson, P.; Wakarchuk, W.; Gray, S.J.; Walia, J.S.; Mark, B.L.; Mahuran, D. Construction of a hybrid β-hexosaminidase subunit capable of forming stable homodimers that hydrolyze GM2 ganglioside in vivo. *Mol. Ther.—Methods Clin. Dev.* **2016**, *3*, 15057. [CrossRef]
12. Corrias, M.V.; Parodi, S.; Haupt, R.; Lacitignola, L.; Negri, F.; Sementa, A.R.; Dau, D.; Scuderi, F.; Carlini, B.; Bianchi, M.; et al. Detection of GD2-positive cells in bone marrow samples and survival of patients with localised neuroblastoma. *Br. J. Cancer* **2008**, *98*, 263–269. [CrossRef]
13. Yi, H.; Lin, Y.; Li, Y.; Guo, Y.; Yuan, L.; Mao, Y. Pan-Cancer Analysis of B4GALNT1 as a Potential Prognostic and Immunological Biomarker. *J. Immunol. Res.* **2022**, *2022*, 4355890. [CrossRef]
14. Yamashiro, S.; Ruan, S.; Furukawa, K.; Tai, T.; Lloyd, K.O.; Shiku, H.; Furukawa, K. Genetic and enzymatic basis for the differential expression of GM2 and GD2 gangliosides in human cancer cell lines. *Cancer Res.* **1993**, *53*, 5395–5400. [PubMed]
15. Chiricozzi, E.; Mauri, L.; Lunghi, G.; Di Biase, E.; Fazzari, M.; Maggioni, M.; Valsecchi, M.; Prioni, S.; Loberto, N.; Pomè, D.Y.; et al. Parkinson's disease recovery by GM1 oligosaccharide treatment in the B4galnt1$^{+/-}$ mouse model. *Sci. Rep.* **2019**, *9*, 19330. [CrossRef] [PubMed]
16. Latour, Y.L.; Yoon, R.; Thomas, S.E.; Grant, C.; Li, C.; Sena-Esteves, M.; Allende, M.L.; Proia, R.L.; Tifft, C.J. Human GLB1 knock-out cerebral organoids: A model system for testing AAV9-mediated GLB1 gene therapy for reducing GM1 ganglioside storage in GM1 gangliosidosis. *Mol. Genet. Metab. Rep.* **2019**, *21*, 100513.
17. Deschenes, N.M.; Cheng, C.; Ryckman, A.E.; Quinville, B.M.; Khanal, P.; Mitchell, M.; Chen, Z.; Sangrar, W.; Gray, S.J.; Walia, J.S. Biochemical Correction of GM2 Ganglioside Accumulation in AB-Variant GM2 Gangliosidosis. *Int. J. Mol. Sci.* **2023**, *24*, 9217. [CrossRef] [PubMed]
18. Vyas, M.; Deschenes, N.M.; Osmon, K.J.L.; Chen, Z.; Ahmad, I.; Kot, S.; Thompson, P.; Richmond, C.; Gray, S.J.; Walia, J.S. Efficacy of Adeno-Associated Virus Serotype 9-Mediated Gene Therapy for AB-Variant GM2 Gangliosidosis. *Int. J. Mol. Sci.* **2023**, *24*, 14611.
19. Coutinho, M.F.; Santos, J.I.; Alves, S. Less Is More: Substrate Reduction Therapy for Lysosomal Storage Disorders. *Int. J. Mol. Sci.* **2016**, *17*, 1065. [CrossRef]
20. Marshall, J.; Nietupski, J.B.; Park, H.; Cao, J.; Bangari, D.S.; Silvescu, C.; Wilper, T.; Randall, K.; Tietz, D.; Wang, B.; et al. Substrate Reduction Therapy for Sandhoff Disease through Inhibition of Glucosylceramide Synthase Activity. *Mol. Ther.* **2019**, *27*, 1495–1506. [CrossRef]
21. Cox, T.; Lachmann, R.; Hollak, C.; Aerts, J.; van Weely, S.; Hrebícek, M.; Platt, F.; Butters, T.; Dwek, R.; Moyses, C.; et al. Novel oral treatment of Gaucher's disease with N-butyldeoxynojirimycin (OGT 918) to decrease substrate biosynthesis. *Lancet* **2000**, *355*, 1481–1485. [CrossRef] [PubMed]
22. Jaskiewicz, E.; Zhu, G.; Bassi, R.; Darling, D.S.; Young, W.W., Jr. Beta1,4-N-acetylgalactosaminyltransferase (GM2 synthase) is released from Golgi membranes as a neuraminidase-sensitive, disulfide-bonded dimer by a cathepsin D-like protease. *J. Biol. Chem.* **1996**, *271*, 26395–26403. [CrossRef] [PubMed]
23. Breton, C.; Fournel-Gigleux, S.; Palcic, M.M. Recent structures, evolution and mechanisms of glycosyltransferases. *Curr. Opin. Struct. Biol.* **2012**, *22*, 540–549. [CrossRef]
24. Boukhris, A.; Schule, R.; Loureiro, J.L.; Lourenço, C.M.; Mundwiller, E.; Gonzalez, M.A.; Charles, P.; Gauthier, J.; Rekik, I.; Acosta Lebrigio, R.F.; et al. Alteration of ganglioside biosynthesis responsible for complex hereditary spastic paraplegia. *Am. J. Hum. Genet.* **2013**, *93*, 118–123. [CrossRef]
25. Harlalka, G.V.; Lehman, A.; Chioza, B.; Baple, E.L.; Maroofian, R.; Cross, H.; Sreekantan-Nair, A.; Priestman, D.A.; Al-Turki, S.; McEntagart, M.E.; et al. Mutations in B4GALNT1 (GM2 synthase) underlie a new disorder of ganglioside biosynthesis. *Brain* **2013**, *136 Pt 12*, 3618–3624. [CrossRef] [PubMed]
26. Alecu, J.E.; Ohmi, Y.; Bhuiyan, R.H.; Inamori, K.I.; Nitta, T.; Saffari, A.; Jumo, H.; Ziegler, M.; de Gusmao, C.M.; Sharma, N.; et al. Functional validation of novel variants in B4GALNT1 associated with early-onset complex hereditary spastic paraplegia with impaired ganglioside synthesis. *Am. J. Med. Genet.* **2022**, *188*, 2590–2598. [CrossRef]

27. Gil-Tommee, C.; Vidal-Martinez, G.; Annette Reyes, C.; Vargas-Medrano, J.; Herrera, G.V.; Martin, S.M.; Chaparro, S.A.; Perez, R.G. Parkinsonian GM2 synthase knock-out mice lacking mature gangliosides develop urinary dysfunction and neurogenic bladder. *Exp. Neurol.* **2019**, *311*, 265–273. [CrossRef]
28. Yamaguchi, T.; Yamauchi, Y.; Furukawa, K.; Ohmi, Y.; Ohkawa, Y.; Zhang, Q.; Okajima, T.; Furukawa, K. Expression of B4GALNT1, an essential glycosyltransferase for the synthesis of complex gangliosides, suppresses BACE1 degradation and modulates APP processing. *Sci. Rep.* **2016**, *6*, 34505. [CrossRef]
29. Dodge, J.C.; Tamsett, T.J.; Treleaven, C.M.; Taksir, T.V.; Piepenhagen, P.; Sardi, S.P.; Cheng, S.H.; Shihabuddin, L.S. Glucosylceramide synthase inhibition reduces ganglioside GM3 accumulation, alleviates amyloid neuropathology, and stabilizes remote contextual memory in a mouse model of Alzheimer's disease. *Alzheimer's Res. Ther.* **2022**, *14*, 19. [CrossRef]
30. Blumenreich, S.; Yaacobi, C.; Vardi, A.; Barav, O.B.; Vitner, E.B.; Park, H.; Wang, B.; Cheng, S.H.; Sardi, S.P.; Futerman, A.H. Substrate reduction therapy using Genz-667161 reduces levels of pathogenic components in a mouse model of neuronopathic forms of Gaucher disease. *J. Neurochem.* **2021**, *156*, 692–701. [CrossRef] [PubMed]
31. Pacuszka, T.; Duffard, R.O.; Nishimura, R.N.; Brady, R.O.; Fishman, P.H. Biosynthesis of bovine thyroid gangliosides. *J. Biol. Chem.* **1978**, *253*, 5839–5846. [CrossRef]
32. Senn, H.J.; Cooper, C.; Warnke, P.C.; Wagner, M.; Decker, K. Ganglioside biosynthesis in rat liver. Characterization of UDP-N-acetylgalactosamine—GM3 acetylgalactosaminyltransferase. *Eur. J. Biochem.* **1981**, *120*, 59–67. [CrossRef]
33. Yamashiro, S.; Haraguchi, M.; Furukawa, K.; Takamiya, K.; Akihito Yamamoto, A. Substrate specificity of β1,4-N-Acetylgalactosaminyltransferase in Vitro and in cDNA-transfected Cells. *J. Biol. Chem.* **1995**, *270*, 6149–6155. [CrossRef]
34. Hidari, J.K.; Ichikawa, S.; Furukawa, K.; Yamasaki, M.; Hirabayashi, Y. Beta 1-4N-acetylgalactosaminyltransferase can synthesize both asialoglycosphingolipid GM2 and glycosphingolipid GM2 in vitro and in vivo: Isolation and characterization of a beta 1-4N-acetylgalactosaminyltransferase cDNA clone from rat ascites hepatoma cell line AH7974F. *Biochem. J.* **1994**, *303 Pt 3*, 957–965.
35. Takamiya, K.; Yamamoto, A.; Furukawa, K.; Yamashiro, S.; Shin, M.; Okada, M.; Fukumoto, S.; Haraguchi, M.; Takeda, N.; Fujimura, K.; et al. Mice with disrupted GM2/GD2 synthase gene lack complex gangliosides but exhibit only subtle defects in their nervous system. *Proc. Natl. Acad. Sci. USA* **1996**, *93*, 10662–10667. [CrossRef]
36. Li, J.; Yen, T.Y.; Allende, M.L.; Joshi, R.K.; Cai, J.; Pierce, W.M.; Jaskiewicz, E.; Darling, D.S.; Macher, B.A.; Young, W.W., Jr. Disulfide bonds of GM2 synthase homodimers. Antiparallel orientation of the catalytic domains. *J. Biol. Chem.* **2000**, *275*, 41476–41486. [CrossRef] [PubMed]
37. Welland, J.W.J.; Barrow, H.G.; Stansfeld, P.J.; Deane, J.E. Conformational dynamics and membrane insertion mechanism of B4GALNT1 in ganglioside synthesis. *Nat. Commun.* **2025**, *16*, 5442. [CrossRef] [PubMed]
38. Li, J.; Rancour, D.M.; Allende, M.L.; Worth, C.A.; Darling, D.S. The DXD motif is required for GM2 synthase activity but is not critical for nucleotide binding. *Glycobiology* **2001**, *11*, 217–229. [CrossRef] [PubMed]
39. Inamori, K.I.; Nakamura, K.; Shishido, F.; Hsu, J.C.; Nagafuku, M.; Nitta, T.; Ikeda, J.; Yoshimura, H.; Kodaira, M.; Tsuchida, N.; et al. Functional evaluation of novel variants of B4GALNT1 in a patient with hereditary spastic paraplegia and the general population. *Front. Neurosci.* **2024**, *18*, 1437668. [CrossRef]
40. Haraguchi, M.; Yamashiro, S.; Furukawa, K.; Takamiya, K.; Shiku, H.; Furukawa, K. The effects of the site-directed removal of N-glycosylation sites from beta-1,4-N-acetylgalactosaminyltransferase on its function. *Biochem. J.* **1995**, *312 Pt 1*, 273–280. [CrossRef]
41. Brockhausen, I.; Dowler, T.; Paulsen, H. Site directed processing: Role of amino acid sequences and glycosylation of acceptor glycopeptides in the assembly of extended mucin type O-glycan core 2. Biochim. *Biophys. Acta* **2009**, *1790*, 1244–1257. [CrossRef] [PubMed]
42. Yang, X.; Tao, S.; Orlando, R.; Brockhausen, I.; Kan, F.W. Structures and biosynthesis of the N- and O-glycans of recombinant human oviduct-specific glycoprotein expressed in human embryonic kidney cells. *Carbohydr. Res.* **2012**, *358*, 47–55. [CrossRef]
43. Gao, Y.; Vlahakis, J.; Szarek, W.; Brockhausen, I. Selective inhibition of glycosyltransferases by bivalent imidazolium salts. *Bioorg. Med. Chem.* **2013**, *21*, 1305–1311. [CrossRef] [PubMed]
44. Brockhausen, I.; Benn, M.; Bhat, J.G.; Marone, S.; Riley, J.G.; Montoya-Peleaz, P.; Vlahakis, J.Z.; Paulsen, H.; Schutzbach, J.S.; Szarek, W.A. UDP-Gal: GlcNAc-R beta1, 4-galactosyltransferase–a target enzyme for drug design. Acceptor specificity and inhibition of the enzyme. *Glycoconj. J.* **2006**, *23*, 525–541. [CrossRef]
45. Gao, Y.; Lazar, C.; Szarek, W.A.; Brockhausen, I. Specificity of β1,4-galactosyltransferase inhibition by 2-naphthyl 2-butanamido-2-deoxy-1-thio-β-D-glucopyranoside. *Glycoconj. J.* **2010**, *27*, 673–684. [CrossRef]
46. Kocev, A.; Melamed, J.; Wang, S.; Kong, X.; Vlahakis, J.Z.; Xu, Y.; Szarek, W.A.; Brockhausen, I. Inhibition of bacterial growth and galactosyltransferase activity of WbwC by α, ω-bis(3-alkyl-1H-imidazolium)alkane salts: Effect of varying carbon content. *Bioorg. Med. Chem.* **2020**, *28*, 115494. [CrossRef]

47. Prabhala, S.V.; Marshall, B.; Galiardi, J.; Fan, Y.; Creamer, E.; Wood, D.W. Highly selective split intein method for efficient separation and purification of recombinant therapeutic proteins from mammalian cell culture fluid. *J. Chromatogr.* **2024**, *1736*, 465430. [CrossRef]
48. Moremen, K.W.; Ramiah, A.; Stuart, M.; Steel, J.; Meng, L.; Forouhar, F.; Moniz, H.A.; Gahlay, G.; Gao, Z.; Chapla, D.; et al. Expression system for structural and functional studies of human glycosylation enzymes. *Nat. Chem. Biol.* **2018**, *14*, 156–162. [CrossRef]
49. Babulic, J.L.; Capicciotti, C.J. Exo-Enzymatic Cell-Surface Glycan Labeling for Capturing Glycan-Protein Interactions through Photo-Cross-Linking. *Bioconjugate Chem.* **2022**, *33*, 773–780. [CrossRef] [PubMed]

molecules

MDPI

*Article*

# Synthesis and Characterization of Glyco-SAMs on Gold Nanoparticles: A Modular Approach Towards Glycan-Based Recognition Studies

**Sebastian Kopitzki † and Joachim Thiem ***

Department of Chemistry, Institute of Organic Chemistry, University of Hamburg, Martin-Luther-King-Platz 6, D-20146 Hamburg, Germany; sebastiankopitzki@web.de
* Correspondence: joachim.thiem@uni-hamburg.de
† Current address: Biosynth GmbH, Robert-Rössle-Straße 10, 13125 Berlin, Germany.

**Abstract**

Within recent years, the interest in tools to investigate carbohydrate-protein (CPI) and carbohydrate-carbohydrate interactions (CCI) has increased significantly. For the investigation of CPI and CCI, several techniques employing different linking methods are available. For mimicking the glycocalyx self-assembled monolayer (SAM) formation of carbohydrate derivatives on gold nanoparticles is most appropriate. In contrast to methods for glyco-SAM formation used previously to analyze CPI/CCI, the novel approach allows for a facile and rapid synthesis to link spacers and carbohydrate derivatives and enhances the binding event by controlling the amount and orientation of ligand. For immobilization on biorepulsive aminooxy functionalized gold nanoparticles by oxime coupling, diverse aldehyde-functionalized glycan structures of mono-, di-, and complex trisaccharides were synthesized, employing several facile steps including olefin metathesis. Effective immobilization and first binding studies are presented for the lectin concanavalin A. This novel and advantageous immobilization method is presently employed in various biomimetic studies of carbohydrates and carbohydrate-based array development for diagnostics and screening.

**Keywords:** carbohydrates; synthesis; self-assembly; monolayers; molecular recognition

## 1. Introduction

Eukaryotic cell membranes are coated by a so-called "glycocalix" consisting of a complex mixture of glycoproteins, glycolipids, complex oligosaccharides, glycoconjugates, and proteoglycans with a size up to 100 nm. Cellular processes, such as bacterial and viral infection, cancer metastasis, modulation and activation of the immune system, tissue differentiation and development, and further intercellular recognition events are largely controlled on the molecular level by these glycoconjugates [1–3]. Cellular recognition events involving carbohydrate derivatives are carbohydrate-protein-interactions (CPI) between sugars and lectins [4] or selectins [5], as well as carbohydrate-carbohydrate-interactions (CCI) [6]. Whereas antigen-antibody interactions range from $10^{-8}$ to $10^{-12}$ M, CPI and CCI are typically weak in in vitro testing, with $K_D$ values in the millimolar or high micromolar range [7–9]. On the other hand, multivalent presentation of carbohydrate recognition units can increase binding affinity considerably [10,11]. Despite intensive studies of the glycocalyx by glycobiology, to date, there is no general mechanistic concept of carbohydrate recognition [12–17].

In order to facilitate studies of carbohydrate-based recognition events, new analytical and synthetic approaches are of interest. These include chemoenzymatic and automated solid phase syntheses of oligosaccharides, and glycomimetics as well as chemical tools, microarrays, glyconanoparticle technology, and molecular modeling [18–21]. Particular advantageous for mimicking the glycocalyx are the self-assembly of thiol-functionalized molecules on gold surfaces, first introduced by Whitesides et al. [22]. The conception of glyco-SAMs proved to be particularly well suited for the investigation of molecular interactions of carbohydrates, allowing characterization as well as control of density and orientation of carbohydrate ligands [23,24]. Furthermore, advantageous studies of glyco-SAMs on gold employed spectroscopic techniques like surface plasmon resonance, ellipsometry, atomic force microscopy, or X-ray photoelectron spectroscopy [25–27].

Gold nanoparticles represent important carrier and delivery materials for biological active compounds in nanomedicine, since they provide both multivalency and multifunctionality. Therefore, their preparation, toxicity, and application for in vivo imaging, diagnosis, and therapy are intensively discussed [28]. Colloidal gold was used for therapeutic and cosmetic properties at an early stage by the Chinese and Egyptians. The first scientific publication about gold nanoparticles (AuNP) by Faraday in 1857 reported and discussed the reason for the intensive red colour of colloidal gold [29]. Employing Maxwell's electromagnetic equation, the visual light absorption of AuNP was described by Mie in 1908 [30]. Nanoscaled AuNPs are typically available in sizes from 1 nm to 120 nm, in a size range of proteins and other biomacromolecules, and show plasmon band absorption. Further, in many other physical properties, the colloidal form differs from the metallic bulk form of gold. These changed properties of AuNP are of interest for optics, catalysis, and material science, but also for biology, biochemistry, and medicine [31]. There are reports about the functionalization of AuNP with proteins, DNA, and RNA [32,33], and increasingly examples in the carbohydrates field are published with AuNP as ideal platforms to mimic the glycocalyx of cell membranes [34–37].

Au-NP are predominantly synthesized by reduction procedures from so-called "gold acid" ($HAuCl_4$). The Turkevitch protocol, introduced already in 1951, is widely used for the preparation of various nanoparticle constructs [38]. Here, an aqueous solution of the $Au^{III}$ salt is reduced by sodium citrate to obtain $Au^0$ colloids [39]. Subsequent ligand exchange reaction with thiols or other thio derivatives on the citrate-stabilised Au-NP provides the individual functionalized Au-NP [40,41]. This ligand exchange results in an elevated stability of Au-NP due to the strength of the Au-S bond (ca. 50 kcal/mol) [42]. Recent developments of the Turkevitch protocol improved the control of size and distribution, thus nowadays monodisperse Au-NP with a controllable size of 9–120 nm can be synthesized [43]. Schiffrin and Brust presented a one-step procedure for thiolated Au-NP. In a phase transfer system of water and toluene, $HAuCl_4$ was reduced with $NaBH_4$ in the present of alkylthiols and tetrabutylammonium bromide [44]. The sizes of the Au-NP strongly dependant on the utilized alkylthiol, however, this method allows AuNP smaller than 9 nm to be synthesized. To obtain uniform Au-NP with different functionalization, it is necessary to apply a second synthetic step, mostly an organochemical reaction or a thiol ligand exchange according to [45]

$$(RS)_n\text{AuNP} + m\,\text{R'SH} \rightarrow (RS)_{n\text{-}m}(\text{R'S})_m\text{AuNP} + n\,\text{RSH}$$

Carbohydrate—functionalized Au-NP were utilized by Penades et al., whose syntheses followed the Brust protocol [46]. Jensen et al. employed syntheses by a modular conception with thiol ligand exchange on citrate-stabilized Au-NP [47]. This modular carbohydrate functionalization of Au-NP was advantageously used throughout this study since citrate-stabilized Au-NP had only to be prepared once and thus allowed for access to many different Au-NP having uniform material, which improves the interpretation of the SPR experiments.

The novel modular approach for immobilization of synthetic glycosides on gold nanoparticle reported herein is based on (1) the control of particles size by bulk preparation of goldnanoparticles by the Tuchevitch protocol, (2) control of ligand density by non-specific binding via tetraethylene glycol tethers, (3) the control of ligand density via tether length, provision of functionalization by preparation of mixed SAMs in different ratio, (4) a facile in situ immobilization by thioalkanes on gold as SAMs and (5) a facile in situ attachment of glycoderivates by reductive amination (Figure 1).

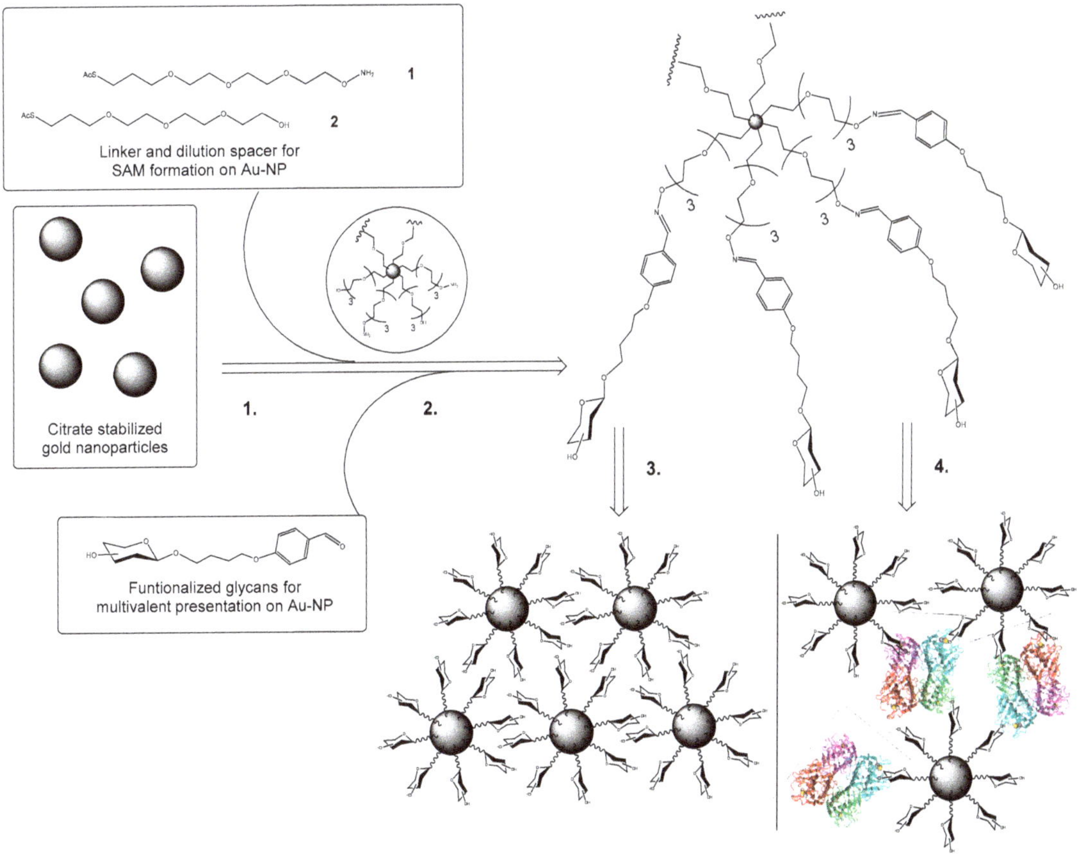


**Figure 1.** Schematic depiction of the modular approach for the preparation of glyco-SAMs on gold nanoparticles: 1. Pre-synthesized, well-defined gold nanoparticles are used for SAM formation with spacer molecules **1** (attachment) and **2** (dilution), 2. Attachment of carbohydrates and application either 3. in carbohydrate-carbohydrate interactions or 4. in carbohydrate-protein interactions.

For control of ligand interdistance, benzaldehyde-functionalized glycoderivatives were employed, in which the carbohydrate head groups were attached to amine-bearing SAMs and diluted with biorepulsive spacers. These biorepulsive compounds essentially require a long alkane chain and a terminal oligoethylene glycol moiety for self-assembly. The alkane chain is needed to form strong van der Waals interactions between spacers, to promote accurate SAM formation [48], and the oligoethylene glycol moiety is employed to exclude non-specific adhesion of biomaterial to the SAM [49]. In addition, by insertion of the amino function, an anchor group is provided.

Whereas conventional glycosylations usually lead to anomeric mixtures and thus require tedious chromatographic separations, the present approach for covalent immobilization via reductive amination of the aldehyde-linking partner facily gives uniform components. Thus, the aldehyde functions are easily introduced to anomerically pure allyl glycosides by cross-metathesis [50,51]. In contrast to ring closing metathesis (RCM) and ring opening metathesis polymerization, cross metathesis (CM), especially in carbohydrate chemistry, is less developed [52]. Based on recent reports, which solved the problems of self-metathesis and could also use aqueous CM [53,54], olefin metathesis was applied to modify carbohydrate derivatives. In fact, synthesis of glyco-SAMs could be successfully accomplished by incubation of plain gold sensor surfaces with amino-functionalized and biorepulsive spacers, following subsequent attachment of glycoderivatives by reductive amination. Evidence of effective immobilization of carbohydrate components could be demonstrated via initial binding experiments of the lectin concanavalin A to modified SAMs carrying α-mannopyranoside structures.

For the present study eighteen benzaldehyde-functionalized glycan structures were employed. These comprise five monosaccharide derivatives **3–9**, four disaccharide derivatives **10–13**, and seven trisaccharide derivatives **14–20** as depicted in Figure 2 (Figure 2).

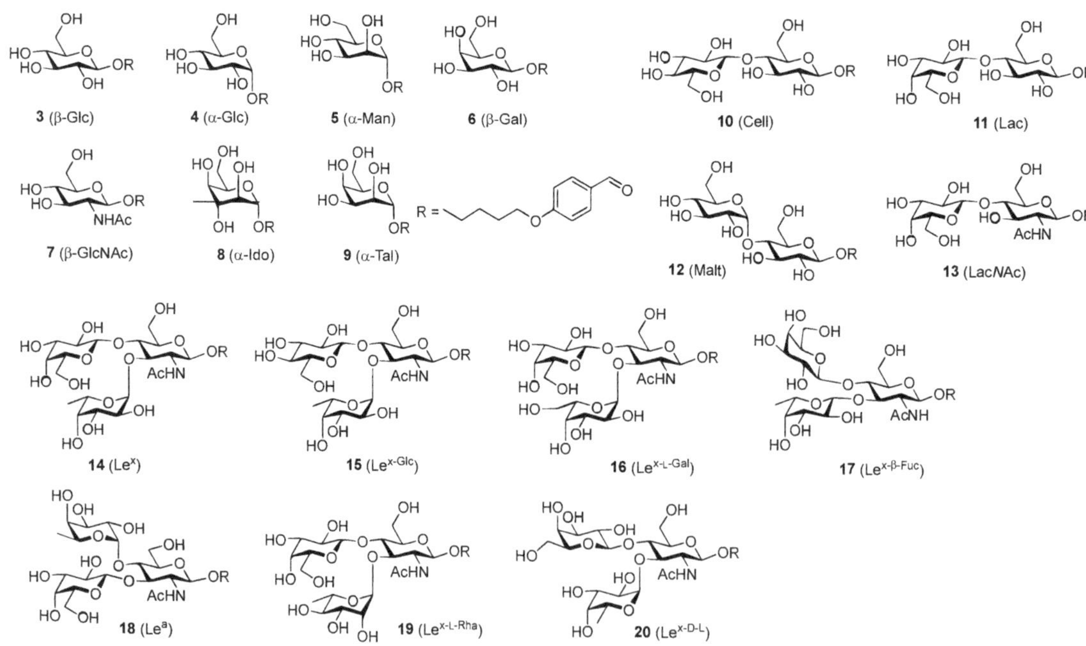


**Figure 2.** Benzaldehyde functionalized glycosides used for the oxime formation on gold nanoparticles. Abbreviated names of carbohydrate head groups: **14** ($Le^x$): Galα(1–4)[Fucα(1–3)]Glc*N*Ac; **15** ($Le^{x\text{-}Glc}$): Glcβ(1–4)[Fuc56(1–3)]Glc*N*Ac; **16** ($Le^{x\text{-}L\text{-}Gal}$): Galβ(1–4)[L-Galα(1–3)]Glc*N*Ac; **17** ($Le^{x\text{-}\beta\text{-}Fuc}$): Galβ(1–4)[Fucβ(1–3)]Glc*N*Ac; **18** ($Le^a$): Galβ(1–3)[Fucα(1–4)]Glc*N*Ac; **19** ($Le^{x\text{-}L\text{-}Rha}$): Galβ(1–4)[Rhaα(1–3)]Glc*N*Ac; **20** ($Le^{x\text{-}D\text{-}L}$): L-Galβ(1–4)[D-Fucα(1–3)]Glc*N*Ac.

## 2. Results and Discussion

### 2.1. Synthesis of Benzaldehyde-Functionalized Carbohydrates

According to previously developed procedures [55] for attachment of glycoderivatives to amino-functionalized SAMs, readily accessible allyl glycosides were chosen as starting components. Thus, allyl 2,3,4,6-tetra-*O*-acetyl-α-D-glucopyranoside (**21**) was treated with para-(allyloxy)benzaldehyde dimethyl acetal (**22**) [55] for olefin metathesis under catalysis with Grubbs-Hoveyda second-generation catalyst **23** [56] to give compound **24** in 91% yield. Employing optimized reaction conditions [57], cross metathesis led exclusively to the (*E*)-configured derivative. Next, deacetalization of the benzaldehyde dimethyl acetal **24** followed by hydrogenation catalyzed employing palladium on charcoal poisoned with diphenyl sulfide and subsequent classical Zemplén deprotection gave the target compound **4** in 91% yield over three steps (Scheme 1).

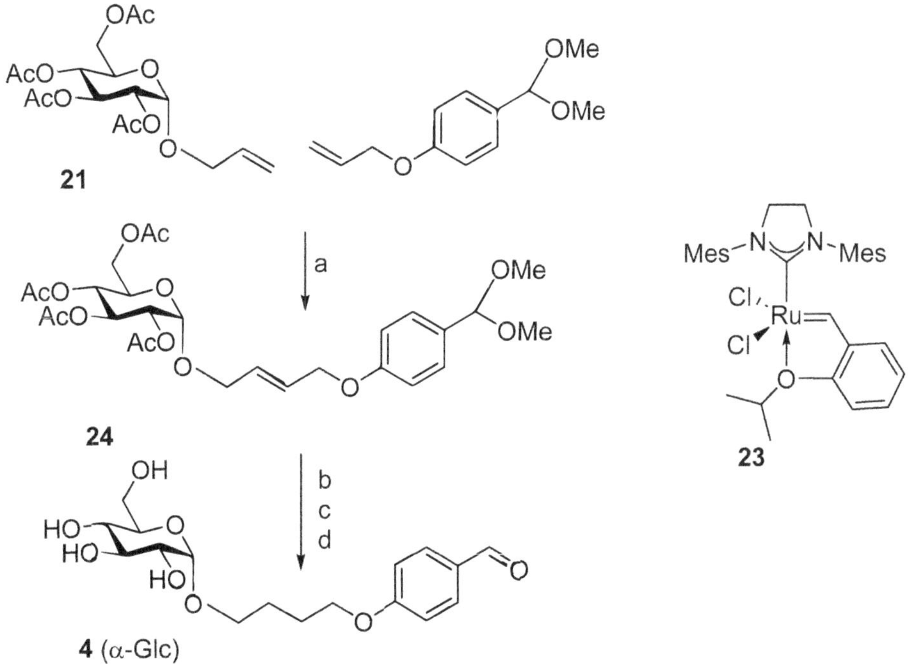

**Scheme 1.** Synthesis of aldehyde-functionalized α-D-glucose. Reagents and conditions: (a) catalyst **23** (10 mol%), **22** (5 eq), 40 °C, 3 h, 91%; (b) $H_2$/Pd, PhSPh, EtOAc, RT, 12 h, quant; (c) TFA/$H_2O$/THF (0.1/9.9/90 *v*/*v*), RT, 6 h, quant.; (d) NaOMe, MeOH, RT, 6 h, 91% (over 3 steps).

Starting with the α-trichloro acetimidate of peracetylated maltose (**25**) [58], a straightforward glycosylation with allyl alcohol gave the crystalline β-allyl derivative **26** in 83% yield. As before, cross metathesis with **22** under catalysis of **23** resulted in the formation of the (*E*)-component **27** in 76% yield. Finally, employing the above reported three step deprotection method gave the crystalline benzaldehyde-functionalized β-maltose derivative 12 in 77% yield (Scheme 2).

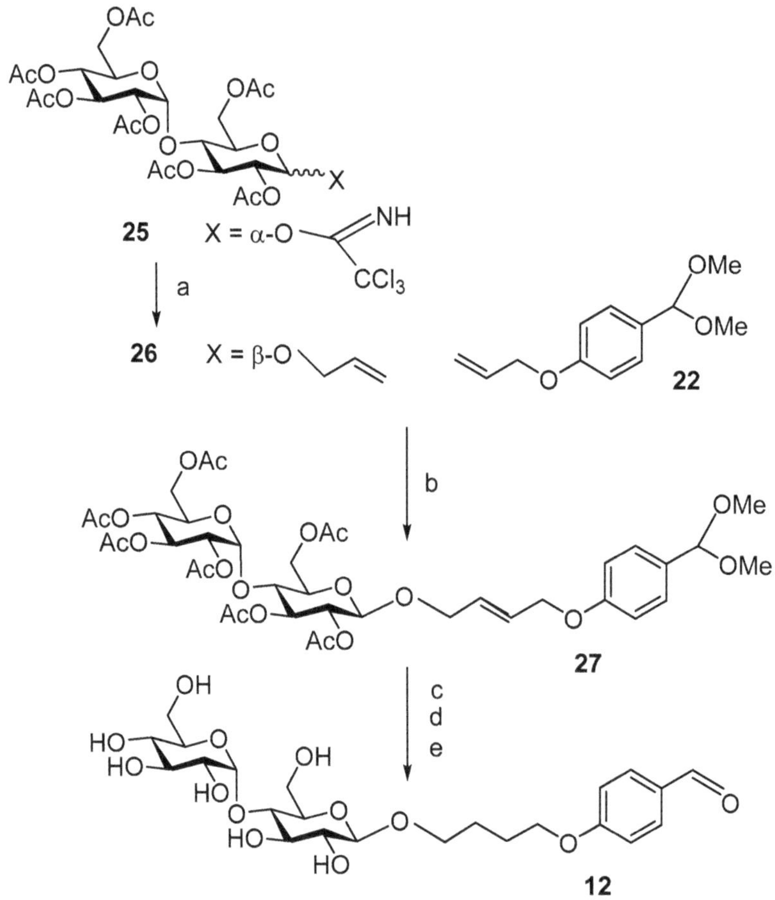

**Scheme 2.** Synthesis of aldehyde-functionalized maltose. Reagents and conditions: (a) AllOH, TSMOTf, 0 °C, 2 h, 83%; (b) catalyst **23** (10 mol%), **22** (5 eq), 40 °C, 3 h, 76%; (c) TFA/$H_2O$/THF (0.1/9.9/90 *v*/*v*), RT, 6 h; (d) $H_2$/Pd, PhSPh, EtOAc, RT, 12 h; (e) NaOMe, MeOH, RT, 6 h, 77% over 3 steps).

### 2.2. Synthesis of Linker and Dilution Spacers

For the preparation of carbohydrate-functionalized nanoparticles with various ligand densities, both an anchor linker and a dilution spacer are required. Oligoethylene glycol spacers are known to suppress nonspecific binding of biomaterials to surfaces. Therefore, aminoxy tetraethylene glycol linker **1** and the tetraethylene glycol spacer **2** were synthesized by a combination of previously published procedures.

Mono-*O*-allylation of tetraethylene glycol **28** was realized with NaH and allyl bromide in 90% yield to give compound **29**, the radical thioacetylation of which gave the required dilution spacer component **2** quantitatively. By the Mitsunobu reaction of **29** with *N*-hydroxy phthalimide compound **30** was obtained in moderate yield. Its hydrazinolysis gave **31**, which again, by radical thioacetylation, led to the quantitative formation of the anchor spacer component **1** (Scheme 3).

For functionalization of saccharides, tetraethylene glycol **28** was treated with *N*-hydroxy phthalimide in a Mitsunobu reaction, employing polymer-bound triphenyl phosphine (PS-$Ph_3P$) to facilitate removal of phosphine oxide. The monosubstituted product **32** was obtained in 37%, corresponding to a statistically adjusted [59] yield of 74%. Under corresponding conditions, the next step used again the Mitsunobu method to give the protected thiol **33** (previously reported by another approach [60]) in 78% yield. Hydrazinolysis gave the aminoxy linker **34** in 97% yield, and for exemplary check, this could be coupled with the β-glucopyranoside component **3** to give the oxime **35** quantitatively in an (*E*/*Z*) ratio of 10:1 (Scheme 4).

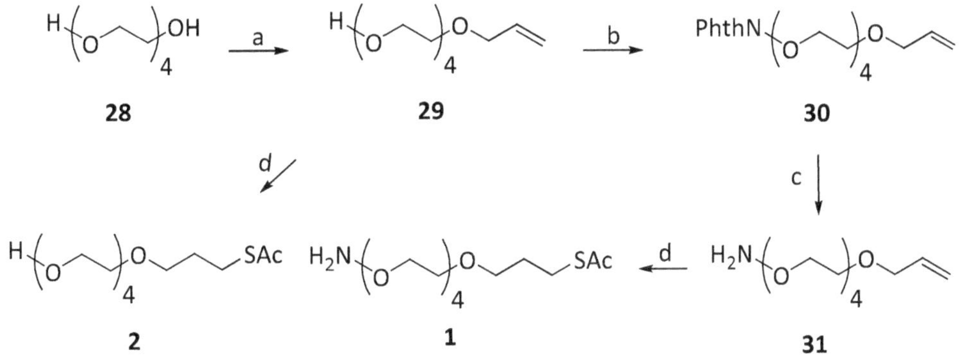

**Scheme 3.** Synthesis of dilution and anchor molecules **1** and **2** for functionalization of Au-NP. Reagents and conditions: (a) NaH, DMF, AllBr, 0°to 20 °C, 12 h, 90%; (b) *N*-hydroxyphthalimide, PS-$PPh_3$, DIAD, THF, 20 °C, 16 h, 56%; (c) hydrazine hydrate, MeCN, 20 °C, 2 h, 75%; (d) AcSH, AIBN, THF, hν, 3 h; **1**: quant., **2**: quant.

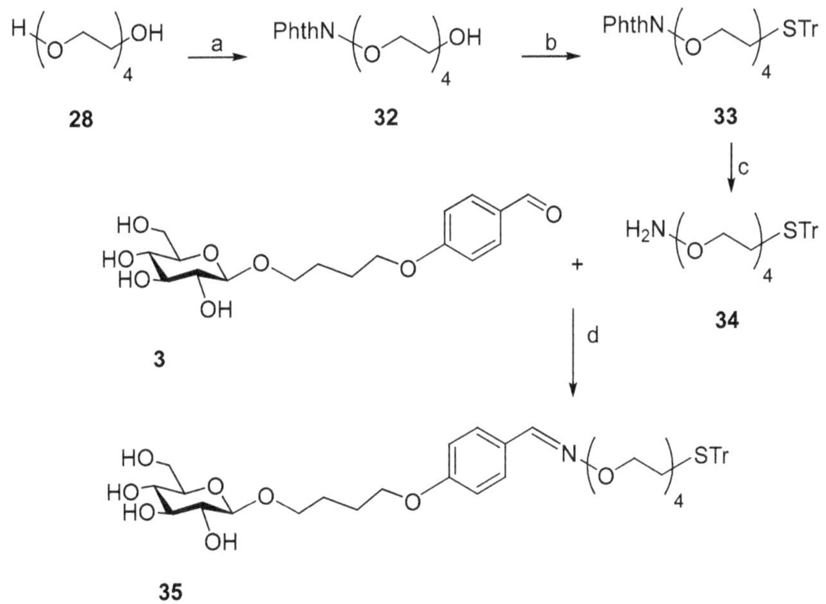

**Scheme 4.** Functionalization of glucose. Reagents and conditions: (a) *N*-hydroxy-phthalimide, PS-$PPh_3$, DIAD, THF, 20 °C, 16 h, 37%; (b) TrSH, PS-$PPh_3$, DIAD, THF, 0° to 20 °C, 16 h, 78%; (c) hydrazine hydrate, MeCN, 20 °C, 2 h, 97%; (d) **3** (1.1 eq), MeCN, $H_2O$, AcOH, 16 h, quant.

Prior to assembling glycoconjugates on aminooxy functionalized gold nanoparticles, the coupling system was tested in solution. As proven in the synthesis of compound **35** their structures could be assumed to be preferentially (*E*)-oximes, which is relevant for CPI and CCI studies with uniform materials (Figure 3).

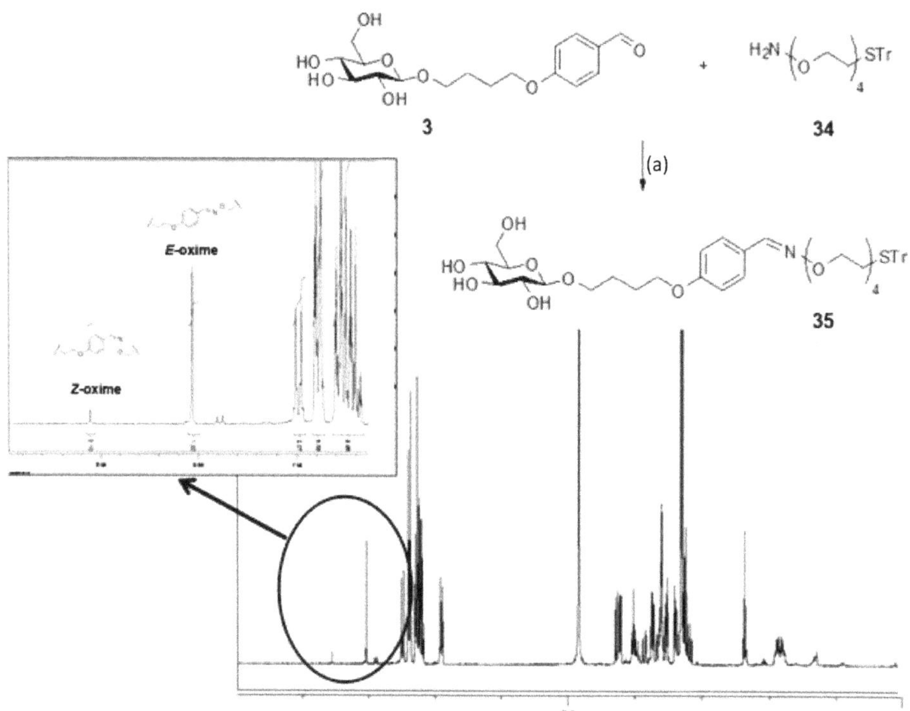


**Figure 3.** Solution test ligation. Benzaldehyde functionalized glucoside **3** was successfully coupled to solution attachment spacer **34** to give component 35. NMR spectra show that (*E*)-oxime was preferentially formed (*E*/*Z* ratio 10:1). Reagents and conditions: (a) 3 (1.1 eq), MeCN, $H_2O$, AcOH, 16 h, quant.

### 2.3. *Synthesis of Glyconanoparticles*

For the binding of free oligosaccharides at their reductive terminal, Jensen et al. [55] employed amino-oxy tetraethylene glycol **34** to give glycoconjugates, which in turn, by their thiol group, formed functionalized Au-NP useful for studies of biological recognition processes. A corresponding approach was to be applied with the novel benzaldehyde-functionalized carbohydrate structures [61,62].

However, to ensure almost homogeneous particle core size for all functionalized gold nanoparticles, a two-step protocol was applied. First, gold nanoparticle formation follows the Turkevitch protocol with subsequent SAM formation using an anchor and dilution spacer. After refluxing $HAuCl_4$ in degassed water for 10 min, a 60 °C warm aqueous solution of sodium citrate was added [63] to give Au-NP with diameters of about 13 nm [64,65].

In this work, the functionalization of AuNP by SAM formation was performed according to a protocol by Jensen et al. [66–68]. Four types of AuNP (AuNP-1–AuNP-4) were synthesized, in which aqueous solutions of citrate-stabilized AuNP were incubated with methanolic solutions of anchor spacer **1** and diluent spacer **2** in different ratios for 16 h (Table 1).

**Table 1.** Synthesis of aminooxy-terminated core-shell gold nanoparticles [a].

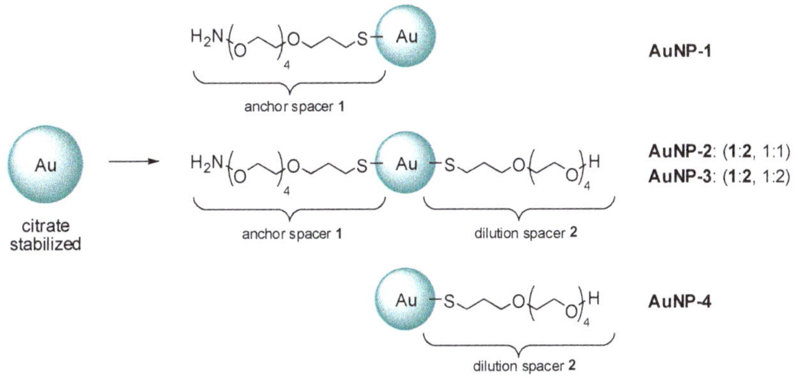


| Entry | Ratio 1:2 | Product | $\Delta\lambda_{SP,max}$ [b] |
|---|---|---|---|
| 1 | 1:0 | **AuNP-1** | ~4 nm |
| 2 | 1:1 | **AuNP-2** | ~3 nm |
| 3 | 1:2 | **AuNP-3** | ~3 nm |
| 4 | 0:1 | **AuNP-4** | ~6 nm |

[a] Au-NP synthesis: $HAuCl_4$, sodium citrate, $H_2O$, 60 °C, 2 h; SAM formation by incubation of aqueous Au-NP solution with methanolic solutions of thioacetates **1** and **2** in different ratios for 16 h. [b] Changes in absorption maxima of surface plasmon band $\lambda_{SP,max}$ occur by SAM formation on gold nanoparticles.

The functionalized Au-NP were transformed by formation of oximes with benzaldehyde glycoconjugates (Figure 2) employing seven monosaccharide derivatives **3**, **5–6** [55], **4**, **8–9** [69], four disaccharide derivatives **10–11**, **13** [55], **12**, and seven trisaccharide derivatives **14–20** [70] into glyconano particles **GNP-1–GNP-19**. Incubation of the Au-NP solution was at pH 4.7 and 40 °C with the corresponding solution of the glycan for 16 h [67].

### *2.4. Characterization of GNPs*

Characterization of GNPs was carried out by TEM and quantification of carbohydrate ligands on their surface. The size of the AuNP of 13 nm determined by UV-Vis spectroscopy was confirmed by TEM images and their evaluation. The measurement of a total of 168 particles resulted in a diameter of 13.3 ± 1.3 nm (Figure 4A). Furthermore, in the case of aminooxy-terminated core-shell gold nanoparticles AuNP-1 and mannose GNP (GNP-3), successful functionalizations were confirmed by TEM images (Figure 4B,C). As expected, the SPR absorption of AuNPs changes after SAM formation employing anchor and/or dilution spacers, and that can also be observed visually (Figure 4D). Both the outside appearance as well as the position of their SPR absorption maximum were identical before and after oxime formation.

Quite accurate quantification of carbohydrate ligands per GNP can be performed employing gas chromatography [71] or elemental analysis [72]. However, these methods require tenfold amounts of material in contrast to colorimetric tests [73], employed for determining the number of carbohydrate ligands per GNP [74,75]. Thus, measuring concentrations of carbohydrates by color reactions and subsequent quantification by UV-VIS spectroscopy was of interest. Sulfuric solutions of anthrone and sugars give green products with an adsorption maximum at about 620 nm [76,77], sufficiently distant from the SPR maximum of the AUNP to exclude any interactions during UV-VIS measurements. The number of carbohydrate ligands per GNP results from the quantity of $n_{\text{sugar}}$ per quantity of $n_{\text{GNP}}$. The amount of carbohydrate in mg per mg of GNP was determined by an anthrone test. The extinction of the color solution was linear proportional to the amount of sugar as determined by balancing via a linear slope. Detailed information on test procedures and the resulting calculations is described in Section 4, and the resulting number of carbohydrate ligands per GNP is listed in Table 2.

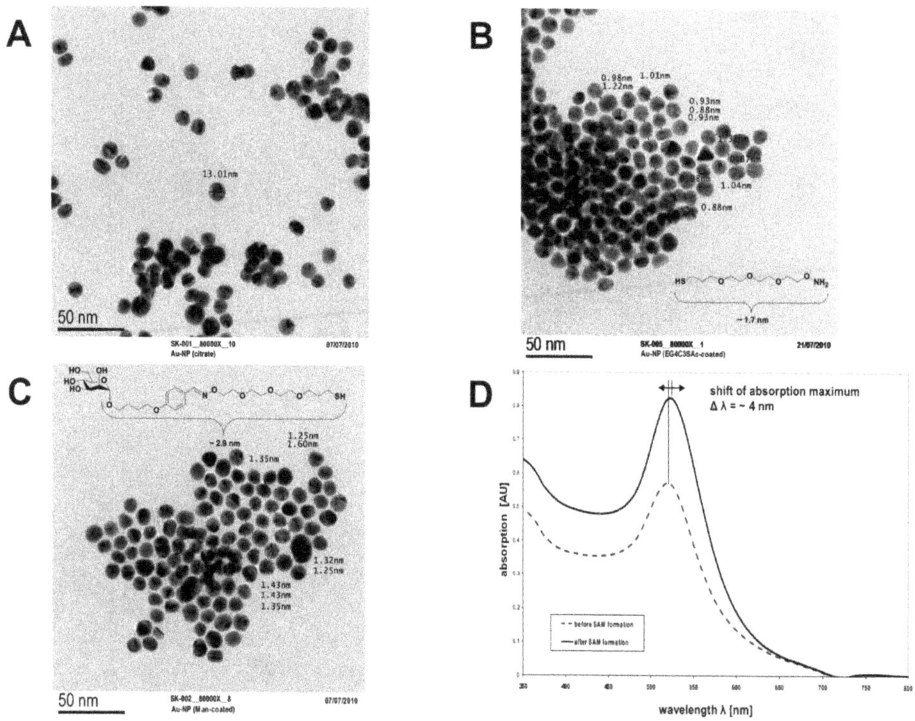


**Figure 4.** Characterization of glyconanoparticles. (**A**) TEM image of citrate-stabilized gold nanoparticles, average diameter ~13 nm; (**B**) TEM image of **AuNP-1**, average inter particle distance ~1 nm; (**C**) TEM image of **GNP-3**, average inter particle distance ~1.4 nm; (**D**) UV-Vis spectra of gold nanoparticles before ($\Delta\lambda_{SP,max}$ = 519 nm) and after ($\Delta\lambda_{SP,max}$ = 523 nm) SAM formation with anchor spacer **1**.

By anthrone tests of the P$^k$ antigen trisaccharide [Galα(1–4)Gal(1–3)Glc] attached to AuNP of 13 nm size, Chien et al. observed 820–1300 ligands attached by short or long spacers, respectively [75]. The amount of ligands found in our case is in the same order of magnitude. Employing quantification by the *trim*-(*trim*ellitoyl) system, Jensen et al. measured 440 ligands on AuNP of 13 nm [67]. Lin et al. used AuNP of 20 nm size and established about 650 mannose-C5- and about 880 mannose tetraethylene glycol spacered ligands [78]. The GNPs formed in our experiments clearly showed more replicants per GNP for smaller carbohydrate ligands: that is, for monosaccharide-functionalized GNPs (**GNP-1** to **GNP-6**), there are 1200–1280 ligands per GNP, for disaccharide-functionalized GNPs (**GNP-7** to

**GNP-10**), there are 980–1060 ligands per GNP, and for trisaccharide-functionalized GNPs (**GNP-11**, **GNP-14** to **GNP-19**), there are only 880–920 ligands per GNP.

**Table 2.** Synthesis of aminooxy-terminated core-shell gold-glyconanoparticles [a].

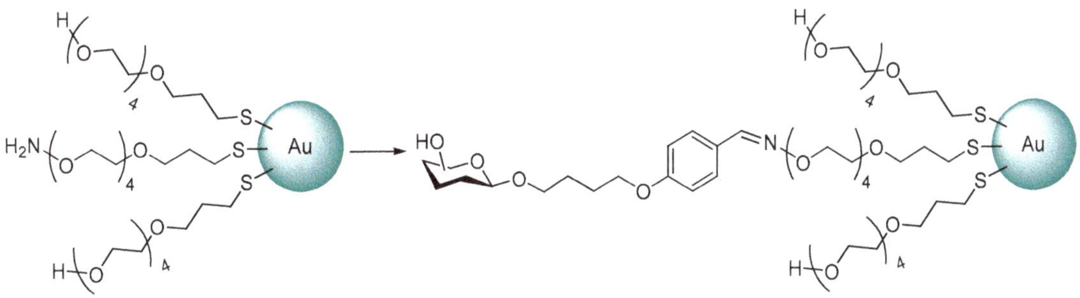

| Entry | Nanoparticle | | Carbohydrate Head Group | | Ligand Density [e] |
|---|---|---|---|---|---|
| 1 | **GNP-1** | **AuNP-1** [b] | β-Glucopyranose | **3** | ~1280 [f] |
| 2 | **GNP-2** | **AuNP-1** [b] | α-Glucopyranose | **4** | ~1160 [f] |
| 3 | **GNP-3** | **AuNP-1** [b] | α-Mannopyranose | **5** | ~1260 [f] |
| 4 | **GNP-4** | **AuNP-1** [b] | β-Galactopyranose | **6** | ~1280 [f] |
| 5 | **GNP-5** | **AuNP-1** [b] | α-Idopyranose | **8** | ~1200 [f] |
| 6 | **GNP-6** | **AuNP-1** [b] | α-Talopyranose | **9** | ~1220 [f] |
| 7 | **GNP-7** | **AuNP-1** [b] | β-Cellobiose | **10** | ~1040 [g] |
| 8 | **GNP-8** | **AuNP-1** [b] | β-Lactose | **11** | ~1060 [g] |
| 9 | **GNP-9** | **AuNP-1** [b] | β-Maltose | **12** | ~1040 [g] |
| 10 | **GNP-10** | **AuNP-1** [b] | β-*N*-Acetyllactosamine | **13** | ~980 [g] |
| 11 | **GNP-11** | **AuNP-1** [b] | Le$^{x}$ | **14** | ~880 [h] |
| 12 | **GNP-12** | **AuNP-2** [c] | Le$^{x}$ | **14** | ~660 [h] |
| 13 | **GNP-13** | **AuNP-3** [d] | Le$^{x}$ | **14** | ~400 [h] |
| 14 | **GNP-14** | **AuNP-1** [b] | Le$^{x\text{-Glc}}$ | **15** | ~900 [h] |
| 15 | **GNP-15** | **AuNP-1** [b] | Le$^{x\text{-L-Gal}}$ | **16** | ~960 [h] |
| 16 | **GNP-16** | **AuNP-1** [b] | Le$^{x\text{-}\beta\text{-Fuc}}$ | **17** | ~880 [h] |
| 17 | **GNP-17** | **AuNP-1** [b] | Le$^{a}$ | **18** | ~880 [h] |
| 18 | **GNP-18** | **AuNP-1** [b] | Le$^{x\text{-L-Rha}}$ | **19** | ~920 [h] |
| 19 | **GNP-19** | **AuNP-1** [b] | Le$^{x\text{-D-L}}$ | **20** | ~900 [h] |

[a] Oxime coupling by incubation of glycans in acetate buffer (pH 4.7) with Au-NP solutions for 16 h at 40 °C; [b] theoretical surface coverage of 100% glycan; [c] theoretical surface coverage of 50% glycan; [d] theoretical surface coverage of 33% glycan; [e] determination by anthrone method; [f] reference: compound **3**; [g] reference: compound **10**; [h] reference: compound **14**.

In case of larger carbohydrate structures, a complete allocation of aminoxy functions on AuNP is precluded due to steric reasons. Apparently, this aspect is confirmed, since the number of Le$^{x}$ ligands per GNP does not linearly correlate with the theoretical functionalization of the surface (Table 2, **GNP-11** to **GNP-13**). In the case of **AuNP-3** with a theoretically 33% functionalization of the surface, an amount of 400 ligands was observed. This would correspond to an equivalent of 1200 ligands, assuming a 100% functionalization of the surface for **AuNP-1**; however, 880 ligands were found, corresponding to a deficit of 25%.

### 2.5. Lectin Binding Experiments of GNPs

Assays by UV-Vis as well as TEM experiments could be used to clearly show these novel GNPs to be suited for specific biological recognition processes [79]. Concanavalin A (ConA), the mannose recognizing and binding lectin, can aggregate with mannose by four binding sites [71]. Thus, **GNP-3** (mannose) dissolved in Hepes buffer, showing a normal UV-Vis spectrum with the SPR absorption maximum at about 520 nm, was treated with a Hepes-buffered ConA solution containing $Ca^{2+}$ and $Mn^{2+}$ ions (1 mM each) for 12 h at room temperature. The subsequent aggregation of the GNP was evident by a distinct shift of the absorption maximum and a considerable flattening of the curve (Figure 5B). Further, this ConA-induced aggregation of **GNP-3** could be observed directly, since the solution showed a shift of color from purple to blue (Figure 5A), and after 2 days, all GNP had precipitated.

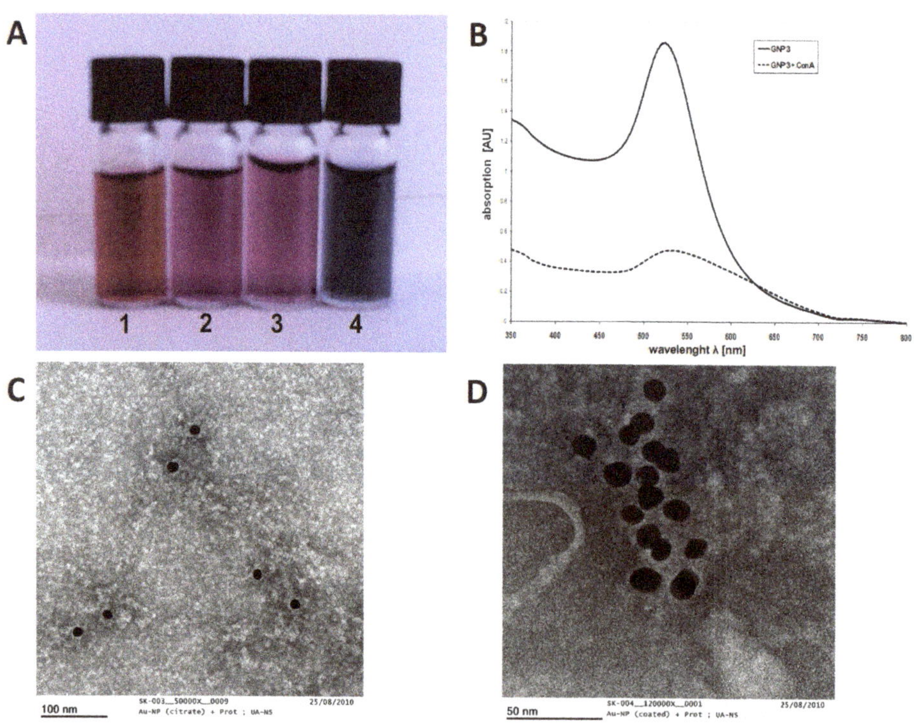


**Figure 5.** Characterization of ConA binding to mannopyranoside functionalized glyconanoparticle GNP-3. ((**A**)1) citrate stabilized AuNP solution, ((**A**)2) GNP-3 solution, ((**A**)3) GNP-3 solution after incubation with BSA, ((**A**)4) GNP-3 solution after incubation with ConA; (**B**) UV-Vis spectra of GNP-3 before (solid line) and after (dotted line) incubation with ConA; (**C**) *negative stain* TEM image of AuNP-4 after incubation with ConA; (**D**) *negative stain* TEM image of GNP-3 after incubation with ConA.

Furthermore, the aggregation of lectin ConA to the mannose GNP could be substantiated by TEM experiments. By negative sampling, this specific interaction could not be observed. In a conventional TEM image, objects such as AuNP are dark since the electron beam is deflected by the atomic nuclei of heavy elements, and does not hit the CCD camera detector. Thus, organic material made only up from light elements cannot be shown this way; however, by employing the so-called "negative staining method", for that purpose, the TEM sample is treated with a solution of heavy metals, such as, for instance, phosphotungstic acid, ruthenium tetroxide, or uranyl acetate [80]. This leads to precipitation of heavy metals and dark spots in areas without organic material, and in contrast to light spots in areas with proteins or carbohydrates, etc. [81].

In the negative staining TEM picture of **GNP-3** plus ConA, proteins are observed as light spots (Figure 5D). These bright spots show a diameter of 8 nm, which corresponds to the size of lectin ConA as given by crystal data. For all the **GNP-3,** a systematic arrangement of proteins was observed by an extended light zone around the GNP. This characteristic arrangement for **GNP-3** was visualized by a red circle. However, if **AuNP-4** (SAM only with dilution spacer **2**) was incubated, the light zones around the AuNP were clearly smaller (Figure 5C). Again, in this sample, the lectins are observed; however, not arranged around the AuNP. The light zone around the AuNP of about 1.5 nm is caused by the organic material of the SAMs of the dilution spacer **2**.

By employing a larger magnification, it could be shown that many lectin units are arranged around one **GNP-3** (Figure 5D), and this attachment is shown by a lighter zone around the GNP. Further, a brightening of the GNP itself is observed due to the spheric mannose SAM attached (Figure 5D). Since ConA disposes of four binding sites, several **GNP-3** could be cross-linked. Figure 5D gives an image of how the ConA-linked GNPs are arranged, and some are even piled. Thus, these negative staining TEM studies unequivocally demonstrate these glyconano particles to represent multiple spherical representations of carbohydrate epitopes to be able to incur specific biological bindings. Apparently, the constitution of both GNP and linkers did not show any negative impact on the specificity of the lectin binding. Therefore, these well-characterized GNPs could be employed to perform CCI experiments by SPR measurements.

## 3. Materials and Methods

### *3.1. General*

Reagents of commercial quality were purchased from Aldrich (St. Louis, MO, USA), Sigma (St. Louis, MO, USA), or Merck (Darmstadt, Germany) and were used without further purification. Solvents were dried according to standard methods. All reactions were carried out under an argon atmosphere with dry solvents under anhydrous conditions, unless otherwise noted. Analytical thin-layer chromatography (TLC) was performed on precoated aluminum plates (Silica Gel 60F254, Merck5554), compound spots were visualized by UV light (254 nm) and by staining with a yellow solution containing $Ce(NH_4)_2(NO_3)_6$ (0.5 g) and $(NH_4)_6Mo_7O_{24}/4\ H_2O$ (24.0 g) in 6% $H_2SO_4$ (500 mL) or with 10% $H_2SO_4$ in ethanol followed by heat treatment. For column chromatography, Silica Gel 60, 230e400 mesh, 40e63 mm (Merck) was used. $^{1}$H NMR and $^{13}$C NMR spectra were recorded on Bruker (Billerica, MA, USA) AMX-400 (400 MHz for $^{1}$H, 100.6 MHz for $^{13}$C) and on Bruker DRX-500 (500 MHz for $^{1}$H, 125.8 MHz for $^{13}$C) at 300 K. Chemical shifts were calibrated to solvent residual peaks ($CDCl_3$: $\delta$ = 7.24 ppm for $^{1}$H and $\delta$ = 77.0 ppm for $^{13}$C; methanol-$d_4$: $\delta$ = 3.35 ppm for $^{1}$H and $\delta$ = 49.30 ppm for $^{13}$C). The signals were assigned by $^{1}$H-$^{1}$H-COSY, HSQC, HMBC, and, if necessary, NOESY experiments. Hydrogen and carbon atoms are indexed as follows: the sugar residue is numbered as usual from 1 to 6, with the anomeric position being number 1, the atoms of the anomeric spacer moiety then receive numbers with index 'bu' for butyl and 'ar' for aromatic by consequent numbering starting from the glycosidic bond. Optical rotations were measured using a Krüss Optronic P8000 (589 nm, Hamburg, Germany) at 20 °C. MALDI-TOF-MS was performed on a Bruker Biflex III with dihydroxybenzoic acid or trihydroxyanthracene as matrices in positive reflector mode.

FAB-HRMS was performed on a Thermo Finnigan MAT95 XL mass spectrometer (San Jose, CA, USA).

### 3.2. Syntheses

#### 3.2.1. (*E*)-4-[4-(Dimethoxymethyl) Phenoxy] But-2-enyl 2,3,4,6-Tetra-*O*-acetyl-α-D-glucopyranoside (**24**)

Allyl α-D-glucopyranoside tetraacetate (**21**, 200 mg, 515 μmol) and *para*-(allyloxy) benzaldehyde dimethyl acetal (**22**, 1.0 g, 4.80 mmol) [55] were dissolved in dry and degassed dichloromethane (30 mL) and placed into a flame-dried flask containing activated molecular sieves (4 Å) by using standard Schlenk techniques. Grubbs–Hoveyda 2nd generation catalyst (**23**, 33 mg, 52 mmol), dissolved in dry and degassed dichloromethane (1 mL), was added by syringe to obtain a 0.02 m solution. The reaction mixture was heated under reflux for 6 h. Conversion of the starting material was monitored by TLC. The solution was concentrated under reduced pressure, and the crude product was directly purified by column chromatography using silica and a petroleum ether/ethyl acetate gradient (4:1–2:1) to give compound **24** (223 mg, 91%) as a colorless syrup. $R_f$ = 0.32 (petroleum ether/ethyl acetate, 1:1). $[\alpha]_D^{20}$ = +5.9 ($c$ = 1.0 in $CHCl_3$). $^1$H-NMR (500 MHz, $CDCl_3$) $\delta$ = 7.32 (d, $J_{Ar}$ = 8.4 Hz, 2 H, H-$1_{arom}$), 6.86 (d, $J_{Ar}$ = 8.4 Hz, 2 H, H-$2_{arom}$), 5.96–5.80 (m, 2 H, H-$2_{bu}$, H-$3_{bu}$), 5.32 (s, 1 H, $H_{acetal}$), 5.18 (dd, $J_{2,3}$ = 2.5, $J_{3,4}$ = 2.1 Hz, 1 H, H-3), 5.06 (dd, $J_{3,4}$ = 9.6, $J_{4,5}$ = 9.9 Hz, 1 H, H-4), 4.99 (dd, $J_{1,2}$ = 7.9, $J_{2,3}$ = 9.4 Hz, 1 H, H-2), 4.53 (d, $J_{1,2}$ = 7.9 Hz, 1 H, H-1), 4.50–4.48 (m, 2 H, H-$4_{bu}$), 4.38–4.32 (m, 1 H, H-$1a_{bu}$), 4.23 (dd, $J_{5,6b}$ = 6.4, $J_{6a,6b}$ = 11.8 Hz, 1 H, H-6b), 4.15–4.03 (m, 2 H, H-$1b_{bu}$, H-6a), 3.65 (ddd, $J_{4,5}$ = 9.9, $J_{5,6a}$ = 2.4 $J_{5,6b}$ = 4.7 Hz, 1 H, H-5), 3.28 (s, 6 H, $OCH_3$), 1.98–2.06 (m, 12 H, 4 $H_{Ac}$); $^{13}$C-NMR (126 MHz, $CDCl_3$) $\delta$ = 170.5, 170.1, 169.3, 169.2 (4 $CH_3CO$), 130.6, 128.4, 128.2, 127.9 (4 $C_{arom}$), 114.2 (C-$2_{bu}$), 114.1 (C-$3_{bu}$), 102.9 ($C_{acetal}$), 99.6 (C-1), 72.8 (C-3), 71.7 (C-5), 71.2 (C-2), 68.8 (C-$4_{bu}$), 68.3 (C-4), 67.5 (C-$1_{bu}$), 61.8 (C-6), 52.5 ($OCH_3$), 20.7, 20.6, 20.5, 20.4 (4 $CH_3CO$). HRMS (FAB): *calcd.* for $C_{27}H_{37}O_{13}$ $[M + H]^+$ 569.2229, found 569.215.

#### 3.2.2. 4-(4-Formylphenoxy) Butyl α-D-Glucopyranoside (**4**)

Compound **24** (500 mg, 879 μmol) was dissolved in a mixture of THF, water, and TFA (90:9.9:0.1, *v*/*v*) to obtain a 0.2 M solution. After stirring for 1 h, the reaction mixture was diluted with $CH_2Cl_2$ (40 mL), and the reaction was stopped by the addition of triethylamine (4 mL), followed by the addition of water (40 mL). The organic phase was separated and dried with $Na_2SO_4$. The solvent was removed under reduced pressure, and the crude product was used in the next transformation without further purification. For hydrogenation, the aldehyde was dissolved in anhydrous ethyl acetate (20 mL) and placed into a flame-dried flask containing palladium (10% on charcoal) and diphenyl sulfide (0.01 equiv.). The suspension was degassed, and after purging with hydrogen, the mixture was stirred for 12 h. Then the suspension was filtered and thoroughly dried before the residue was redissolved in methanolic sodium methoxide solution (40 mL, 0.1 M). The solution was stirred at room temperature for 6 h, and then the mixture was neutralized with Amberlite IR 120 ($H^+$) resin. After filtration and evaporation of the solvent, the crude product was purified by flash chromatography using silica and dichloromethane/methanol (5:1) to give **4** (285 mg, 91%) as a colorless syrup. $R_f$ = 0.46 ($CH_2Cl_2$/MeOH, 5:1). $[\alpha]_D^{20}$ = +36.0 ($c$ = 1.0 in MeOH). $^1$H-NMR (500 MHz, $CDCl_3$) δ = 9.81 (s, 1H, $H_{ald}$), 7.83 (d, $J_{Ar}$ = 8.7 Hz, 1H, H-$1_{arom}$), 7.00 (d, $J_{Ar}$ = 8.7 Hz, 1H, H-$2_{arom}$), 4.12 (t, $J_{3bu,4bu}$ = 6.7 Hz, 2H, H-$4_{bu}$), 4.09 (d, $J_{1,2}$ = 3.6 Hz, 1H, H-1), 4.01–3.95 (m, 1H, H-$1a_{bu}$), 3.90–3.83 (m, 1H, H-$1b_{bu}$), 3.55–3.23 (m, 4H, H-3, H-4, H-6ab), 3.20 (dd, $J_{1,2}$ = 3.6, $J_{2,3}$ = 9.4 Hz, 1H, H-2), 1.96–1.89 (m, 2H, H-$3_{bu}$), 1.83–1.75 (m, 2H, H-$2_b$u); $^{13}$C-NMR (126 MHz, $CDCl_3$) δ = 192.0 ($C_{ald}$), 133.1, 116.0 (2

$C_{arom}$), 99.7 (C-1), 77.9 (C-4), 77.7 (C-5), 74.1 (C-2), 71.7 (C-$1_{bu}$), 70.31 (C-$4_{bu}$), 69.9 (C-3), 62.9 (C-6), 27.2 (C-$3_{bu}$), 26.9 (C-$2_{bu}$). HRMS (FAB): *calcd.* for $C_{17}H_{25}O_8$ $[M + H]^+$ 357.1544, found 357.1601.

3.2.3. Allyl 2,3,6-Tri-*O*-acetyl-4-*O*-(2,3,4,6-tetra-*O*-acetyl-α-D-glucopyranosyl)-β-D-glucopyranoside (**26**)

A solution of 2,3,6-tri-*O*-acetyl-4-*O*-(2,3,4,6-tetra-*O*-acetyl-α-D-glucopyranosyl)-α-D-glucopyranosyl-trichloroacetimidate (**24**, 781mg, 1.00 mmol) [58] and allyl alcohol (190 μL, 3.00 mmol) in anhydrous dichloromethane (20 mL) under argon was stirred with powdered molecular sieves (4A, 1.0 g) for 2h at room temperature. The mixture was cooled to −20 °C and treated dropwise with a solution of trimethylsilyl triflate (18 μL, 1.0 mmol) in anhydrous $CH_2Cl_2$ (5 mL) and monitored by thin-layer chromatography. After 30 min, the reaction was quenched with triethylamine (1 mL), diluted with dichloromethane (20 mL), and filtered via Celite. The filtrate was washed with HCl (1 N, 20 mL), water (20 mL), and saturated NaCl solution (20 mL). The organic phase was dried via $Na_2SO_4$, evaporated and the sirupy residue purified by column chromatography on silica gel with dichloromethane-methanol (30:1) to give **26** (560 mg, 83%) as colorless solid, $R_f$ = 0.32 (PE/EE 1:1); mp 113 °C; $[\alpha]_D^{20}$ = +45.3 (*c* = 1.0 in $CHCl_3$); $^1$H-NMR (400 MHz, $CDCl_3$) δ = 5.84 (ddd, $J_{2,3all}$ = 5.5, $J_{cis}$ = 10.7, $J_{trans}$ = 16.7 Hz, 1H, $H_{All-2}$), 5.40 (d, $J_{1',2'}$ = 3.8 Hz, 1H, H-1′), 5.35 (dd, *J* = 8.25, 8.25 Hz, 1H, H-3′), 5.30–5.16 (m, 3H, H-3, HAll-3), 5.04 (t, *J* = 9.7, 9.7 Hz, 1H, H-4′), 4.91–4.80 (m, 2H, H-2, H-2′), 4.57 (d, $J_{1,2}$ = 7.7 Hz, 1H, H-1), 4.47 (dd, $J_{6ab}$ = 12.1 Hz, 1H H-6a), 4.34–4.16 (m, 3H, $H_{All-1a}$), 4.15–3.89 (m, 5H, H-5′, $H_{All-1b}$), 3.75–3.58 (m, 1H, H-5), 2.14–2.00 (7 × s, 7 × 3H, $H_{Ac}$); MALDI-TOF: *m*/*z calcd.* for $C_{29}H_{40}O_{18}$ 676.22, found: 699.5 $[M + Na]^+$.

3.2.4. (*E*)-4-(4-Dimethoxymethylphenoxy)-but-2-enyl 2,3,6-Tri-*O*-acetyl-4-*O*-(2,3,4,6-tetra-*O*-acetyl-α-D-glucopyranosyl)-β-D-glucopyranoside (**27**)

Under Schlenk conditions, allyl glycoside **26** (412 mg, 609 μmol) and para-allyloxy-benzaldehyde dimethylacetal **22** (1.27 g, 6.10 mmol) were dissolved in anhydrous and degassed dichloromethane (50 mL), and freshly activated molecular sieves 4A (500 mg) were added. Methathesis catalyst Grubbs-Hoveyda 2nd generation **23** (39 mg, 61 mmol) dissolved in dry and degassed dichloromethane (1 mL), was added by syringe to obtain a 0.02 m solution. The reaction mixture was heated under reflux for 6 h. Conversion of the starting material was monitored by TLC. The solution was concentrated under reduced pressure, and the crude product was directly purified by column chromatography using silica and a petroleum ether/ethyl acetate gradient to give **27** (396 mg, 76%) as colorless syrup; $R_f$ = 0.30 (PE/EE 1:1); $[\alpha]_D^{20}$ = +23.8 (*c* = 1.0 in $CHCl_3$); $^1$H-NMR (400 MHz, $CDCl_3$) δ = 7.34 (d, $J_{Ar}$ = 8.5 Hz, 2H, H-$1_{arom}$), 6.88 (d, $J_{Ar}$ = 8.5 Hz, 2H, H-$2_{arom}$), 5.95–5.80 (m, 2H, H-$2_{bu}$, H-$3_{bu}$), 5.40 (d, $J_{1',2'}$ = 3.8 Hz, 1H, H-1′), 5.36–5.33 (m, 2H, H-3′, $H_{acetal}$), 5.20 (dd, $J_{2,3}$ = 9.8, $J_{3,4}$ = 7.8 Hz, 1H, H-3), 5.04 (dd, $J_{3',4'}$ = 9.8, $J_{4',5'}$ = 9.9 Hz, 1H, H-4′), 4.91–4.80 (m, 2H, H-2, H-2′), 4.59 (d, $J_{1,2}$ = 7.9 Hz, 1H, H-1), 4.47 (dd, $J_{6ab}$ = 12.1 Hz, 1H H-6a), 4.43–4.40 (m, 2H, H-$1_{bu}$), 4.34–4.16 (m, 2H, H-$4_{bu}$), 4.15–3.89 (m, 3H, H-6′ab, H-5′), 3.66 (ddd, $J_{4,5}$ = 9.8, $J_{5,6a}$ = 3.4, $J_{5,6b}$ = 6.9 Hz, 1H, H-5), 3.30 (s, 6H, $OCH_3$), 2.14–2.00 (7 × s, 7 × 3H, HAc); $^{13}$C-NMR (126 MHz, $CDCl_3$) δ = 170.4, 170.3, 170.2, 169.8, 169.5, 169.2, 169.0 (7 × s 7 × 3H, $CH_3$CO), 132.0, 129.3, 128.0, 126.9 (4 × s 4 × 1H, $C_{arom}$), 114.9 (C-$2_{bu}$), 114.5 (C-$3_{bu}$), 105.5 ($C_{acetal}$), 102.7 (C-1′), 98.3 (C-1), 76.1 (C-4), 72.9 (C-5), 72.7 (C-3′), 72.1 (C-2), 72.0 (C-5′), 71.6 (C-2′), 71.6 (C-3), 68.0 (C-$1_{bu}$) 67.7 (C-4′) 67.5 (C-$4_{bu}$), 61.8 (C-6), 61.5 (C-6′), 52.0 ($OCH_3$), 20.8, 20.7, 20.6, 20.5 (4 × 1C, $CH_3$CO); HRMS (FAB): *m/z calcd.* for $C_{39}H_{53}O_{21}^+$: 857.3074, found: 857.3100 $[M + H]^+$.

3.2.5. 4-(4-Formylphenoxy)-butyl 4-*O*-(α-D-Glucopyranosyl)-β-D-glucopyranoside (**12**)

The metathesis product **27** (300 mg, 350 μmol) was dissolved in a mixture of THF, water, and TFA (90:9.9:0.1 *v*/*v*) to obtain a 0.2 M solution. After stirring for 1 h, the reaction mixture was diluted with dichloromethane (20 mL), and the reaction was stopped by the addition of triethylamine (2 mL) followed by the addition of water (20 mL). The organic layer was separated and dried over $Na_2SO_4$. The solvent was removed under reduced pressure, and the crude product was used directly without further purification. For hydrogenolysis, the aldehyde was dissolved in anhydrous ethyl acetate (10 mL) in a flame-dried flask containing palladium (10%) on charcoal and diphenylsulfide (0.01 eq). The suspension was degassed, and after purging with hydrogen, the mixture was stirred for 12 h. Then the suspension was filtered and thoroughly dried before the residue was redissolved in dry THF (10 mL) and cooled to 0° C, followed by the addition of acetic acid (5 eq). After warming to room temperature, 1 M TBAF solution in THF (1.2 eq) was added and the mixture stirred overnight. After termination (TLC control), the reaction was quenched by the addition of brine and extracted with ethyl acetate (4 × 100 mL). The combined organic layers were concentrated, redissolved in methanolic sodium methoxide solution (20 mL, 0.1 M), and stirred for 6h at room temperature. The mixture was neutralized with Amberlite IR 120 ($H^+$) resin, filtered, and evaporated. The crude product was purified by flash chromatography on silica with dichloromethane/methanol (3:1) to give product **12** (139 mg, 77%) as a colorless solid. $R_f$ = 0.20 ($CH_2Cl_2$/MeOH 3:1); mp: 140 °C; $[\alpha]_D^{20}$ = −9.8 (*c* = 1.0 in MeOH); $^1$H-NMR (500 MHz, $CDCl_3$): δ = 9.84 (s, 1H, $H_{ald}$), 7.88 (d, $J_{Ar}$ = 8.6 Hz, 1H, H-$1_{arom}$), 7.10 (d, $J_{Ar}$ = 8.6 Hz, 1H, H-$2_{arom}$), 4.56 (d, $J_{1,2}$ = 7.9 Hz, 1H, H-1), 4.43 (d, $J_{1,2}$ = 7.9 Hz, 1H, H-1′) 4.20 (t, $J_{3bu,4bu}$ = 6.6 Hz, 2H, H-$4_{bu}$), 3.86–3.50 (m, 6H, H-$1_{bu}$, H-6, H-6′), 3.50–3.30 (m, 8H, H-2, H-3, H-4, H-5, H-2′, H-3′, H-4′, H-5′), 1.90–1.85 (m, 2H, H-$3_{bu}$), 1.84–1.76 (m, 2H, H-$2_{bu}$); $^{13}$C-NMR (126 MHz, $CDCl_3$): δ =192.9 ($C_{ald}$), 129.0, 122.3, 115.1 (3C, $C_{arom}$), 102.2 (C-1), 96.3 (C-1′), 79.7 (C-4), 77.1 (C-3′), 77.0 (C-5′) 75.0 (C-2′), 74.0 (C-5), 73.8 (C-3′), 73.5 (C-2), 71.2 (C-4′), 68.2 (C-$1_{bu}$), 67.6 (C-$4_{bu}$), 63.9 (C-6′), 63.0 (C-6), 27.3 (C-$2_{bu}$), 27.0 (C-$3_{bu}$); HRMS (FAB): *m*/*z calcd.* for $C_{23}H_{35}O_{13}^+$: 519.2072, found: 519.2075 $[M + H]^+$.

3.2.6. Allyl-tetraethylene-glycol (**29**)

To a solution of tetraethylene glycol (**28**, 3.41 g, 17.6 mmol) in anhydrous DMF (20 mL), sodium hydride (510 mg, 21.1 mmol) was added slowly. After stirring for 1 h at room temperature, the suspension was cooled to 0 °C, and allyl bromide (760 μL, 8.80 mmol) was added. After warming to room temperature, the mixture was stirred for 12 h. The reaction was quenched with saturated aqueous $NH_4Cl$ (10 mL) and diluted with ethyl acetate (20 mL). The aqueous layer was extracted with ethyl acetate (3 × 20 mL), the combined organic layers were washed with saturated NaCl solution, and dried over $Na_2SO_4$. After filtration and evaporation, the crude product was purified by column chromatography on silica gel (dichloromethane/ methanol 20:1) to yield **29** as a yellowish oil (1.86 g, 90%). $R_f$ = 0.38 ($CH_2Cl_2$/MeOH 15:1); $^1$H- NMR (400 MHz, $CDCl_3$) δ = 5.90 (tdd, 1H, $J_{1a,2}$ = 17.4, $J_{1b,2}$ = 10.3, $J_{2,3}$ = 4.9 Hz, H-2), 5.26 (d, 1H, $J_{1a,2}$ = 17.4 Hz, H-1a), 5.20 (d, 1H, $J_{1b,2}$ = 10.3 Hz, H-1b), 4.00 (d, 2H, $J_{2,3}$ = 4.9 Hz, H-3), 3.69–3.63 (m, 2H, H-4), 3.63–3.57 (m, 10H, H-5,-6,-7,-8,-9), 3.56–3.51 (m, 4H, H-10, -11), 2.67 (s, 1H, OH); $^{13}$C- NMR (100 MHz, $CDCl_3$) δ = 134.6 (C-2), 117.0 (C-1), 72.4 (C-3), 72.1 (C-4), 71.1 -69.9 (C-6,-7,-8,-9,-10), 69.3 (C-5), 61.7 (C-11).

3.2.7. Allyl-(ω-*O*-phthalimido)-tetraethylene-glycol (**30**)

Under argon atmosphere, derivative **29** (1.10 g, 4.70 mmol), triphenylphosphine (1.85 g, 7.05 mmol), and *N*-hydroxy-phthalimide (1.15 g, 7.05 mmol) dissolved in anhydrous THF

(80 mL) were treated with diisopropyl azodicarboxylate (1.43 g, 7.05 mmol) in anhydrous THF (20 mL). After stirring for 4 h, filtration, evaporation, and purification by column chromatography on silica gel (petroleum ether/ethyl acetate 1:1), **30** was obtained as a colorless oil (1.50 g, 56%). $R_f$ = 0.31 (petroleum ether/ethyl acetate 1:1); $^1$H-NMR (400 MHz, $CDCl_3$) δ = 7.87–7.80 (m, 2H, Phth-ortho), 7.76–7.70 (m, 2H, Phth-meta), 5.97 (tdd, 1H, $J_{1a,2}$ = 17.1, $J_{1b,2}$ = 10.5, $J_{2,3}$ = 5.4 Hz, H-2), 5.24 (d, 1H, $J_{1a,2}$ = 17.1 Hz, H-1a), 5.17 (d, 1H, $J_{1b,2}$ = 10.5 Hz, H-1b), 4.40–4.37 (m, 2H, H-11), 4.00 (d, 2H, $J_{2,3}$ = 5.4 Hz, H-3), 3.89–3.84 (m, 2H, H-10), 3.68–3.57 (m, 12H, H-4,-5,-6,-7,-8,-9); $^{13}$C-NMR (100 MHz, $CDCl_3$) δ = 163.4 (C-Carboxy), 134.7 (C-2), 133.9 (C-Phth-meta), 128.9 (C-Phth-ipso), 123.9 (C-Phth-ortho), 117.0 (C-1), 77.2 (C-11), 72.0–69.9 (C-3,4,5,6,7,8,9), 68.9 (C-10). MALDI-TOF *m/z calcd* for $C_{19}H_{25}NO_7$: 379.16, found: 402.50 $[M + Na]^+$.

### 3.2.8. Allyl-(ω-hydroxylamino)-tetraethylene-glycol (**31**)

Compound **30** (1.45 g, 3.82 mmol) dissolved in anhydrous acetonitrile (30 mL) was treated with hydrazine hydrate (880 mL, 19.1 mmol). After stirring for 2h at room temperature, the white suspension was evaporated and purified by column chromatography on silica gel (dichloromethane/ methanol 10:1) to give **31** as a colorless oil (750 mg, 75%). $R_f$ = 0.31 (petroleum ether/ ethyl acetate 1:1); $^1$H-NMR (400 MHz, $CDCl_3$) δ = 5.99 (tdd, 1H, $J_{1a,2}$ = 10.3, $J_{1b,2}$ = 17.2, $J_{2,3}$ = 5.1 Hz, H-2), 5.45 (s, 2H, $ONH_2$), 5.28 (d, 1H, $J_{1b,2}$ = 17.2 Hz, H-1b), 5.19 (d, 1H, $J_{1a,2}$ = 10.3 Hz, H-1a), 4.00 (d, 2H, $J_{2,3}$ = 5.1 Hz, H-3), 3.84–3.79 (m, 2H, H-11), 3.70–3.64 (m, 2H, H-10), 3.62–3.50 (m, 12H, H-4,-5,-6,-7,-8,-9); $^{13}$C-NMR (100 MHz, $CDCl_3$) δ = 133.9 (C-2), 117.0 (C-1), 74.2 (C-11), 72.0–68.9 (C-3,-4,-5,-6,-7,-8,-9,-10); MALDI-TOF: *m/z calcd* for $C_{11}H_{23}NO_5$: 249.16, found: 272.9 $[M + Na]^+$.

### 3.2.9. Aminooxy-(ω-3-thioacetylpropyl)-tetraethylene-glycol (**1**)

To a stirred solution of alkene **31** (600 mg, 2.41 mmol) in anhydrous THF (10 mL), thioacetic acid (850 μL, 12.0 mmol) and a catalytic amount of AIBN were added. The mixture was irradiated for 3 h with UV light. Then the solvent was removed under reduced pressure and the yellowish crude was purified by column chromatography on silica gel ($CH_2Cl_2$/ methanol 20: 1). Compound **1** was obtained as a yellowish oil (780 mg, quant.). $R_f$ = 0.28 ($CH_2Cl_2$/ MeOH 10:1); $^1$H-NMR (400 MHz, $CDCl_3$) δ = 5.44 (s, 2H, $ONH_2$), 3.74–3.70 (m, 2H, H-11), 3.68–3.51 (m, 16H, H-3,-4,-5,-6,-7,-8,-9,-10), 3.08 (t, 2H, $J_{1,2}$ =6.4 Hz, H-1), 2.33 (s, 3H, Sac), 2.13–1.98 (m, 2h, H-2); $^{13}$C-NMR (100 MHz, $CDCl_3$) δ = 72.3 (C-3), 71.3–68.9 (C-4,-5,-6,-7,-8,-9,-10), 74.2 (C-11), 30.5 (C-1), 28.6 (Sac), 21.2 (C-2). MALDI-TOF *m/z calcd* for $C_{11}H_{23}NO_5$: 325.16, found: 348.0 $[M + Na]^+$.

### 3.2.10. 3-Thioacetylpropyl-tetraethylene-glycol (**2**)

According to the synthesis of compound **1**, alkene **29** (680 mg, 2.90 mmol) was reacted with thioacetic acid (1.03 mL, 14.5 mmol) and AIBN (450 mg, 2.74 mmol) in anhydrous THF (10 mL). Compound **2** was obtained after column chromatography using silica gel ($CH_2Cl_2$/ methanol 20:1) as a yellowish oil (905 mg, quant.) $R_f$ = 0.40 ($CH_2Cl_2$/ MeOH 15:1); $^1$H-NMR (400 MHz, $CDCl_3$) δ = 3.75–3.71 (m, 2H, H-11), 3.69–3.52 (m, 16H, H-3,-4,-5,-6,-7,-8,-9,-10), 3.09 (t, 2H, $J_{1,2}$ = 6.4 Hz, H-1), 2.57 (s, 1H, OH), 2.33 (s, 3H, SAc), 2.14–1.97 (m, 2H, H-2). $^{13}$C- NMR (101 MHz, $CDCl_3$) δ = 72.4 (C-3), 71.5–68.8 (C-4,-5,-6,-7,-8,-9,-10), 61.7 (C-11), 30.5 (C-1), 28.7 (SAc), 21.2 (C-2).

### 3.2.11. *O*-Phthalimido-tetraethylene glycol (**32**)

Under an argon atmosphere, tetraethylene glycol **28** (187 mg, 0.97 mmol), triphenylphosphine (5.07 g, 7.5 mmol), and *N*-hydroxy-phthalimide (0.82 g, 5.5 mmol) dissolved in anhydrous THF (80 mL) were treated dropwise with a solution of diisopropyl azodicarboxylate (1.52 g, 7.5 mmol) in anhydrous THF (20 mL). Stirring for 4 h, filtration, evaporation, and purification by column chromatography on silica gel (ethyl acetate) gave **32** as a colorless oil (622 mg, 37%). $^1$H-NMR (300 MHz, $CDCl_3$) δ = 7.88–7.80 (m, 2H, Phth-ortho), 7.78–7.71 (m, 2H, Phth-meta), 4.41–4.36 (m, 2H, $CH_2ON$), 3.89–3.84 (m, 2H, $OCH_2CH_2ON$), 3.74–3.66 (m, 4H, $OCH_2CH_2O$), 3.66–3.57 (m, 8H, $OCH_2CH_2O$), 2.47 (t, 1H, $J_{CH2,OH}$ = 6.2 Hz, OH); $^{13}$C-NMR (75 MHz, $CDCl_3$) δ = 163.4 (C = O), 134.4 (C-Phth-meta), 128.9 (C-Phth-ipso), 123.5 (C-Phth-ortho), 77.2 ($CH_2ON$), 72.4, 70.8, 70.6, 70.4, 70.3, ($OCH_2CH_2OH$), 69.2 ($OCH_2CH_2ON$), 61.7 ($CH_2OH$); HRMS (ES): *m*/*z calcd* for $C_{16}H_{22}NO_7$: 340.1396 [M + H], found: 340.1378 [M + H].

#### 3.2.12. *O*-Phthalimido-(ω-thiotrityl)-tetraethylene glycol (**33**)

Under an argon atmosphere, diisopropyl azodicarboxylate (607 mg, 3.00 mmol) and a suspension of polymer-bound triphenylphosphine (2.03 g, 3.0 mmol, 1.48 mmol/g) in anhydrous THF (25 mL) were mixed and stirred for 30 min at 0 °C. Then compound **32** (509 mg, 1.5 mmol) and triphenylmethanethiol (829 mg, 3.0 mmol) dissolved in anhydrous THF (20 mL) were added dropwise at 0 °C. Under stirring for 4 h, the mixture was gradually warmed to 20 °C, the resin filtered via Celite, the residue evaporated, and the material purified on silica gel by chromatography (diethyl ether-hexane, gradient 1:1 to 1:0). Compound **33** resulted as a colorless oil (699 mg, 78%). $^1$H-NMR (300 MHz, $CDCl_3$) δ = 7.86–7.79 (m, 2H, o-ArH), 7.76–7.69 (m, 2H, m-ArH), 7.44–7.38 (m, 6H, H-trityl), 7.31–7.16 (m, 9H, H-trityl), 4.38–4.33 (m, 2H, $\mathbf{CH_2}ON$), 3.87–3.81 (m, 2H, $O\mathbf{CH_2}CH_2ON$), 3.66–3.60 (m, 2H), 3.57–3.51 (m, 2H, $OCH_2CH_2O$), 3.51–3.45 (m, 2H, $OCH_2CH_2O$), 3.43–3.36 (m, 2H, $OCH_2CH_2O$), 3.28 (t, 2H, $J_{OCH2,CH2STr}$ = 6.7 Hz, $O\mathbf{CH_2}CH_2STr$), 2.41 (t, 2H, $J_{OCH2,CH2STr}$ = 6.7 Hz, $OCH_2\mathbf{CH_2}STr$); $^{13}$C-NMR (75 MHz, $CDCl_3$) δ = 163.3 (C = O), 144.7 (C-trityl-ipso), 134.3 (C-Phth-meta), 129.5 (C-trityl), 129.0 (C-Phth-ipso), 127.8 (C-trityl), 126.6 (C-trityl-para), 123.4 (C-Phth-ortho), 77.1 ($\mathbf{CH_2}ON$), 70.7, 70.4, 70.3, 70.0, 69.5 ($OCH_2CH_2O$), 69.2 ($O\mathbf{CH_2}CH_2STr$), 66.5 (C-trityl-quaternary), 31.6 ($OCH_2\mathbf{CH_2}STr$); HRMS (ES): *m*/*z calcd* for $C_{35}H_{36}NO_6S$: 620.2083, [*M* + Na] found: 620.2089 [*M* + Na].

#### 3.2.13. Aminooxy-(ω-thiotrityl)-tetraethylene-glycol (**34**)

Compound **33** (598 mg, 1.0 mmol) was dissolved in acetonitrile (20 mL) and stirred with hydrazine-hydrate (230 μL, 5.0 mmol) for 2 h at room temperature. The white suspension was evaporated, the residue suspended in dichloromethane, and filtered via Celite. Evaporation gave compound **34** as a colorless oil (454 mg, 97%). $^1$H-NMR (300 MHz, $CDCl_3$) δ = 7.44–7.38 (m, 6H, H-trityl), 7.31–7.16 (m, 9H, H-trityl), 5.48 (s, 2H, $ON\mathbf{H_2}$), 3.84–3.79 (m, 2H, $CH_2ON$), 3.69–3.64 (m, 2H, $O\mathbf{CH_2}CH_2ON$), 3.63 (m, 4H, $OCH_2CH_2O$), 3.60–3.55 (m, 2H, $OCH_2CH_2O$), 3.48–3.43 (m, 2H, $OCH_2CH_2O$), 3.31 (t, 2H, $J_{OCH2,CH2STr}$ = 6.9 Hz, $O\mathbf{CH_2}CH_2STr$), 2.43 (t, 2H, $J_{OCH2,CH2STr}$ = 6.9 Hz, $OCH_2\mathbf{CH_2}STr$); $^{13}$C-NMR (75 MHz, $CDCl_3$) δ = 144.8 (C-trityl-ipso), 129.6 (C-trityl), 127.8 (C-trityl), 126.6 (C-trityl-para), 74.8 ($\mathbf{CH_2}ON$), 70.6, 70.5, 70.4, 70.1, 69.7, 69.6 ($\mathbf{CH_2}O$), 66.6 (C-trityl-quaterny), 31.6 ($\mathbf{CH_2}STr$); HRMS (ES): *m*/*z calcd* for $C_{27}H_{34}NO_4S$: 468.2209 [M + H], found: 468.2199.

#### 3.2.14. 4-[4-(ω-Thiotrityl)-tetraethylene-glycolyl-benzaldehyde-oxime]-butyl β-D-glucopyranoside (**35**)

The aminooxy linker **34** (150 mg, 0.32 mmol) and glycoside **3** (124 mg, 0.35 mmol) [53] dissolved in acetonitrile (5 mL) were stirred with glacial acetic acid (150 μL) for 16 h at 20 °C. Following evaporation, purification was by chromatography on silica gel (dichloromethane-methanol 20:1) to give compound **35** as a colorless syrup (260 mg, quant). $^1$H-NMR

(300 MHz, $CDCl_3$) δ = 8.03 (s, 1H, CH = NO), 7.49 (d, 2H, $J_{ar}$ = 8.8 Hz, H-1ar), 7.43–7.34 (m, 6H, H-trityl), 7.32–7.13 (m, 9H, H-trityl), 6.90 (d, 2H, $J_{ar}$ = 8.8 Hz, H-2ar), 4.31–4.16 (m, 3H, H-1), 4.08–3.91 (m, 3H), 3.87 (dd, 1H, $J$ = 11.9, 1.80 Hz), 3.80–3.72 (m, 3H), 3.67–3.58 (m, 6H), 3.57–3.50 (m, 2H), 3.45–3.36 (m, 1H), 3.36–3.12 (m, 18H), 2.37 (2 t, 4H, $J_{OCH2,CH2STr}$ = 6.7, $J_{OCH2,CH2STr}$ = 6.7 Hz, $OCH_2CH_2STr$ and $OCH_2CH_2STr$), 1.96–1.71 (m, 4H, H-2bu,-3bu); $^{13}$C-NMR (75 MHz, $CDCl_3$) δ = 150.0 (CH = NO), 146.3 (C-trityl-ipso), 130.8 (C-trityl), 129.6 (C-arom), 128.9 (C-trityl), 127.8 (C-trityl), 126.1 (C-arom), 115.8 (C-arom), 104.4 (C-1), 78.2 (C-4), 78.0 (C-5), 75.2 (C-2), 74.3 ($CH_2ON$), 71.7, 71.6, 71.5, 71.4, 71.2, 70.7, 70.6, 70.3 (C-1bu, -2bu, $CH_2$-OEG), 68.9 (C-3), 62.8 (C-6), 32.8 ($CH_2$-STr), 27.4 (C-2bu), 27.0 (C-3bu); MALDI-TOF: $m/z$ *calcd* for $C_{44}H_{55}NO_{11}S$: 805.35, found: 828.5 $[M + Na]^+$.

### *3.3. Formation of Saccharide-Functionalized Gold Nanoparticles*

#### 3.3.1. Synthesis and Characterization of Citrate Stabilized Gold Nanoparticles

Prior to use, all glassware was cleaned thoroughly with aqua dest, aqua regia, and finally with degassed Millipore water. A solution of gold (III) chloride trihydrate ($HAuCl_4$ * $3H_2O$, 158 mg, 400 μmol) in degassed Millipore water (400 mL) was heated to reflux. To this pale yellow solution, a warm (~50–60 °C) solution of sodium citrate dihydrate (447 mg, 1.52 mmol) in degassed Millipore water (40 mL) was added under reflux and vigorous stirring. The solution was stained deep red and stirred under reflux for 30 min. Then the solution was cooled to room temperature and filtered through a syringe filter unit (0.2 μm). The average diameter of the gold nanoparticles was determined by size measurements in TEM images (Ø 13.3 ± 1.3 nm) and UV-Vis spectroscopy (13.0 ± 0.4 nm). The concentration of the gold nanoparticle solution was 10.9 nM.

#### 3.3.2. Self-Assembly of Tetraethylene Glycol Linkers **1** and **2** on Gold Nanoparticles (AuNP)

The linker solution (0.06 mmol: 1. **1** 100% = 19.5 mg; 2. **1/2** 1:1 = 9.8 mg + 9.3 mg; 3. **1/2** 1:2 = 6.5 mg + 12.4 mg) in methanol (5 mL) was added to a vigorously stirred 11 nM solution of ~13 nm citrate-stabilized gold nanoparticles (120 mL, 1.32 nmol). After the addition, a color change from red to purple was visible. The mixture was stirred for 16 h at room temperature. Subsequently, the coated **Au-NPs** were purified via centrifuge filtration (Millipore Amicon Ultra, Darmstadt, Germany, 50 kDa cutoff) in 30 mL portions by dissolving in Millipore water (5 mL) and filtration (five times). After the final filtration, the nanoparticles were diluted with Millipore water (330 μL:200 μL filtration residue + 130 μL addition) to obtain 1.0 μM solutions of SAM-coated gold nanoparticles **Au-NP 1, Au-NP 2, Au-NP 3,** and **Au-NP 4** (total volume each 1.32 mL). These solutions could be stored at 4 °C for at least four weeks.

#### 3.3.3. Oxime Coupling of Benzaldehyde Functionalized Glycosides **3–20** to Aminooxy Coated Gold Nanoparticles

To a 1 M stock solution of glycosides **3–11**, **13–15**, and **17–20** (100 μL, 100 μmol) for **12** and **16** (50 μL, 50 μmol) in 10 mM acetate buffer (pH 4.7), a 1.0 μM solution of aminooxy-coated AuNPs (400 μL, 0.400 nmol) in Millipore water was added. These transparent deep red reaction mixtures were shaken at 40 °C for 16 h. Then the glyconanoparticles **GNP-1 –GNP 19** were purified by ten times repeating centrifuge filtration (Millipore Amicon Ultra, 50 kDa cutoff) and dilution with Millipore water. After the final filtration, the GNPs were diluted with Millipore water (400 μL: 200 μL filtration residue + 200 μL addition) to obtain 1.0 μM solutions (400 μL) of glyconanoparticles **GNP-1–GNP-19**. These solutions were storable at −20 °C for at least four weeks (storage in sterile filtrated, $NaN_3$ containing Millipore water to avoid bacterial contamination).

3.3.4. Determination of Carbohydrate Concentration on Gold Nanoparticles (Anthrone Test)

The amount of sugar on the GNPs can be easily determined by an anthrone-sulfuric acid assay. For this purpose, standard curves were recorded with the glycoconjugates **4** (for monosaccharides, MZ = 356.37 g/mol), **10** (for the disaccharides, MZ = 518.51 g/mol), and **14** (for trisaccharides, MZ = 705.70 g/mol). The stock solutions were diluted 1:100 (**4**: 3.56 µg/µL; **10**: 5.18 µg/µL; **14**: 7.05 µg/µL). Aliquots were taken from the dilutions to obtain a content of 7–176 µg per mL solution (see Table S1 in Supplementary Materials). The color reaction was carried out by adding 2.5 mL of a freshly prepared solution of 0.5% anthrone (wt%) in 95% sulfuric acid to 1 mL of carbohydrate solution in a test tube. The reaction mixture was carefully heated for 10 min in a boiling water bath. After cooling, the absorbance was immediately measured in the UV-Vis spectrometer at 620 nm (double determination). A blank sample was prepared without carbohydrate. To determine the content of carbohydrates on the nanoparticles, 1 µM GNP solution (40 µL/538 µg) was diluted to 1 mL and treated with anthrone-sulfuric acid as described for the preparation of the standard curves. The absorbance was measured at 620 nm. The formulas in Table S1 can be used to calculate the carbohydrate contents. The number of ligands per nanoparticle is given by the corresponding equation. The results are summarized in Table S2 in the Supplementary Materials.

3.3.5. ConA Binding Study Using TEM

Transmission electron microscopy was performed on a JEOL JEM 2100F (Tokyo, Japan) equipped with a 2 k × 2 k CCD camera and an EDXS system at 200 kV. For the measurements, a drop of the respective nanoparticle solutions was applied to a carbon-coated copper carrier and dried in air. For ConA binding studies, the nanoparticle solution was incubated with ConA solution (50 µM containing 10 mM $Ca^{2+}$ and $Mn^{2+}$ ions) for 12 h. Additionally, a drop of uranyl acetate solution was applied. The carrier was previously hydrophilized in an oxygen plasma (discharge grid) so that the drop of the aqueous solutions was distributed homogeneously over the sample carrier. The composition of the particles was determined using energy dispersive X-ray spectroscopy (EDXS). The particle sizes were determined, and the images of the nanoparticles were processed using DigitalMicrograph software (version 1.83) from Gatan Inc. (Pleasanton, CA, USA).

## 4. Conclusions

For studies of carbohydrate-protein and carbohydrate-carbohydrate interactions (CCI/CPI), powerful tools are required. Here, we present a novel modular approach. Facile and rapid syntheses for linking spacers and carbohydrate derivatives could be developed, and enhanced binding events were realized by controlling the amount and orientation of the ligand. For immobilization on biorepulsive aminoxy functionalized gold nanoparticles by oxime formation, aldehyde-functionalized glycan structures of mono-, di-, and complex trisaccharides were synthesized, employing several facile steps including olefin metathesis. Uniform gold nanoparticles were synthesized and subsequently functionalized. Glyconanoparticles were well characterized by TEM and the anthrone method. Effective immobilization and binding studies are exemplary presented for the lectin concanavalin A. This novel method showed advantageous perspectives to be employed in various biomimetic studies of carbohydrates and carbohydrate-based array development for diagnostics and screening.

**Supplementary Materials:** The following supporting information can be downloaded at: https://www.mdpi.com/article/10.3390/molecules30183765/s1, Table S1. Data for determination of standard curves and formula for calculation of carbohydrate content. Figure S1. Standard curves for calculation of carbohydrate content. Table S2. Absorption after Treatment of GNPs with Anthrone Method and Calculated Ligands pro GNP.

**Author Contributions:** Investigation, original draft preparation, S.K.; writing, review, and editing, J.T. All authors have read and agreed to the published version of the manuscript.

**Funding:** Financial support of this work by the Deutsche Forschungsgemeinschaft (SFB 470, A5) is gratefully acknowledged.

**Institutional Review Board Statement:** Not applicable.

**Informed Consent Statement:** Not applicable.

**Data Availability Statement:** The original contributions presented in this study are included in the article/Supplementary Material. Further inquiries can be directed to the corresponding authors.

**Acknowledgments:** We thank Marco Moeller, electro microscopy, CICbiomaGUNE, San Sebastian, Spain, for providing the TEM images. Furthermore, we are grateful to K.J. Jensen, Department of Natural Sciences, Faculty of Life Sciences, Centre for Carbohydrate Recognition and Signalling, University of Copenhagen, Thorvaldsensvej 40, 1871 Frederiksberg, Denmark, for providing a personal research sabbatical to SK.

**Conflicts of Interest:** Author Sebastian Kopitzki was employed by the company Biosynth GmbH. The remaining authors declare that the research was conducted in the absence of any commercial or financial relationships that could be construed as a potential conflict of interest.

## References


1. Ratner, D.M.; Adams, E.W.; Disney, M.D.; Seeberger, P.H. Tools for glycomics: Mapping interactions of carbohydrates in biological systems. *ChemBioChem* **2004**, *15*, 1375–1383. [CrossRef]
2. Lindhorst, T.K. Struktur und Funktion von Kohlenhydraten. *Chem. Unserer Zeit* **2000**, *34*, 38–52. [CrossRef]
3. Varki, A. Biological roles of oligosaccharides: All of the theories are correct. *Glycobiology* **1993**, *3*, 97–130. [CrossRef] [PubMed]
4. Ito, S. The enteric surface coat on cat intestinal microvilli. *J. Cell Biol.* **1965**, *27*, 475–491. [CrossRef]
5. Bennett, H.S. Morphological Aspects of Extracellular Polysaccharides. *J. Histochem. Cytochem.* **1963**, *11*, 14–23. [CrossRef]
6. Ito, S. Form and function of the glycocalyx on free cell surfaces. *Philos. Trans. B. Biol. Sci.* **1974**, *268*, 55–66.
7. Marth, J.D.; Grewal, P.K. Mammalian glycosylation in immunity. *Nat. Rev.* **2008**, *8*, 874–887. [CrossRef]
8. Nuenke, R.L.; Cunningham, L.W. Preparation and structural studies of ovalbumin glycopeptides. *J. Biol. Chem.* **1961**, *236*, 2452–2460. [CrossRef] [PubMed]
9. Schachter, H. The joys of HexNAc. The synthesis and function of *N*- and *O*-glycan branches. *Glycoconj. J.* **2000**, *17*, 465–483. [CrossRef]
10. Haltiwanger, R.S.; Lowe, J.B. Role of Glycosylation in Development. *Annu. Rev. Biochem.* **2004**, *73*, 491–537. [CrossRef]
11. Low, M.G.; Ferguson, M.A.; Silman, I. Cell-Surface Anchoring of Proteins via Glycosyl-Phosphatidylinositol Structures. *Annu. Rev. Biochem.* **1988**, *57*, 285–320.
12. Ohtsubo, K.; Marth, J.D. Glycosylation in cellular mechanisms of health and disease. *Cell* **2006**, *126*, 855–867. [CrossRef]
13. Dwek, R.A. Glycobiology: Toward Understanding the Function of Sugars. *Chem. Rev.* **1996**, *96*, 683–720. [CrossRef]
14. Ono, M.; Hakomori, S. Glycosylation defining cancer cell motility and invasiveness. *Glycoconj. J.* **2004**, *20*, 71–78.
15. Lis, H.; Sharon, N. Lectins: Carbohydrate-Specific Proteins That Mediate Cellular Recognition. *Chem. Rev.* **1998**, *98*, 637–674. [CrossRef] [PubMed]
16. Simanek, E.E.; McGarvey, G.J.; Jablonowski, J.A.; Wong, C.-H. Selectin minus sign Carbohydrate Interactions: From Natural Ligands to Designed Mimics. *Chem. Rev.* **1998**, *98*, 833–862. [CrossRef] [PubMed]
17. Hakomori, S.-I. Carbohydrate-carbohydrate interaction as an initial step in cell recognition. *Pure Appl. Chem.* **1991**, *63*, 473–482. [CrossRef]
18. Hernaiz, M.J.; de la Fuente, J.M.; Barrientos, A.G.; Penades, S. A Model System Mimicking Glycosphingolipid Clusters to Quantify Carbohydrate Self-Interactions by Surface Plasmon Resonance. *Angew. Chem. Int. Ed.* **2002**, *41*, 1554–1557. [CrossRef]

19. Lundquist, J.J.; Toone, E.J. The Cluster Glycoside Effect. *Chem. Rev.* **2002**, *102*, 555–578. [CrossRef]
20. Kiessling, L.L.; Pohl, N.L. Strength in numbers: Non-natural polyvalent carbohydrate derivatives. *Chem. Biol.* **1996**, *3*, 71–77. [CrossRef]
21. Garcia-Manyes, S.; Bucior, I.; Ros, R.; Anselmetti, D.; Sanz, F.; Burger, M.M.; Fernandez-Busquets, X. Proteoglycan mechanics studied by single-molecule force spectroscopy of allotypic cell adhesion glycans. *J. Biol. Chem.* **2006**, *280*, 5992–5999. [CrossRef]
22. Bain, C.D.; Whitesides, G.M. Modeling organic surfaces with self-assembled monolayers. *Angew. Chem. Int. Ed. Engl.* **1989**, *28*, 506–528. [CrossRef]
23. Marradi, M.; Chiodo, F.; Garcia, I.; Penades, S. Glyconanoparticles as multifunctional and multimodal carbohydrate systems. *Chem. Soc. Rev.* **2013**, *42*, 4728–4745. [CrossRef]
24. Kim, Y.; Hyun, J.Y.; Shin, I. Multivalent glycans for biological and biomedical applications. *Chem. Soc. Rev.* **2021**, *50*, 10567–10593. [CrossRef] [PubMed]
25. Svedhem, S.; Oehberg, L.; Borrelli, S.; Valiokas, R.; Andersson, M.; Oscarson, S.; Svensson, S.C.T.; Liedberg, B.; Konradsson, P. Synthesis and Self-Assembly of Globotriose Derivatives: A Model System for Studies of Carbohydrate-Protein Interactions. *Langmuir* **2002**, *18*, 2848–2858. [CrossRef]
26. Dhayal, M.; Ratner, D.M. XPS and SPR Analysis of Glycoarray Surface Density. *Langmuir* **2009**, *25*, 2181–2187. [CrossRef]
27. Seo, J.H.; Adachi, K.; Lee, B.K.; Kang, D.G.; Kim, Y.K.; Kim, K.R.; Lee, H.Y.; Kawai, T.; Cha, H.J. Facile and Rapid Direct Gold Surface Immobilization with Controlled Orientation for Carbohydrates. *Bioconjugate Chem.* **2007**, *18*, 2197–2201. [CrossRef]
28. Boisselier, E.; Astruc, D. Gold nanoparticles in nanomedicine: Preparations, imaging, diagnostics, therapies and toxicity. *Chem. Soc. Rev.* **2009**, *38*, 1759–1782. [CrossRef] [PubMed]
29. Faraday, M. The Bakerian Lecture: Experimental Relations of Gold (and Other Metals) to Light. *Philos. Trans. R. Soc.* **1857**, *147*, 145–181.
30. Mie, G. Beiträge zur Optik trüber Medien, speziell kolloidaler Metallösungen. *Ann. Phys.* **1908**, *25*, 377–445. [CrossRef]
31. Daniel, M.-C.; Astruc, D. Gold nanoparticles: Assembly, supramolecular chemistry, quantum-size-related properties, and applications toward biology, catalysis, and nanotechnology. *Chem. Rev.* **2004**, *104*, 293–346. [CrossRef]
32. Niemeyer, C.M. Nanoparticles, Proteins, and Nucleic Acids: Biotechnology Meets Materials Science. *Angew. Chem. Int. Ed.* **2001**, *40*, 4128–4158. [CrossRef]
33. Katz, E.; Willner, I. Integrated Nanoparticle–Biomolecule Hybrid Systems: Synthesis, Properties, and Applications. *Angew. Chem. Int. Ed.* **2004**, *33*, 6042–6108. [CrossRef] [PubMed]
34. de la Fuente, J.M.; Penades, S. Glyconanoparticles: Types, synthesis and applications in glycoscience, biomedicine and material science. *Biochim. Biophys. Acta* **2006**, *1760*, 636–651. [PubMed]
35. Chiodo, F.; Marradi, M.; Park, J. Galactofuranose-coated gold nanoparticles elicit a pro-inflammatory response in human monocyte-derived dendritic cells and are recognized by DC-SIGN. *ACS Chem Biol.* **2014**, *9*, 383–389. [CrossRef]
36. Latxague, L.; Gaubert, A.; Barthelemy, P. Recent advances in the chemistry of glycoconjugate amphiphiles. *Molecules* **2018**, *23*, 89. [CrossRef]
37. Vijayakumar, S.; Chen, J.; Kalaiselvi, V.; Divya, M.; González-Sánchez, Z.I.; Durán-Lara, E.F.; Vaseeharan, B. Antibacterial and antibiofilm activities of marine polysaccharide laminarin formulated gold nanoparticles: An ecotoxicity and cytotoxicity assessment. *J. Env. Chem. Eng.* **2021**, *9*, 105514. [CrossRef]
38. Turkevitch, J.; Stevenson, P.C.; Hillier, J. A study of the nucleation and growth processes in the synthesis of colloidal gold. *Disuss. Faraday Soc.* **1951**, *11*, 55–75. [CrossRef]
39. Enüstün, E.J.; Turkevitch, J. Coagulation of Colloidal Gold. *J. Am. Chem. Soc.* **1963**, *85*, 3317–3328. [CrossRef]
40. Sajanlal, P.R.; Khatun, E.; Pradeep, T. *Kirk-Othmer Encyclopedia of Chemical Technology*; Wiley: Hoboken, NJ, USA, 2019.
41. Broadbent, M.; Chadwick, S.J.; Brust, M.; Volk, M. Gold Nanoparticles for Photothermal and Photodynamic Therapy. *ACS Omega* **2024**, *9*, 44846–44859. [CrossRef]
42. Giersig, M.; Mulvaney, P. Preparation of ordered colloid monolayers by electrophoretic deposition. *Langmuir* **1993**, *9*, 3404–3413. [CrossRef]
43. Kimling, J.; Maier, M.; Okenve, B.; Kotaidis, V.; Ballot, H.; Plech, A. Turkevich method for gold nanoparticle synthesis revisited. *J. Phys. Chem. B* **2006**, *110*, 5700–5757. [CrossRef]
44. Brust, M.; Walker, M.; Bethell, D.; Schiffrin, D.J.; Whyman, R. Synthesis of thiol-derivatised gold nanoparticles in a two-phase Liquid–Liquid system. *J. Chem. Soc. Chem. Commun.* **1994**, 801–802. [CrossRef]
45. Templeton, A.C.; Wuelfing, W.P.; Murray, R.W. Monolayer-protected cluster molecules. *Acc. Chem. Res.* **2001**, *33*, 27–36. [CrossRef]
46. Rojo, J.; Diaz, V.; de la Fuente, J.M.; Segura, I.; Barrientos, A.G.; Riese, H.H.; Bernad, A.; Penades, S. Gold glyconanoparticles as new tools in antiadhesive therapy. *ChemBioChem* **2004**, *5*, 291–297. [CrossRef]

47. Larsen, L.; Thygesen, M.B.; Guillaumie, F.; Willats, W.G.T.; Jensen, K.J. Solid-phase chemical tools for glycobiology. *Carbohydr. Res.* **2006**, *341*, 1209–1234. [CrossRef] [PubMed]
48. Love, J.C.; Estroff, L.A.; Kriebel, J.K.; Nuzzo, R.G.; Whitesides, G.M. Self-Assembled Monolayers of Thiolates on Metals as a Form of Nanotechnology. *Chem. Rev.* **2005**, *105*, 1103–1169. [CrossRef]
49. Prime, K.L.; Whitesides, G.M. Adsorption of proteins onto surfaces containing end-attached oligo(ethylene oxide): A model system using self-assembled monolayers. *J. Am. Chem. Soc.* **1993**, *115*, 10714–10721. [CrossRef]
50. Fürstner, A. Olefin metathesis and beyond. *Angew. Chem. Int. Ed.* **2000**, *39*, 3012–3043. [CrossRef]
51. Hoveyda, A.H.; Zhugralin, A.R. The remarkable metal-catalysed olefin metathesis reaction. *Nature* **2007**, *450*, 243–251. [CrossRef]
52. Connon, S.J.; Blechert, S. Recent developments in olefin cross-metathesis. *Angew. Chem. Int. Ed.* **2003**, *42*, 1900–1923. [CrossRef] [PubMed]
53. Lin, Y.A.; Chalker, J.M.; Floyd, N.; Bernardes, G.J.L.; Davis, B.G. Allyl Sulfides are Privileged Substrates in Aqueous Cross-Metathesis: Application to Site-Selective Protein Modification. *J. Am. Chem. Soc.* **2008**, *130*, 9642–9643. [CrossRef]
54. Ritter, T.; Hejl, A.; Wenzel, A.G.; Funk, T.W.; Grubbs, R.H. A standard system of characterization for olefin metathesis catalysts. *Organometallics* **2006**, *25*, 5740–5745. [CrossRef]
55. Kopitzki, S.; Jensen, K.S.; Thiem, J. Synthesis of Benzaldehyde Functionalized Glycans: Novel Approach Towards Glyco-SAMs as Tools for Surface Plasmon Resonance Studies. *Chem. Eur. J.* **2010**, *16*, 7017–7029. [CrossRef] [PubMed]
56. Morgan, J.P.; Morill, C.; Grubbs, R.H. Selective Ring Opening Cross Metathesis of Cyclooctadiene and Trisubstituted Cycloolefins. *Org. Lett.* **2002**, *4*, 67–70. [CrossRef]
57. Plettenburg, O.; Mui, C.; Bodmer-Narkevitch, V.; Wong, C.-H. Rapid preparation of glycolipid libraries by cross metathesis. *Adv. Synth. Catal.* **2002**, *344*, 622–626. [CrossRef]
58. Schmidt, R.R.; Michel, J.; Roos, M. Glycosylimidate, 12 Direkte Synthese von O-α-und O-β-Glycosyl-imidaten. *Liebigs Ann. Chem.* **1984**, 1343–1357. [CrossRef]
59. Thygesen, M.B.; Sauer, J.; Jensen, K.J. Chemoselective capture of glycans for analysis on gold nanoparticles: Carbohydrate oxime tautomers provide functional recognition by proteins. *Chem. Eur. J.* **2009**, *15*, 1649–1660. [CrossRef]
60. Jin, S.; Miduturu, C.V.; McKinney, D.C.; Silverman, S.K. Synthesis of Amine- and Thiol-Modified Nucleoside Phosphoramidites for Site-Specific Introduction of Biophysical Probes into RNA. *J. Org. Chem.* **2005**, *70*, 4284–4299. [CrossRef]
61. Sperling, R.A.; Rivera Gil, P.; Zhang, F.; Zanella, M.; Parak, W.J. Biological applications of gold nanoparticles. *Chem. Soc. Rev.* **2008**, *37*, 1896–1908. [CrossRef]
62. Zhou, W.; Gao, X.; Liu, D.; Chen, X. Gold Nanoparticles for In Vitro Diagnostics. *Chem. Rev.* **2015**, *115*, 10575–10636. [CrossRef]
63. Levy, R.; Thanh, N.T.K.; Doty, R.C.; Hussain, I.; Nichols, R.J.; Schiffrin, D.J.; Brust, M.; Fernig, D.G. Rational and Combinatorial Design of Peptide Capping Ligands for Gold Nanoparticles. *J. Am. Chem. Soc.* **2004**, *126*, 10076–10084. [CrossRef]
64. Grabar, K.C.; Freeman, R.G.; Hommer, M.B.; Natan, M.J. Preparation and Characterization of Au Colloid Monolayers. *J. Anal. Chem.* **1995**, *67*, 735–743. [CrossRef]
65. Frens, G. Controlled Nucleation for the Regulation of the Particle Size in Monodisperse Gold Suspensions. *Nat. Phys. Sci.* **1973**, *241*, 20–22. [CrossRef]
66. Carvalho de Souza, A.; Halkes, K.M.; Meeldijk, J.M.; Verkleij, A.J.; Vliegenthart, J.F.G.; Kamerling, J.P. Gold glyconanoparticles as probes to explore the carbohydrate-mediated self-recognition of marine sponge cells. *ChemBioChem* **2005**, *6*, 828–831. [CrossRef] [PubMed]
67. Thygesen, M.B.; Sørensen, K.K.; Clo, E.; Jensen, K.J. Direct chemoselective synthesis of glyconanoparticles from unprotected reducing glycans and glycopeptide aldehydes. *Chem. Commun.* **2009**, 6367–6369. [CrossRef] [PubMed]
68. Thygesen, M.B.; Jensen, K.J. Carbohydrate-modified gold nanoparticles. *Carbohydr. Nanotechnol.* **2016**, 79–98.
69. Kopitzki, S.; Thiem, J. Short Synthetic Route to Benzaldehyde—Functionalized Idose and Talose Derivatives by Acetoxonium Ion Rearrangements. *Eur. J. Org. Chem.* **2013**, *2013*, 4008–4016. [CrossRef]
70. Kopitzki, S.; Dilmaghani, K.A.; Thiem, J. Synthesis of benzaldehyde-functionalized Lewis$^{X}$ trisaccharide analogs for glyco-SAM formation. *Tetrahedron* **2013**, *69*, 10621–10636. [CrossRef]
71. Wittmann, V.; Pieters, R.J. Bridging lectin binding sites by multivalent carbohydrates. *Chem. Soc. Rev.* **2013**, *42*, 4492–4503. [CrossRef]
72. de Souza, A.C.; Halkes, K.M.; Meeldijk, J.M.; Verkleij, A.J.; Vliegenthart, J.F.G.; Kamerling, J.P. Synthesis of gold glyconanoparticles: Possible probes for the exploration of carbohydrate-mediated self-recognition of marine sponge cells. *Eur. J. Org. Chem.* **2004**, *2004*, 4323–4339. [CrossRef]
73. Mrksich, M. A surface chemistry approach to studying cell adhesion. *Chem. Soc. Rev.* **2000**, *29*, 267–273. [CrossRef]
74. Scherz, H.; Bonn, G. *Analytical Chemistry of Carbohydrates*; Thieme: Stuttgart, Germany, 1998.

75. Chien, Y.-Y.; Jan, M.-D.; Adak, A.K.; Tzeng, H.-C.; Lin, Y.-P.; Chen, Y.-J.; Wang, K.-T.; Chen, C.-T.; Chen, C.-C.; Lin, C.C. Globotriose-functionalized gold nanoparticles as multivalent probes for Shiga-like toxin. *ChemBioChem* **2008**, *9*, 1100–1109. [CrossRef] [PubMed]
76. Dubios, M.; Gilles, K.A.; Hamilton, J.K.; Rebers, P.A.; Smith, F. Colorimetric method for determination of sugars and related substances. *Anal. Chem.* **1956**, *28*, 350–356. [CrossRef]
77. Koehler, L.H. Differentiation of carbohydrates by anthrone reaction rate and color intensity. *Anal. Chem.* **1952**, *24*, 1576–1579. [CrossRef]
78. Lin, C.C.; Yeh, Y.C.; Yang, C.Y.; Chen, G.F.; Chen, Y.C.; Wu, Y.C.; Chen, C.C. Quantitative analysis of multivalent interactions of carbohydrate-encapsulated gold nanoparticles with concanavalin A. *Chem. Commun.* **2003**, 2920–2921. [CrossRef] [PubMed]
79. de la Fuente, J.M.; Eaton, P.; Barrientos, A.G.; Menendez, M.; Penades, S. Thermodynamic Evidence for $Ca^{2+}$-Mediated Self-Aggregation of Lewis X Gold Glyconanoparticles. A Model for Cell Adhesion via Carbohydrate-Carbohydrate Interaction. *J. Am. Chem. Soc.* **2005**, *127*, 6192–6197. [CrossRef]
80. Penney, D.P.; Powers, J.M.; Frank, M.; Churukian, C. Analysis and testing of biological stains—The biological stain comission procedures. *Biotech. Histochem.* **2002**, *77*, 237–275. [CrossRef]
81. de Carlo, S.; Harris, J.R. Negative staining and cryo-negative staining of macromolecules and viruses for TEM. *Micron* **2011**, *42*, 117–131. [CrossRef]

*molecules*

MDPI

*Article*

# Synthesis and Docking Studies of Glycolipids Inspired by *Bacteroides fragilis* Lipid A

**Davie Kenneth [1,†], Cristina Manuela Santi [1,2,†], Francesca Tanda [1], Alessia Izzo [1], Monica Civera [1], Giuseppe D'Orazio [1,*] and Luigi Lay [1,*]**

1 Department of Chemistry, Università degli Studi di Milano, Via C. Golgi 19, 20133 Milan, Italy; davie.kenneth@unimi.it (D.K.); cristina.santi@iit.it (C.M.S.); francesca.tanda@unimi.it (F.T.); alessia.izzo96@gmail.com (A.I.); monica.civera@unimi.it (M.C.)

2 Laboratory for Polymers and Biomaterials, Fondazione Istituto Italiano di Tecnologia, 16163 Genova, Italy

* Correspondence: giuseppe.dorazio@unimi.it (G.D.); luigi.lay@unimi.it (L.L.)

† These authors equally contributed to this work.

**Abstract**

*Bacteroides fragilis*, a prominent commensal of the human gut microbiota, plays a vital role in immune system regulation through its capsular polysaccharide A (PSA), which requires a glycolipid anchor structurally reminiscent of lipid A. While canonical *Escherichia coli* lipid A acts as a potent TLR4 agonist contributing to septic shock and inflammatory disorders, certain *B. fragilis*-derived glycolipids demonstrate antagonistic effects, offering potential as anti-inflammatory agents. In this study, we report the synthesis and preliminary computational evaluation of a library of glycolipids inspired by *B. fragilis* lipid A. Three lipid As, including a tetra-acylated 1-phosphoryl lipid A analog (Tetra C-1), were synthesized and assessed using molecular docking simulations targeting the human TLR4/MD-2 complex. Docking results reveal that Tetra C-1 exhibits more favorable antagonist binding characteristics compared to the well-studied TLR4 antagonist Eritoran. This work highlights a microbiota-informed strategy for the development of novel TLR4 antagonists, potentially enabling targeted modulation of innate immunity for therapeutic applications in inflammatory diseases and as vaccine adjuvants.

**Keywords:** lipid A; glycolipids; *Bacteroides fragilis*; immunomodulation; glycoderivatives; anti-inflammatory agents; molecular docking

## 1. Introduction

The innate immune system is highly sensitive to microbial signatures of invaders, with lipopolysaccharides (LPS)—particularly their lipid A component—playing a pivotal role in triggering the immune response [1]. Lipid A, composed of a mono- or di-phosphorylated di-glucosamine core with multiple fatty acid chains, anchors LPS into the bacterial membrane and is largely responsible for the toxicity of Gram-negative bacteria [2]. As the bioactive core of LPS, bacterial lipid As play a dual/Janus-like role. On one side, they are vital immunostimulants, able to modulate humans' immune response through their interaction with key components of the innate immune system named Toll-like receptors (TLRs). Some particular lipid As have been proven to have a low-toxicity profile, stimulating a mild activation of the innate immune system capable of boosting the immune response to co-administered antigens. On the other side, other lipid A structures lead to excessive immune reaction and, upon being lysed by the immune system, their fragments are released into the bloodstream, causing dysregulated inflammatory processes and endotoxin-related pathologies such as fever, diarrhea, and possible fatal endotoxic shock [3,4]. This paradox highlights the need for precise modulation of innate immunity and emphasizes the therapeutic potential of targeting Toll-like receptor 4 (TLR4)—the central sensor for LPS in several types of immune cells like monocytes, neutrophils, dendritic cells, and macrophages [5–7] as well as intestinal epithelial cells [8,9].

*Escherichia coli* lipid A (Figure 1), the reference structure of this family of glycolipids, strongly activates the TLR4/MD-2 (Myeloid Differentiation factor 2) complex, initiating a potent inflammatory cascade, which strictly depends on the lipid A/LPS structure and concentration [10]. Structure-activity relationship studies reveal that the endotoxicity of lipid A is influenced by its phosphorylation level and the number, length, and spatial arrangement of acyl chains [11]. Some over- or under-acylated lipid A variants behave as TLR4 antagonists, suggesting that structural modifications can provide lipid A with an immunomodulatory activity [12]. Although several antagonists have been developed, such as Eritoran (based on *Rhodobacter sphaeroides* lipid A, Figure 1) [13], clinical outcomes have been limited, reflecting the need for better-designed molecules with enhanced efficacy and safety profiles. Specific lipid A derivatives also hold promise in vaccine adjuvant development. Many subunit vaccines rely on adjuvants to improve immune response strength and duration, lower the required dosage, and enhance the generation of memory B cells [14]. Adjuvants can also boost immune responses and accelerate initial vaccine efficacy [15].

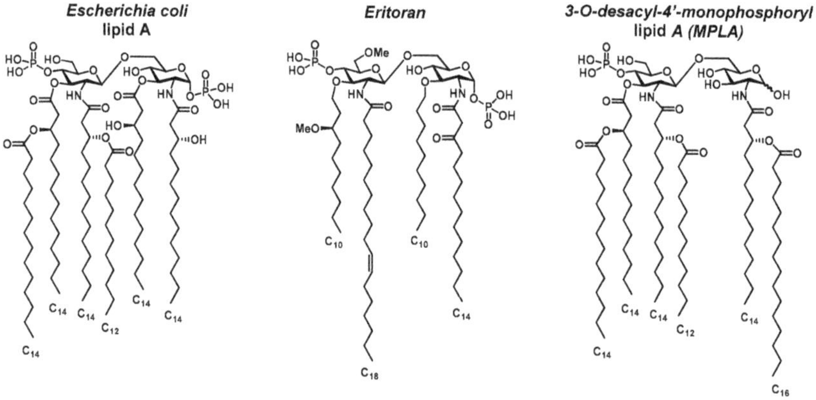


**Figure 1.** Structures of different lipid A from *E. coli*, Eritoran, and MPLA.

To uncouple the toxicity of lipid A from its immune-stimulating properties, Masihi et al. developed monophosphoryl lipid A (MPLA) in 1986 [16]. Derived from lipid A of *Salmonella minnesota* R595, MPLA was obtained by removing one acyl chain and one phosphate group, reducing toxicity by 100-fold while preserving immunostimulatory effects (Figure 1). MPLA acts as a TLR4 agonist, promoting dendritic cell maturation and Th1 CD4+ T cell polarization [17–19]. MPL®, an MPLA-based adjuvant, is the only TLR4 ligand approved for human vaccines and is included in Cervarix® and Fendrix® (by GSK, London, UK), targeting HPV and Hepatitis B, respectively [18].

A promising frontier in developing new and safer lipid A-based immunomodulatory agents lies in the exploration of commensal bacteria [20] and their unique glycolipid architectures—especially those with immunomodulatory rather than pro-inflammatory properties. The human immune system's tolerance toward symbiotic bacteria makes them attractive candidates for novel, safe vaccine adjuvants. Among these, *Bacteroides fragilis*, a Gram-negative bacterium inhabiting the human gastrointestinal tract, has emerged as a model organism.

*Bacteroides fragilis* is known for its capsular polysaccharide A (PSA), which promotes immune tolerance and anti-inflammatory cytokine production, particularly IL-10, through interactions with dendritic cells and regulatory T cells [21–23]. In 2019, the Kasper group revealed that the lipid A moiety anchoring PSA plays a crucial immunomodulatory role [24]. The authors demonstrated that upon the glycolipid-mediated activation of the TLR1-TLR2 heterodimer and polysaccharide-based Dectin-1 engagement, PSA processing and its presentation to CD4+ T cells were induced, ultimately leading to production of IL-10, indicating an anti-inflammatory response which did not occur when PSA lacked its lipid anchor, hence highlighting the importance of the glycolipid moieties. By means of NMR and LC/MS-MS, the authors characterized *B. fragilis* lipid A as having a di-glucosamine backbone with variable acylation (C15–C17 chains) and phosphorylation patterns [24], along with tri- and tetra-acylated variants. Compared to *E. coli* lipid A, *B. fragilis* lipid A shows greater structural diversity, which may explain its distinct TLR2/TLR4-mediated immune responses [25].

Interestingly, as early as 1989, Weintraub et al. proposed a lipid A structure for *B. fragilis*, characterized by a distinct pattern of fatty acid chains, including terminally-branched chains [26] (Penta-C1*, Figure 2), the same features that were later acknowledged in the more recent work of Kasper [24]. Furthermore, this structure was recently synthesized and immunologically evaluated [27]. The compound exhibited an antagonistic response towards the production of pro-inflammatory cytokines induced by *E. coli* lipid A, which is a natural ligand of the TLR4/MD-2 heterodimer [27].

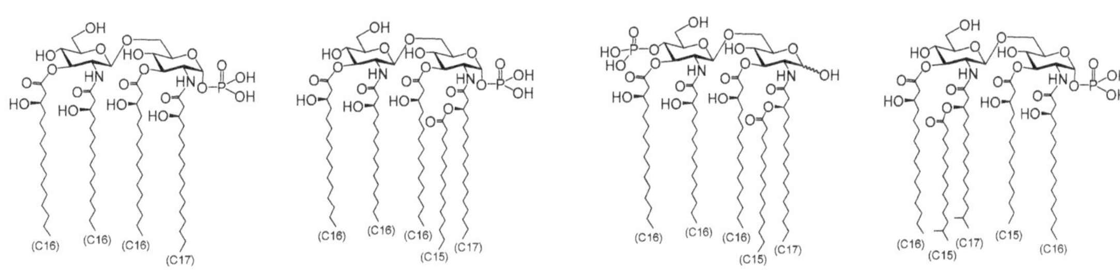


**Figure 2.** Structures of the lipid A analogues object of this work.

Despite its biological relevance, the detailed structure of *B. fragilis* lipid A remains only partially understood. Due to gaps in structural characterization, our study is focused on the synthesis and computational evaluation of glycolipid compounds inspired by *B. fragilis* lipid A, considering the various structural hypotheses proposed to date.

In this work, we report the synthesis of three lipid A structures (**Tetra-C1**, **Penta-C1**, and **Penta-C4′**, Figure 2) along with preliminary molecular docking analysis to assess their binding affinity to the TLR4/MD-2 complex. The compounds panel included both newly synthesized structures and the one previously reported by Weintraub and Boons [26,27] (**Penta-C1***, Figure 2). We evaluated the antagonistic potential of synthesized lipid A analogs against the TLR4/MD-2 complex through in silico methods, comparing their inhibitory efficacy to that of the known antagonist Eritoran, and aimed to highlight the distinctive functional properties of *Bacteroides fragilis*-derived lipid A to support its further biological validation as a promising immunomodulatory candidate.

## 2. Results and Discussion

### 2.1. Library Design

Inspired by literature, particularly the studies conducted in 2019 [24], in 1989 [26], and more recently in 2024 [27]—we designed a set of lipid A analogues featuring three novel structures: **Tetra-C1**, **Penta-C1**, and **Penta-C4′** (Figure 2). These compounds contain either four or five fatty acid chains, including C16 and C17 chains bearing 3-hydroxy groups with (*R*)-configuration. Each chain is directly attached to the hydroxyl or amino functionalities of the di-glucosamine disaccharide backbone. In both **Penta-C1** and **Penta-C4′**, the 3-hydroxy group of the C17 chain connected to the amino group of the reducing-end glucose is further acylated with a linear C15 fatty acid. Phosphorylation occurs at the anomeric position of the reducing-end glucose in **Tetra-C1** and **Penta-C1**, whereas in **Penta-C4′** the phosphate group is positioned at C4′.

The library also includes the lipid A analog **Penta-C1***, originally proposed by Weintraub [26] and subsequently synthesized and evaluated by Boons in 2024 [27]. This compound exhibits a distinct fatty acid profile, including C15 and C17 branched chains, setting it apart from the newly designed structures. All three novel analogues were synthesized from a common disaccharide intermediate. The complete set—comprising both the newly developed and the previously proposed structures—was subjected to molecular docking simulations with the TLR4/MD-2 heterodimer to evaluate binding affinities. The results were compared to reference ligands, including Eritoran, a well-known TLR4 antagonist [13,28], and lipid IVa from *E. coli*, a canonical TLR4 antagonist [29–31].

### 2.2. Synthesis of Target Compounds-General Approach

The synthesis of the lipid A analogues (**Tetra-C1**, **Penta-C1**, and **Penta-C4′**) relied on two key structural components: a protected di-glucosamine disaccharide (**3**) and three fatty acid chains (**4**, **6**, and **7**), derived from commercially available natural fatty acids. In our approach (Scheme 1), the di-glucosamine disaccharide **3**, obtained via glycosylation of donor **1** and acceptor **2**, served as the carbohydrate scaffold. Protective groups on the donor and acceptor were selected to enable stereoselective β-glycosylation and facilitate selective deprotection at the sites intended for functionalization with fatty acid chains. To introduce 16-carbon chains at positions 3, 2′, and 3′ of the disaccharide, phthalimide and acetyl groups were employed to protect the amine at 2′ and hydroxy groups at 3 and 3′, respectively, due to their susceptibility to base-mediated cleavage. In order to install the fatty acid chain **7** on the amine at position 2, this group was temporarily masked as an azide, allowing for selective reduction and functionalization post-chain **4** introduction. Selective phosphorylation at the anomeric position (C1)—required for synthesizing **Tetra-C1**—was achieved by 1-O-silyl ether deprotection using fluoride-based reagents. Additionally, positions 4′ and 6′ were protected as a benzylidene acetal, which was removed in the final step of the **Penta-C1** synthesis alongside benzyl groups. Alternatively, regioselective reductive ring opening of the benzylidene acetal revealed a free hydroxy group at C4′, which was subsequently phosphorylated to generate compound **Penta-C4′**.

### 2.3. Synthesis of Disaccharide 3

The disaccharide backbone **3** was synthesized via glycosylation with thioglucoside donor **1** of acceptor **2**. The preparation of acceptor **2** began from D-glucosamine hydrochloride, which was transformed into the azido derivative through a diazotransfer reaction using the safe and stable reagent imidazole-1-sulfonyl azide hydrogen sulfate [32], with subsequent O-acetylation to afford intermediate **9**. Selective removal of the acetyl group at the anomeric position, followed by 1-O-silylation, yielded compound **11**. This intermediate was then deacetylated, regioselectively protected at positions 4 and 6 with a benzylidene acetal (giving alcohol **12**), and finally acetylated at 3-OH to generate compound **13**. Reductive opening of the benzylidene acetal using phenylboron dichloride as Lewis acid and triethylsilane as the reducing agent [33] afforded acceptor **2** with an overall yield of 56% across seven synthetic steps.

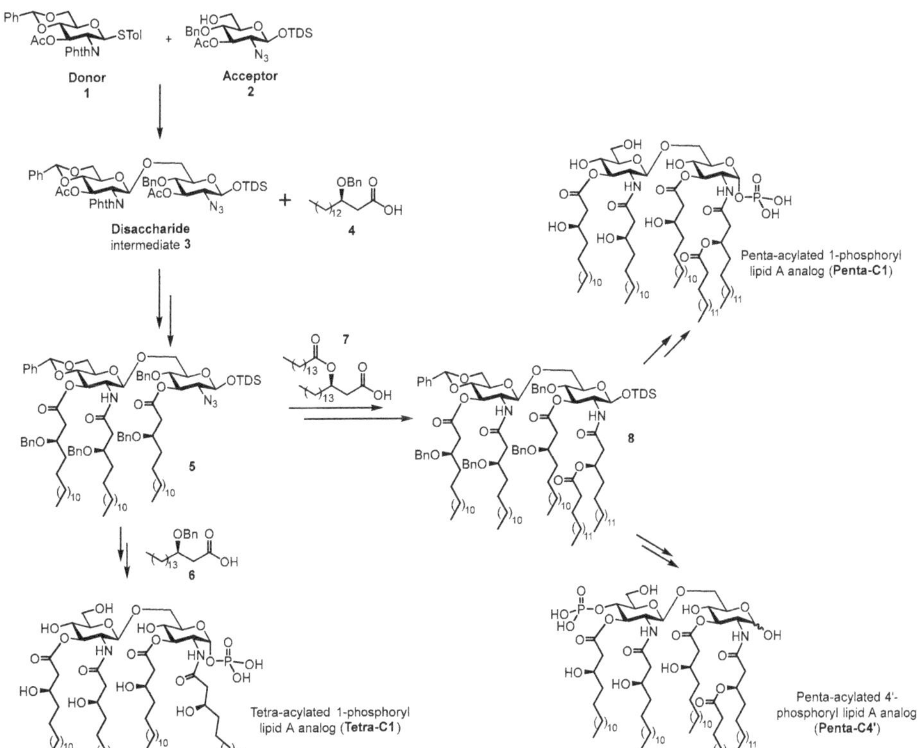


**Scheme 1.** General scheme for the synthesis of lipid A compounds **Tetra-C1**, **Penta-C1**, and **Penta-C4′**.

Donor **1** was prepared according to the route reported by Liu and Wei in 2012 [34], employing phthalimide as the amino-protecting group. This choice enabled its removal under basic conditions using hydrazine hydrate, which also allowed simultaneous deprotection of the acetyl groups at positions C3′ and C3 of the disaccharide—thereby unveiling the attachment sites for the C16 fatty acid chain derivative **4**. Glucosamine hydrochloride was first reacted with phthalic anhydride to afford the corresponding imide, which was then fully O-acetylated. Thioglycoside formation, subsequent O-deacetylation, and benzylidene protection at 4-OH and 6-OH furnished compound **14** [34]. Final acetylation at 3-OH yielded donor **1**, with an overall yield of 52% (Scheme 2).

The final disaccharide backbone was synthesized by glycosylation of acceptor **2** with donor **1**, promoted by TMSOTf/NIS at –20 °C, affording intermediate **3** in 71% yield (Scheme 2). $^{1}$H-NMR spectroscopy confirmed the β-configuration of the newly formed glycosidic linkage, as evidenced by the measured $J_{1\text{-}2}$ coupling constant of 8.4 Hz for H1′, typical of 1,2-trans glycosides.

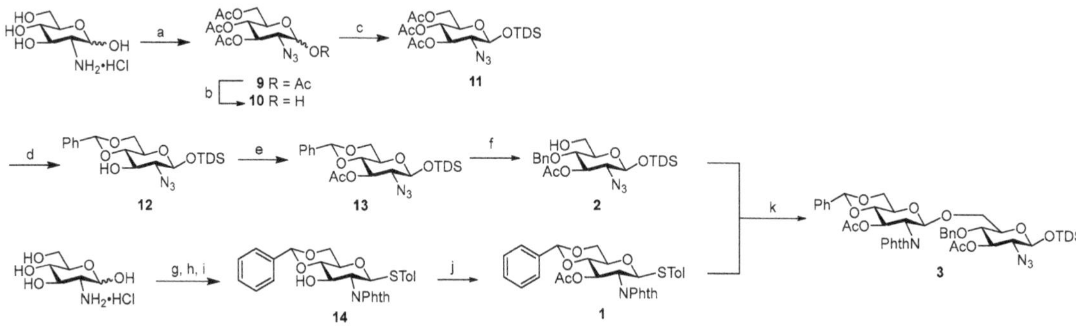

**Scheme 2.** Synthesis of disaccharide **2**. Reagents and conditions. (a) 1. Imidazole-1-sulfonyl azide hydrogen sulfate, $K_2CO_3$, MeOH, 0 °C to r.t., 4 h; 2. $Ac_2O$, DMAP, Py, $CH_2Cl_2$, 0 °C to r.t., 12 h 84% over two steps; (b) hydrazine hydrate, AcOH, DMF, 0 °C to r.t., 4 h; (c) TDSCl, imidazole, $CH_2Cl_2$, 0 °C to r.t., 12 h, 92% over two steps; (d) 1. NaOMe, MeOH, r.t., 1 h; 2. PhCH $(OMe)_2$, *p*TSA, $CH_3CN$, r.t., 8 h; (e) $Ac_2O$, DMAP, Py, $CH_2Cl_2$, 0 °C to r.t., 4 h, 80% over three steps; (f) $PhBCl_2$, $Et_3SiH$, $CH_2Cl_2$, −78 °C, 3 h, 93%, (g) 1. Phthalic anhydride, MeONa, MeOH, $Et_3N$, r.t., 24 h, 76% [34]; 2. $Ac_2O$, Py, 0 °C to r.t., 48 h, 90% [34]; (h) p-thiocresol, $BF_3$ $OEt_2$, $CH_2Cl_2$, 0 °C to r.t., 24 h, 91% [34]; (i) 1. NaOMe, MeOH, r.t., 1 h, quant. [34]; 2. PhCH $(OMe)_2$, *p*TSA, THF, reflux, 10 h, 84% [34]; (j) $Ac_2O$, DMAP, Py, $CH_2Cl_2$, 0 °C to r.t., 2 h, 95%; (k) NIS, TMSOTf, 4 Å MS, dry $CH_2Cl_2$, −20 °C, 2 h, 71%.

### 2.4. Synthesis of the Fatty Acid Chain Derivatives

The synthesis of fatty acid chain derivatives **4**, **6**, and **7** followed a common route starting from two commercially available fatty acids—tetradecanoic and pentadecanoic acid. These precursors were elongated into the corresponding β-ketoesters using a method reported by Brinkerhoff et al. in 2014 [35]. The approach involved initial condensation with Meldrum's acid and DCC to form enol derivatives, which were subsequently converted into the desired β-ketoesters via acid-mediated reaction with the appropriate alcohol. Using this strategy, we obtained three β-ketoesters (**15**, **16**, and **17**) with yields ranging from 80% to 90%. Compound **17**, which incorporated a tert-butyl group as the alcohol component, proved to be particularly advantageous for the synthesis of the double-chain fatty acid derivative **7**. The tert-butyl group could be selectively removed under mild acidic conditions, thereby preserving the ester functionality at the 3-OH position.

Subsequent efforts focused on the stereoselective reduction of the ketone functionality to achieve the desired (*R*)-configuration at the β-carbon. The application of Noyori reduction conditions, employing the freshly prepared dibromodiphosphineruthenium(II) complex [(R)-BINAP]$RuBr_2$ [36], proved highly efficient—yielding (R)-β-hydroxy fatty acids **18**–**20** in good overall yields. The absolute configuration of the resulting β-hydroxy esters was confirmed via Mosher's ester analysis. This involved performing $^1$H-NMR experiments on the corresponding (*S*)- and (*R*)-MTPA esters [37] (see Supporting Information), as well as complementary NMR studies using the chiral lanthanide shift reagent $Eu(hfc)_3$ [38] (see Supporting Information). To prevent racemization, protection of the 3-hydroxy group in compounds **18** and **19** was carried out under carefully selected conditions. Specifically, acid-catalyzed reductive benzylation of compounds **18** and **19** utilizing benzaldehyde [39] and subsequent hydrolysis of the ester groups furnished the final benzyl-protected β-hydroxy fatty acids **4** and **6**. For the synthesis of the double-chain fatty acid derivative 7, initial condensation with pentadecanoic acid provided compound **21** in quantitative yield. The final fatty acid derivative **7** was obtained by acidic hydrolysis of the tert-butyl ester moiety (Scheme 3).

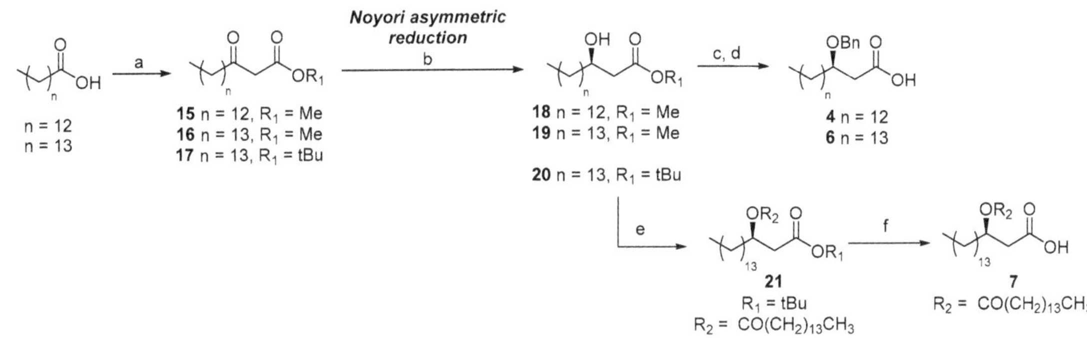


**Scheme 3.** Synthesis of the fatty acid chain derivatives. Reagents and conditions. (a) i. DCC, $CH_2Cl_2$, DMAP, r.t., 45 min, then Meldrum's acid, Pyridine, $CH_2Cl_2$, r.t., 2 h; ii. $R_1OH$, $H_2SO_4$ (cat.), reflux, 16 h, 64–74% yields over three steps; (b) [(R)-BINAP]$RuBr_2$ (10 mol%), dry MeOH, $H_2$ (balloon), 50 °C, 18 h, **18** (89%), **19** (85%), **20** (73%); (c) PhCHO, $TMS_2O$, TMSOTf, THF, 0 °C, 3 h, then $Et_3SiH$, 0 °C, 2 h; (d) LiOH, THF/MeOH/$H_2O$ (3/2/1), rt., 5 h, **4** (80% over two steps), **6** (82% over two steps); (e) pentadecanoic acid, DIC, DMAP, $CH_2Cl_2$, rt, 6 h, quant.; (f) TFA/$CH_2Cl_2$, r.t., 12 h, 93%.

### *2.5. Fatty Acid Chains Condensation and Final Deprotections*

The final phase of the synthetic work comprised the following key steps (Scheme 4): (1) deprotection of the phthalimido group together with the cleavage of acetyl esters at the C3 and C3′ hydroxyls; (2) condensation of fatty acid derivative **4** on the newly liberated hydroxyl and amino sites; (3) reduction of the azido group, followed by condensation with fatty acid derivatives **6** or **7**; (4) phosphorylation at either C1 or C4′, followed by final global deprotection. The phthalimide and acetyl groups were removed using hydrazine hydrate in ethanol at 70 °C, and the resulting crude disaccharide **22** was directly used in the subsequent fatty acid coupling step. This stage proved particularly challenging, as the introduction of multiple long-chain and bulky fatty acids required extensive optimization to identify ideal conditions for satisfactory yields. After screening different conditions (Table 1), optimal coupling was achieved using four equivalents of fatty acid **4** and 10 equivalents of EDC hydrochloride. Reduction of the azide at C2 using zinc in acetic acid, followed by condensation with benzyl-protected 3-hydroxy fatty acids **6** and **7**, furnished intermediates **24** and **27** in excellent yields (87% and 85% over two steps, respectively).

Removal of the TDS group from compounds **24** and **27** by treatment with HF·pyridine afforded hemiacetals **25** and **28**, which were phosphorylated using tetrabenzyl pyrophosphate and LiHDMS, yielding protected precursors **26** and **29**, respectively. Final compounds **Tetra-C1** and **Penta-C1** were obtained through hydrogenolysis to remove benzylidene acetal and benzyl groups. For the synthesis of **Penta-C4′**, in which the phosphate group is located at the C4′ hydroxyl, regioselective ring opening of the benzylidene acetal on intermediate **27** generated derivative **30**, delivering the C4′ hydroxyl. The phosphorylation conditions used for compounds **26** and **29** proved to be ineffective in this case. Then, dibenzyl *N,N*-diisopropyl phosphoramidite was employed as an alternative phosphorylating agent, affording the C4′-dibenzylphosphate intermediate **31** in excellent yield. Cleavage of the silyl ether at C1 using previously described conditions, followed by exhaustive protecting group cleavage, furnished the final **Penta-C4′** compound.

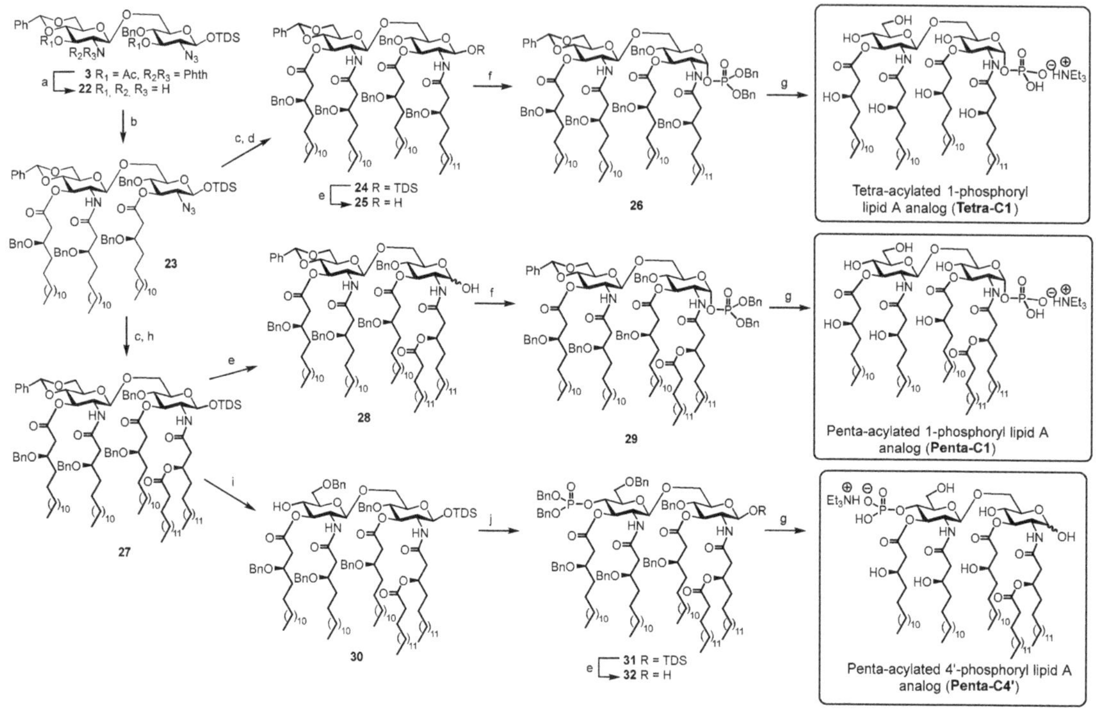

**Scheme 4.** Condensation of fatty acid chains and final deprotection. Reagents and conditions. (a) Hydrazine hydrate, EtOH, 70 °C, 2 h; (b) **4**, $CH_2Cl_2$/DMF (5:1), 45 °C, 12 h, Table 1; (c) Zn, 1,4-dioxane/AcOH (10:1), r.t., 12 h; (d) **6**, HATU, DIPEA, $CH_2Cl_2$/DMF (5:1), rt, 7 h, 87% over two steps; (e) HF·pyridine, THF/Pyridine, r.t., 8 h, 84% (for **25**), 75% (for **28**), 73% (for **32**); (f) tetrabenzyl pyrophosphate, LiHDMS, THF, −70 °C to 0 °C, 1 h, 63% (for **26**), 65% (for **29**); (g) $Pd(OH)_2$/C, $H_2$ (balloon), $CH_2Cl_2$/MeOH/AcOH (10/10/1), 2 d, then $Et_3N$, quant. (for **Tetra-C1**, **Penta-C1** and **Penta-C4′**); (h) **7**, HATU, DIPEA, $CH_2Cl_2$/DMF (5:1), r.t., 4 h, 85% over two steps; (i) $Et_3SiH$, TfOH, MS4Å, $CH_2Cl_2$, −78 °C, 3 h, 70%; (j) dibenzyl *N,N*-diisopropyl phosphoramidite, 1*H*-tetrazole, $CH_2Cl_2$/$CH_3CN$ (5:1), then *m*CPBA, rt, 2.5 h, 87%.

**Table 1.** Screening of conditions for condensation of fatty acid **4** with partially deprotected disaccharide **22**.

| Entry | Reaction Reagents (Equivalents) | Equivalents of Fatty Acids | Reaction Concentration (M) | Yield (%) |
|---|---|---|---|---|
| 1 | HBTU (6.0 eq.), DIPEA (10 eq.), DMAP (3.0 eq.) | 4.5 | 0.05 | 63 |
| 2 | HATU (6.0 eq.), DIPEA (10 eq.), DMAP (3.0 eq.) | 4.5 | 0.05 | 64 |
| 3 | EDC HCl (6.0 eq.), DMAP (1.5 eq.) | 4.5 | 0.05 | 66 |
| 4 | EDC HCl (10. eq.), DMAP (1.5 eq.) | 4.0 | 0.1 | 75 |

### 2.6. Docking of Compounds 1–4 in hTLR4/MD-2 Complex

To investigate the binding features of the four compounds of Figure 2, we performed docking calculations in the agonist and antagonist models of TLR4/MD-2 dimer, generated starting from the corresponding X-ray structures (see Section 3 for details). The protein, the sugar core, and the first two torsions of each acyl chain were kept rigid during the docking calculations using AutoDock Vina [40,41]. For each ligand, we run docking simulations in triplicate and build two ligand conformations, one based on the crystallographic diglucosamine backbone of lipid A bound to MD-2 (agonist-like structure, PDB ID: 3FXI [10]) and the other from lipid IVa (antagonist-like structure, PDB ID: 2E59 [30]) complex. *E. coli* lipid A is the active portion of LPS and acts as an agonist; lipid IVa is a tetra-acylated precursor of lipid A that binds to the same hMD-2/TLR4 complex without triggering the biological response. In the X-ray structures, lipid A and lipid IVa share a similar disaccharide conformation but exhibit an inverted orientation of the entire ligand within the MD-2 pocket, corresponding to a 180° horizontal rotation (Figure S1). In the agonist complex, the sugar moiety of lipid A interacts with residues S118 and S120 at the MD-2 surface, whereas in the antagonist structure, contacts are established with S120 and K122 residues. In both complexes, one phosphate group establishes an electrostatic contact with R264 on TLR4 (Figure 3).

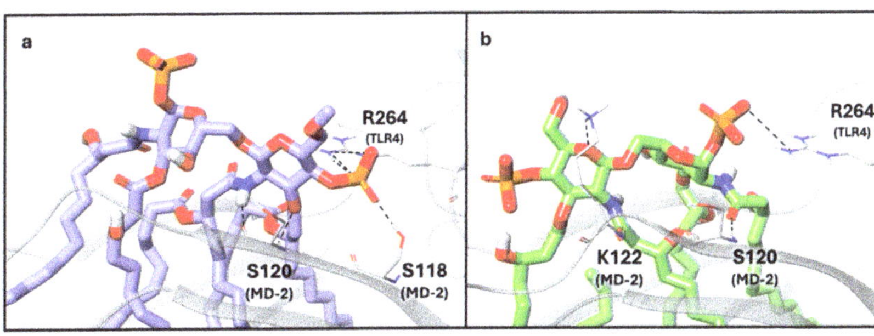


**Figure 3.** (**a**) X-ray structure (PDB ID: 3FXI) of the lipid A (purple carbon, blue nitrogen, red oxygen, orange sulfur and white polar hydrogen atoms) and (**b**) generated structure (PDB ID: 2E59) of the lipid IVa (green carbon, blue nitrogen, red oxygen, orange sulfur and white polar hydrogen atoms) bound to MD-2 and TLR-4 (grey ribbon representation). Key residues involved in the interactions with the disaccharide unit are shown in tube representation (grey for carbon, blue for nitrogen, red for oxygen and white for polar hydrogen atoms). For lipid A, we highlighted the salt bridge and the hydrogen bond between the ligand phosphate group and R264 (TLR4) and the side chain of S118 (MD2), respectively. For lipid IVa, we highlighted the salt bridge between the ligand phosphate group and R264 (TLR4), the hydrogen bond between the ligand amide and the backbone of S120 (MD-2), and between the sugar ring oxygen and the side chain of K122 (MD-2).

Beyond the sugar region, more pronounced structural differences are observed in the orientation of the acyl chains. In the agonist state, five of the six lipid A acyl chains bind within the hydrophobic cavity of MD-2, while the sixth chain extends outward from the pocket to interact with a hydrophobic surface on a second TLR4* unit (Figure 4a—the asterisk indicates the TLR4 of the other heterodimer involved in tetramer formation).

This interface, mainly formed by TLR4* residues (Q436, F440, L444, and F463, Figure 4a) and the F126 of the MD-2 loop, has a pivotal role in the recruitment of the second TLR4*·MD-2*-LPS complex and in the formation of the signaling-competent multimer. Residue F126 is located at the entrance of the MD-2 binding pocket and, after the binding of the ligand, reorients its side chain and creates, together with one LPS chain, the hydrophobic pocket for the interaction with the second TLR4 protein. Site-directed mutagenesis studies confirmed that F126 is essential for receptor dimerization [28].

By contrast, in the antagonist structure, the human lipid IVa is fully embedded within the MD-2 pocket, with all four acyl chains buried inside the pockets and no hydrophobic chain available for interaction with another MD-2/TLR4 unit (Figure 4b). F126 adopts an open conformation (Figure S2) that prevents its side chain from engaging the ligand lipid chain (no contacts between the ligand and F126 are formed).

To evaluate whether our synthesized ligands behave as agonist or antagonist ligands, docking poses were compared to the crystallographic structures of lipid A, for the results in the agonist model, and of lipid IVa, for the antagonist model.

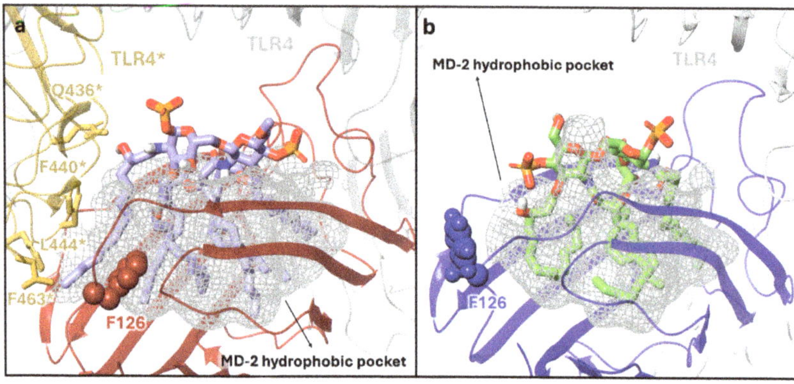


**Figure 4.** (**a**) X-ray structure (PDB ID: 3FXI) of the lipid A (purple carbon, blue nitrogen, red oxygen, orange sulfur and white polar hydrogen atoms) in complex with MD-2 (red) and two TLR4 proteins (grey and yellow). The MD-2 hydrophobic pocket is represented as a grey meshed surface. MD-2 is shown in red, with residue F126 highlighted in CPK representation in its "closed" conformation, forming hydrophobic interactions with one of the acyl chains of lipid A. One TLR4 chain is shown in grey, the second TLR4* in yellow. The key residues of the TLR4* (yellow)–MD-2 interface are also shown. (**b**) Generated structure of the lipid IV*a* (green carbon, blue nitrogen, red oxygen, orange sulfur and white polar hydrogen atoms) in complex with MD-2 and TLR4 dimer. MD-2 is shown in blue, and F126 is displayed in CPK representation in its "open" conformation, not interacting with the acyl chains of lipid IV*a*. TLR4 is shown in grey.

For each ligand, after clustering the docking results based on volume overlap, we first quantified the displacement of the disaccharide moiety from the X-ray structure by calculating the heavy-atom RMSD for each representative binding mode. Low RMSD scores (<2 Å, see Table S1 for the atom selection) indicate a good superimposition of the ligand pose to the experimental structure. In addition, we compared the interactions formed with MD-2 and TLR4 proteins (shown in Figure 3) and the overlay of the acyl chains in the MD-2 pockets to lipid A (agonist docking results) or to lipid IVa (antagonist docking results) by visual inspection. We also considered contacts between the outer lipophilic chain and Phe126 side chain (distance cutoff < 5.5 Å) as indicative of agonist-like behavior. According to these criteria, in the agonist MD-2/TLR4 model, none of the compound geometries reproduces the binding mode of lipid A (Figures S4–S7). With the exception of **Penta-C1***, the disaccharide core in all ligands deviates significantly from the crystallographic reference, showing RMSD values greater than 5.84 Å. Even if **Penta-C1*** in the top-ranked pose is well-aligned to the sugar moiety of lipid A (RMSD = 1.22 Å), none of the X-ray protein–ligand interactions are formed, and the acyl chains are entirely buried within the pocket, and no contacts with Phe126 side chain are present, failing to create the hydrophobic interface required for dimerization (Figure S7a). Taken together, these in silico analyses suggest that none of the compounds can bind to MD-2 like the agonist lipid A.

Conversely, the analysis of docking poses generated in the antagonist model, besides showing good convergence among the three independent docking runs, highlighted that only the **Tetra-C1** ligand can bind to the MD-2 pocket like the crystallographic ligand. Compound **Tetra-C1** features an antagonist-like binding mode with low RMSD scores in the top-ranked and most populated cluster (50% populated, RMSD range values 0.25–1.05 Å, Table S5 and Figures S8 and 5a,b). All lipid chains are properly accommodated within the MD-2 pocket, and the salt bridge with R264 (TLR4) and hydrogen bonds with S120 and K122 (MD-2) are formed. For **Penta-C1***, only a small subset of docking poses shows an antagonist-like positioning of the sugar moiety (8% populated, RMSD range values 0.53–1.76 Å, Table S6) and the interactions with R264 on TLR4 and K122, and S120 residues on MD-2. One acyl chain of **Penta-C1*** is partially solvent-exposed and poorly aligned to the lipid IVa reference structure (Figure S9). Docking poses of **Penta-C4′** clustered into two main binding modes: the first (12% populated) shows a good sugar core alignment (RMSD range values 0.29–0.98 Å, while the second (57% populated) has a significantly higher displacement (Table S7). In both binding modes, one lipid chain points outside the MD-2 pocket and does not overlap with the antagonist binding mode (Figure S10). Compound **Penta-C1** (Figure S11) adopts two binding modes, both with the disaccharide alignment deviating from the crystallographic pose of lipid IVa (Table S8) and one acyl chain protruding from the MD-2 pocket.

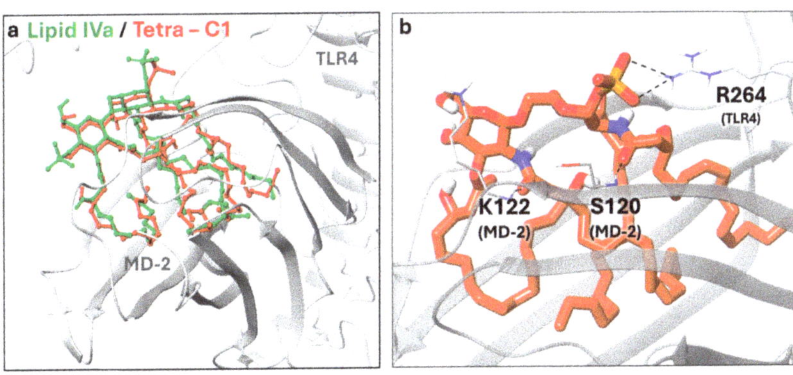


**Figure 5.** (**a**) Top-ranked pose of the most populated cluster of **Tetra-C1** ligand (red) in the antagonist models overlaid to the crystallographic ligand (green). Both TLR4 and MD-2 are shown in grey. The RMSD range values of the cluster for the di-glucosamine backbone are 0.25–1.05 Å (**b**). Close-up view of the **Tetra-C1** (orange carbon, blue nitrogen, red oxygen, orange sulfur and white polar hydrogen atoms) binding in the MD-2 pocket. Key residues involved in the interactions with the disaccharide unit are shown in tube representation (grey for carbon, blue for nitrogen, red for oxygen and white for polar hydrogen atoms).The salt bridge between the ligand phosphate group and R264 of TLR4, the hydrogen bond between the ligand amide and the backbone of S120 of MD-2, and between the sugar ring oxygen and the side chain of K122 of MD-2 are shown.

To compare our results with a well-characterized TLR4 antagonist, we applied our docking protocol to Eritoran, selecting its crystallographic conformation bound to human MD-2 as the starting point (PDB ID: 2Z65 [28]) and both the antagonist and agonist MD-2/TLR4 models. Despite the structural differences with lipid IVa, the two antagonists bind in a similar way to MD-2, inserting the lipid chains in the hydrophobic pocket. Their di-glucosamine backbones conformation is superimposable (they share a similar conformation) with a 3.25 Å shift (Figure S10) with the MD-2 site. As expected, in the docking results of the agonist model, Eritoran adopted two binding modes, both with poor alignment of the disaccharide core to the lipid A (RMSD range values 7.29–9.31 Å, Table S9, Figure S13). Conversely, in the antagonist model, the top-ranked cluster of docking poses (23% populated), all the acyl chains are properly buried inside the MD-2 hydrophobic pocket, and the interactions with residues K122 (MD-2) and the salt bridge with R264 (TLR4) are formed. The disaccharide core is still not well aligned to lipid IVA, as in the X-ray structures (average RMSD values 4.35 Å, Table S10). The remaining two clusters of docking poses are poorly aligned to both the lipid IVa and Eritoran X-ray structures, with one of the acyl chains partially solvent-exposed and mispositioned outside the MD-2 pocket (Figure 6).

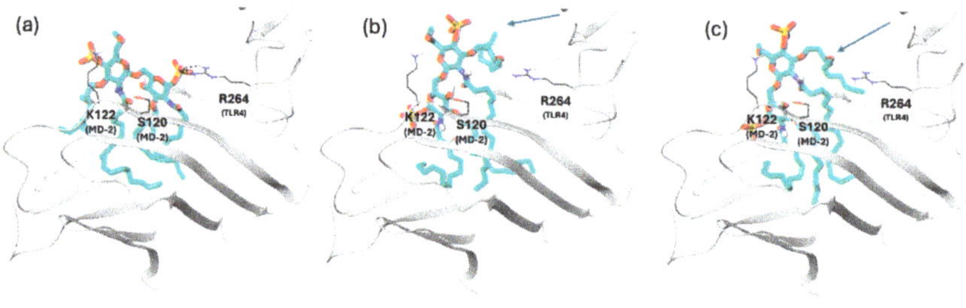


**Figure 6.** Representative docking poses of Eritoran (orange) in the antagonist MD-2/TLR4 model from cluster analysis. Both TLR4 and MD-2 are shown in grey and Eritoran with cyan carbon, blue nitrogen, red oxygen, orange sulfur and white polar hydrogen atoms. Key residues involved in the interactions with the disaccharide unit are shown in tube representation (grey for carbon, blue for nitrogen, red for oxygen and white for polar hydrogen atoms). (**a**) Cluster 1: moderate disaccharide alignment with the lipid IVa X-ray reference; all acyl chains accommodated in the MD-2 pocket; salt bridge with R264 observed, and the hydrogen bond with K122, consistent with an antagonist-like binding mode. (**b**) Cluster 2: poor disaccharide alignment; one acyl chain protrudes from the pocket as indicated by the arrow. (**c**) Cluster 3: misaligned disaccharide; one acyl chain protrudes toward TLR4 as indicated by the arrow.

Finally, to provide a consensus ranking that complements the docking scores, docking poses from the antagonist model were re-scored using MM-GBSA calculations [42]. This method estimates the free energies of binding by calculating the difference between the solvation energy of the bound complex and the energy of the unbound protein and ligand using implicit solvent. This method, though it neglects configurational entropy and depends on the quality of force field parameters and solvation model [43], provides a more accurate estimation of binding free energy compared to the docking score, allowing for a better ranking of compounds [44]. Kaus et al. [45] have reported that the MM-GBSA method performs better than standard scoring functions in ranking binding poses. Similarly, Rastelli et al. [46] have shown that both MM-GBSA and MM-PBSA rescoring methodologies improve the affinity ranking estimated by the AutoDock scoring function. Based on this estimation, the complex with the lowest MM-GBSA energy should have the strongest binding affinity. Notably, within the same cluster of docking poses, MM-GBSA values show greater variability compared to docking scores or RMSD values (Tables S5–S8, S10, and S12), reflecting the intrinsic sensitivity of MM-GBSA calculations to small ligand conformational differences. When focusing only on agonist-like binding modes, the reference compounds, Eritoran and lipid IVa, and the **Tertra-C1,** showed comparable affinities based on docking scores, whereas the MM-GBSA values of **Tertra-C1** were higher than those of the reference ligands (Table 2). Experimental validation will be essential to confirm these predictions and to guide further refinement of the model. In addition, advanced MD-based free-energy calculations could be performed in the future to provide more robust and quantitatively reliable estimates of binding free energies.

**Table 2.** MM-GBSA ranking of the antagonist-like binding mode of Tetra-C1, Eritoran, and lipid IVa. Autodock Vina score and disaccharide RMSD value are also reported.

| Compound | RMSD * (Å) | Autodock Score (kcal/mol) | MM-GBSA (kcal/mol) |
|---|---|---|---|
| Tetra-C1 | 0.52 | −7.65 | −129.79 |
| lipid IVa | 0.80 | −7.84 | −126.09 |
| Eritoran | 3.51 | −7.69 | −83.74 |

* reference structure = lipid IVa X-ray structure PDB ID 2E59.

## 3. Materials and Methods

### 3.1. Docking Studies

#### 3.1.1. Ligands Preparation

The 3D structures of lipid A and lipid IVa were obtained from the corresponding X-ray complexes (PDB ID 3FXI [10] and PDB ID 2E59 [30], respectively). To build the 3D conformations of compounds **1–4**, both lipid A and lipid IVa were used as templates, and their hydrophobic side was modified using the "*building panel*" of the Maestro interface. Only the newly added and modified atoms of the aliphatic chains were locally minimized using the "Minimize Selected Atoms" function of Maestro (OPLS4 force field [47]).

#### 3.1.2. Protein Structures Preparation

For the agonist model, the X-ray structure of 2:2:2 hTLR4, hMD-2, and *E. coli* LPS complex (PDB ID 3FXI [10]) was selected. After removing all the water molecules and ions, the Protein Preparation tool of Maestro (Schrödinger Release 2024-3, version 14.1.138) was used to add hydrogen atoms and determine the protonation states of the protein residues at pH 7 $\pm$ 2. The system was relaxed by a restrained energy minimization step with the OPLS3 force field [48] and a heavy atom RMSD convergence set to the default value of 0.3 Å. Chain A of TLR4 and chain C of MD-2 were considered for the final agonist heterodimer model for docking calculations. Since the crystallographic structure of the hTLR4/MD-2 complex in the agonist-bound state is not available, we generated the antagonist model starting from the structure of hMD-2 bound to lipid IVa (PDB ID 2E59, [30]) and adding the TLR4 protein by superimposition to the agonist model, following the procedure described in this work by Lembo-Fazio et al. [49]. We prepared the system for docking calculation by applying the Protein Preparation workflow used for the agonist model. The final structure was then used to generate the grids for docking calculations.

#### 3.1.3. Docking Workflow

Docking calculations were performed with AutoDock Vina [40,41] using a grid spacing of 1 Å and a box size of 60 Å. The grid box (33.00 $\times$ 40.50 $\times$ 35.25 Å) was centered at the TLR4/MD-2 interface, specifically between the center of mass of R264 (TLR4), R90 (MD-2), and K122 (MD-2) for LPS. For lipid IVa, K122 was replaced by R96 (MD-2), following the protocol reported in the work by Sestito et al. [50]. Gasteiger charges for ligands and receptors were calculated with AutoDockTools 1.5.6 [51], after merging non-polar hydrogens on the receptors. The receptor structures were kept rigid, while ligands were treated as partially flexible. Ligand conformations were generated as described in the ligand preparation section. We considered the saccharide core as rigid, and we introduced dihedral constraints on lipid chains. To determine the number and position of the constraints, the X-ray crystallographic ligands were used as references. For the agonist model, lipid A structure (i.e., the phosphorylated di-glucosamine core and lipid chains) was selected, and for the antagonist model, the lipid IVa. Docking calculations were performed with different combinations of dihedral constraints to identify the setup that best reproduced the experimental binding mode: RMSD of the di-glucosamine core < 2 Å (according to the atoms selection shown in Figure S3); the formation of the relevant polar interactions shown in Figure 3; and all the lipidic chains properly inserted in the MD-2 pocket examined through visual inspection of the docking poses with respect to the X-ray structures and through the formation of ligand–Phe126 contacts (within a 5.5 Å distance cutoff). The best results were obtained by constraining the first two dihedral angles linked to the sugar ring in each hydrophobic chain (dihedral values reported in Table S13). For each ligand, we run three independent docking calculations, and we collected the top-ranked 20 poses, yielding a total of 60 poses (20 poses $\times$ 3 runs). These poses were then clustered according to a volume-based overlap criterion (using the utility "phase_volCalc" available in Maestro 2024-3, version 14.1.138) to identify the relevant binding modes of each ligand. We selected all ligand atoms to capture the full spatial shape, enabling a rigorous rigid alignment and grouping based on normalized volume overlap. For the most representative clusters (i.e., those accounting for more than 10% of the total population), docking scores and RMSD values of the saccharide core relative to the reference structures were calculated (Tables S1–S12). The most representative clusters were then evaluated for their capability of reproducing the binding modes of the agonist/antagonist X-ray ligands based on the validation criteria. If all criteria were satisfied, ligand binding modes were classified as agonist-like when the outer acyl chain interacted with the Phe126 side chain (distance < 5.5 Å), and as antagonist-like when no such contact was observed. For each ligand docked in the antagonist model, the most representative clusters were re-scored using the MM-GBSA tool of Prime version 3.0 [42] using the VSGB2.0 implicit water model [52] and the OPLS4 force field [47] after the addition of non-polar hydrogens to both ligands and protein (Tables S5–S8 for compounds **1–4**, Table S10 for Eritoran and Table S12 for lipidIVa).

### 3.2. *Synthesis—General Remarks*

All reagents were purchased from commercial suppliers and were used as supplied without further purification. Dry solvents such as DCM, MeOH, pyridine, toluene, DMF, $CH_3CN$, and THF were purchased from Merck© (Darmstadt, Germany) or Thermo Fisher Scientific© (Waltham, MA, USA) and were used without further purification. Non-anhydrous solvents were used directly as supplied from commercial vendors. Analytical thin-layer chromatography (TLC) was performed on Merck© precoated 60F254 plates (0.25 mm), visualized by UV light at 254 nm and/or dipping the plates into stains: molybdic solution (21 g of $(NH_4)_6Mo_7O_{24}$, 1 g of $Ce(SO_4)_2$, 31 mL of $H_2SO_4$ 98%, 970 mL $H_2O$) sulfuric acid solution (50 mL of $H_2SO_4$ 98%, 450 mL of MeOH, 450 mL $H_2O$), ninhydrin (2.7 g of 2,2-dihydroxyindane-1,3-dione, 27 mL of AcOH, 900 mL of EtOH) with detection by charring at 300 °C. For some reactions, High Performance Thin Layer Chromatography (HPTLC) (Merck© precoated 60 F254 plates 0.20 mm) was used and correspondently pursued. Flash chromatography was performed using Silica gel ($SiO_2$, high-purity grade (Merck Grade 9385), pore size 60 Å, 230–400 mesh particle size) from Merck©. Biotage® (Biotage Sweden AB™, Uppsala, Sweden) Isolera SP1 system was used to purify some compounds. Normal phase Biotage cartridges (sizes from 3 g to 340 g, standard 50 μm silica) were employed. Freeze-drying of aqueous solutions was performed using the Lio5P Lypholizer. $^1H$, $^{13}C$, and $^{31}P$-NMR spectra were recorded at 25 °C, unless otherwise stated, on a Bruker© Avance TM NEO 400 MHz (Billerica, MA, USA) (400, 100.6 MHz, and 162 MHz for $^1$H, $^{13}$C, and $^{31}$P, respectively). All NMR measurements were performed at room temperature. The samples were prepared using deuterated solvents ($CDCl_3$, $CD_3COCD_3$, $D_2O$, and $CD_3OD$ from Merck© or Eurisotop®). Chemical shifts are reported in ppm and coupling constants (J) in Hz. Multiplicities are abbreviated as br (broad), s (singlet), d (doublet), t (triplet), hept (heptet), m (multiplet), or combinations thereof. $^1$H-NMR spectra were recorded for all the compounds. For unknown structures, characterization is reported by $^1$H-NMR, $^{13}$C-NMR, and $^{31}$P-NMR. Low resolution mass analysis was recorded in negative or positive mode on a Thermo© (Waltham, MA, USA) Finnigan LCQAdvantage equipped with an ESI source. High resolution mass spectra (HR-MS) were acquired on a Synapt G2-Si QTof mass spectrometer (Waters™, Milford, MA, USA) equipped with a Zspray ESI-probe (Waters) for electrospray ionization in full scan mode. Data were processed using MassLynx v4.2 software (Waters).

### 3.3. *Synthesis—Synthetic Procedures*

#### 3.3.1. 1,3,4,6-Tetra-*O*-acetyl-2-deoxy-2-azido-α/β-D-glucopyranose **9**

Glucosamine hydrochloride (7.95 g, 36.9 mmol) was dissolved in MeOH (185 mL). Anhydrous $K_2CO_3$ (15.3 g, 110.7 mmol) was added slowly at 0 °C. After 15 min, imidazole-1-sulfonyl azide hydrogen sulfate (12 g, 44.3 mmol) was added in portions, followed by $CuSO_4{\cdot}5H_2O$ (92 mg, 0.37 mmol). The reaction progress was monitored through TLC (DCM/MeOH 7:3). After 4 h, the starting material was consumed, and the solvent was removed under reduced pressure and dried under vacuum. The crude 2-azido-2-deoxy-glucopyranose was then suspended in dichloromethane (185 mL) and pyridine (48 mL), then cooled to 0 °C. DMAP (225 mg, 1.85 mmol) and $Ac_2O$ (27.9 mL, 295 mmol) were added, and the reaction was stirred at r.t. for 12 h. The reaction was followed by TLC (DCM/MeOH 7:3 and hexane/EtOAc 7:3). The reaction was diluted with DCM, then washed three times with 5% HCl solution, twice with saturated solution of $NaHCO_3$, and finally with brine. The organic phase was dried with sodium sulfate, filtered, and evaporated. Purification by filtration on silica (hexane/EtOAc 6:4) led to **9** (13 g, 35 mmol) in 95% yield over 2 steps as a 2:1 mixture of the α and β anomers. The spectroscopic data are in agreement with those reported in the literature [53].

#### 3.3.2. 3,4,6-Tri-*O*-acetyl-2-deoxy-2-azido-β-D-glucose **10**

Hydrazine hydrate (1.8 mL, 37.2 mmol) was dissolved in MeOH (30 mL). AcOH (1.77 mL, 31 mmol) was slowly added at 0 °C, and the solution was stirred for 20 min to give hydrazine acetate. The freshly prepared hydrazine acetate solution was slowly added to a solution of **9** (7.7 g, 20.66 mmol) in DMF (40 mL). When the reaction was completed, indicated by TLC (hexane/EtOAc 5:5), the reaction mixture was diluted with $Et_2O$ and washed 1x $H_2O$ and 1x brine. The organic phase was dried over $Na_2SO_4$, filtered, and the solvent was removed under reduced pressure, obtaining 8 g of crude compound whose purity was checked by $^1$H-NMR analysis. The spectroscopic data are in agreement with those reported in the literature [54].

#### 3.3.3. 1-*O*-Thexyldimethylsilyl-3,4,6-tri-*O*-acetyl-2-deoxy-2-azido-β-D-glucopyranoside **11**

Crude compound **10** (8 g, 20.66 mmol) was dissolved in DCM (200 mL). Imidazole (3.1 g, 45.45 mmol) was added, and the solution was stirred for 10 min at r.t. The solution was cooled to 0 °C, and the thexyldimethylsilyl chloride (4.9 mL, 24.8 mmol) was added. The reaction was followed by TLC (hexane/EtOAc 7:3). After 12 h, TLC indicated the consumption of the starting material; the reaction mixture was then washed with $H_2O$ and brine. The organic phase was dried over $Na_2SO_4$, filtered, and the solvent was removed under reduced pressure. The crude was purified by flash chromatography (hexane/EtOAc 7:3), leading to **11** (9.04 g, 19 mmol) with 92% yield over two steps. The spectroscopic data are in agreement with those reported in the literature [53].

3.3.4. 1-*O*-Thexyldimethylsilyl-2-deoxy-2-azido-4,6-*O*-benzylidene-β-D-glucopyranoside **12**

Compound **11** (9.01 g, 19.04 mmol) was dissolved in MeOH (170 mL), then NaOMe (308 mg, 5.71 mmol) was added. After 1 h, as indicated by TLC (hexane/EtOAc 3:7), the reaction was completed. Amberlite IR-120 H+ form was added until pH 6, then the mixture was filtered, and the solution was evaporated to dryness. The residue was dried under vacuum, then redissolved in acetonitrile (175 mL); benzaldehyde dimethylacetal (6.3 mL, 41.9 mmol) and a catalytic amount of *p*TSA (724 mg, 3.8 mmol) were added to the reaction solution. After 8 h, TLC (hexane/EtOAc 3:7) indicated the completion of the reaction. Triethylamine (1.5 mL) was added, and the reaction mixture was evaporated. The purity of the compound, checked by $^1$H-NMR, was found to be adequate for direct use in the subsequent reaction. A sample was purified by flash chromatography hexane/EtOAc 7:3) for characterization.

$^1$H-NMR (400 MHz, $CDCl_3$) δ 7.34–7.10 (m, 5H, H-Arom), 5.33 (s, 1H, CH benzylidene), 4.43 (d, J = 7.7 Hz, 1H, H-1), 4.09 (dd, J = 10.5, 5.0 Hz, 1H, 6-a), 3.58 (t, J = 10.3 Hz, 1H, 6-b), 3.42 (dd, J = 9.3, 1.5 Hz, 1H, H-3), 3.36 (t, J = 9.1 Hz, 1H, H-4), 3.19 (ddd, J = 10.1, 9.1, 5.0 Hz, 1H, H-5), 3.15–3.04 (m, 1H, H-2), 2.52 (brs, 1H, OH), 1.49 (h, J = 7.1 Hz, 1H, CH-TDS), 0.72 (s, 3H, $CH_3$-TDS), 0.70 (s, 9H, $CH_3$-TDS), 0.02 (s, 3H, $CH_3$Si-TDS), 0.00 ppm (s, 3H, $CH_3$Si-TDS).

$^{13}$C-NMR (101 MHz, $CDCl_3$) δ 136.87-129.36-128.38-126.29 (C-Arom), 102.00 (CH benzylidene), 97.40 (C-1), 80.76 (C-4), 71.92 (C-3), 69.13 (C-2), 68.58 (C-6), 66.26 (C-5), 33.88 (CH-TDS), 24.81 (Cquat-TDS), 19.98 ($CH_3$-TDS), 19.82 ($CH_3$-TDS), 18.50 ($CH_3$-TDS), 18.40 ($CH_3$-TDS), −2.11 ($CH_3$Si-TDS), −3.18 ppm ($CH_3$Si-TDS).

$C_{21}H_{33}N_3O_5Si$; calcd. mass 435.22; ESI-MS: *m*/*z* 458.43 $[M+Na]^+$.

3.3.5. 1-*O*-Thexyldimethylsilyl-2-deoxy-2-azido-3-*O*-acetyl-4,6-*O*-benzylidene-2-deoxy-β-D-glucopyranoside **13**

Crude compound **12** (10.5 g) was acetylated using a procedure analogous to that reported for compound **9** (3.6 mL of acetic anhydride, 117 mg of DMAP, 6.2 mL of pyridine in 90 mL DCM). The crude compound after extraction was purified by filtration on silica gel (hexane/EtOAc, 7:3), to give compound **13** (7.27 g, 15.2 mmol) with an 80% yield over three steps.

$^1$H-NMR (400 MHz, $CDCl_3$) δ7.48-7.30 (m, 5H, H-Arom), 5.48 (s, 1H, CH benzylidene), 5.13 (t, J = 9.8 Hz, 1H, H-3), 4.69 (d, J = 7.5 Hz, 1H, H-1), 4.29 (dd, J = 10.3, 5.0 Hz, 1H, H-6a), 3.78 (t, J = 10.3 Hz, 1H, H-6b), 3.63 (t, J = 9.5 Hz, 1H, H-4), 3.48 (dd, J = 5.0, 10.3 Hz, 1H, H-5), 3.43-3.36 (dd, J = 7.5, 9.8 Hz, 1H, H-2), 2.13 (s, 3H, $CH_3$CO), 1.67 (hept, J = 6.8, 1H, CH-TDS), 0.96-0.78 (m, 12 H, 4x$CH_3$-TDS), 0.22 (s, 3H, $CH_3$Si-TDS), 0.21 ppm (s, $CH_3$Si-TDS).

$^{13}$C-NMR (101 MHz, $CDCl_3$) δ169.88 ($CH_3$CO), 136.98-129.25-128.38-126.28 (C-Arom), 101.66 (C-benzylidene), 97.61 (C-1), 78.85 (C-4), 71.29 (C-3), 68.68 (C-6), 67.45 (C-2), 66.66 (C-5), 34.01 (CH-TDS), 21.02 ($CH_3$CO), 20.07-19.94-18.64-18.53 ($CH_3$-TDS), −2.03 ($CH_3$Si-TDS), −3.09 ppm ($CH_3$Si-TDS).

$C_{23}H_{35}N_3O_6Si$; calcd. mass 477.23; ESI-MS: *m*/*z* 500.31 $[M+Na]^+$.

### 3.3.6. 1-*O*-Thexyldimethylsilyl-2-deoxy-2-azido-3-*O*-acetyl-4-*O*-benzyl-2-deoxy-β-D-glucopyranoside **2**

Compound **13** (2.3 g, 4.85 mmol) was dissolved in DCM (95 mL), then freshly activated 4Å MS were added under an Ar atmosphere. The mixture was stirred for 1 h at −78 °C before the addition of $Et_3SiH$ (2.7 mL, 17 mmol). The mixture was stirred for an additional 15 min, then $PhBCl_2$ (0.69 mL, 5.34 mmol) was added. The reaction progress was monitored by TLC (hexane/EtOAc 8:2). When the reaction was completed (3 h), it was quenched at −55 °C by the addition of $Et_3N$ (1.5 mL) and MeOH (5 mL). The MS were removed by filtration over celite, the organic phase was washed once with NaHCO3 (saturated aqueous solution), thenH2O, brine, and dried over Na2SO4. After filtration, the crude was concentrated in vacuo, then purified by flash chromatography (hexane/EtOAc 8:2) to give pure **2** (2.1 g, 4.5 mmol) in 93% yield.

$^1$H-NMR (400 MHz, $CDCl_3$) δ 7.42-7.25 (m, 5H, CH Ar), 5.06 (dd, J = 10.4, 9.3 Hz, 1H, H-3), 4.46 (d, J = 7.7, 1H, H-1), 4.63 (d, J = 3.7 Hz, 2H, $CH_2Ph$), 3.88 (ddd, J = 12.2, 5.3, 2.7 Hz, 1H, H-6a), 3.74 (ddd, J = 11.9, 8.3, 3.9, 1H, H-6b), 6.65 (t, J = 9.5 Hz, 1H, H-4), 3.42 (ddd, J = 9.8, 4.1, 2.6 Hz, 1H, H-5), 3.30 (dd, J = 10.4, 7.7 Hz, 1H, H-2), 2.05 (s, 3H, $CH_3CO$), 1.80 (dd, J = 8.4, 5,4, 1H, OH), 1.67 (hept, J = 6.8, 1H, CH-TDS), 0.92 (s, 3H, $CH_3$-TDS), 0.90 (s, 9H, $CH_3$-TDS), 0.22 (s, 3H, $CH_3Si$-TDS), 0.21 ppm (s, 3H, $CH_3Si$-TDS).

$^{13}$C-NMR (101 MHz, $CDCl_3$) δ169.98 ($CH_3CO$), 137.65 (CquatAr), 128.70, 128.16, 127.94 (CAr), 97.01 (C-1), 75.72 (C-4), 75.40 (C-5), 74.72 ($CH_2Ph$), 74.14 (C-3), 67.01 (C-2), 61.89 (C-6), 34.07 (CH-TDS), 25.30 (Cquat), 21.07 ($CH_3CO$), 20.11-20.10-18.57-18.54 ($CH_3$-TDS), 1.96 ($CH_3Si$-TDS), -3.00 ppm ($CH_3Si$-TDS).

$C_{23}H_{37}N_3O_6Si$; calcd. mass 479.25; ESI-MS: $m/z$ 502.49 $[M+Na]^+$.

### 3.3.7. 1-*S*-(4-Methylphenyl)-2-deoxy-2-phthalimido-3-*O*-acetyl-4,6-*O*-benzylidene-β-D-glucopyranoside **1**

Compound **14** [34] (3.05 g, 6.06 mmol) was dissolved in DCM (30 mL), then DMAP (37 mg, 0.05 mmol) and pyridine (1.96 mL, 24 mmol) were added; the reaction was cooled to 0 °C then acetic anhydride (1.15 mL, 12.1 mmol) was added slowly, then the reaction was raised to r.t. After 2 h TLC (hexane/EtOAc 75:25) indicated the disappearance of the starting material. The reaction was worked up as for compounds **9** and **13**; the product was purified by flash chromatography, using hexane/EtOAc 75:25 as eluent, affording compound **1** (3.1 g) with a yield of 95%.

$^1$H NMR (400 MHz, $CDCl_3$) δ 7.91–7.83 (m, 2H, CHAr), 7.80–7.70 (m, 2H, CHAr), 7.47–7.40 (m, 2H, CHAr), 7.40–7.32 (m, 3H, CHAr), 7.32–7.24 (m, 2H, CHAr), 7.08 (d, J = 8.0 Hz, 2H, CHAr), 5.88 (t, J = 9.4 Hz, 1H, H-3), 5.76 (d, J = 10.5 Hz, 1H, H-1), 5.53 (s, 1H, CH benzylidene), 4.46–4.38 (m, 1H, H-5), 4.33 (dd, J = 10.6, 9.9 Hz, 1H, H-2), 3.87–3.69 (m, 3H, H-6, H-4), 2.32 (s, 3H, $CH_3$Ph), 1.87 (s, 3H, $CH_3$CO).

$^{13}$C NMR (101 MHz, $CDCl_3$) δ 170.32 ($CH_3CO$), 168.01 (C(O)N Phth), 167.41 (C(O)N Phth), 138.87 (Cquat Ar), 137.01 (Cquat Ar), 134.58 (CAr), 134.35 (CAr), 133.86 (CAr), 131.86 (Cquat Ar), 131.37 (Cquat Ar), 129.90 (CAr), 129.33 (CAr), 128.40 (CAr), 127.35 (Cquat Ar), 126.40 (CAr), 123.87 (CAr), 123.75 (CAr), 101.80 (CHPh), 84.15 (C-1), 79.16 (C-4), 70.79 (C-3), 70.66 (C-5), 68.75 (C-6), 54.50 (C-2), 21.33 ($CH_3$-STol), 20.71 ($CH_3CO$).

$C_{30}H_{27}NO_7S$; calcd. mass 545.15; ESI-MS: $m/z$ 568,32 $[M+Na]^+$.

3.3.8. 6-*O*-(2-Deoxy-2-phthalimido-3-*O*-acetyl-4,6-*O*-benzylidene-β-D-glucopyranosyl)-β-1-(*O*-thexyldimethylsilyl)-2-deoxy-2-azido-3-*O*-acetyl-4-*O*-benzyl-β-D-glucopyranoside **3**

Compound **1** (3.195 g, 5.85 mmol) and compound **2** (2.58 g, 5.38 mmol) were co-evaporated three times with toluene and stored under vacuum overnight. Then, activated 4Å MS were added under an Ar atmosphere, 50 mL of dry DCM was added, and the mixture was stirred at r.t. for 1 h. Then the reaction was cooled to −20 °C; NIS (1.6 g, 7 mmol) and after 3 min TMSOTf (0.1 mL, 0.538 mmol) were added. The reaction was stirred at this temperature for 2 h, when TLC (hexane/EtOAc 7:3) showed complete consumption of the acceptor. The reaction was quenched by adding 5 mL of triethylamine at −20 °C, then raised to r.t.; the mixture was filtered over celite, and the filtrate concentrated to dryness. The residue was purified by silica gel chromatography (hexane/EtOAc 8:2), affording disaccharide **3** (3.44 g) in a yield of 71%.

$^1$H NMR (400 MHz, $CDCl_3$) δ 7.77–7.72 (m, 2H, CHAr), 7.67–7.58 (m, 2H, CHAr), 7.48–7.40 (m, 2H, CHAr), 7.40–7.28 (m, 3H, CHAr), 7.24–7.15 (m, 4H, CHAr), 7.00–6.91 (m, 2H, CHAr), 5.82 (dd, J = 10.3, 9.1 Hz, 1H, H-3′), 5.52 (s, 1H, CHPh), 5.49 (d, J = 8.4 Hz, 1H, H-1′), 4.90 (dd, J = 10.5, 8.6 Hz, 1H, H-3), 4.49 (d, J = 7.6 Hz, 1H, H-1), 4.39–4.31 (m, 2H, H-2′, H-6a′), 4.31–4.21 (m, 2H, $CH_2Ph$), 3.98 (dd, J = 10.7, 1.7 Hz, 1H, H-6b), 3.85–3.73 (m, 2H, H-4′, H-6b′), 3.73–3.63 (m, 2H, H-6a, H-5′), 3.47–3.33 (m, 2H, H-4, H-5), 3.18 (dd, J = 10.4, 7.7 Hz, 1H, H-2), 1.95 (s, 3H, $CH_3CO$), 1.87 (s, 3H, $CH_3CO$), 1.63–1.55 (m, 1H, CH-TDS), 0.83 (dd, J = 6.8, 1.3 Hz, 6H, $CH_3$-TDSx2), 0.79 (s, 6H, $CH_3$-TDSx2), 0.08 (s, 3H, $CH_3Si$-TDS), 0.00 (s, 3H, $CH_3Si$-TDS).

$^{13}$C NMR (101 MHz, $CDCl_3$) δ 170.35 ($CH_3CO$), 169.87 ($CH_3CO$), 137.43 (CquatAr), 137.06 (CquatAr), 134.37 (CquatAr), 131.57 (CquatAr), 129.34 (CAr), 128.50 (CAr), 128.41 (CAr), 128.03 (CAr), 127.75 (CAr), 126.40 (CAr), 123.70 (CAr), 101.86 (CHPh), 98.42 (C-1′), 97.01 (C-1), 79.34 (C-4′), 76.30 (C-4), 74.52 ($CH_2Ph$), 74.27 (C-5), 73.95 (C-3), 70.11 (C-3′), 68.81 (C-6′), 68.13 (C-6), 66.72 (C-2), 66.43 (C-5′), 55.43 (C-2′), 34.01 (CH-TDS), 24.89 (Cquat-TDS), 21.00 ($CH_3CO$), 20.69 ($CH_3CO$), 19.9954 ($CH_3$-TDS), 19.9854 ($CH_3$-TDS), 18.5954 ($CH_3$-TDS), 18.5554 ($CH_3$-TDS), −1.97 ($CH_3Si$-TDS), −3.39 ($CH_3Si$-TDS).

$C_{46}H_{56}N_4O_{13}Si$: calcd. mass 900.36; ESI-MS: *m*/*z* 901.76 $[M+H]^+$, 923.64 $[M+Na]^+$.

3.3.9. General Procedure for the Synthesis of β-Ketoesters

The appropriate carboxylic acid (10 mmol) was dissolved in DCM (100 mL); DCC (11 mmol) and DMAP (3 mmol) were added at r.t. and the reaction was stirred at this temperature for 1 h. A white precipitate formed during the reaction. Then Meldrum's acid (20 mmol) was dissolved in a DCM/pyridine solution (100 + 3 mL) and stirred at r.t. for 45 min, during which the solution turned pink. This solution was then slowly added to the first flask, and the reaction was stirred at r.t. After 3 h, TLC (hexane/EtOAc 7:3) showed the disappearance of the starting material; the reaction was then filtered, the filtrate washed with 5% HCl and then brine. The organic phase was dried with sodium sulfate, filtered, and evaporated. The crude residue was dissolved in the appropriate alcohol (100 mL, MeOH for **15** and **16** or tBuOH for **17**), then 5% mol of sulfuric acid was added, and the solution was refluxed for 15 h. The reaction was then evaporated, and the crude purified on silica gel (hexane/EtOAc 9:1 to 8:2).

Methyl 3-oxohexadecanoate **15**. Yield: 74%. The spectroscopic data are in agreement with those reported in the literature [55].

Methyl 3-oxoheptadecanoate **16**. Yield: 64%.

$^1$H-NMR (400 MHz, Chloroform-*d*) δ 3.73 (s, 3H, $OCH_3$), 3.44 (s, 2H, CO-$CH_2$-$CO_2$Me), 2.52 (t, J = 7.4 Hz, 2H, CO-$CH_2$-$CH_2$-Chain), 1.64–1.53 (m, 2H, CO-$CH_2$-$CH_2$-Chain),

1.34–1.17 (m, 22H, chain), 0.88 (t, J = 6.7 Hz, 3H, $CH_3$). $^{13}C$ NMR (101 MHz, $CDCl_3$) δ 202.98 (CO), 167.83 ($CO_2Me$), 52.44 ($CO_2CH_3$), 49.14, 43.22, 32.06, 29.79, 29.72, 29.57, 29.48, 29.32, 29.14, 23.61, 22.82 ($CH_2$ chain), 14.24 ($CH_3$). $C_{18}H_{34}O_3$: calcd. mass 298.25; ESI-MS: *m/z* 321.47 $[M+Na]^+$.

t-Butyl 3-oxoheptadecanoate **17**. Yield: 73%.

$^1H$-NMR (400 MHz, $CDCl_3$) δ 3.26 (s, 2H, CO-$CH_2$-$CO_2$tBu), 2.44 (t, J = 7.4 Hz, 2H, $CH_2$), 1.50 (q, J = 7.0 Hz, 2H, $CH_2$), 1.41 (d, J = 7.3 Hz, 9H, tBu), 1.19 (d, J = 7.8 Hz, 26H, $CH_2$ chain), 0.88–0.73 (m, 3H, $CH_3$). $^{13}C$ NMR (101 MHz, $CDCl_3$) δ 203.21 (CO), 166.48 ($CO_2$tBu), 50.61, 42.89, 31.94, 29.71, 29.69, 29.67, 29.62, 29.50, 29.47, 29.43, 29.40, 29.37, 29.21, 29.14, 29.09, 27.95 ($CH_3$ tBu), 23.50, 22.69. $C_{21}H_{40}O_3$: calcd. mass 340.30; ESI-MS: *m/z* 363.29 $[M+Na]^+$.

#### 3.3.10. General Procedure for the Stereoselective Reduction of β-Ketoesters to (*R*)-β-Hydroxyesters

The stereoselective reductions of β-ketoesters were performed employing a slightly modified procedure reported by Ratovelomanana et al. [36]. Briefly, the β-ketoester (3.5 mmol) was dissolved in degassed MeOH (7 mL, 0.5 M) and added to the freshly prepared [(*R*)-BINAP]$RuBr_2$ catalyst (0.07 mmol) [36]. The mixture was placed under argon, and then the atmosphere was replaced with hydrogen (from a balloon). The reaction was heated to 50 °C for 16 h, and the progress of the reaction was checked by TLC (hexane/EtOAc 7:3). The reaction was evaporated, and the crude product was purified by flash chromatography (hexane/EtOAc 8:2), affording the respective (R)-β-hydroxyesters.

Methyl 3-hydroxyhexadecanoate **18**. Yield: 89%. The spectroscopic data are in agreement with those reported in the literature [36]. Mosher's esters analysis [37] and the $^1H$-NMR experiment employing the chiral lanthanide shift reagent $Eu(Hfc)_3$ [38] confirmed the stereoselective reduction, affording the (*R*)-enantiomers (see Supporting Information).

Methyl 3-hydroxyheptadecanoate **19**. Yield: 85%.

$^1H$ NMR (400 MHz, $CDCl_3$) δ 4.07–3.93 (m, 1H, H-3), 3.71 (s, 3H, O$CH_3$), 2.83 (d, J = 4.0 Hz, 1H, OH), 2.52 (dd, J = 16.4, 3.1 Hz, 1H, H-2a), 2.41 (dd, J = 16.4, 9.0 Hz, 1H, H-2b), 1.48–1.38 (m, 2H, H-4), 1.26 (s, 22H, $CH_2$ chain), 0.88 (t, J = 6.7 Hz, 3H, $CH_3$).

$^{13}C$ NMR (101 MHz, $CDCl_3$) δ 173.63 (COOMe), 68.16 (C-3), 51.84 ($OCH_3$), 41.25 (C-2), 36.68 (C-4), 32.05 ($CH_2$ chain), 29.80 ($CH_2$ chain), 29.71 ($CH_2$ chain), 29.65 ($CH_2$ chain), 29.49 ($CH_2$ chain), 29.41 ($CH_2$ chain), 25.61 ($CH_2$ chain), 22.82 ($CH_2$ chain), 14.24 ($CH_3$). $C_{18}H_{36}O_3$: calcd. mass 300.27; ESI-MS: *m/z* 300.99 $[M+H]^+$.

t-Butyl-3-hydroxyheptadecanoate **20**. Yield: 73%.

$^1H$ NMR (400 MHz, $CDCl_3$) δ 3.98–3.88 (m, 1H, H-3), 3.09 (s, 1H, OH), 2.41 (dd, J = 16.3, 3.2 Hz, 1H, H-2a), 2.30 (dd, J = 16.3, 8.9 Hz, 1H, H-2b), 1.45 (s, 9H, $CH_3$ tBu), 1.43–1.34 (m, 2H, H-4), 1.33–1.18 (m, 24H, $CH_2$ chain), 0.91–0.83 (m, 3H, $CH_3$). $^{13}C$ NMR (101 MHz, $CDCl_3$) δ 172.71 (COOtBu), 81.30 (Cquat tBu), 68.04 (C-3), 42.42 (C-2), 36.62 (C-4), 32.07 ($CH_2$ chain), 29.83 ($CH_2$ chain), 29.80 ($CH_2$ chain), 29.78 ($CH_2$ chain), 29.72 ($CH_2$ chain), 29.70 ($CH_2$ chain), 29.49 ($CH_2$ chain), 28.24 ($CH_3$ tBu), 25.60 ($CH_2$ chain), 22.80 ($CH_2$ chain), 14.23 ($CH_3$). $C_{21}H_{42}O_3$: calcd. mass 342.31; ESI-MS: *m/z* 343.65 $[M+H]^+$.

#### 3.3.11. General Procedure for the Reductive Benzylation of the 3-Hydroxyl Group of Fatty Acids and Base-Mediated Ester Hydrolysis

Respective Methyl 3-hydroxyhexadecanoate (**18**) or heptadecanoate (**19**) (0.67 mmol) was dissolved in dry THF (1.3 mL), and the solution was cooled to 0 °C. After the addition of benzaldehyde (2.0 mmol) and hexamethyldisiloxane (4.0 mmol), trimethylsilyl trifluoromethanesulfonate (1.3 mmol) was added dropwise, and the reaction mixture was stirred for 3 h at 0 °C. Then, triethylsilane (2.0 mmol) was added, and after stirring for 2 h at 0 °C, the reaction mixture was diluted with EtOAc, washed with saturated aqueous $NaHCO_3$ solution and brine; dried over $Na_2SO_4$, filtered, and concentrated.

To the crude was added THF (3 mL), MeOH (2 mL), and $H_2O$ (1 mL). To the solution was added LiOH.$H_2O$ (2.7 mmol), and the reaction was stirred at room temperature for 5 h. The reaction mixture was then diluted with EtOAc, washed with aqueous 1 N HCl, deionized $H_2O$, and brine; dried over $Na_2SO_4$, filtered, and concentrated. The crude was purified by flash silica gel column chromatography (hexane/EtOAc 7:3), affording the respective (*R*)-3-(benzyloxy) fatty acids.

(*R*)-3-(benzyloxy)-hexadecanoic acid **4**; Yield: 80%. The spectroscopic data are in agreement with those reported in the literature [27].

(*R*)-3-(benzyloxy)-heptadecanoic acid **6**; Yield: 82%.

$^1$H NMR (400 MHz, $CDCl_3$) δ 7.28–7.17 (m, 5H, CHAr), 4.56–4.45 (m, 2H, $CH_2$Ph), 3.81 (ddd, J = 12.2, 6.8, 5.5 Hz, 1H, H-3), 2.58 (dd, J = 15.4, 7.1 Hz, 1H, H-2a), 2.48 (dd, J = 15.4, 5.2 Hz, 1H, H-2b), 1.66–1.43 (m, 2H, H-4), 1.40–1.14 (m, 24H, $CH_2$ chain), 0.82 (t, J = 6.6 Hz, 3H, $CH_3$). $^{13}$C NMR (101 MHz, $CDCl_3$) δ 177.47 (COOMe), 138.29 (Cquat Ar), 128.52 (CAr), 127.98 (CAr), 127.84 (CAr), 75.87 (C-3), 71.69 ($CH_2$Ph), 39.69 (C-2), 34.29 (C-4), 32.07 ($CH_2$ chain), 29.83 ($CH_2$ chain), 29.82 ($CH_2$ chain), 29.80 ($CH_2$ chain), 29.79 ($CH_2$ chain), 29.73 ($CH_2$ chain), 29.70 ($CH_2$ chain), 29.51 ($CH_2$ chain), 25.26 ($CH_2$ chain), 22.84 ($CH_2$ chain), 14.26 ($CH_3$). $C_{24}H_{40}O_3$: calcd. mass 376.30; ESI-MS: *m/z* 399.70 $[M+Na]^+$.

#### 3.3.12. t-Butyl (R)-3-(Pentadecanoyloxy)-Heptadecanoate **21**

To the solution of *t*-Butyl (*R*)-3-hydroxy-heptadecanoate **20** (1.04 g, 2.92 mmol) and pentadecanoic acid (0.92 g, 3.79 mmol) in dry $CH_2Cl_2$ (29 mL) was added DIC (904 μL, 5.84 mmol) and DMAP (0.36 g, 2.92 mmol). The reaction was stirred at room temperature for 6 h, and then it was diluted with $CH_2Cl_2$. The diisopropylurea (DIU) precipitate was filtered off, and the filtrate was washed with saturated aqueous $NH_4Cl$ solution and brine, dried over $Na_2SO_4$, filtered, and concentrated. The crude was purified by flash silica gel column chromatography (hexane/EtOAc = 19:1) to give **21** (1.7 g, quant.)

$^1$H NMR (400 MHz, $CDCl_3$) δ 5.19 (tt, *J* = 7.3, 5.7 Hz, 1H, H-3), 2.53–2.37 (m, 2H, H-2), 2.29–2.21 (m, 2H, H-2′), 1.58 (qd, J = 11.5, 6.1 Hz, 4H, H-4, H-3′), 1.42 (s, 9H, $CH_3$ tBu), 1.25 (d, J = 5.1 Hz, 46H, $CH_2$ chains), 0.91–0.82 (m, 6H, $CH_3$ × 2).

$^{13}$C NMR (101 MHz, $CDCl_3$) δ 173.19 (COOtBu), 169.84 (COO C15 chain), 80.82 (Cquat tBu), 70.64 (C-3), 40.75 (C-2), 34.65 (C-2′), 34.19 (C-4), 32.05 ($CH_2$ chains), 29.84 ($CH_2$ chains), 29.82 ($CH_2$ chains), 29.81 ($CH_2$ chains), 29.79 ($CH_2$ chains), 29.78 ($CH_2$ chains), 29.75 ($CH_2$ chains), 29.68 ($CH_2$ chains), 29.63 ($CH_2$ chains), 29.62 ($CH_2$ chains), 29.53 ($CH_2$ chains), 29.50 ($CH_2$ chains), 29.43 ($CH_2$ chains), 29.31 ($CH_2$ chains), 28.15 ($CH_3$ tBu), 25.25 ($CH_2$ chains), 25.17 ($CH_2$ chains), 22.83 ($CH_2$ chains), 14.24 ($CH_3$). $C_{36}H_{70}O_4$: calcd. mass 566.53; ESI-MS: *m/z* 589.63 $[M+Na]^+$.

#### 3.3.13. (*R*)-3-(Pentadecanoyloxy)-Heptadecanoic Acid **7**

Compound **21** (1.6 g, 2.82 mmol) was dissolved in $CH_2Cl_2$ (23 mL), and then TFA (6 mL) was added, and the reaction mixture was stirred at room temperature for 12 h. The reaction was diluted with $CH_2Cl_2$, washed with saturated aqueous $NaHCO_3$ solution and brine; dried over $Na_2SO_4$, filtered, and concentrated. The crude was purified by flash silica gel column chromatography (hexane/EtOAc = 6:4) to give **7** (1.3 g, 93%).

$^1$H NMR (400 MHz, $CDCl_3$) δ 5.20 (t, J = 6.5 Hz, 1H, H-3), 2.67–2.51 (m, 2H, H-2), 2.27 (t, J = 7.5 Hz, 2H, H-2′), 1.60 (h, J = 7.2 Hz, 4H, H-4, H-3′), 1.46–1.04 (m, 46H, $CH_2$ chains), 0.94–0.80 (m, 6H, $CH_3$ x 2). $^{13}$C NMR (101 MHz, $CDCl_3$) δ 176.69 (COOH), 173.44 (COO C15 chain), 70.14 (C-3), 39.06 (C-2), 34.62 (C-2′), 34.11 (C-4), 32.07 ($CH_2$ chains), 29.86 ($CH_2$

chains), 29.85 ($CH_2$ chains), 29.82 ($CH_2$ chains), 29.81 ($CH_2$ chains), 29.78 ($CH_2$ chains), 29.70 ($CH_2$ chains), 29.64 ($CH_2$ chains), 29.51 ($CH_2$ chains), 29.49 ($CH_2$ chains), 29.43 ($CH_2$ chains), 29.26 ($CH_2$ chains), 25.25 ($CH_2$ chains), 25.15 ($CH_2$ chains), 22.85 ($CH_2$ chains), 14.25 ($CH_3$). $C_{32}H_{62}O_4$: calcd. mass 510.46; ESI-MS: $m/z$ 533.46 $[M+Na]^+$.

#### 3.3.14. 6-*O*-(2-Deoxy-2-((*R*)-3-(benzyloxy)-hexadecanoylamido)-3-*O*-((*R*)-3-(benzyloxy)-hexadecanoyl)-4,6-*O*-benzylidene-β-D-glucopyranosyl)-1-(*O*-thexyldimethylsilyl)-2-deoxy-2-azido-3-*O*-((*R*)-3-(benzyloxy)-hexadecanoyl)-4-*O*-benzyl-β-D-glucopyranoside **23**

Disaccharide **3** (620 mg, 0.688 mmol) was dissolved in absolute ethanol (20 mL) with sonication to obtain a cloudy solution. Then, hydrazine monohydrate (2 mL) was added, and the reaction mixture was stirred at 70 °C for 2 h. After cooling to room temperature, the reaction was diluted with $CH_2Cl_2$, washed with deionized $H_2O$ and brine; dried over $Na_2SO_4$, filtered, and concentrated.

A mixture of the obtained crude compound **22** and benzyl-protected fatty acid **4** (1.0 g, 2.75 mmol) was co-evaporated with toluene twice and dried in vacuo for 30 min. To the mixture was added dry $CH_2Cl_2$ (5.8 mL) and dry DMF (1.2 mL), and stirred at room temperature. Then EDC.HCl (1.3 g, 6.88 mmol) and DMAP (126 mg, 1.03 mmol) were added, and the reaction was stirred at 45 °C for 12 h. After dilution with $CH_2Cl_2$, the reaction mixture was washed with saturated aqueous $NH_4Cl$ solution and brine, dried over $Na_2SO_4$, filtered, and concentrated. The crude was purified by flash silica gel column chromatography (Toluene/EtOAc = 19:1) to give **23** (0.9 g, 75%).

$^1$H NMR (400 MHz, $CDCl_3$) δ 7.52–7.18 (m, 25H, CHAr), 6.32 (d, J = 9.0 Hz, 1H, N*H*), 5.42 (s, 1H, H-7′), 5.31 (t, J = 9.8 Hz, 1H, H-3′), 5.03 (dd, J = 10.4, 9.1 Hz, 1H, H-3), 4.62–4.35 (m, 11H), 4.27 (dd, J = 10.4, 4.9 Hz, 1H), 4.03–3.61 (m, 10H), 3.56 (t, J = 9.4 Hz, 1H), 3.41 (dtd, J = 11.7, 5.8, 3.9 Hz, 2H), 3.26 (dd, J = 10.4, 7.6 Hz, 1H, H-2), 2.73–2.53 (m, 2H), 2.51–2.40 (m, 2H), 2.34–2.22 (m, 2H), 1.73–1.66 (m, 1H, CH TDS), 1.62–1.38 (m, 8H), 1.38–1.10 (m, 70H, $CH_2$ chains), 0.96–0.87 (m, 21H, $CH_3$ (chains + TDS)), 0.22 (d, J = 3.1 Hz, 6H, $CH_3$ TDS). $^{13}$C NMR (101 MHz, $CDCl_3$) δ 171.38, 171.30, 170.80, 138.75, 138.70, 138.46, 137.92, 137.03, 129.15, 128.73, 128.56, 128.54, 128.39, 128.37, 128.29, 128.27, 127.96, 127.89, 127.88, 127.82, 127.57, 127.55, 126.25, 101.61, 101.56, 97.06, 79.04, 76.26, 76.14, 75.88, 75.69, 74.62, 74.30, 74.11, 71.76, 71.63, 71.27, 70.95, 68.78, 67.93, 66.90, 66.53, 54.62, 41.36, 39.88, 34.66, 34.51, 34.07, 33.84, 32.07, 29.85, 29.83, 29.81, 29.79, 29.77, 29.74, 29.73, 29.50, 25.28, 25.26, 25.23, 25.21, 24.98, 22.82, 20.13, 20.08, 18.65, 18.60, 14.23, −1.63, −3.17.

$C_{103}H_{158}N_4O_{15}Si$: calcd. mass 1719.15; ESI-MS: $m/z$ 1742.42 $[M+Na]^+$.

#### 3.3.15. General Procedure for Azide Reduction

To a solution of compound **23** (27 mg, 0.016 mmol) in 1,4-dioxane (1 mL) and AcOH (0.1 mL) was added Zn powder (150 mg), and after stirring the reaction for 12 h at room temperature, it was diluted with $CH_2Cl_2$, filtered over celite, washed with saturated aqueous $NaHCO_3$ solution and brine; dried over $Na_2SO_4$, filtered and concentrated to obtain the amine as crude which was used in the next reaction without further purification.

#### 3.3.16. 6-*O*-(2-Deoxy-2-((*R*)-3-(benzyloxy)-hexadecanoylamido)-3-*O*-((*R*)-3-(benzyloxy)-hexadecanoyl)-4,6-*O*-benzylidene-β-D-glucopyranosyl)-1-(*O*-thexyldimethylsilyl)-2-deoxy-2-((*R*)-3-(benzyloxy)-heptadecanoylamido)-3-*O*-((*R*)-3-(benzyloxy)-hexadecanoyl)-4-*O*-benzyl-β-D-glucopyranoside **24**

The crude amine obtained after azide reduction (0.019 mmol) was co-evaporated with toluene and dried in vacuo. To the crude was added benzyl-protected fatty acid

**6** (11 mg, 0.029 mmol) and HATU (15 mg, 0.038 mmol), and the mixture was dissolved in dry $CH_2Cl_2$ (1 mL) and DMF (0.2 mL). After adding DIPEA (7 μL), the reaction was stirred at room temperature for 7 h. The reaction mixture was then diluted with $CH_2Cl_2$, washed with saturated aqueous $NH_4Cl$ solution and brine, dried over $Na_2SO_4$, filtered, and concentrated. The crude was purified by size exclusion chromatography (Sephadex® LH-20) ($CHCl_3$/MeOH = 1:1) to give **24** (34 mg, 87%).

$^1$H NMR (400 MHz, $CDCl_3$) δ 7.50–7.18 (m, 25H, CHAr), 6.38 (d, J = 9.0 Hz, 1H, NH), 6.24 (d, J = 9.3 Hz, 1H, NH), 5.42 (s, 1H, H-7′), 5.34 (t, J = 9.8 Hz, 1H, H-3′), 5.14 (dd, J = 10.4, 8.8 Hz, 1H, H-3), 4.63–4.37 (m, 13H), 4.30 (dd, J = 10.5, 4.9 Hz, 1H), 4.02–3.93 (m, 1H), 3.86 (ddt, J = 18.9, 11.9, 6.1 Hz, 4H), 3.78–3.54 (m, 5H), 3.47–3.38 (m, 2H), 2.73–2.50 (m, 2H), 2.49–2.24 (m, 6H), 1.67–1.42 (m, 9H), 1.41–1.10 (m, 105H, $CH_2$ chains), 0.96–0.81 (m, 25H, $CH_3$ (chains + TDS)), 0.17 (s, 3H, $CH_3$ TDS), 0.10 (s, 3H, $CH_3$ TDS).

$^{13}$C NMR (101 MHz, $CDCl_3$) δ 171.54, 171.38, 171.37, 171.00, 138.70, 138.61, 138.41, 138.39, 138.00, 137.01, 128.69, 128.61, 128.47, 128.36, 128.34, 128.23, 127.91, 127.89, 127.85, 127.83, 127.82, 127.81, 127.79, 127.77, 127.55, 127.52, 126.21, 101.48, 101.44, 96.19, 78.97, 76.27, 76.10, 76.00, 75.88, 75.81, 75.62, 75.54, 74.67, 74.43, 74.16, 71.97, 71.72, 71.63, 71.61, 71.44, 71.21, 70.97, 70.65, 68.72, 68.29, 66.41, 55.95, 54.61, 41.27, 41.13, 39.81, 39.46, 37.15, 34.60, 34.37, 34.35, 34.05, 33.84, 33.76, 32.07, 29.87, 29.86, 29.83, 29.81, 29.80, 29.78, 29.74, 29.71, 29.69, 29.68, 29.66, 29.48, 25.34, 25.26, 25.24, 25.16, 24.80, 22.80, 22.78, 20.21, 20.19, 18.70, 18.69, 14.23, −1.31, −3.16.

$C_{127}H_{198}N_2O_{17}Si$: calcd. mass 2051.45; ESI-MS: $m/z$ 2074.86 $[M+Na]^+$.

#### 3.3.17. 6-*O*-(2-Deoxy-2-((*R*)-3-(benzyloxy)-hexadecanoylamido)-3-*O*-((*R*)-3-(benzyloxy)-hexadecanoyl)-4,6-*O*-benzylidene-β-D-glucopyranosyl)-1-(*O*-thexyldimethylsilyl)-2-deoxy-2-((*R*)-3-(pentadecanoyloxy)-heptadecanoylamido)-3-*O*-((*R*)-3-(benzyloxy)-hexadecanoyl)-4-*O*-benzyl-β-D-glucopyranoside **27**

The crude amine obtained after azide reduction (0.016 mmol) was co-evaporated with toluene and dried in vacuo. To the crude was added double fatty acid **7** (12 mg, 0.024 mmol) and HATU (12 mg, 0.031 mmol), and the mixture was dissolved in dry $CH_2Cl_2$ (1 mL) and DMF (0.2 mL). After adding DIPEA (5.5 μL), the reaction was stirred at room temperature for 4 h. The reaction mixture was then diluted with $CH_2Cl_2$, washed with saturated aqueous $NH_4Cl$ solution and brine, dried over $Na_2SO_4$, filtered, and concentrated. The crude was purified by size exclusion chromatography (Sephadex® LH-20) ($CHCl_3$/MeOH = 1:1) to give **27** (29 mg, 85%).

$^1$H NMR (400 MHz, $CDCl_3$) δ 7.44–7.13 (m, 25H, CHAr), 6.29 (d, J = 8.9 Hz, 1H, NH), 5.75 (d, J = 9.1 Hz, 1H, NH), 5.39 (s, 1H, H-7′), 5.34–5.25 (m, 1H, H-3′), 5.11 (dd, J = 10.4, 8.5 Hz, 1H, H-3), 5.08–5.00 (m, 1H), 4.69 (d, J = 7.6 Hz, 1H, H-1), 4.60–4.33 (m, 10H), 4.31–4.20 (m, 1H), 3.98–3.74 (m, 5H), 3.73–3.51 (m, 4H), 3.51–3.43 (m, 1H), 3.43–3.34 (m, 1H), 2.64 (dd, J = 15.0, 6.4 Hz, 1H), 2.53 (dd, J = 15.9, 7.0 Hz, 1H), 2.47–2.18 (m, 8H), 1.67–1.39 (m, 4H), 1.39–1.07 (m, 101H, $CH_2$ chains), 0.92–0.79 (m, 31H, $CH_3$ (chains + TDS)), 0.16 (s, 3H, $CH_3$ TDS), 0.13 (s, 3H, $CH_3$ TDS).

$^{13}$C NMR (101 MHz, $CDCl_3$) δ 173.92, 171.71, 171.39, 171.32, 169.30, 138.69, 138.66, 138.49, 138.04, 137.05, 128.73, 128.55, 128.53, 128.51, 128.42, 128.38, 128.36, 128.29, 128.27, 128.25, 127.91, 127.86, 127.83, 127.81, 127.78, 127.76, 127.63, 127.55, 126.27, 126.25, 126.23, 101.59, 101.49, 96.04, 79.00, 76.33, 76.03, 75.67, 75.59, 74.84, 74.55, 74.23, 71.76, 71.43, 71.26, 71.01, 70.91, 68.76, 68.34, 66.49, 56.36, 54.66, 41.87, 41.36, 39.86, 39.69, 34.68, 34.65, 34.35, 34.14, 33.89, 32.07, 32.05, 29.92, 29.91, 29.90, 29.88, 29.86, 29.83, 29.81, 29.79, 29.77, 29.75, 29.74, 29.71, 29.69, 29.58, 29.51, 29.49, 29.38, 25.45, 25.35, 25.28, 25.20, 25.18, 25.15, 24.88, 22.83, 20.24, 20.22, 18.75, 18.73, 18.70, 18.68, 14.25, −1.37, −3.15.

$C_{135}H_{220}N_2O_{18}Si$: calcd. mass 2185.61; ESI-MS: *m*/*z* 2208.58 $[M+Na]^+$.

3.3.18. General Procedure for Desilylation

The respective TDS-protected substrate (0.017 mmol) was dissolved in dry THF (1.2 mL) and dry pyridine (0.4 mL). To the solution was added HF-pyridine (0.12 mL), and the reaction was stirred at room temperature for 8 h. The reaction mixture was then diluted with $CH_2Cl_2$, washed with saturated aqueous $NH_4Cl$ solution and brine, dried over $Na_2SO_4$, filtered, and concentrated. The crude was purified by flash silica gel column chromatography (Toluene/EtOAc = 8:2) to give the corresponding desilylated products.

6-*O*-(2-deoxy-2-((*R*)-3-(benzyloxy)-hexadecanoylamido)-3-*O*-((*R*)-3-(benzyloxy) -hexadecanoyl)-4,6-*O*-benzylidene-β-D-glucopyranosyl)-2-deoxy-2-((*R*)-3-(benzyloxy) -heptadecanoylamido)-3-*O*-((*R*)-3-(benzyloxy)-hexadecanoyl)-4-*O*-benzyl-β-D-glucopyranose **25**. Yield: 84%.

$^1$H NMR (400 MHz, $CDCl_3$) δ 7.40–7.02 (m, 30H, CHAr), 6.30 (d, J = 8.3 Hz, 1H, NH), 6.20 (d, J = 9.4 Hz, 1H, NH), 5.33 (s, 1H, H-7′), 5.27–5.18 (m, 2H, H-3, H-3′), 4.84 (d, J = 3.5 Hz, 1H, H-1), 4.62 (d, J = 8.4 Hz, 1H, H-1′), 4.52–4.19 (m, 11H), 4.06 (td, J = 9.8, 3.5 Hz, 1H), 3.90 (dd, J = 11.7, 1.9 Hz, 1H), 3.83–3.63 (m, 5H), 3.56 (t, J = 9.4 Hz, 1H), 3.45–3.33 (m, 2H), 3.22 (t, J = 9.6 Hz, 1H), 2.63–2.42 (m, 2H), 2.39–2.04 (m, 5H), 1.53–1.30 (m, 8H, $CH_2$ chains), 1.29–1.01 (m, 99H, CH, $CH_2$ chains), 0.86–0.73 (m, 12H, $CH_3$ chains).

$^{13}$C NMR (101 MHz, $CDCl_3$) δ 172.00, 171.97, 171.51, 171.46, 138.76, 138.72, 138.70, 137.76, 137.65, 137.06, 128.76, 128.64, 128.61, 128.56, 128.53, 128.48, 128.45, 128.42, 128.40, 128.35, 128.33, 128.00, 127.98, 127.96, 127.94, 127.91, 127.90, 127.87, 127.85, 127.79, 127.61, 127.59, 126.30, 102.28, 101.64, 91.51, 79.04, 76.60, 76.56, 75.69, 75.66, 74.39, 73.50, 71.94, 71.65, 71.55, 71.53, 71.39, 71.20, 70.64, 69.41, 68.81, 66.71, 55.36, 52.75, 42.10, 41.74, 40.00, 39.85, 34.64, 34.50, 34.37, 33.94, 32.11, 29.92, 29.90, 29.89, 29.85, 29.84, 29.81, 29.55, 25.40, 25.38, 25.35, 25.31, 22.87, 14.29.

$C_{119}H_{180}N_2O_{17}$: calcd. mass 1909.33; ESI-MS: *m*/*z* 1932.42 $[M+Na]^+$.

6-*O*-(2-deoxy-2-((*R*)-3-(benzyloxy)-hexadecanoylamido)-3-*O*-((*R*)-3-(benzyloxy) -hexadecanoyl)-4,6-*O*-benzylidene-β-D-glucopyranosyl)-2-deoxy-2-((*R*)-3-(pentadecanoyloxy) -heptadecanoylamido)-3-*O*-((*R*)-3-(benzyloxy)-hexadecanoyl)-4-*O*-benzyl-D-glucopyranose **28**; Yield: 75%.

$^1$H NMR (400 MHz, $CDCl_3$) δ 7.47–7.12 (m, 25H. CHAr), 6.38 (d, J = 8.5 Hz, 1H, NH), 5.92 (d, J = 9.2 Hz, 1H, NH), 5.42 (s, 1H, H-7′), 5.35–5.24 (m, 2H, H-3, H-3′), 5.12 (dd, J = 7.4, 4.9 Hz, 1H), 4.93 (d, J = 3.5 Hz, 1H, H-1), 4.71 (d, J = 8.4 Hz, 1H, H-1′), 4.63–4.27 (m, 8H, $CH_2$Ph), 4.12–3.98 (m, 2H), 3.95–3.72 (m, 4H), 3.65 (t, J = 9.4 Hz, 1H), 3.55–3.40 (m, 2H), 3.29 (t, J = 9.5 Hz, 1H), 2.71–2.53 (m, 2H), 2.48–2.19 (m, 6H), 1.96 (s, 1H, OH), 1.70–1.40 (m, 10H, $CH_2$ chains), 1.39–1.04 (m, 106H, $CH_2$ chains), 0.89 (t, J = 6.8 Hz, 15H, $CH_3$ chains).

$^{13}$C NMR (101 MHz, $CDCl_3$) δ 173.30, 172.30, 171.86, 171.39, 169.80, 138.64, 137.63, 137.57, 136.98, 128.74, 128.51, 128.45, 128.37, 128.34, 128.29, 127.94, 127.88, 127.86, 127.84, 127.82, 127.58, 127.56, 126.24, 102.26, 101.59, 91.25, 78.98, 77.48, 77.16, 76.85, 76.77, 76.53, 75.63, 75.55, 74.47, 73.53, 71.86, 71.43, 71.23, 71.15, 71.13, 71.01, 70.65, 69.32, 68.75, 66.65, 55.23, 52.88, 41.55, 39.93, 39.80, 34.64, 34.57, 34.43, 34.37, 33.83, 32.06, 32.04, 29.87, 29.86, 29.84, 29.82, 29.80, 29.77, 29.75, 29.73, 29.71, 29.59, 29.53, 29.50, 29.38, 25.40, 25.35, 25.28, 25.26, 25.20, 22.82, 14.25.

$C_{127}H_{202}N_2O_{18}$: calcd. mass 2043.50; ESI-MS: *m*/*z* 2066.72 $[M+Na]^+$.

6-*O*-(2-deoxy-2-((*R*)-3-(benzyloxy)-hexadecanoylamido)-3-*O*-((*R*)-3-(benzyloxy) -hexadecanoyl)-4-(*O*-dibenzylphosphoryl)-6-*O*-benzyl-β-D-glucopyranosyl)-2-deoxy-2-((*R*) -3-(pentadecanoyloxy)-heptadecanoylamido)-3-*O*-((*R*)-3-(benzyloxy)-hexadecanoyl)-4-*O* -benzyl-D-glucopyranose **32**. Yield: 73%.

$^{1}$H NMR (400 MHz, $CDCl_3$) δ 7.38–7.12 (m, 35H, CHAr), 6.22 (d, J = 7.9 Hz, 1H, NH), 5.87 (d, J = 9.3 Hz, 1H, NH), 5.38 (t, J = 9.7 Hz, 1H, H-3′), 5.28 (t, J = 9.7 Hz, 1H, H-3), 5.17–5.08 (m, 1H), 4.99–4.83 (m, 6H, H-1, H-1′, $(CH_2PhO)_2PO$-), 4.60–4.31 (m, 11H, $CH_2Ph$, H-4′), 4.12–4.02 (m, 1H, H-2), 3.99–3.90 (m, 2H), 3.87–3.69 (m, 5H), 3.68–3.45 (m, 4H), 3.25 (t, J = 9.4 Hz, 1H, H-4), 2.62–2.24 (m, 8H), 2.22–2.14 (m, 2H), 1.80 (s, 2H), 1.65–1.40 (m, 3H), 1.25 (d, J = 9.5 Hz, 148H, $CH_2$ chains), 0.88 (t, J = 6.7 Hz, 15H, $CH_3$ chains).

$^{13}$C NMR (101 MHz, $CDCl_3$) δ 173.27, 172.24, 172.04, 171.30, 169.73, 138.78, 138.70, 138.11, 137.83, 137.62, 135.71, 128.67, 128.53, 128.46, 128.38, 128.31, 128.16, 128.11, 127.96, 127.90, 127.86, 127.83, 100.58, 91.28, 76.34, 75.57, 74.52, 74.31, 73.83, 73.60, 71.45, 69.74, 69.66, 68.84, 55.57, 52.85, 41.53, 39.95, 39.00, 34.65, 34.42, 34.30, 34.01, 32.08, 29.87, 29.82, 29.52, 25.36, 22.83, 14.25.

$^{31}$P NMR (162 MHz, $CDCl_3$) δ −2.14.

$C_{141}H_{217}N_2O_{21}P$: calcd. mass 2305.57; ESI-MS: $m/z$ 2328.67 $[M+Na]^+$.

### 3.3.19. General Procedure for Phosphorylation of the Anomeric Position

To the respective substrate with a free anomeric hydroxyl group (0.01 mmol) was added dry THF (1 mL), and the solution was cooled to −70 °C. Then, LiHMDS (0.03 mmol) was added, and the reaction was stirred at −70 °C for 30 min. Thereafter, tetrabenzyl pyrophosphate (0.02 mmol) was added, and stirring continued for 1 h while warming from −70 °C to 0 °C. The reaction mixture was diluted with $CH_2Cl_2$, washed with saturated aqueous $NH_4Cl$ solution and brine, dried over $Na_2SO_4$, filtered, and concentrated. The crude was purified by flash silica gel column chromatography (Toluene/EtOAc = 9:1) to give the corresponding phosphorylated products.

6-*O*-(2-deoxy-2-((*R*)-3-(benzyloxy)-hexadecanoylamido)-3-*O*-((*R*)-3-(benzyloxy) -hexadecanoyl)-4,6-*O*-benzylidene-β-D-glucopyranosyl)-1-*O*-(dibenzylphosphoryl)-2-deoxy -2-((*R*)-3-(benzyloxy)-heptadecanoylamido)-3-*O*-((*R*)-3-(benzyloxy)-hexadecanoyl)-4-*O* -benzyl-α-D-glucopyranoside **26**. Yield: 63%.

$^{1}$H NMR (400 MHz, $CDCl_3$) δ 7.44–7.19 (m, 40H, CHAr), 7.02 (d, J = 8.9 Hz, 1H, NH), 6.24 (d, J = 8.7 Hz, 1H, NH), 5.68 (dd, J = 5.4, 3.3 Hz, 1H, H-1), 5.40 (s, 1H, H-7′), 5.34–5.23 (m, 2H, H-3, H-3′), 5.08–4.89 (m, 5H), 4.70 (d, J = 8.4 Hz, 1H, H-1′), 4.61–4.24 (m, 17H), 4.08–3.95 (m, 1H), 3.87–3.67 (m, 5H), 3.62 (t, J = 9.5 Hz, 1H), 3.54 (t, J = 9.6 Hz, 1H), 3.40–3.32 (m, 1H), 2.66 (dd, J = 15.0, 6.6 Hz, 1H), 2.54 (dd, J = 16.1, 7.3 Hz, 1H), 2.46–2.11 (m, 6H), 2.03–1.93 (m, 1H), 1.86–1.68 (m, 2H), 1.60–1.41 (m, 2H), 1.37–1.07 (m, 95H, $CH_2$ chains), 0.89 (t, J = 6.9 Hz, 12H, $CH_3$ chains).

$^{13}$C NMR (101 MHz, $CDCl_3$) δ 172.04, 171.71, 171.45, 171.36, 138.78, 138.72, 138.59, 138.55, 137.41, 137.10, 128.94, 128.92, 128.84, 128.83, 128.62, 128.57, 128.55, 128.53, 128.51, 128.46, 128.41, 128.38, 128.37, 128.34, 128.32, 128.29, 128.26, 128.21, 128.13, 128.10, 127.94, 127.91, 127.86, 127.83, 127.80, 127.69, 127.64, 127.49, 126.24, 101.49, 100.99, 79.16, 76.09, 75.67, 75.43, 75.34, 74.92, 71.54, 71.27, 71.09, 71.04, 70.11, 70.06, 70.01, 66.56, 54.37, 52.32, 41.46, 41.24, 39.94, 39.81, 34.70, 34.35, 34.08, 32.07, 29.87, 29.85, 29.83, 29.81, 29.78, 29.51, 25.29, 22.83, 14.27.

$^{31}$P NMR (162 MHz, $CDCl_3$) δ −3.10.

$C_{133}H_{193}N_2O_{20}P$: calcd. mass 2169.39; ESI-MS: $m/z$ 2192.68 $[M+Na]^+$.

6-*O*-(2-deoxy-2-((*R*)-3-(benzyloxy)-hexadecanoylamido)-3-*O*-((*R*)-3-(benzyloxy) -hexadecanoyl)-4,6-*O*-benzylidene-β-D-glucopyranosyl)-1-(*O*-dibenzylphosphoryl)-2-deoxy -2-((*R*)-3-(pentadecanoyloxy)-heptadecanoylamido)-3-*O*-((*R*)-3-(benzyloxy)-hexadecanoyl)-4-*O*-benzyl-α-D-glucopyranose **29**. Yield: 65%.

$^{1}$H NMR (400 MHz, $CDCl_3$) δ 7.43–7.12 (m, 35H, CHAr), 6.99 (d, J = 9.0 Hz, 1H, NH-1), 5.99 (d, J = 8.6 Hz, 1H NH-2), 5.63 (dd, J = 5.5, 3.2 Hz, 1H, H-1), 5.38 (s, 1H, H-7′), 5.33–5.18

(m, 2H, H-3′, H-3), 5.15–4.97 (m, 5H, (($CH_2$Ph)O)$_2$PO-), H-3′ C17 chain), 4.67 (d, J = 8.4 Hz, 1H, H-1′), 4.60–4.34 (m, 8H, $CH_2$Ph), 4.28 (dd, J = 10.5, 4.9 Hz, 1H, H-6′a), 4.19 (tt, J = 8.4, 4.3 Hz, 1H, H-2), 4.05–3.97 (m, 2H, H-5, H-2′), 3.87–3.68 (m, 6H, H-6′b, H-6a,b, H-3′ C16 chains), 3.60 (t, J = 9.4 Hz, 1H, H-4′), 3.52 (t, J = 9.6 Hz, 1H, H-4), 3.36 (dt, J = 9.7, 4.9 Hz, 1H, H-5′), 2.65 (dd, J = 15.0, 6.6 Hz, 1H), 2.55 (dd, J = 16.0, 7.3 Hz, 1H), 2.50–2.35 (m, 3H), 2.34–2.18 (m, 4H), 2.17–2.07 (m, 1H), 1.66–1.37 (m, 5H, $CH_2$ chains), 1.36–0.99 (m, 112H, $CH_2$ chains), 0.88 (t, J = 6.8 Hz, 15H, $CH_3$ chains).

$^{13}$C NMR (101 MHz, $CDCl_3$) δ 173.35, 172.47, 171.71, 171.34, 169.96, 138.74, 138.69, 138.54, 137.41, 137.07, 135.56, 135.50, 128.95, 128.94, 128.85, 128.83, 128.61, 128.53, 128.41, 128.33, 128.24, 128.18, 128.14, 127.91, 127.89, 127.84, 127.81, 127.65, 127.48, 126.22, 101.47, 100.99, 96.05, 79.13, 76.07, 75.65, 75.40, 75.20, 74.97, 73.64, 72.46, 71.92, 71.44, 71.25, 71.01, 70.58, 70.14, 70.11, 70.09, 70.05, 68.74, 66.53, 54.35, 52.60, 52.51, 41.41, 41.24, 39.92, 39.73, 34.68, 34.50, 34.32, 34.27, 33.94, 32.05, 29.87, 29.85, 29.84, 29.80, 29.78, 29.76, 29.75, 29.50, 25.33, 25.27, 25.13, 22.82, 14.27.

$^{31}$P NMR (162 MHz, $CDCl_3$) δ −3.05.

$C_{141}H_{215}N_2O_{21}P$: calcd. mass 2303.56; ESI-MS: *m*/*z* 2326.52 $[M+Na]^+$.

3.3.20. 6-*O*-(2-Deoxy-2-((*R*)-3-(benzyloxy)-hexadecanoylamido)-3-*O*-((*R*)-3-(benzyloxy)-hexadecanoyl)-6-*O*-benzyl-β-D-glucopyranosyl)-1-*O*-thexyldimethylsilyl-2-deoxy-2-((*R*)-3-(pentadecanoyloxy)-heptadecanoylamido)-3-*O*-((*R*)-3-(benzyloxy)-hexadecanoyl)-4-*O*-benzyl-β-D-glucopyranoside **30**

Dry $CH_2Cl_2$ (4 mL) was added to a mixture of compound **27** (87 mg, 0.040 mmol) with activated MS4Å, and the suspension was stirred at room temperature for 3 h before cooling to -78 °C. Then triethylsilane (51 μL, 0.32 mmol) and triflic acid (35 μL, 0.40 mmol) were added successively to the reaction mixture. After stirring the reaction for 3 h at -78 °C, triethylamine (58 μL) and MeOH (0.4 mL) were added, and stirring continued for 10 min. The reaction was then warmed to room temperature, diluted with $CH_2Cl_2$, washed with saturated aqueous $NaHCO_3$ solution and brine, dried over $Na_2SO_4$, filtered, and concentrated. The crude was purified by flash silica gel column chromatography (Toluene/EtOAc = 9:1) to give **30** (62 mg, 70%).

$^1$H NMR (400 MHz, $CDCl_3$) δ 7.39–7.12 (m, 25H, CHAr), 6.16 (d, J = 8.8 Hz, 1H, NH-1′), 5.68 (d, J = 9.2 Hz, 1H, NH-1), 5.07 (m, 3H, H-3′, H-3, H-3′ C17 chain), 4.64 (d, J = 7.8 Hz, 1H, H-1), 4.59–4.38 (m, 10H), 3.91–3.59 (m, 8H), 3.55 (t, J = 9.1 Hz, 1H), 3.46 (dt, J = 9.6, 4.2 Hz, 2H), 2.62 (dd, J = 14.8, 7.6 Hz, 1H), 2.56–2.16 (m, 8H), 1.68–1.47 (m, 3H), 1.25 (dd, J = 7.3, 3.5 Hz, 125H, $CH_2$ chains), 0.94–0.77 (m, 27H, $CH_3$ chains, $CH_3$ TDS), 0.14 (s, 3H, $CH_3$ TDS), 0.11 (s, 3H, $CH_3$ TDS).

$^{13}$C NMR (101 MHz, $CDCl_3$) δ 173.88, 172.41, 171.74, 171.23, 169.25, 138.69, 138.53, 138.34, 138.04, 137.99, 128.68, 128.57, 128.47, 128.01, 127.93, 127.87, 127.83, 127.81, 127.78, 127.61, 101.09, 96.08, 76.32, 76.12, 76.02, 75.96, 75.56, 75.05, 74.70, 74.22, 74.20, 73.81, 71.41, 71.30, 71.08, 70.97, 70.89, 70.53, 68.10, 56.44, 53.77, 41.82, 41.54, 39.91, 39.65, 34.68, 34.35, 34.30, 34.12, 33.98, 32.07, 29.88, 29.86, 29.85, 29.83, 29.81, 29.80, 29.78, 29.51, 25.45, 25.34, 25.29, 25.15, 24.85, 22.83, 20.20, 18.71, 14.30, −1.35, −3.17.

$C_{135}H_{222}N_2O_{18}Si$: calcd. mass 2187.63; ESI-MS: *m*/*z* 2210.42 $[M+Na]^+$.

3.3.21. 6-*O*-(2-Deoxy-2-((*R*)-3-(benzyloxy)-hexadecanoylamido)-3-*O*-((*R*)-3-(benzyloxy)-hexadecanoyl)-4-(*O*-dibenzylphosphoryl)-6-*O*-benzyl-β-D-glucopyranosyl)-1-(*O*-thexyldimethylsilyl)-2-deoxy-2-((*R*)-3-(pentadecanoyloxy)-heptadecanoylamido)-3-*O*-((*R*)-3-(benzyloxy)-hexadecanoyl)-4-*O*-benzyl-D-glucopyranose **31**

The mixture of Compound **30** (13 mg, 0.006 mmol) and *1H*-tetrazole (4 mg, 0.06 mmol) was dissolved in dry $CH_2Cl_2$ (0.6 mL) and MeCN (0.1 mL), and the solution was cooled to 0 °C. Dibenzyl *N*,*N*-diisopropylphosphoramidite (5 µL, 0.015 mmol) was added, and the reaction was stirred at room temperature for 2 h. Then, *m*CPBA (4 mg, 0.023 mmol) was added and continued stirring for 30 min. The reaction mixture was diluted with $CH_2Cl_2$, washed with saturated aqueous $NaHCO_3$ solution and brine; dried over $Na_2SO_4$, filtered, and concentrated. The crude was purified by size exclusion chromatography (Sephadex® LH-20) ($CHCl_3$/MeOH = 1:1) to give **31** (13 mg, 87%).

$^1$H NMR (400 MHz, $CDCl_3$) δ 7.38–7.13 (m, 35H, CHAr), 6.06 (d, J = 8.2 Hz, 1H, NH), 5.64 (d, J = 9.3 Hz, 1H, NH), 5.42 (t, J = 9.7 Hz, 1H, H-3′), 5.16–5.00 (m, 2H, H-3, H-3′ C17 chain), 4.92–4.84 (m, 4H, $((CH_2Ph)O)_2PO$-)), 4.65 (d, J = 8.3 Hz, 1H, H-1′), 4.61 (d, J = 7.7 Hz, 1H, H-1), 4.55–4.31 (m, 12H, $CH_2Ph$), 3.91–3.43 (m, 8H), 2.60–2.10 (m, 9H), 1.81–1.67 (m, 2H), 1.67–1.38 (m, 4H), 1.38–1.03 (m, 95H, $CH_2$ chains), 0.97–0.73 (m, 32H, $CH_3$ chains, $CH_3$ TDS), 0.15 (s, 3H, $CH_3$ TDS), 0.10 (s, 3H, $CH_3$ TDS).

$^{13}$C NMR (101 MHz, $CDCl_3$) δ 173.87, 171.73, 171.50, 171.39, 169.23, 138.82, 138.72, 138.66, 138.45, 137.97, 135.80, 128.65, 128.44, 128.39, 128.14, 128.07, 127.98, 127.83, 127.71, 127.64, 127.59, 100.60, 96.12, 76.25, 76.15, 75.56, 75.18, 74.64, 74.32, 74.24, 73.53, 72.78, 71.41, 71.07, 71.00, 70.92, 69.75, 69.69, 69.62, 69.57, 68.88, 68.63, 56.50, 55.24, 41.85, 41.55, 39.66, 39.08, 34.69, 34.37, 34.15, 32.07, 29.87, 29.82, 29.51, 25.46, 25.40, 25.35, 25.16, 22.83, 20.23, 18.72, 14.24, −1.36, −3.15.

$^{31}$P NMR (162 MHz, $CDCl_3$) δ −2.11.

$C_{149}H_{235}N_2O_{21}PSi$: calcd. mass 2447.69; ESI-MS: *m*/*z* 2470.80 $[M+Na]^+$.

3.3.22. General Procedure for Hydrogenolysis

The respective substrate (0.027 mmol) was dissolved in a mixture of $CH_2Cl_2$ (1.2 mL) and MeOH (1.2 mL), and to the solution was added AcOH (0.12 mL) and $Pd(OH)_2/C$ catalyst (120 mg), and the reaction was stirred for 2 days under a hydrogen gas atmosphere using a balloon at room temperature. The reaction mixture was neutralized with $Et_3N$ (0.36 mL), and the palladium was filtered off using a micropore membrane filter. The filtrate was co-evaporated with toluene and concentrated in a rotary evaporator. The crude was purified by size exclusion chromatography (Sephadex® LH-20) ($CHCl_3$/MeOH = 1:1) to obtain the corresponding final fully deprotected lipid A compound.

Tetra C-1; quantitative yield.

$^1$H NMR (400 MHz, $CDCl_3$/MeOD) δ 5.37–5.28 (m, 1H), 5.21 (s, 1H), 5.08–4.93 (m, 4H), 4.89–4.75 (m, 1H), 4.66 (d, J = 8.5 Hz, 1H), 4.38 (dd, J = 14.6, 9.5 Hz, 1H), 3.68–3.59 (m, 1H), 3.46 (t, J = 9.5 Hz, 1H), 3.39–3.28 (m, 9H), 3.26–3.17 (m, 5H), 3.01 (q, J = 7.3 Hz, 1H), 2.87 (q, J = 7.2 Hz, 16H), 2.43–1.99 (m, 10H), 1.95–1.79 (m, 1H), 1.50–0.89 (m, 90H), 0.74 (t, J = 6.8 Hz, 15 H).

$^{13}$C NMR (101 MHz, $CDCl_3$/MeOD) δ 173.60, 173.27, 172.76, 172.58, 129.88, 129.61, 75.75, 68.85, 68.75, 68.37, 68.29, 68.15, 67.99, 62.42, 61.18, 52.52, 42.06, 36.98, 31.81, 29.65, 29.62, 29.59, 29.56, 29.54, 29.52, 29.47, 29.45, 29.26, 29.24, 29.22, 29.18, 25.49, 25.45, 22.55, 13.88, 10.81, 7.21.

$^{31}$P NMR (162 MHz, $CDCl_3$/MeOD) δ 2.81.

HR-MS: $C_{77}H_{147}N_2O_{20}P$: calcd. mass 1451.03; ESI-qTOF MS: *m*/*z* 1450.0225 $[M-H]^-$.

Penta C-1; quantitative yield.

$^{1}$H NMR (400 MHz, $CDCl_3$) δ 5.27–5.13 (m, 1H), 5.05–4.93 (m, 2H), 4.79 (t, J = 9.8 Hz, 2H), 4.58 (d, J = 8.5 Hz, 1H), 3.90–3.70 (m, 3H), 3.68–3.57 (m, 2H), 3.53 (d, J = 12.9 Hz, 2H), 3.44 (t, J = 9.5 Hz, 3H), 3.37–3.26 (m, 1H), 3.25–3.18 (m, 9H), 2.99 (q, J = 7.5 Hz, 6H), 2.85 (q, J = 7.2 Hz, 21H), 2.39–1.96 (m, 21H), 1.86 (d, J = 5.1 Hz, 2H), 1.52–1.00 (m, 80H), 0.72 (t, J = 6.7 Hz, 15H).

$^{13}$C NMR (101 MHz, $CDCl_3$) δ 173.73, 173.54, 173.01, 172.80, 171.19, 162.64, 130.06, 129.77, 71.03, 68.80, 68.67, 68.35, 68.09, 51.95, 51.79, 51.54, 46.08, 42.54, 42.05, 41.34, 37.47, 37.17, 36.64, 35.88, 34.54, 34.18, 34.14, 34.07, 31.99, 31.97, 29.85, 29.83, 29.81, 29.78, 29.75, 29.73, 29.72, 29.70, 29.44, 29.42, 27.25, 27.21, 25.94, 25.80, 25.73, 25.61, 25.52, 25.37, 25.12, 24.99, 24.98, 22.73, 14.12, 14.10, 11.03, 8.58.

$^{31}$P NMR (162 MHz, $CDCl_3$) δ 1.99.

HR-MS: $C_{92}H_{175}N_2O_{21}P$: calcd. mass 1675.24; ESI-qTOF MS: $m/z$ 1674.2378 $[M-H]^-$.

Penta C-4′; quantitative yield.

$^{1}$H NMR (400 MHz, $CDCl_3$) δ 5.09–5.04 (m, 1H), 4.96–4.84 (m, 2H), 4.77 (d, J = 3.5 Hz, 0H), 4.11 (s, 1H), 3.96 (q, J = 10.0 Hz, 1H), 3.91–3.82 (m, 1H), 3.82–3.74 (m, 2H), 3.74–3.66 (m, 1H), 3.56–3.43 (m, 1H), 3.32–3.16 (m, 1H), 2.89 (q, J = 7.3 Hz, 7H), 2.74 (q, J = 7.3 Hz, 5H), 2.63–2.53 (m, 1H), 2.26–2.08 (m, 3H), 2.02 (q, J = 6.3 Hz, 3H), 1.94 (t, J = 7.7 Hz, 2H), 1.82–1.67 (m, 3H), 1.43–1.26 (m, 6H), 1.26–0.79 (m, 120H), 0.61 (t, J = 6.7 Hz, 15H).

$^{13}$C NMR (101 MHz, $CDCl_3$) δ 173.52, 173.05, 172.81, 172.11, 170.69, 101.53, 91.10, 75.49, 74.04, 73.78, 70.81, 70.25, 70.05, 68.84, 68.35, 68.22, 60.12, 53.73, 51.81, 43.49, 42.17, 41.92, 41.75, 40.80, 36.98, 36.87, 34.14, 33.80, 31.61, 29.38, 29.03, 25.32, 25.14, 24.91, 24.74, 22.33, 13.53, 10.56.

$^{31}$P NMR (162 MHz, $CDCl_3$) δ 4.90.

HR-MS: $C_{92}H_{175}N_2O_{21}P$: calcd. mass 1675.24; ESI-qTOF MS: $m/z$ 1674.2343 $[M-H]^-$.

## 4. Conclusions

In this work, we underscore the therapeutic potential of microbiota-inspired glycolipids as modulators of innate immunity. By synthesizing and computationally evaluating a panel of *Bacteroides fragilis*-derived lipid A analogs, we identified **Tetra C-1** (Figure 2) as a promising TLR4 antagonist with superior binding characteristics compared to Eritoran, a well-established antagonist that we used as a benchmark. This indicates a potentially superior antagonistic profile, meriting further investigation. As a next step, we will evaluate our designed *Bacteroides fragilis*-derived glycolipids both in vitro and in vivo, using established cell line models and murine models of LPS-induced inflammation—systems previously employed to assess the immunomodulatory effects of *B. fragilis*-derived lipids. These models will enable quantification of cytokine levels (e.g., TNF-α, IL-6), assessment of epithelial barrier integrity, and analysis of innate immune modulation. A direct comparison with Eritoran and other known TLR4 ligands as reference compounds in these models could validate our findings and support the development of *B. fragilis*-derived glycolipids as therapeutic agents for inflammatory diseases and immunomodulation. Our findings pave the way for the rational design of novel anti-inflammatory agents and vaccine adjuvants, leveraging the immunomodulatory properties of commensal-derived molecules to target TLR4-mediated pathways in inflammatory diseases. This study marks a further advance toward the development of novel therapeutics targeting TLR4 in pursuit of next-generation immunomodulators and vaccine adjuvants inspired by the human microbiome.

**Supplementary Materials:** The following supporting information can be downloaded at https://www.mdpi.com/article/10.3390/molecules30193927/s1. NMR and MS spectra of synthesized compounds.

**Author Contributions:** Conceptualization, L.L., M.C. and G.D.; methodology, C.M.S., D.K., F.T. and A.I.; investigation, G.D. and M.C.; writing—original draft preparation, G.D. and M.C.; writing—reviewing and editing, L.L., M.C. and G.D.; supervision, M.C., L.L. and G.D.; funding acquisition, L.L. and M.C. All authors have read and agreed to the published version of the manuscript.

**Funding:** This research was supported by EU funding within the NextGenerationEU-MUR PNRR Extended Partnership initiative on Emerging Infectious Diseases (Project no. PE00000007, INF-ACT). This project has received funding from the European Union's Horizon 2020 research and innovation programme under the Marie Curie Sklodowska–Curie European Training Network (ETN) "GLYCOVAX", grant agreement No 675671 (http://glycovax.eu/, (accessed on 24 September 2025)). This research was also supported by Next Generation EU "Missione 4—Componente 1-CUP: G53D23003320006 (GLYCONANOVACS)" and by the University of Milan Research Grant PSR2025 (PSR2025_DIP_005_MCIVE).

**Data Availability Statement:** Data are contained within this article or in Supplementary Materials.

**Acknowledgments:** We acknowledge the mass spectrometry facility of Unitech COSPECT at the University of Milan (Italy) for HRMS analysis.

**Conflicts of Interest:** The authors declare no conflicts of interest.

## Abbreviations

LPS, lipopolysaccharides; TLRs, Toll-like receptors; TLR1-2-4, Toll-like receptor 1, 2 or 4; MD-2, Myeloid Differentiation factor 2; MPLA, monophosphoryl lipid A; Th1, T-helper; CD4, Cluster of Differentiation 4; PSA, polysaccharide A; IL-10, Interleukin 10; NMR, Nuclear Magnetic Resonance; LC, Liquid chromatography; MS, Mass Spectrometry; DMAP, Dimethylaminopyridine; Py, Pyridine; AcOH, Acetic Acid; DMF, Dimethylformamide; TDSCl, Thexyl-dimethyl silyl chloride; pTSA, para-Toluensulfonic acid; THF, Tetrahydrofuran; NIS, N-Iodosuccinimide; TMSOTf, Trimethylsilyl trifluoromethanesulfonate; DCC, Dicyclohexylcarbodiimide; BINAP, 2,2′-bis(diphenylphosphino)-1,1′-binaphthyl; MTPA, α-methoxy-α-trifluoromethylphenylacetic acid; DIC, Diisopropylcarbodiimide; TFA, trifluoroacetic acid; LiHDMS, Lithium bis(trimethylsilyl)amide; HBTU, Hexafluorophosphate Benzotriazole Tetramethyl Uronium; DIPEA, N,N-Diisopropylethylamine; HATU, 1-[Bis(dimethylamino)methylene]-1H-1,2,3-triazolo [4,5-b]pyridinium 3-oxide hexafluorophosphate; EDC, 1-Ethyl-3-(3-dimethylaminopropyl)carbodiimide; PDB, Protein DataBank; RMSD, Root mean square deviation; MM-GBSA, Molecular Mechanics-Generalized Born Surface Area; EtOAc, Ethyl acetate; DCM, dichloromethane; TLC, Thin-Layer Chromatography; ESI, ElectroSpay Ionization; mCPBA, meta-chloroperbenzoic acid.

## References

1. Di Lorenzo, F.; Duda, K.A.; Lanzetta, R.; Silipo, A.; De Castro, C.; Molinaro, A. A Journey from Structure to Function of Bacterial Lipopolysaccharides. *Chem. Rev.* **2022**, *122*, 15767–15821. [CrossRef]
2. Raetz, C.R.H.; Whitfield, C. Lipopolysaccharide Endotoxins. *Annu. Rev. Biochem.* **2002**, *71*, 635–700. [CrossRef]
3. Galanos, C.; Lüderitz, O.; Rietschel, E.T.; Westphal, O.; Brade, H.; Brade, L.; Freudenberg, M.; Schade, U.; Imoto, M.; Yoshimura, H.; et al. Synthetic and Natural *Escherichia Coli* Free Lipid A Express Identical Endotoxic Activities. *Eur. J. Biochem.* **1985**, *148*, 1–5. [CrossRef]
4. Cohen, J. The Immunopathogenesis of Sepsis. *Nature* **2002**, *420*, 885–891. [CrossRef]
5. Yethon, J.; Whitfield, C. Lipopolysaccharide as a Target for the Development of Novel Therapeutics in Gram-Negative Bacteria. *Curr. Drug Target-Infect. Disord.* **2001**, *1*, 91–106. [CrossRef] [PubMed]
6. Miller, S.I.; Ernst, R.K.; Bader, M.W. LPS, TLR4 and Infectious Disease Diversity. *Nat. Rev. Microbiol.* **2005**, *3*, 36–46. [CrossRef] [PubMed]
7. Akira, S.; Uematsu, S.; Takeuchi, O. Pathogen Recognition and Innate Immunity. *Cell* **2006**, *124*, 783–801. [CrossRef]
8. La Ferla, B.; Spinosa, V.; D'Orazio, G.; Palazzo, M.; Balsari, A.; Foppoli, A.A.; Rumio, C.; Nicotra, F. Dansyl C -Glucoside as a Novel Agent Against Endotoxic Shock. *ChemMedChem* **2010**, *5*, 1677–1680. [CrossRef]

9. Cardani, D.; Sardi, C.; La Ferla, B.; D'Orazio, G.; Sommariva, M.; Marcucci, F.; Olivero, D.; Tagliabue, E.; Koepsell, H.; Nicotra, F.; et al. Sodium Glucose Cotransporter 1 Ligand BLF501 as a Novel Tool for Management of Gastrointestinal Mucositis. *Mol. Cancer* **2014**, *13*, 23. [CrossRef] [PubMed]
10. Park, B.S.; Song, D.H.; Kim, H.M.; Choi, B.-S.; Lee, H.; Lee, J.-O. The Structural Basis of Lipopolysaccharide Recognition by the TLR4–MD-2 Complex. *Nature* **2009**, *458*, 1191–1195. [CrossRef]
11. Komaniecka, I.; Choma, A.; Lindner, B.; Holst, O. The Structure of a Novel Neutral Lipid A from the Lipopolysaccharide of *Bradyrhizobium Elkanii* Containing Three Mannose Units in the Backbone. *Chem.–A Eur. J.* **2010**, *16*, 2922–2929. [CrossRef]
12. Coats, S.R.; Pham, T.-T.T.; Bainbridge, B.W.; Reife, R.A.; Darveau, R.P. MD-2 Mediates the Ability of Tetra-Acylated and Penta-Acylated Lipopolysaccharides to Antagonize *Escherichia Coli* Lipopolysaccharide at the TLR4 Signaling Complex. *J. Immunol.* **2005**, *175*, 4490–4498. [CrossRef]
13. Shirey, K.A.; Lai, W.; Scott, A.J.; Lipsky, M.; Mistry, P.; Pletneva, L.M.; Karp, C.L.; McAlees, J.; Gioannini, T.L.; Weiss, J.; et al. The TLR4 Antagonist Eritoran Protects Mice from Lethal Influenza Infection. *Nature* **2013**, *497*, 498–502. [CrossRef] [PubMed]
14. Edelman, R. The Development and Use of Vaccine Adjuvants. *Mol. Biotechnol.* **2002**, *21*, 129–148. [CrossRef]
15. Coffman, R.L.; Sher, A.; Seder, R.A. Vaccine Adjuvants: Putting Innate Immunity to Work. *Immunity* **2010**, *33*, 492–503. [CrossRef]
16. Noel Masihi, K.; Lange, W.; Brehmer, W.; Ribi, E. Immunobiological Activities of Nontoxic Lipid A: Enhancement of Nonspecific Resistance in Combination with Trehalose Dimycolate against Viral Infection and Adjuvant Effects. *Int. J. Immunopharmacol.* **1986**, *8*, 339–345. [CrossRef]
17. Chen, C.; Zhang, C.; Li, R.; Wang, Z.; Yuan, Y.; Li, H.; Fu, Z.; Zhou, M.; Zhao, L. Monophosphoryl-Lipid A (MPLA) Is an Efficacious Adjuvant for Inactivated Rabies Vaccines. *Viruses* **2019**, *11*, 1118. [CrossRef] [PubMed]
18. Coler, R.N.; Bertholet, S.; Moutaftsi, M.; Guderian, J.A.; Windish, H.P.; Baldwin, S.L.; Laughlin, E.M.; Duthie, M.S.; Fox, C.B.; Carter, D.; et al. Development and Characterization of Synthetic Glucopyranosyl Lipid Adjuvant System as a Vaccine Adjuvant. *PLoS ONE* **2011**, *6*, e16333. [CrossRef] [PubMed]
19. Komai-Koma, M.; Ji, Y.; Cao, H.; Liu, Z.; McSharry, C.; Xu, D. Monophosphoryl Lipid A Directly Regulates Th1 Cytokine Production in Human CD4+ T-Cells through Toll-like Receptor 2 and 4. *Immunobiology* **2021**, *226*, 152132. [CrossRef]
20. Erturk-Hasdemir, D.; Kasper, D.L. Resident Commensals Shaping Immunity. *Curr. Opin. Immunol.* **2013**, *25*, 450–455. [CrossRef]
21. Dasgupta, S.; Erturk-Hasdemir, D.; Ochoa-Reparaz, J.; Reinecker, H.-C.; Kasper, D.L. Plasmacytoid Dendritic Cells Mediate Anti-Inflammatory Responses to a Gut Commensal Molecule via Both Innate and Adaptive Mechanisms. *Cell Host Microbe* **2014**, *15*, 413–423. [CrossRef]
22. Mazmanian, S.K.; Round, J.L.; Kasper, D.L. A Microbial Symbiosis Factor Prevents Intestinal Inflammatory Disease. *Nature* **2008**, *453*, 620–625. [CrossRef] [PubMed]
23. Ochoa-Repáraz, J.; Mielcarz, D.W.; Wang, Y.; Begum-Haque, S.; Dasgupta, S.; Kasper, D.L.; Kasper, L.H. A Polysaccharide from the Human Commensal *Bacteroides Fragilis* Protects against CNS Demyelinating Disease. *Mucosal Immunol.* **2010**, *3*, 487–495. [CrossRef] [PubMed]
24. Erturk-Hasdemir, D.; Oh, S.F.; Okan, N.A.; Stefanetti, G.; Gazzaniga, F.S.; Seeberger, P.H.; Plevy, S.E.; Kasper, D.L. Symbionts Exploit Complex Signaling to Educate the Immune System. *Proc. Natl. Acad. Sci. USA* **2019**, *116*, 26157–26166. [CrossRef]
25. Alhawi, M.; Stewart, J.; Erridge, C.; Patrick, S.; Poxton, I.R. *Bacteroides Fragilis* Signals through Toll-like Receptor (TLR) 2 and Not through TLR4. *J. Med. Microbiol.* **2009**, *58*, 1015–1022. [CrossRef]
26. Weintraub, A.; Zähringer, U.; Wollenweber, H.W.; Seydel, U.; Rietschel, E.T. Structural Characterization of the Lipid A Component of *Bacteroides Fragilis* Strain NCTC 9343 Lipopolysaccharide. *Eur. J. Biochem.* **1989**, *183*, 425–431. [CrossRef]
27. Verpalen, E.C.J.M.; Ehlers, A.M.; van Wingaarden, A.C.A.; Brouwer, A.J.; Boons, G.-J. Synthesis and Biological Evaluation of Lipid A Derived from Commensal *Bacteroides*. *Org. Biomol. Chem.* **2024**, *22*, 8793–8800. [CrossRef] [PubMed]
28. Kim, H.M.; Park, B.S.; Kim, J.-I.; Kim, S.E.; Lee, J.; Oh, S.C.; Enkhbayar, P.; Matsushima, N.; Lee, H.; Yoo, O.J.; et al. Crystal Structure of the TLR4-MD-2 Complex with Bound Endotoxin Antagonist Eritoran. *Cell* **2007**, *130*, 906–917. [CrossRef]
29. Takasawa, K.; Kano, R.; Maruyama, H.; Hasegawa, A.; Kamata, H. The Antagonist Activity of Lipid IVa on the Stimulation by Lipid A of TNF-Alpha Production from Canine Blood Mononuclear Cells. *Vet. Immunol. Immunopathol.* **2011**, *143*, 167–169. [CrossRef]
30. Ohto, U.; Fukase, K.; Miyake, K.; Satow, Y. Crystal Structures of Human MD-2 and Its Complex with Antiendotoxic Lipid IVa. *Science* **2007**, *316*, 1632–1634. [CrossRef]
31. Ohto, U.; Fukase, K.; Miyake, K.; Shimizu, T. Structural Basis of Species-Specific Endotoxin Sensing by Innate Immune Receptor TLR4/MD-2. *Proc. Natl. Acad. Sci. USA* **2012**, *109*, 7421–7426. [CrossRef]
32. Potter, G.T.; Jayson, G.C.; Miller, G.J.; Gardiner, J.M. An Updated Synthesis of the Diazo-Transfer Reagent Imidazole-1-Sulfonyl Azide Hydrogen Sulfate. *J. Org. Chem.* **2016**, *81*, 3443–3446. [CrossRef]

33. Hénault, J.; Quellier, P.; Mock-Joubert, M.; Le Narvor, C.; Alix, A.; Bonnaffé, D. Regioselective Reductive Opening of Benzylidene Acetals with Dichlorophenylborane/Triethylsilane: Previously Unreported Side Reactions and How to Prevent Them. *J. Org. Chem.* **2022**, *87*, 963–973. [CrossRef]
34. Liu, R.; Wei, A. Solid-Phase Synthesis of 2-Aminoethyl Glucosamine Sulfoforms. *J. Carbohydr. Chem.* **2012**, *31*, 384–419. [CrossRef] [PubMed]
35. Brinkerhoff, R.C.; Tarazona, H.F.; de Oliveira, P.M.; Flores, D.C.; Montes D'Oca, C.D.R.; Russowsky, D.; Montes D'Oca, M.G. Synthesis of β-Ketoesters from Renewable Resources and Meldrum's Acid. *RSC Adv.* **2014**, *4*, 49556–49559. [CrossRef]
36. Ratovelomanana-Vidal, V.; Girard, C.; Touati, R.; Tranchier, J.P.; Hassine, B.B.; Genêt, J.P. Enantioselective Hydrogenation of Β-Keto Esters Using Chiral Diphosphine-Ruthenium Complexes: Optimization for Academic and Industrial Purposes and Synthetic Applications. *Adv. Synth. Catal.* **2003**, *345*, 261–274. [CrossRef]
37. Hoye, T.R.; Jeffrey, C.S.; Shao, F. Mosher Ester Analysis for the Determination of Absolute Configuration of Stereogenic (Chiral) Carbinol Carbons. *Nat. Protoc.* **2007**, *2*, 2451–2458. [CrossRef]
38. Ghosh, I.; Zeng, H.; Kishi, Y. Application of Chiral Lanthanide Shift Reagents for Assignment of Absolute Configuration of Alcohols. *Org. Lett.* **2004**, *6*, 4715–4718. [CrossRef]
39. Fukase, K.; Fukase, Y.; Oikawa, M.; Liu, W.-C.; Suda, Y.; Kusumoto, S. Divergent Synthesis and Biological Activities of Lipid A Analogues of Shorter Acyl Chains. *Tetrahedron* **1998**, *54*, 4033–4050. [CrossRef]
40. Trott, O.; Olson, A.J. AutoDock Vina: Improving the Speed and Accuracy of Docking with a New Scoring Function, Efficient Optimization, and Multithreading. *J. Comput. Chem.* **2010**, *31*, 455–461. [CrossRef] [PubMed]
41. Eberhardt, J.; Santos-Martins, D.; Tillack, A.F.; Forli, S. AutoDock Vina 1.2.0: New Docking Methods, Expanded Force Field, and Python Bindings. *J. Chem. Inf. Model.* **2021**, *61*, 3891–3898. [CrossRef] [PubMed]
42. Hou, T.; Wang, J.; Li, Y.; Wang, W. Assessing the Performance of the Molecular Mechanics/Poisson Boltzmann Surface Area and Molecular Mechanics/Generalized Born Surface Area Methods. II. The Accuracy of Ranking Poses Generated from Docking. *J. Comput. Chem.* **2011**, *32*, 866–877. [CrossRef] [PubMed]
43. El Khoury, L.; Santos-Martins, D.; Sasmal, S.; Eberhardt, J.; Bianco, G.; Ambrosio, F.A.; Solis-Vasquez, L.; Koch, A.; Forli, S.; Mobley, D.L. Comparison of Affinity Ranking Using AutoDock-GPU and MM-GBSA Scores for BACE-1 Inhibitors in the D3R Grand Challenge 4. *J. Comput. Aided Mol. Des.* **2019**, *33*, 1011–1020. [CrossRef] [PubMed]
44. Zhang, X.; Perez-Sanchez, H.; Lightstone, F.C. A Comprehensive Docking and MM/GBSA Rescoring Study of Ligand Recognition upon Binding Antithrombin. *Curr. Top. Med. Chem.* **2017**, *17*, 1631–1639. [CrossRef]
45. Kaus, J.W.; Harder, E.; Lin, T.; Abel, R.; McCammon, J.A.; Wang, L. How To Deal with Multiple Binding Poses in Alchemical Relative Protein–Ligand Binding Free Energy Calculations. *J. Chem. Theory Comput.* **2015**, *11*, 2670–2679. [CrossRef]
46. Rastelli, G.; Rio, A.D.; Degliesposti, G.; Sgobba, M. Fast and Accurate Predictions of Binding Free Energies Using MM-PBSA and MM-GBSA. *J. Comput. Chem.* **2010**, *31*, 797–810. [CrossRef]
47. Lu, C.; Wu, C.; Ghoreishi, D.; Chen, W.; Wang, L.; Damm, W.; Ross, G.A.; Dahlgren, M.K.; Russell, E.; Von Bargen, C.D.; et al. OPLS4: Improving Force Field Accuracy on Challenging Regimes of Chemical Space. *J. Chem. Theory Comput.* **2021**, *17*, 4291–4300. [CrossRef]
48. Harder, E.; Damm, W.; Maple, J.; Wu, C.; Reboul, M.; Xiang, J.Y.; Wang, L.; Lupyan, D.; Dahlgren, M.K.; Knight, J.L.; et al. OPLS3: A Force Field Providing Broad Coverage of Drug-like Small Molecules and Proteins. *J. Chem. Theory Comput.* **2016**, *12*, 281–296. [CrossRef]
49. Lembo-Fazio, L.; Billod, J.-M.; Di Lorenzo, F.; Paciello, I.; Pallach, M.; Vaz-Francisco, S.; Holgado, A.; Beyaert, R.; Fresno, M.; Shimoyama, A.; et al. Bradyrhizobium Lipid A: Immunological Properties and Molecular Basis of Its Binding to the Myeloid Differentiation Protein-2/Toll-Like Receptor 4 Complex. *Front. Immunol.* **2018**, *9*, 1888. [CrossRef]
50. Sestito, S.E.; Facchini, F.A.; Morbioli, I.; Billod, J.-M.; Martin-Santamaria, S.; Casnati, A.; Sansone, F.; Peri, F. Amphiphilic Guanidinocalixarenes Inhibit Lipopolysaccharide (LPS)- and Lectin-Stimulated Toll-like Receptor 4 (TLR4) Signaling. *J. Med. Chem.* **2017**, *60*, 4882–4892. [CrossRef]
51. Morris, G.M.; Huey, R.; Lindstrom, W.; Sanner, M.F.; Belew, R.K.; Goodsell, D.S.; Olson, A.J. AutoDock4 and AutoDockTools4: Automated Docking with Selective Receptor Flexibility. *J. Comput. Chem.* **2009**, *30*, 2785–2791. [CrossRef]
52. Li, J.; Abel, R.; Zhu, K.; Cao, Y.; Zhao, S.; Friesner, R.A. The VSGB 2.0 Model: A next Generation Energy Model for High Resolution Protein Structure Modeling. *Proteins Struct. Funct. Bioinform.* **2011**, *79*, 2794–2812. [CrossRef] [PubMed]
53. Morelli, L.; Lay, L. Synthesis of Neisseria Meningitidis X Capsular Polysaccharide Fragments. *Arkivoc* **2012**, *2013*, 166–184. [CrossRef]

54. Melnyk, J.E.; Mohanan, V.; Schaefer, A.K.; Hou, C.-W.; Grimes, C.L. Peptidoglycan Modifications Tune the Stability and Function of the Innate Immune Receptor Nod2. *J. Am. Chem. Soc.* **2015**, *137*, 6987–6990. [CrossRef] [PubMed]
55. Sieben, D.; Santana, A.; Nowka, P.; Weber, S.; Funke, K.; Hüttenhain, S.H. Preparation of the Even-Numbered 3-Oxo Fatty Acid Nicotinyl Esters from C6:0 to C18:0. *Tetrahedron Lett.* **2016**, *57*, 808–810. [CrossRef]

molecules

MDPI

*Article*

# Exploring the Scope of Functionalized *N*-Acylneuraminic Acid β-Methyl Glycosides as Inhibitors of *Neisseria meningitidis* CMP-Sialic Acid Synthetase

**Pradeep Chopra [1], Jana Führing [2], Preston Ng [1], Thomas Haselhorst [1], Jeffrey C. Dyason [1], Faith J. Rose [1], Robin J. Thomson [1], Rita Gerardy-Schahn [2], I. Darren Grice [1], Michael P. Jennings [1,*], Anja K. Münster-Kühnel [2,*] and Mark von Itzstein [1,*]**

1 Institute for Biomedicine and Glycomics, Griffith University, Gold Coast, QLD 4222, Australia; pchopra@uga.edu (P.C.); t.haselhorst@griffith.edu.au (T.H.); r.thomson@griffith.edu.au (R.J.T.); d.grice@griffith.edu.au (I.D.G.)

2 Institut für Klinische Biochemie, Medizinische Hochschule, Carl-Neuberg-Straße 1, 30625 Hannover, Germany; fuehring.jana@mh-hannover.de (J.F.); gerardy-schahn.rita@mh-hannover.de (R.G.-S.)

* Correspondence: m.jennings@griffith.edu.au (M.P.J.); muenster.anja@mh-hannover.de (A.K.M.-K.); m.vonitzstein@griffith.edu.au (M.v.I.)

**Abstract**

Cell surface sialylation is utilized by a number of pathogenic bacteria to evade the host immune system through molecular mimicry of host sialoglycoconjugates. Human pathogen *Neisseria meningitidis* serotype B (*Nm*B) expresses both sialylated capsule and surface lipooligosaccharides as pivotal virulence factors. An essential enzyme in the sialylation pathway of *Nm*B is CMP-sialic acid synthetase (CSS), which produces the activated nucleotide sugar necessary for sialic acid transfer. In this work, novel C-4, -5, -7, and -9 functionalized derivatives of neuraminic acid β-methyl glycoside (Neuβ2Me) were synthesized as candidate CSS inhibitors. A number of these were found to reduce the activity of *Nm*B CSS in vitro. The highest inhibition of *Nm*B CSS, in a mixed mode manner, was observed with a Neu5Acβ2Me C-9 serine carboxamide. Direct interaction with the enzyme was confirmed by Saturation Transfer Difference (STD) NMR. Supplementation of growth media with this compound reduced lipooligosaccharide (LOS) sialylation of living *N. meningitidis*, thus providing an interesting starting point for the development of specific *Nm*B CSS inhibitors as an alternative treatment strategy to fight bacterial infections.

**Keywords:** sialic acids; CMP-sialic acid; CMP-sialic acid synthase; inhibitors; *Neisseria meningitidis*

## 1. Introduction

*Neisseria meningitidis* serogroup B (*Nm*B) is an obligate human pathogen and a leading cause of meningitis and septicemia in developed countries [1]. Vaccines are currently available for protection against *N. meningitidis* serogroups A, B, C, W, and Y [2,3]. The Men A, C, W, and Y vaccines are based on bacterial capsular polysaccharide antigens [2–4]. In contrast, the MenB vaccine uses subcapsular membrane proteins and outer membrane vesicle proteins [2,5] due to the similarity of its capsular polysaccharides to human neuronal glycans [6–8]. *Nm*B expresses sialic acid (Sia) on its surface, both in its capsule, which is composed of an α2,8-linked homopolymer of Sia (polySia), and as the terminal sugar on lipooligosaccharides (LOS) [9,10]. Sialic acids are 9-carbon acidic sugars [11], incorporated into glycoconjugates as α-linked sialosides, which are negatively charged under physiological conditions. The highly charged polySia capsule of serogroup B meningococci, as well as LOS terminated by Sia, are pivotal virulence factors [12]. Sia not only mediates interactions with the host [13] but also promotes survival and dissemination of this pathogen. Sialylated LOS has been shown to be involved in immune evasion of *Nm*B, resistance against phagocytosis by human dendritic cells [14,15], and hindrance of complement attack by interfering with the alternative pathway of complement activation [16–18]. Genetic interruption of the *Nm*B sialylation pathway abolishes capsule formation as well as LOS sialylation, and results in rapid bacterial killing by human serum [19]. Interference with *Nm*B cell surface sialylation may therefore provide an attractive antimicrobial strategy.

A prerequisite for incorporation of Sia into sialylated glycans in both pro- and eukaryotes is the enzymatic conversion of free Sia to the activated nucleotide sugar CMP-Sia. Exclusively CMP-Sia is used by sialyltransferases for the transfer of Sia onto nascent sialoglycoconjugates. The sialic acid present in *Nm*B polysaccharides is *N*-acetylneuraminic acid, Neu5Ac **1**. Activation of Neu5Ac to CMP-Neu5Ac **2** through reaction with cytidine 5′-triphosphate (CTP) [20,21] is catalyzed by CMP-sialic acid synthetase (*N*-acylneuraminate cytidylyltransferase, CSS, EC 2.7.7.43) (Figure 1A) [20–25]. Interference with CSS activity in mammalian cells abolishes cell surface sialylation [26], suggesting the enzyme as a potential target for manipulation of cell surface sialylation in *Nm*B.

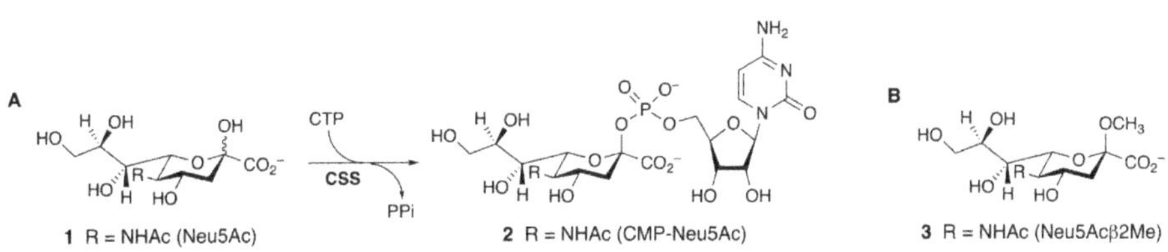


**Figure 1.** (**A**). Activation of the Sia *N*-acetylneuraminic acid (Neu5Ac) **1** to the nucleotide sugar form CMP-Neu5Ac **2**, catalyzed by CMP-sialic acid synthetase (CSS). (**B**). Neu5Acβ2Me (**3**), an inhibitor of mammalian CSS [27,28].

Tolerance of *Nm*B CSS to structural variation on the substrate Neu5Ac at carbon atoms C-5 [29–32], C-7 [33], C-8 [34], and C-9 [30,31,35], and indeed to replacement of the glycerol side chain [36], has been reported. This substrate tolerance has been exploited for the generation of modified sialoglycans [29,37], including in multi-enzyme and chemoenzymatic syntheses [30–33,36,38,39]. In contrast, comparatively little research has to date been reported on the inhibition of CSS. Only a relatively small number of studies have examined the inhibitory effect of modified sialic acids [27,28,40,41] or nucleotide isosteres [42] on CSS activity. Removal of the C-2 hydroxyl group of Neu5Ac (the proposed nucleophile in coupling Neu5Ac **1** to CTP [43,44]) was found to abolish recognition by mammalian CSS [27]. Retaining the C-2 oxygen but blocking it with a methyl group—giving the methyl glycoside in the β-configuration to mimic the β-configuration of CMP-Neu5Ac **2**—however, led to inhibition of mammalian CSS, albeit at millimolar concentration [Neu5Acβ2Me (**3**): $K_i$ = 2.5 mM [27]; 15 mM [28]. As part of our continuing work in sialic acid and CSS biology [13,18,24,45–48], we sought to explore the potential for inhibition of CSS from pathogenic *Nm*B using novel Neu5Ac derivatives. As the inhibitor template, we chose the above-mentioned β-methyl glycoside of Neu5Ac (Neu5Acβ2Me **3**, Figure 1B), based on its known inhibitory effect on CSS activity [27,28].

## 2. Results and Discussion

### 2.1. Structure Analysis of NmB CSS

To decide on a rational basis which positions of Neu5Acβ2Me (**3**) would be most promising for modification, to potentially increase binding affinity and inhibitory effect, we began our study with an analysis of the available X-ray structural data for CSS. At the beginning of our studies, the X-ray crystal structure of *Nm*B CSS in complex with CDP as a substrate mimetic (PDB: 1EYR) was available [49]. This structure showed the enzyme as an asymmetric homodimer, with the active site present at the interface of the core domain of one subunit and the dimerization domain of the other subunit [49]. In 2020, X-ray crystal structures of *Nm*B CSS in complex with substrate CTP, and with the product of reaction CMP-Neu5Ac **2** (PDB: 6CKM; Figure 2) were reported [44], providing further insights into the catalytic cycle of the enzyme. From kinetic studies with active-site mutants and the X-ray crystallographic studies, it is apparent that the CSS active site switches between an open form for substrate binding (and product release), and a catalytically active 'closed' form [43,44,49].

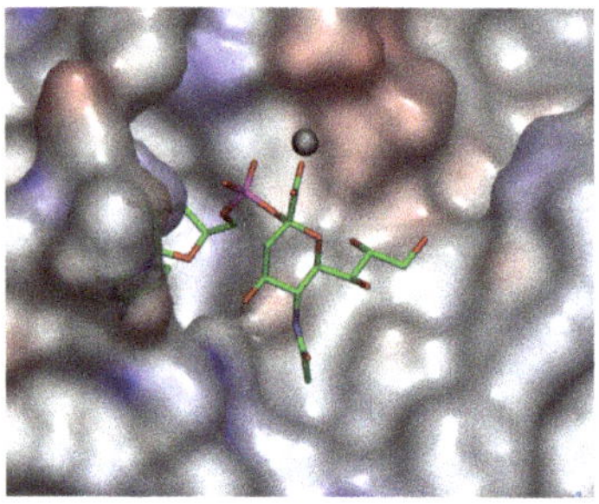

**Figure 2.** X-Ray crystal structure of CMP-Neu5Ac **2** in the active site of *Nm* CSS in an open form (PDB 6CKM) [44]. Solvent accessible surfaces are coloured according to charge (red, negative charge; blue, positive charge; grey, neutral regions). CMP-Neu5Ac is shown in stick representation (green). Bound $Ca^{2+}$ (grey sphere). Visualized using OpenAstexViewer (Version 3.0) [50].

It is clear that in the open form of the enzyme, there would be significant space around bound Neu5Ac. Substitutions on the Neu5Ac template, however, would no doubt have implications for the enzyme being able to close to give the correct positioning of the sialic acid and CTP for catalytic activity. The apparent extensive flexibility of the protein as it performs its function may limit a traditional molecular docking study. Instead, more complex techniques would need to be employed that would take into account protein flexibility, e.g., molecular dynamics simulations [51,52]. Our approach was therefore to systematically explore changes in spatial and functional group characteristics on the Neu5Ac template, based on analysis of contacts around Neu5Ac in the available models and crystal structures.

In our original modelling study, and subsequently in the *Nm*B CSS–CMP-Neu5Ac (**2**) complex [44] (Figure 2), the glycerol side-chain (C-7 to C-9 hydroxyl groups) of the Neu5Ac moiety is seen to project into a spacious channel or cleft. The potential binding environment of the Neu5Ac glycerol side chain contains a number of charged amino acid residues, including positively charged residues on one side of the channel. This suggested that the introduction of a negatively charged substituent at the terminus of the glycerol side chain was of particular interest. The scope for this modification could be explored through the synthesis of a C-9 carboxylic acid derivative that could then be further elaborated with various simple neutral amines and with amino acids to generate a series of C-9 carboxamide derivatives of Neu5Acβ2Me (Figure 3, Series I).

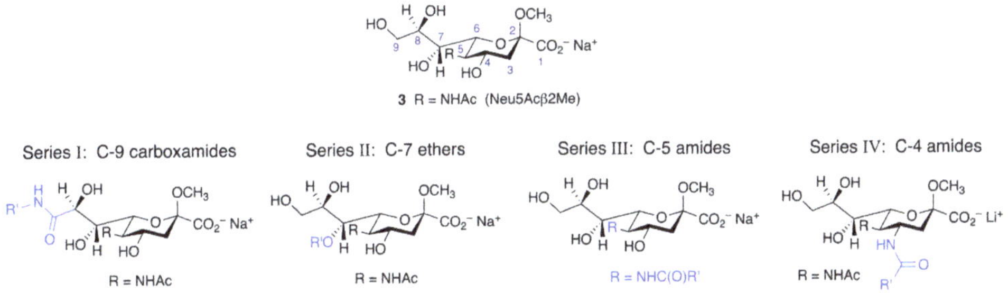


**Figure 3.** Overview of modifications incorporated onto the Neu5Acβ2Me (**3**) scaffold to explore interactions with *Nm* CSS.

Activation of 7-deoxy-Neu5Ac by *Nm* CSS [33] indicates that the C-7 hydroxyl group is not essential for substrate binding. Inspection of *Nm* CSS crystal structures revealed that a phenylalanine residue is in proximity to the Neu5Ac C-7 hydroxyl group, suggesting that an opportunity for possible π-π interaction exists in this region. With this in mind, we chose to synthesize a number of 7-*O*-aryl (e.g., benzyl) and unsaturated-alkyl (e.g., allyl) ethers, as well as the ethyl and propyl ethers, to explore this possible binding domain (Figure 3, Series II).

Apparent spatial constraints around C-4 and C-5 of Neu5Ac in *Nm* CSS structures seem to imply there is little potential for further functionalization at these positions. However, there are extensive protein dynamics involved in the function of CSS [44,45,53], and reports of successful activation of various *N*-acylated neuraminic acids by *Nm*B CSS [29–32], and a number of C-4 modified Neu5Ac derivatives (including 4-acetamido-4-deoxy-Neu5Ac [54]) by mammalian CSS. Together, these indicate that some structural variation at positions C-5 and C-4 can be tolerated. We therefore chose to introduce separately at each position, amides with varying functionality, ranging from small alkyl chains to large hydrophobic groups, in order to investigate structure–activity relationships around these positions (Figure 3, Series III and IV). In summary, four series of novel functionalized Neu5Ac derivatives, as the β-methyl glycosides (Figure 3), were proposed to probe interactions with *Nm* CSS, introducing carboxamides at C-9, ethers at C-7, and amides at C-5 and C-4.

### 2.2. *Synthesis of Functionalized Neu5Acβ2Me Derivatives*

#### 2.2.1. Synthesis of C-9 Functionalized Neu5Acβ2Me Derivatives (Series I)

The starting point for the current studies was the synthesis of the β-methyl glycoside of *N*-acetylneuraminic acid methyl ester [(Neu5Ac1,β2$Me_2$, **4**) from Neu5Ac **1** (Scheme 1). Methyl esterification of **1** and concomitant formation of the β-methyl glycoside can be achieved by heating **1** under reflux [55], or under microwave irradiation [56], with dry acidic ion exchange resin in anhydrous methanol. We have found [56] that while both approaches give comparable yields, microwave irradiation provides a more rapid reaction, and is more economical in terms of solvent and resin requirement than reaction under conventional heating. Saponification of **4** provided the benchmark β-methyl glycoside derivative, Neu5Acβ2Me, **3** [27,57,58].

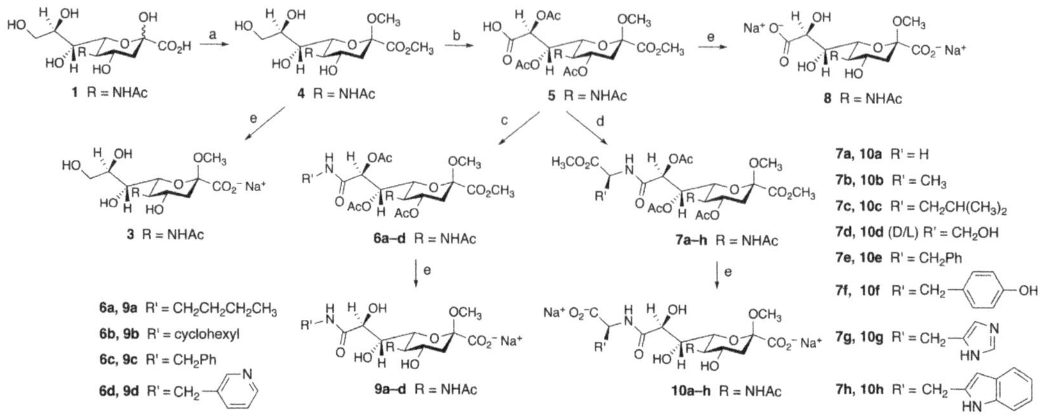

**Scheme 1.** Synthesis of C-9 carboxamide derivatives of Neu5Acβ2Me. Reagents and conditions: (a) MeOH, $H^+$ ion exchange resin, (i) reflux, 48 h (62%), or (ii) MW (100 W), 120 °C, 15 min (63%); (b) i. TEMPO, DCM, sat. aq. $NaHCO_3$, KBr, $Bu_4NBr$, aq. NaOCl, 0 °C to 5 °C, 1 h; ii. $Ac_2O$, pyridine, rt, 16 h (66% over 2 steps) [59]; (c) $R'NH_2$, HOBt, EDC, DIPEA, DCM/DMF (4:1), rt, 16 h (**6a** 60%, **6b** 60%, **6c** 56%, **6d** 55%); (d) $R'C(NH_2)CO_2CH_3$, HOBt, EDC, DIPEA, DCM/DMF (4:1), rt, 16 h (**7a** 55%, **7b** 60%, **7c** 60%, **7d** 62%, **7e** 55%, **7f** 60%, **7g** 40%, **7h** 45%); (e) NaOH, MeOH/$H_2O$ (1:1), pH 13, 0 °C-rt, 16 h.

To pursue modification at the C-9 position, we took advantage of the efficient synthesis of the C-9 carboxylic acid derivative of Neu5Ac2βMe [59], which could then be elaborated to generate a series of C-9 carboxamides. Regioselective TEMPO-mediated oxidation of the primary C-9 hydroxyl group of Neu5Ac2βMe methyl ester **4** was followed by peracetylation to assist in isolation and purification (Scheme 1). This gave the key intermediate 9-carboxy derivative **5** [59] in 66% yield over two steps. Compound **5** ultimately provides an interesting probe of CSS in its own right (see **8**), while the C-9 carboxy group of **5** provided us with the opportunity to further modify this position through coupling with amines of varying functionality. Elaboration of the carboxylic acid was carried out by reaction with neutral amines, as well as carboxy-protected amino acids. The latter would, upon deprotection, provide a negative charge at the end of the glycerol side-chain that may engage binding with positively charged protein residues.

The amide coupling was achieved using a slight modification of an established method for coupling the C-1 carboxyl group of Neu5Ac to amino acid esters [60,61], with the addition of 1-hydroxybenztriazole (HOBt) to minimize racemization and side reactions. 1-Ethyl-3-(3-dimethylaminopropyl) carbodiimide (EDC), which gives a water-soluble urea byproduct that can be easily removed by aqueous work-up, was used in place of (benzotriazol-1-yloxy)-tris(dimethylamino)phosphonium hexafluorophosphate (BOP reagent) because of the potential hazard of HMPA formation associated with coupling using the latter reagent [62]. 9-Carboxy derivative **5** was coupled with a series of aliphatic amines, and with a range of L-amino acid methyl esters (with the exception of D-/L-serine methyl ester), to give carboxamides **6a–d** and **7a–h**, respectively, in moderate yields (Scheme 1). De-*O*-acetylation followed by saponification yielded the 1,9-dicarboxylic acid derivative **8** [59], and carboxamides **9a–d** and **10a–h**. For the serine amino acid carboxamide **10d**, prepared from racemic D-/L-serine methyl ester, two isomers were separable by HPLC (denoted **10d-1** and **10d-2**); however, the relative stereochemistries at the serine a-carbon were not determined.

### 2.2.2. Synthesis of C-7 Functionalized Neu5Acβ2Me Derivatives (Series II)

For the synthesis of the target Neu5Acβ2Me C-7 ethers (Series II), we utilized a route analogous to that reported by Masuda and coworkers [63–65] (Scheme 2). To isolate the relatively unreactive C-7 hydroxyl group, Neu5Ac1,β2Me$_2$ **4** was reacted with 2,2-dimethoxypropane under acid catalysis to give the 8,9-*O*-isopropylidenated derivative **11** [66] in 90% yield. Regioselective monosilylation of the C-4 hydroxyl group then afforded 4-*O*-*t*-butyldimethylsilyl-8,9-*O*-isopropylidene derivative **12** [67]. The C-7 hydroxyl group of **12** was successfully alkylated using ethyl iodide (**13a**), allyl bromide (**13b**), propargyl bromide (**13d**), and benzyl bromide (**13e**) in the presence of NaH in DMF, to afford the corresponding alkyl ethers in moderate yield (36–60%). Use of ethyl iodide gave a comparable yield of **13a** to that reported when using diethylsulfate as the alkylating agent [64]. Higher equivalents of base were used to overcome the low reactivity of the C-7 hydroxyl group, which resulted in partial to complete transesterification in each reaction. Advantage was taken of the 7-*O*-allyl derivative **13b**, to prepare the 7-*O*-propyl derivative **13c** [63,64] by hydrogenation. Finally, simultaneous deprotection of both acid-labile groups (isopropylidene and TBDMS) of compounds **13a–e** by heating with 80% acetic acid at 80 °C, followed by saponification, gave the target 7-*O*-alkylated Neu5Acβ2Me derivatives **14a–e**.

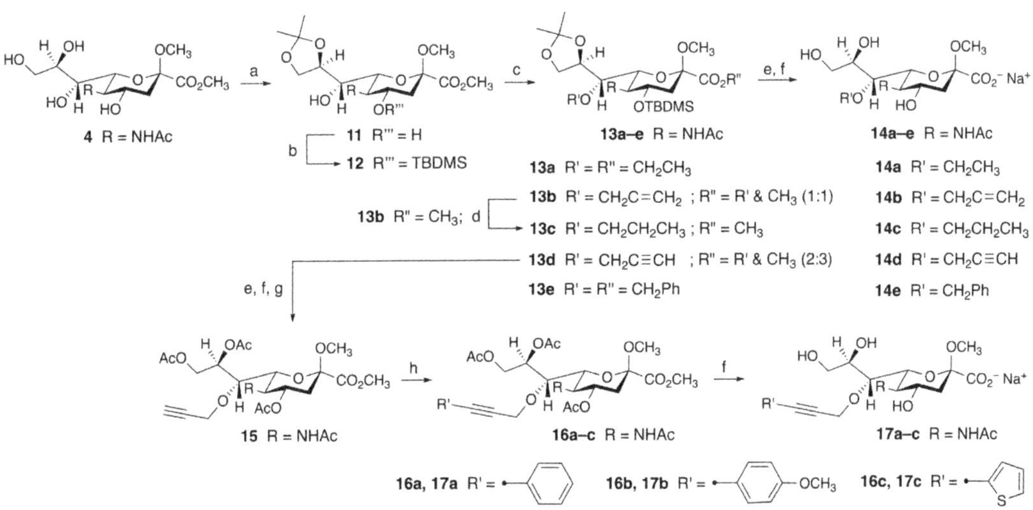


**Scheme 2.** Synthesis of 7-*O*-alkylated derivatives of Neu5Acβ2Me. Reagents and conditions: (a) 2,2-dimethoxypropane, H$^+$ ion exchange resin, acetone, rt, 16 h (90%); (b) *t*-BuMe$_2$SiCl, pyridine, 0 °C-rt, 16 h (58%); (c) R$^1$X, NaH, TBAI, DMF, 0 °C-rt, 16 h (**13a** 56%, **13b** 60%, **13d** 60%, **13e** 36%); (d) H$_2$, 10% Pd/C, MeOH, rt, 16 h (**13c** 87%); (e) 80% aq. AcOH, 80 °C, 2 h; (f) NaOH, MeOH/H$_2$O (1:1), pH 13, 0 °C-rt, 16 h; (g) (i) MeOH, H$^+$ ion exchange resin, MW (100 W), 80 °C, 15 min; (ii) Ac$_2$O, pyridine, 16 h (88% over 4 steps from **13d**); (h) R'I, PdCl$_2$(PPh$_3$)$_2$, CuI, Et$_3$N, CH$_3$CN, rt, 2–6 h (**16a** 60%, **16b** 87%, **16c** 85%). In **16,17a–c**, the sphere indicates the point of attachment of R' to the alkyne carbon.

The C-7 propyne ether derivative **13d** provided us with the opportunity to utilize Sonogashira coupling to further diversify the C-7 substituent. In order to achieve coupling selectively at the C-7 propyne ether, a sequence of deprotection and re-protection was employed on **13d** to remove the propyne ester functionality, and subsequently produce the acetylated methyl ester derivative **15** (Scheme 2). Following the method of Sato and coworkers [68], compound **15** was treated with aryl iodides in the presence of PdCl$_2$(PPh$_3$)$_2$,

CuI, and $Et_3N$ in anhydrous $CH_3CN$ to yield the disubstituted alkyne derivatives **16a–c** in good yield (60–87%). Upon saponification, these afforded the target compounds **17a–c**.

### 2.2.3. Synthesis of C-5 Functionalized Neuβ2Me Derivatives (Series III)

For the synthesis of compounds of Series III, which have alternative acyl groups on the amine at C-5, basic hydrolysis of the C-5 acetamido group of **4** was utilized to obtain key 5-amino derivative **18** (Scheme 3). Instead of using conventional heating, which requires long reaction times [69,70], in this work, we took advantage of reaction under microwave irradiation to provide a method for rapid de-*N*-acetylation of **4** [71]. Under microwave irradiation (max. 100 W) at 120 °C, NaOH-promoted de-*N*-acetylation of **4** was complete in only 15 min, with the 5-amino derivative **18** obtained cleanly in 80% yield. The carboxylate group of **18** was then re-esterified under acidic conditions to give methyl ester **19**. *N*-Acylation of the C-5 amino group of **19** was performed with acid chlorides of varying functionality using triethylamine as base in water/dioxane (1:5) to provide the desired C-5 amide derivatives **20a–i** in high yields (78–91%). In contrast to an earlier reported method [72], no bis-acylation was observed. Finally, saponification afforded the target variously *N*-acylated Neuβ2Me derivatives **21a–i**.

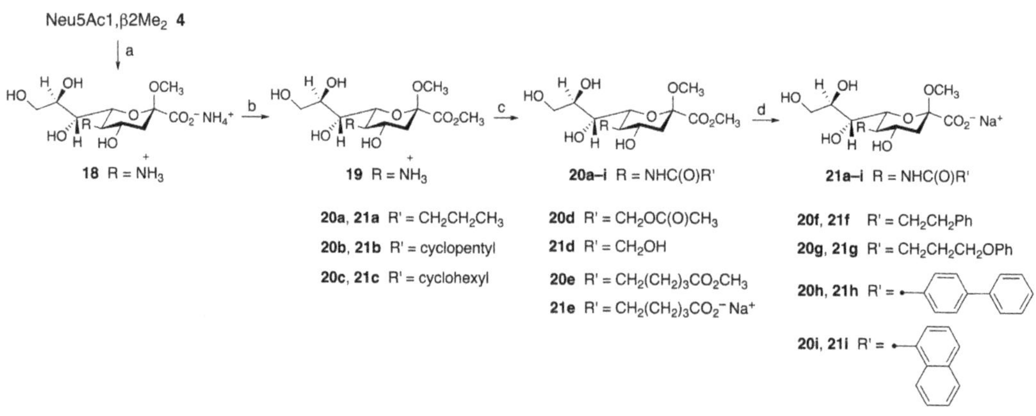


**Scheme 3.** Synthesis of C-5 amide derivatives of Neuβ2Me. Reagents and conditions: (a) NaOH (2.0 M), MW (100 W), 120 °C, 15 min (80%) [71]; (b) $SOCl_2$, MeOH, 0 °C-rt, 16 h (92%); (c) R′COCl, $Et_3N$, 1,4-dioxane/water (5:1), 0 °C–40 °C, 2 h (**20a** 87%, **20b** 90%, **20c** 91%, **20d** 91%, **20e** 78%, **20f** 80%, **20g** 82%, **20h** 80% and **20i** 88%); d) NaOH, MeOH/$H_2O$ (1:1), pH 13, 0 °C-rt, 16 h. In **20,21h,i**, the sphere indicates the point of attachment of R' to the amide carbon.

### 2.2.4. Synthesis of C-4 Functionalized Neu5Acβ2Me Derivatives (Series IV)

The synthesis of the C-4 amide derivatives of Neu5Acβ2Me (Series IV) required the introduction of an amino group at C-4 in the equatorial configuration. Nitrogen functionality can be introduced at C-4 on a suitably protected Neu5Ac template via a 3-step sequence of oxidation of the equatorially substituted C-4 hydroxyl group, reduction in the 4-oxo derivative to the 4-*epi*-OH derivative, and finally introduction of azide in the equatorial configuration via a Mitsunobu reaction using $HN_3$.[54] An alternative approach is to proceed via the 2,3-unsaturated sialic acid, Neu5Ac2en, scaffold, where the 4-azido derivative **22** [73–75] can be produced in good yield over four steps from Neu5Ac **1** [75]. Having **22** at hand, we chose the second approach (Scheme 4). Bromo-methoxylation [76] of the 2,3 double bond of 4-azido-4-deoxy-Neu5Ac2en derivative **22** gives rise to an approximately 1:1 isomeric mixture of 2,3-diaxial (**23**) and 2,3-diequatorial (**24**) products [77]. Treatment of the 2,3-di-axial bromomethoxylated isomer (the β-methyl glycoside) **23** with tributyltin hydride in the presence of AIBN [77] resulted in both debromination at C-3 and reduction in the C-4 azide, to give 4-amino Neu5Acβ2Me derivative **25**. Treatment of **25** with a range of acid chlorides afforded the C-4 amide derivatives **26a–i**, which, after global deprotection, gave the target 4-acylamino-4-deoxy-Neu5Acβ2Me derivatives **27a–i**.

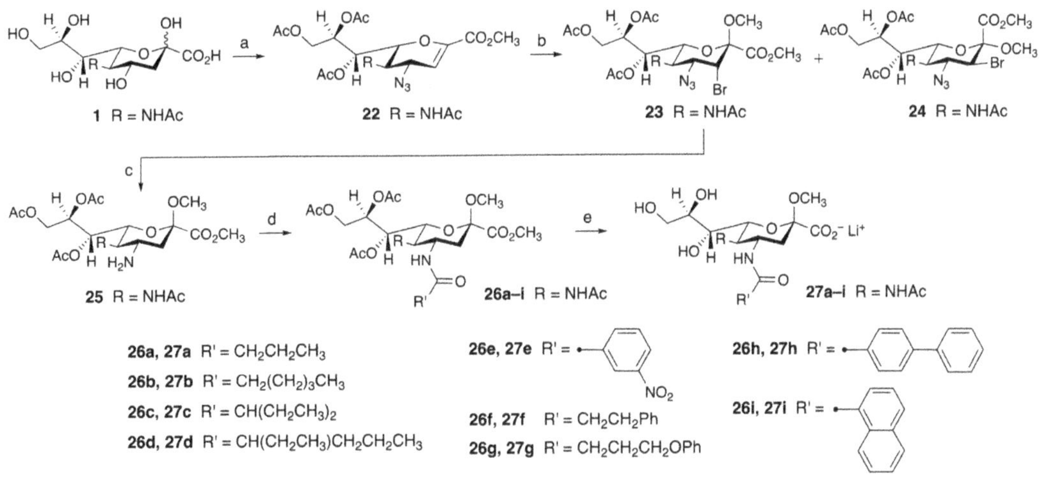

**Scheme 4.** Synthesis of C-4 amide derivatives of 4-deoxy-Neu5Acβ2Me. Reagents and conditions: (a) over 4 steps according to the reported procedures: (i) MeOH, $H^+$ ion exchange resin, rt, 16 h; (ii) $Ac_2O$, pyridine, rt, 16 h; (iii) TMSOTf, EtOAc, 50 °C, 2 h; (iv) $TMSN_3$, *t*-BuOH, reflux, 6 h [75]; (b) according to the reported procedure: NBS, MeOH, 0–5 °C, 72 h (**23 + 24**, 87%) [77]; (c) $Bu_3SnH$, AIBN, dioxane, 65 °C, 16 h (77%) [77]; (d) R′COCl, $Et_3N$, DCM, 0 °C-rt, 2 h (**26a** 55%, **26b** 67%, **26c** 60%, **26d** 65%, **26e** 65%, **26f** 65%, **26g** 65%, **26h** 60%, **26i** 65%); (e) LiOH, MeOH/$H_2O$ (1:1), pH 13, 0 °C-rt, 16 h. In **26,27e,h,i**, the sphere indicates the point of attachment of R' to the amide carbon.

### *2.3. Screening of Functionalized Neu5Acβ2Me Derivatives for Inhibition of NmB CSS*

To analyze the effect of the compounds on *Nm*B CSS activity in vitro, we recombinantly expressed and purified the enzyme to homogeneity. The EnzChek™ pyrophosphate assay kit (Life Technologies, Darmstadt, Germany) was used to determine *Nm*B CSS activity in the forward reaction (i.e., formation of CMP-Neu5Ac and PPi), revealing $K_m$ values of 22 and 37 μM for CTP and Neu5Ac, respectively. After confirming that the test compounds did not show any inhibitory effect on the assay components, we screened all functionalized Neu5acylβ2Me derivatives and the parent compound **3** at a concentration of 100 μM (Figure 4). Interestingly, the benchmark parent compound, unsubstituted β-methyl glycoside **3**, previously reported [27] to be a modest inhibitor of mammalian CSS ($K_i$ = 2.5 mM), showed no significant inhibitory effect on *Nm*B CSS in this in vitro assay (Figure 4). This difference might be attributed to differences in the employed CSS species and/or in assay conditions: while Zbiral and coworkers [27] utilized the Warren assay [78] under CTP-limiting conditions, we provided both substrates in saturating concentrations of 1 mM each. Three of the C-9 modified Neu5Acβ2Me derivatives—parent 9-carboxy-Neu5Acβ2Me **8** [59], benzamide **9c**, and serine carboxamide **10d-1**—and one of the C-7 ethers—7-*O*-benzyl-Neu5Acβ2Me **14e**, however, showed inhibition of *Nm*B CSS activity by more than 20% in this screen. Detailed kinetic analysis of these four compounds suggests a mixed-mode inhibition that most closely resembles the non-competitive mode, as indicated by α-values close to 1 and supported by the appearance of Lineweaver-Burk plots (Supplementary Figure S1). As all compounds were designed to target the active site and compete with the natural substrate Neu5Ac **1** for binding, the finding that none of the four characterized inhibitors display competitive kinetics may be somewhat unexpected, but does not necessarily indicate binding outside of the catalytic center. Indeed, multiple mechanisms exist that may account for non-competitive or "mixed-mode mimicking" inhibition patterns of inhibitors that in fact bind to the active site [79,80]. One of these mechanisms applies to multisubstrate/-product enzymes that follow a sequential ordered mechanism, which is the case for CSS, where CTP binds prior to Neu5Ac [44]. With inhibitory constants ($K_i$) in the range between 100 and 500 μM, these compounds appear to be stronger inhibitors than the previously reported sulfone-based nucleotide isosteres, which inhibited *Nm*B CSS activity up to 55% at a concentration of 1 mM [42].

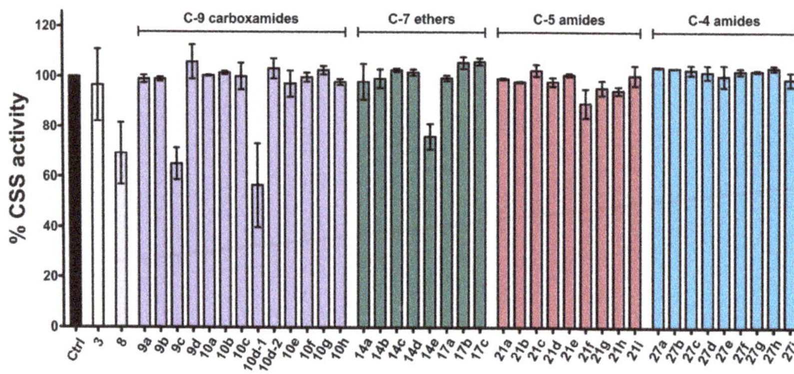


**Figure 4.** Relative activity of *Nm*B CSS in the presence of C-9, C-7, C-5, and C-4 functionalized Neu5acylβ2Me derivatives (at 100 μM) in the EnzChek® pyrophosphate assay (Life Technologies, Darmstadt, Germany). Bar graphs represent the means of two individual experiments, with error bars representing the standard deviation. Activity of *Nm*B CSS in the absence of functionalized compounds was defined as 100% (Ctrl).

### *2.4. Interaction of Neu5Acβ2Me C-9 Serine Carboxamide 10d-1 with NmB CSS as Determined by STD NMR*

To confirm direct binding and to investigate the interactions of the most promising C9 carboxamide derivative **10d-1** with CSS, we used Saturation Transfer Difference (STD) NMR spectroscopy [46,81,82]. This powerful tool allows evaluation of ligand–biomolecule interactions by mapping the binding epitope of the ligand, essentially through screening the protons of the ligand for their proximity to the protein upon binding.

The binding epitope of **10d-1** in complex with *Nm*B CSS (Figure 5iii) was determined from the relative strength of transferred saturation (with the effect at the C-5 NHAc methyl group set to 100%). It is evident from the STD NMR spectrum (Figure 5ii) that the carbohydrate ring of **10d-1** is in close contact with the enzyme. Relatively strong STD NMR signals were observed for the equatorial proton at C-3 (H-3eq 217%), with more moderate signal intensities observed for the axial C-3 proton (H-3ax 165%) and the glycoside methyl group (126%). Interaction of the serine 'tail' could not be unequivocally determined due to the signal overlap, for example, of H-2′ with H-5; however, any interaction would appear to be only weak.

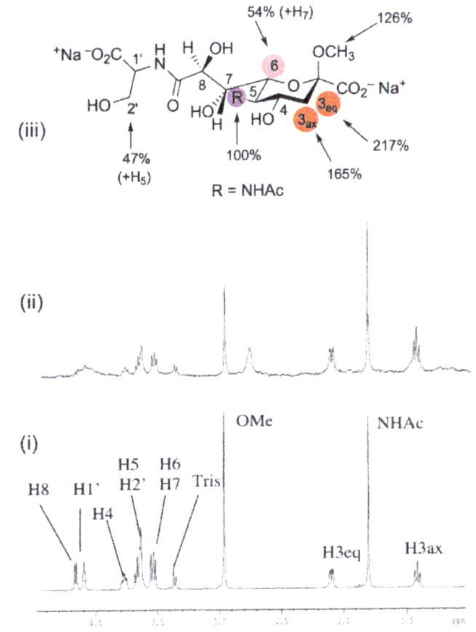


**Figure 5.** [1]H NMR and STD NMR spectra of Neu5Acβ2Me C-9 serine carboxamide **10d-1** with *Nm*B CSS. **(i)** [1]H NMR spectrum of **10d-1**; **(ii)** STD NMR of **10d-1** with *Nm*B CSS (ligand to protein ratio of 100:1); **(iii)** epitope map of **10d-1**.

### *2.5. Neu5Acβ2Me C-9 Serine Carboxamide **10d-1** as a Modulator of NmB LOS Sialylation*

To determine whether Neu5Acβ2Me C-9 serine carboxamide **10d-1** was capable of inhibiting CSS (encoded by the gene *neuA*) in living bacteria, we established an ELISA-based assay utilizing a sialic acid-recognizing lectin to detect LOS sialylation on the cell surface of *Nm*B. The capsule-deficient *N. meningitidis* strain ¢3 (¢3 WT) [13] expressed sialylated LOS at high levels (defined as 100%, Figure 6). To define the background signal in the absence of *neuA* activity, we generated a corresponding *neuA* knockout strain (*¢3neuA::kan*) and observed a reduction in LOS sialylation to about 2%. Complementation of the *neuA* knockout mutant by reintroducing the *neu*A gene (*¢3neuA::kan*, NeuA+) confirmed that loss of signal in the assay was due to inactivity of CSS and thus, loss of sialylation in this mutant (Supplementary Figure S2). To evaluate whether the results of this assay can be interpreted in a quantitative manner, we mixed ¢3 wild type strain and *¢3neuA::kan* knockout mutant in a 1:1 ratio. This resulted in a ~44% ± 6% reduction in signal intensity, indicating a corresponding reduction in LOS sialylation of wild-type *Nm*, which was close to ~50%. Given that not all ¢3 wild-type LOS molecules are sialylated, a value of under 50% is reasonable (Supplementary Figure S2). We then tested the effect of C-9 serine carboxamide **10d-1** on LOS sialylation by supplementing the growth medium with 1 mM of the compound. The ELISA results indicated that bacterial growth in the presence of **10d-1** resulted in a reduction in *Nm*B LOS sialylation to ~66% of ¢3 wild wild-type level (Figure 6). This suggests that inhibitor **10d-1** is able to cross the cell membranes and act on CSS, leading to a decrease in cell surface sialylation of whole cells.

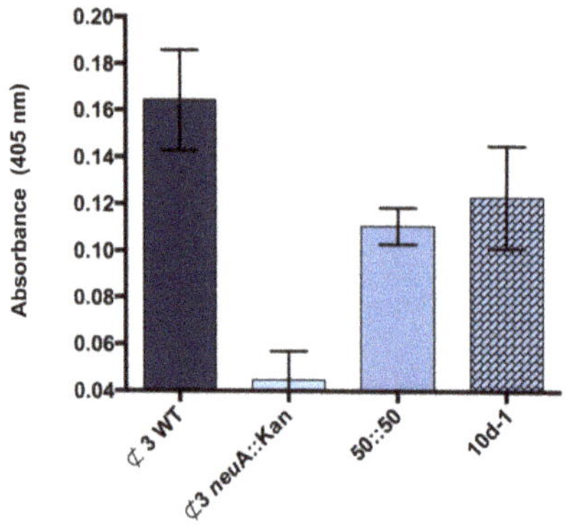


**Figure 6.** ELISA to measure the effect of **10d-1** on *N. meningitidis* LOS sialylation. ¢3 WT (capsule-deficient *N. meningitidis* wild type), *¢3neuA::kan*: (*CSS-deficient* ¢3), and a 1:1 ratio of ¢3 WT and *¢3neuA::kan* (50::50) were coated on microtiter plates and surface sialic acid detected photometrically (405 nm) with the sialic-acid-specific lectin MAA-AP conjugate. Similarly, ¢3 wild type was incubated with **10d-1** at 1 mM final concentration for 4 h, after which surface sialic acid was detected. Absorbance at 405 nM was measured to detect *p*-nitrophenol released from the alkaline phosphatase substrate *p*-nitrophenyl phosphate (PNPP). Experiments were performed in triplicate.

## 3. Materials and Methods

### 3.1. Modelling of Neu5Ac 3 in the Active Site of NmB CSS

The PyMOL Molecular Graphics System (Version 1.7.x Schrödinger, LLC, New York, NY, USA) was used for modelling.

### 3.2. Chemical Synthesis—Materials and Methods

*N*-Acetylneuraminic acid (5-acetamido-3,5-dideoxy-D-*glycero*-D-*galacto*-non-2-ulosonic acid) (**1**) was obtained from Jülich Chiral Solutions GmbH (Jülich, Germany) and Carbosynth Ltd. (Compton, UK). Reagents and dry solvents purchased from commercial sources were used without further purification. Anhydrous reactions were carried out under an atmosphere of nitrogen or argon, using oven-dried glassware. Microwave reactions were conducted using a CEM Discover® SP Explorer Hybrid-12 (CEM Corporation, Matthews, NC, USA) microwave system with a single-mode cavity, and were carried out in a 10 mL pressure tube, sealed with a Teflon septum.

Reactions were monitored using thin-layer chromatography (TLC) on aluminum plates precoated with Silica Gel 60 F254 (E. Merck, Darmstadt, Germany). Developed plates were observed under UV light at 254 nm and then visualized after application of a solution of $H_2SO_4$ in EtOH (5% *v*/*v*) or ninhydrin in EtOH (0.2% *v*/*v*), as appropriate, and heating. Flash chromatography was performed on silica gel 60 (0.040–0.063 mm) or using a Reveleris® flash chromatography system (Grace Davison, Columbia, MD, USA) (as indicated) using distilled solvents.

$^{1}H$ and $^{13}C$ NMR spectra were recorded either at 600 or 300 MHz and 150 or 75.5 MHz, respectively, on a Bruker Avance 600 or 300 MHz spectrometer (Bruker, Rheinstetten, Germany; as indicated). Low-resolution mass spectra (LRMS) were recorded, in electrospray ionization mode, on a Bruker Daltonics Esquire 3000 ESI spectrometer (Bruker, Bremen, Germany), using positive or negative mode (as indicated). High-resolution mass spectrometry (HRMS) was carried out by the Griffith University FTMS Facility on a Bruker Daltonics Apex III 4.7e Fourier Transform MS, fitted with an Apollo ESI source, or by the

University of Queensland MS Facility on a Bruker MicrOTOF-Q with a Bruker ESI source (Bruker, Bremen, Germany).

HPLC purification was performed on an Agilent HP1100 instrument (Agilent, Santa Clara, CA, USA) using a Phenomenex Aqua 5 μ C18 124 Å column (250 × 10 mm) (Phenomenex, Torrance, CA, USA) at a flow rate of 3 mL/min and column temperature of 40 °C using isocratic elution with solvents as indicated. The purities of all synthetic intermediates after chromatographic purification were judged to be >90% by $^{1}$H and $^{13}$C NMR. The purity of tested compounds was ≥95% by HPLC analysis or by $^{1}$H and $^{13}$C NMR.

Detailed synthetic methods for all test compounds and intermediates [83] are provided in the Supporting Information—Part 1. NMR spectra ($^{1}$H and $^{13}$C) of all test compounds and intermediates [83] are provided in the Supporting Information—Part 2.

### 3.3. Protein Expression and Purification (For Screening Assay and STD NMR)

A modified pET22b-Strep vector (IBA Lifesciences, Göttingen, Germany) allowing the prokaryotic expression of *Nm*B CSS (accession No. M95053) was kindly provided by PD Dr. Martina Mühlenhoff, MHH, Germany. Recombinant *Nm*B CSS with an N-terminal StrepII-tag was expressed in *E. coli* BL21(DE3) (Novagen, registered trademark of Merck, Darmstadt, Germany) at 15 °C and purified by affinity chromatography utilizing the StrepII-Tag. Peak fractions were desalted (HiPrep 26/10, GE Healthcare/now Cytiva, Freiburg, Germany) and concentrated in buffer containing 50 mM Tris-HCl, pH 8, 20 mM $MgCl_2$, 150 mM NaCl, and 1 mM DTT to 2.5 mg/mL. Protein concentrations were determined by measuring the absorption at 280 nm under consideration of the specific extinction coefficients calculated at https://web.expasy.org/protparam/ (accessed on 15 January 2020). Purified protein samples were flash-frozen in liquid nitrogen and stored at −80 °C until required.

### 3.4. In Vitro Inhibition Assay and Kinetic Characterization

For initial screening of the synthesized compounds, CSS in vitro activity in the absence or presence of synthesized compounds was measured in 96-well half-area plates (Greiner Bio-One, Frickenhausen, Germany) in a reaction volume of 100 μL, utilizing the EnzChek pyrophosphate assay kit (Life Technologies, Darmstadt, Germany). Briefly, the inorganic pyrophosphate (PPi), produced by CSS as a byproduct, is cleaved by inorganic pyrophosphatase (IPP), yielding two molecules of phosphate (Pi), which subsequently react with 2-amino-6-mercapto-7-methylpurine ribonucleoside (MESG) in a reaction catalyzed by purine nucleoside phosphorylase (PNP). Detection of the reaction product 2-amino-6-mercapto-7-methylpurine at 360 nm allows quantitation of the initially generated Pi. The assay reaction mixture contained 50 mM Tris-HCl, pH 8.0, 25 mM $MgCl_2$, 1 mM Neu5Ac (Nacalai Tesque, Kyoto, Japan), 1 mM CTP (Sigma-Aldrich, Taufkirchen, Germany), as well as 0.22 mM MESG, 1 U/mL PNP, and 0.03 U/mL IPP (all from Life Technologies), and 0.1 mM of the compounds to be tested. To exclude an effect of the compounds on the coupling enzymes PNP and IPP, a control assay without Neu5Ac, CTP, and CSS was performed, and the reaction was instead initiated by the addition of inorganic pyrophosphate (Sigma-Aldrich) at 20 μM final concentration. For CSS activity measurements, the reaction was initiated by the addition of purified *Nm*B CSS at a final concentration of 157.5 ng/mL. The enzyme was diluted in 50 mM Tris-HCl, pH 8.0, 25 mM $MgCl_2$, and briefly pre-incubated with the compounds before transferring the CSS/compound mixture to the reaction mixture with a 96-well pipette. Product formation was monitored at 360 nm in a 96-well plate reader at 25 °C for 4 min. As a reference, CSS activity without compounds was measured and defined as 100% activity. Activity in the presence of each compound was calculated as a percentage of the reference activity.

For kinetic characterization, the CSS in vitro activity assay was carried out as described above, but with Neu5Ac and effector compound concentrations varied between 0.01 and 1.0 mM and 0–0.4 mM, respectively, while CTP concentration was maintained at 1 mM. All measurements were performed in duplicate. Uninhibited CSS activity was used to calculate kinetic parameters for the *Nm*B enzyme by nonlinear regression according to the Michaelis–Menten model. Inhibitory constants were determined by plotting CSS activity in the presence of different compound concentrations against Neu5Ac concentration, and performing nonlinear regression curve fitting according to different inhibition modes (competitive, non-competitive, uncompetitive, and mixed-mode). Lineweaver–Burk plots were generated by plotting the reciprocal values of enzyme activity (1/v) against substrate concentration (1/[s]). The displayed linear regression lines with x- and y-intercepts at $-1/K_m$ and $1/V_{max}$, respectively, were constructed based on the corresponding kinetic parameters obtained from nonlinear regression analysis of the original data. All kinetic analyses were carried out using GraphPad Prism version 5.01 for Windows (GraphPad Software, Boston, MA, USA, www.graphpad.com).

### 3.5. STD NMR Spectroscopy

Saturation Transfer Difference (STD) NMR experiments were performed on a Bruker Avance 600 MHz NMR spectrometer, equipped with a 5 mm TXI probe with triple-axis gradient at 298 K [83]. The protein was saturated on-resonance at −1 ppm and off-resonance at 33 ppm with a cascade of 40 selective Gaussian-shaped pulses, of 50 ms duration with a 100 μs delay between each pulse in all STD NMR experiments. The total duration of the saturation time was set to 2 s. For the STD NMR experiments, 23.3 μM *Nm*B CSS in deuterated 50 mM $TRIS_{D5}$, 50 mM NaCl, and 5 mM $MgCl_2$ at pH 8.0 was used. The test compound was added in a molecular protein:ligand ratio of 1:100. A total of 1024 scans per STD NMR experiment were acquired, and a WATERGATE sequence was used to suppress the residual HDO signal. A spin-lock filter with a strength of 5 kHz and a duration of 10 ms was applied to suppress protein background.

### 3.6. ELISA-Based Assay to Monitor Sialylation of NmB LOS

Bacterial strains, plasmids, media, and growth conditions: *Neisseria* strains were grown on brain heart infusion (BHI, Oxoid; Thermo Scientific, Scoresby, Australia) agar plates made with 1% agar and supplemented with 10% Levinthal base [84] at 37 °C with 5% $CO_2$. *E. coli* strain DH5α was grown on Luria–Bertani agar plate or broth [85]. Antibiotics ampicillin (amp) and kanamycin (kan) were used at a final concentration of 100 μg/mL where appropriate.

Construction of mutant strains: A PCR product encompassing 1 kb upstream and downstream of the sequence flanking the CMP-Neu5Ac synthase gene (*neuA*, NMB0069) was amplified from *Neisseria* strain MC58 using primers containing *Neisseria* uptake sequence; NeuAMC58_FOR (5′-GCCGTCTGAACAGAACCTACAAGGAAGTAAC-3′) and NeuAMC58_REV (5′- GACGCT GAAGTCTCCATTG-3′) and cloned into pGEM-T-Easy (Promega, Madison, WI, USA). Expression of *neuA* was disrupted through the insertion of a kanamycin resistance cassette ($Kan^R$) from the pUC4Kan plasmid (Cytiva (formerly GE Healthcare, Amersham Biosciences), USA) into a *Bam*HI restriction site. The resulting mutant construct named pGEM-T::*neuA*::$Kan^R$ was linearized with *Nco*I-HF and transformed into *Neisseria* strain MC58 ¢3 as described previously [86]. The new construct was named (¢3*neuA::kan*). Orientation of the insert into the vector was verified through sequencing and the generation of a unique PCR product with primers NeuABMC58_REV and KanUP_OUT (5′-AGACGTTTCCCGTTGAATATGGTCAT-3′), a primer located within the Kan$^R$ cassette.

Complementation of the siaB gene: The *neuA* mutant was complemented as described previously [87]. A PCR product of *siaB* with 300 bp upstream and downstream flanking region was amplified using primers with added random bases and restriction enzyme sequence; NeuA_COMP_FOR (5′-*ACATCG***CTTAAG** GCAACTCAAGTGCAGGTATTAG-3′) (*Afl*II restriction enzyme sequence) and NeuA_COMP_REV (5′-*GTCGTA***CCCGGG**CTT-CTTCATTCAGGGCGCAAC-3′) (*Sma*I restriction enzyme sequence). This insert was cloned into the corresponding *Afl*II and *Sma*I digested pCTS32 vector under the control of the vector promoter. The subsequent construct, named *¢3neuA::kan*, NeuA+, was linearized with *Sal*I and transformed into *¢3neuA::kan*. Clones were selected on BHI plates supplemented with 10% Levinthal base and antibiotics ampicillin (100 μg/mL) and spectinomycin (50 μg/mL), respectively. Correct recombination of *siaB* into the chromosome was verified by PCR.

ELISA-based LOS sialylation assay: *Neisseria* strains were plated out on BHI plates (supplemented with antibiotics where necessary) and incubated at 37 °C + 5% $CO_2$ overnight. Cell suspension was prepared from overnight growth in 2X BHI supplemented with levinthal, and normalized to an OD600 of 1.0.

Test compound was suspended in 300 μL of sterile $dH_2O$ to give a final concentration of 1 mM and filter sterilized. Standing cultures were grown with the test compound at a starting OD of 0.1 for 4 h at 37 °C, shaken at 200 rpm. After incubation, cells were washed and resuspended in PBS and normalized to an OD600 of 0.2. Cells were heat-killed for 1 h at 60 °C before being used for ELISA. A total of 50 μL of normalized heat-killed cells was added per well (Nunc, Thermo Scientific, Waltham, MA, USA), and the plates were dried overnight at room temperature. Wells were blocked with 5% BSA + 1× TBST overnight, followed by 1 h incubation with 100 μL of *Maackia amurensis* lectin–alkaline phosphatase conjugate (MAA-AP; EY Laboratories, San Mateo, CA, USA) in 1/1400 dilution. Wells were washed at 5 min intervals with 1× TBST, repeated 4 times, and developed with substrate *p*-nitrophenyl phosphate (PNPP, Thermo Scientific) for 30 min. Reaction was stopped with 50 μL of 1N NaOH, and absorbance was read at 405 nm. Experiments were performed in triplicate.

## 4. Conclusions

We have functionalized the C-9, -7, and -4 positions of Neu5Acβ2Me and varied the C-5 *N*-acyl substituent on Neuβ2Me to produce a series of compounds with varied substitution around the sialic acid framework to probe interaction with *N. meningitidis* CSS. Evaluation of the compounds for inhibition of *Nm*B CSS activity in vitro identified four compounds that showed an increased inhibition of the recombinant enzyme ($K_i$ range between 100 and 500 μM), compared to the unsubstituted parent compound **3**. Importantly, the Neu5Acβ2Me C-9 serine carboxamide **10d-1** was shown to directly interact with *Nm*B CSS, to exhibit mixed mode inhibition of the enzyme, and to reduce the level of sialylation of *N. meningitidis* LOS. This data provides an intriguing starting point for further development of *Nm*B CSS inhibitors as novel and specific anti-microbial agents for the treatment of bacterial meningitis.

**Supplementary Materials:** The following supporting information can be downloaded at: https://www.mdpi.com/article/10.3390/molecules30224329/s1: Supporting Information-S1, which includes: Figure S1, Kinetic analysis of test compounds with *Nm*B CSS; Figure S2, ELISA to measure *N. meningitidis* LOS sialylation; Detailed experimental procedures for synthesis of the reported functionalized sialic acids and their characterization, including NMR data. Supporting Information-S2: NMR spectra.

**Author Contributions:** Conceptualization, M.v.I., R.G.-S., J.F., A.K.M.-K.; structure analysis, P.C., J.C.D., A.K.M.-K.; compound synthesis and characterization, P.C., I.D.G.; compound purification, F.J.R.; STD NMR and analysis, P.C., T.H.; protein production and enzymatic assays, J.F., A.K.M.-K.; *Nm* in vitro assay development and assay analysis, P.N., M.P.J.; *Nm* assays P.N.; supervision, I.D.G., R.J.T., M.v.I., M.P.J.; writing—original draft preparation, P.C., P.N., J.F., A.K.M.-K.; writing—review and editing, P.C., J.F., P.N., T.H., J.C.D., R.J.T., R.G.-S., I.D.G., M.P.J., A.K.M.-K., M.v.I. All authors have read and agreed to the published version of the manuscript.

**Funding:** M.v.I. thanks the Australian Research Council (A10017093, FF0345363) and the National Health and Medical Research Council (APP1071659) for financial support. P.C. thanks the Institute for Glycomics and Griffith University for the award of scholarships. R.G.-S. and A.K.M.-K. acknowledge the German Research Foundation (DFG) for financial support (MU-1849/2-1), also in the framework of DFG Research Unit 548 (Ge801/10-2), and J.F. acknowledges the Hannover Medical School for intramural funding (HILF). M.P.J. thanks the National Health and Medical Research Council for a Principal Research Fellowship (1138466).

**Institutional Review Board Statement:** This study does not involve humans or animals.

**Informed Consent Statement:** Not applicable.

**Data Availability Statement:** The data presented in this study are included in the article and Supplementary Material. Further inquiries can be directed to the corresponding authors.

**Acknowledgments:** This project is part of the Fraunhofer International Consortium for Antimicrobial Research (*i*CAIR) initiative. We thank Paul Madge and Ryan Bennet (Institute for Glycomics) for the preparation of synthetic intermediates. We thank Martina Mühlenhoff (MHH) for kindly providing the *Nm*B CSS expression vector and Wiebke Schaper and Friedrich Freiberger (MHH) for their contribution to activity assay design. We thank Plabon Das (IBG) for his contribution to the preparation of figures. Aspects of the work reported in this paper, including the synthesis of functionalized Neu5Acβ2Me derivatives described in Section 2.1, and the STD NMR study described in Section 2.3, are drawn from the doctoral thesis of P.C. [83].

**Conflicts of Interest:** The authors declare no conflicts of interest.

## Abbreviations

The following abbreviations are used in this manuscript:

| | |
|---|---|
| CSS | CMP-sialic acid synthetase |
| LOS | lipooligosaccharide |
| *Nm*B | *Neisseria meningitidis* serotype B |
| STD | Saturation Transfer Difference (STD) NMR |

## References

1. Harrison, L.H.; Trotter, C.L.; Ramsay, M.E. Global epidemiology of meningococcal disease. *Vaccine* **2009**, *27* (Suppl. 2), B51–B63. [CrossRef]
2. Dull, P.M.; McIntosh, E.D. Meningococcal vaccine development—From glycoconjugates against MenACWY to proteins against MenB—Potential for broad protection against meningococcal disease. *Vaccine* **2012**, *30*, B18–B25. [CrossRef]
3. Berti, F.; Romano, M.R.; Micoli, F.; Adamo, R. Carbohydrate based meningococcal vaccines: Past and present overview. *Glycoconj. J.* **2021**, *38*, 401–409. [CrossRef] [PubMed]
4. Basu, N.; Mukherjee, M.M.; Ghosh, R. Recent advances on the syntheses of oligosaccharides and the corresponding glycoconjugates towards development of vaccine candidates against invasive *Neisseria meningitidis* strains. *Carbohydr. Res.* **2025**, *555*, 109545. [CrossRef] [PubMed]

5. Serruto, D.; Bottomley, M.J.; Ram, S.; Giuliani, M.M.; Rappuoli, R. The new multicomponent vaccine against meningococcal serogroup B, 4CMenB: Immunological, functional and structural characterization of the antigens. *Vaccine* **2012**, *30*, B87–B97. [CrossRef]
6. Finne, J.; Bitter-Suermann, D.; Goridis, C.; Finne, U. An IgG monoclonal antibody to group B meningococci cross-reacts with developmentally regulated polysialic acid units of glycoproteins in neural and extraneural tissues. *J. Immunol.* **1987**, *138*, 4402–4407. [CrossRef]
7. Häyrinen, J.; Jennings, H.; Raff, H.V.; Rougon, G.; Hanai, N.; Gerardy-Schahn, R.; Finne, J. Antibodies to polysialic acid and its *N*-propyl derivative: Binding properties and interaction with human embryonal brain glycopeptides. *J. Infect. Dis.* **1995**, *171*, 1481–1490. [CrossRef]
8. Granoff, D.M.; Bartoloni, A.; Ricci, S.; Gallo, E.; Rosa, D.; Ravenscroft, N.; Guarnieri, V.; Seid, R.C.; Shan, A.; Usinger, W.R.; et al. Bactericidal monoclonal antibodies that define unique meningococcal B polysaccharide epitopes that do not cross-react with human polysialic acid. *J. Immunol.* **1998**, *160*, 5028–5036. [CrossRef] [PubMed]
9. Frosch, M.; Weisgerber, C.; Meyer, T.F. Molecular characterization and expression in *Escherichia coli* of the gene complex encoding the polysaccharide capsule of *Neisseria meningitidis* group B. *Proc. Natl. Acad. Sci. USA* **1989**, *86*, 1669–1673. [CrossRef]
10. Mandrell, R.E.; Kim, J.J.; John, C.M.; Gibson, B.W.; Sugai, J.V.; Apicella, M.A.; Griffiss, J.M.; Yamasaki, R. Endogenous sialylation of the lipooligosaccharides of *Neisseria meningitidis*. *J. Bacteriol.* **1991**, *173*, 2823–2832. [CrossRef]
11. Schauer, R.; Kamerling, J.P. Exploration of the sialic acid world. *Adv. Carbohydr. Chem. Biochem.* **2018**, *75*, 1–213. [CrossRef] [PubMed]
12. Vogel, U.; Hammerschmidt, S.; Frosch, M. Sialic acids of both the capsule and the sialylated lipooligosaccharide of *Neisseria meningitis* serogroup B are prerequisites for virulence of meningococci in the infant rat. *Med. Microbiol. Immunol.* **1996**, *185*, 81–87. [CrossRef]
13. Virji, M.; Makepeace, K.; Peak, I.R.A.; Ferguson, D.J.P.; Jennings, M.P.; Moxon, E.R. Opc- and pilus-dependent interactions of meningococci with human endothelial cells: Mechanisms and modulation by surface polysaccharides. *Mol. Microbiol.* **1995**, *18*, 741–754. [CrossRef]
14. Unkmeir, A.; Kämmerer, U.; Stade, A.; Hübner, C.; Haller, S.; Kolb-Mäurer, A.; Frosch, M.; Dietrich, G. Lipooligosaccharide and polysaccharide capsule: Virulence factors of *Neisseria meningitidis* that determine meningococcal interaction with human dendritic cells. *Infect. Immun.* **2002**, *70*, 2454–2462. [CrossRef]
15. Kurzai, O.; Schmitt, C.; Claus, H.; Vogel, U.; Frosch, M.; Kolb-Mäurer, A. Carbohydrate composition of meningococcal lipopolysaccharide modulates the interaction of *Neisseria meningitidis* with human dendritic cells. *Cell Microbiol.* **2005**, *7*, 1319–1334. [CrossRef]
16. Jarvis, G.A.; Vedros, N.A. Sialic acid of group B *Neisseria meningitidis* regulates alternative complement pathway activation. *Infect. Immun.* **1987**, *55*, 174–180. [CrossRef]
17. Jarvis, G.A. Recognition and control of neisserial infection by antibody and complement. *Trends Microbiol.* **1995**, *3*, 198–201. [CrossRef]
18. Hammerschmidt, S.; Hilse, R.; van Putten, J.P.; Gerardy-Schahn, R.; Unkmeir, A.; Frosch, M. Modulation of cell surface sialic acid expression in *Neisseria meningitidis* via a transposable genetic element. *EMBO J.* **1996**, *15*, 192–198. [CrossRef] [PubMed]
19. Kahler, C.M.; Martin, L.E.; Shih, G.C.; Rahman, M.M.; Carlson, R.W.; Stephens, D.S. The (alpha 2–>8)-linked polysialic acid capsule and lipooligosaccharide structure both contribute to the ability of serogroup B *Neisseria meningitidis* to resist the bactericidal activity of normal human serum. *Infect. Immun.* **1998**, *66*, 5939–5947. [CrossRef]
20. Warren, L.; Blacklow, R.S. The biosynthesis of cytidine 5′-monophospho-*N*-acetylneuraminic acid by an enzyme from *Neisseria meningitidis*. *J. Biol. Chem.* **1962**, *237*, 3527–3534. [CrossRef] [PubMed]
21. Roseman, S. Enzymatic synthesis of cytidine 5′-mono-phospho-sialic acids. *Proc. Natl. Acad. Sci. USA* **1962**, *48*, 437–441. [CrossRef]
22. Kean, E.L. Sialic acid activation. *Glycobiology* **1991**, *1*, 441–447. [CrossRef] [PubMed]
23. Masson, L.; Holbein, B.E. Physiology of sialic acid capsular polysaccharide synthesis in serogroup B *Neisseria meningitidis*. *J. Bacteriol.* **1983**, *154*, 728–736. [CrossRef]
24. Münster-Kühnel, A.K.; Tiralongo, J.; Krapp, S.; Weinhold, B.; Ritz-sedlacek, V.; Jacob, U.; Gerardy-Schahn, R. Structure and function of vertebrate CMP-sialic acid synthetase. *Glycobiology* **2004**, *14*, 43R–51R. [CrossRef]
25. Severi, E.; Hood, D.W.; Thomas, G.H. Sialic acid utilization by bacterial pathogens. *Microbiology* **2007**, *153*, 2817–2822. [CrossRef]
26. Abeln, M.; Borst, K.M.; Cajic, S.; Thiesler, H.; Kats, E.; Albers, I.; Kuhn, M.; Kaever, V.; Rapp, E.; Münster-Kühnel, A.; et al. Sialylation is dispensable for early murine embryonic development in vitro. *ChemBioChem* **2017**, *18*, 1305–1316. [CrossRef] [PubMed]

27. Schmid, W.; Christian, R.; Zbiral, E. Synthesis of both epimeric 2-deoxy-*N*-acetylneuraminic acids and their behaviour towards CMP-sialate synthetase—A comparison with 2-β-methylketoside of *N*-acetylneuraminic acid. *Tetrahedron Lett.* **1988**, *29*, 3643–3646. [CrossRef]
28. Petrie, C.R.I.; Sharma, M.; Simmons, O.D.; Korytnyk, W. Synthesis of analogs of *N*-acetylneuraminic acid and their effect on CMP-sialate synthase. *Carbohydr. Res.* **1989**, *186*, 326–334. [CrossRef]
29. Knorst, M.; Fessner, W.-D. CMP-Sialate synthetase from *Neisseria meningitidis*—Overexpression and application to the synthesis of oligosaccharides containing modified sialic acids. *Adv. Synth. Catal.* **2001**, *343*, 698–710. [CrossRef]
30. Yu, H.; Yu, H.; Karpel, R.; Chen, X. Chemoenzymatic synthesis of CMP-sialic acid derivatives by a one-pot two-enzyme system: Comparison of substrate flexibility of three microbial CMP-sialic acid synthetases. *Bioorg. Med. Chem.* **2004**, *12*, 6427–6435. [CrossRef]
31. He, N.; Yi, D.; Fessner, W.-D. Flexibility of substrate binding of cytosine-5′-monophosphate-*N*-acetylneuraminate synthetase (CMP-Sialate Synthetase) from *Neisseria meningitidis*: An enabling catalyst for the synthesis of neo-sialoconjugates. *Adv. Synth. Catal.* **2011**, *353*, 2384–2398. [CrossRef]
32. Yi, D.; He, N.; Kickstein, M.; Metzner, J.; Weiß, M.; Berry, A.; Fessner, W.-D. Engineering of a cytidine 5′-monophosphate-sialic acid synthetase for improved tolerance to functional sialic acids. *Adv. Synth. Catal.* **2013**, *355*, 3597–3612. [CrossRef]
33. Khedri, Z.; Li, Y.; Muthana, S.; Muthana, M.M.; Hsiao, C.-W.; Yu, H.; Chen, X. Chemoenzymatic synthesis of sialosides containing C7-modified sialic acids and their application in sialidase substrate specificity studies. *Carbohydr. Res.* **2014**, *389*, 100–111. [CrossRef] [PubMed]
34. Morley, T.J.; Withers, S.G. Chemoenzymatic synthesis and enzymatic analysis of 8-modified cytidine monophosphate-sialic acid and sialyl lactose derivatives. *J. Am. Chem. Soc.* **2010**, *132*, 9430–9437. [CrossRef] [PubMed]
35. Rauvolfova, J.; Venot, A.; Boons, G.-J. Chemo-enzymatic synthesis of C-9 acetylated sialosides. *Carbohydr. Res.* **2008**, *343*, 1605–1611. [CrossRef] [PubMed]
36. Mertsch, A.; Poschenrieder, S.; Fessner, W.-D. Semi-synthetic sialic acid probes for challenging the substrate promiscuity of enzymes in the sialoconjugation pathway. *Adv. Synth. Catal.* **2020**, *362*, 5485–5495. [CrossRef]
37. Mizanur, R.M.; Pohl, N.L. Bacterial CMP-sialic acid synthetases: Production, properties, and applications. *Appl. Microbiol. Biotechnol.* **2008**, *80*, 757–765. [CrossRef]
38. Yu, H.; Chokhawala, H.A.; Huang, S.; Chen, X. One-pot three-enzyme chemoenzymatic approach to the synthesis of sialosides containing natural and non-natural functionalities. *Nat. Protoc.* **2006**, *1*, 2485–2492. [CrossRef]
39. Malekan, H.; Fung, G.; Thon, V.; Khedri, Z.; Yu, H.; Qu, J.; Li, Y.; Ding, L.; Lam, K.S.; Chen, X. One-pot multi-enzyme (OPME) chemoenzymatic synthesis of sialyl-Tn-MUC1 and sialyl-T-MUC1 glycopeptides containing natural or non-natural sialic acid. *Bioorg. Med. Chem.* **2013**, *21*, 4778–4785. [CrossRef]
40. Hartmann, M.; Christian, R.; Zbiral, E. Structural variations of *N*-acetylneuraminic acid, 14. Synthesis of the β-methyl ketosides of 4-oxo-, 7-oxo-, and 8-oxo-N-acetylneuraminic acid and the corresponding 7,7- and 8,8-dimethoxy derivatives their behaviour towards CMP-sialate synthase. *Liebigs Ann. Chem.* **1990**, 83–91. [CrossRef]
41. Hartmann, M.; Christian, R.; Zbiral, E. Structural variations of *N*-acetylneuraminic acid. Part 19. Synthesis of both epimeric pairs of the 4-*C*-methyl- and 4-deoxy-4-*C*-methyl- as well as of the β-methylketoside of 4-deoxy-4-*C*-methylene-*N*-acetylneuraminic acid. Behavior towards CMP-sialate synthase. *Monatsh. Chem.* **1991**, *122*, 111–125. [CrossRef]
42. Wong, J.H.; Sahni, U.; Li, Y.; Chen, X.; Gervay-Hague, J. Synthesis of sulfone-based nucleotide isosteres: Identification of CMP-sialic acid synthetase inhibitors. *Org. Biomol. Chem.* **2009**, *7*, 27–29. [CrossRef]
43. Horsfall, L.E.; Nelson, A.; Berry, A. Identification and characterization of important residues in the catalytic mechanism of CMP-Neu5Ac synthetase from *Neisseria meningitidis*. *FEBS J.* **2010**, *277*, 2779–2790. [CrossRef] [PubMed]
44. Matthews, M.M.; McArthur, J.B.; Li, Y.; Yu, H.; Chen, X.; Fisher, A.J. Catalytic cycle of *Neisseria meningitidis* CMP-sialic acid synthetase illustrated by high-resolution protein crystallography. *Biochemistry* **2020**, *59*, 3157–3168. [CrossRef] [PubMed]
45. Krapp, S.; Münster-Kühnel, A.K.; Kaiser, J.T.; Huber, R.; Tiralongo, J.; Gerardy-Schahn, R.; Jacob, U. The crystal structure of murine CMP-5-*N*-acetylneuraminic acid synthetase. *J. Mol. Biol.* **2003**, *334*, 625–637. [CrossRef]
46. Haselhorst, T.; Münster-Kühnel, A.K.; Stolz, A.; Oschlies, M.; Tiralongo, J.; Kitajima, K.; Gerardy-Schahn, R.; von Itzstein, M. Probing a CMP-Kdn synthetase by $^{1}$H, $^{31}$P, and STD NMR spectroscopy. *Biochem. Biophy. Res. Commun.* **2005**, *327*, 565–570. [CrossRef]
47. Sellmeier, M.; Weinhold, B.; Münster-Kühnel, A. CMP-Sialic acid synthetase: The point of constriction in the sialylation pathway. In *SialoGlyco Chemistry and Biology I: Biosynthesis, Structural Diversity and Sialoglycopathologies*; Gerardy-Schahn, R., Delannoy, P., von Itzstein, M., Eds.; Springer: Berlin, Heidelberg, 2015; pp. 139–167.

48. Jen, F.E.-C.; Ketterer, M.R.; Semchenko, E.A.; Day, C.J.; Seib, K.L.; Apicella, M.A.; Jennings, M.P. The Lst sialyltransferase of *Neisseria gonorrhoeae* can transfer keto-deoxyoctanoate as the terminal sugar of lipooligosaccharide: A glyco-achilles heel that provides a new strategy for vaccines to prevent Gonorrhea. *mBio* **2021**, *12*, e03666-20. [CrossRef] [PubMed]
49. Mosimann, S.C.; Gilbert, M.; Dombroswki, D.; To, R.; Wakarchuk, W.; Strynadka, N.C.J. Structure of a sialic acid-activating synthetase, CMP-acylneuraminate synthetase in the presence and absence of CDP. *J. Biol. Chem.* **2001**, *276*, 8190–8196. [CrossRef]
50. Hartshorn, M.J. AstexViewer: A visualisation aid for structure-based drug design. *J. Comp. Aided Mol. Des.* **2002**, *16*, 871–881. [CrossRef]
51. Yuan, Y.; Bleile, D.W.; Wen, X.; Sanders, D.A.R.; Itoh, K.; Liu, H.-w.; Pinto, B.M. Investigation of binding of UDP-Galf and UDP-[3-F]Galf to UDPgalactopyranose mutase by STD-NMR spectroscopy, molecular dynamics, and CORCEMA-ST calculations. *J. Am. Chem. Soc.* **2008**, *130*, 3157–3168. [CrossRef]
52. Winger, M.; von Itzstein, M. Exposing the flexibility of human parainfluenza virus hemagglutinin-neuraminidase. *J. Am. Chem. Soc.* **2012**, *134*, 18447–18452. [CrossRef]
53. Bose, S.; Purkait, D.; Joseph, D.; Nayak, V.; Subramanian, R. Structural and functional characterization of CMP-*N*-acetylneuraminate synthetase from *Vibrio cholerae*. *Acta Crystallogr. D Struct. Biol.* **2019**, *75*, 564–577. [CrossRef]
54. Zbiral, E.; Schreiner, E.; Christian, R. Synthesis of the 4-acetamido-4-deoxy analogue of *N*-acetylneuraminic acid and its behaviour towards CMP-sialate synthase. *Carbohydr. Res.* **1989**, *194*, c15–c18. [CrossRef]
55. Kuhn, R.; Lutz, P.; MacDonald, D.L. Syntheses anomerer sialinsaure-methylketoside. *Chem. Ber.* **1966**, *99*, 611–617. [CrossRef]
56. Chopra, P.; Thomson, R.J.; Grice, I.D.; von Itzstein, M. Rapid and clean microwave-assisted synthesis of *N*-acetylneuraminic acid methyl ester and its β-methyl glycoside. *Tetrahedron Lett.* **2012**, *53*, 6254–6256. [CrossRef]
57. Byramova, N.E.; Tuzikov, A.B.; Bovin, N.V. A simple procedure for the synthesis of the methyl and benzyl glycosides of Neu5Ac and 4-epi-Neu5Ac. Conversion of the benzyl and methyl glycosides of Neu5Ac into *N*-trifluoroacetylneuraminic acid benzyl glycosides. *Carbohydr. Res.* **1992**, *237*, 161–175. [CrossRef]
58. Czarniecki, M.F.; Thornton, E.R. Carbon-13 nuclear magnetic resonance spin-lattice relaxation in the *N*-acylneuraminic acids. Probes for internal dynamics and conformational analysis. *J. Am. Chem. Soc.* **1977**, *99*, 8273. [CrossRef]
59. Kiefel, M.J.; Chopra, P.; Madge, P.D.; Szyczew, A.; Thomson, R.J.; Grice, I.D.; von Itzstein, M. Synthesis of C-9 oxidised *N*-acetylneuraminic acid derivatives as biological probes. *Tetrahedron Lett.* **2011**, *52*, 98–100. [CrossRef]
60. Ramamoorthy, P.S.; Gervay, J. Solution phase synthesis of amide-linked *N*-acetyl neuraminic acid, α-amino acid, and sugar amino acid conjugates. *J. Org. Chem.* **1997**, *62*, 7801–7805. [CrossRef]
61. Gregar, T.Q.; Gervay-Hague, J. Synthesis of oligomers derived from amide-linked neuraminic acid analogues. *J. Org. Chem.* **2003**, *69*, 1001–1009. [CrossRef]
62. Valeur, E.; Bradley, M. Amide bond formation: Beyond the myth of coupling reagents. *Chem. Soc. Rev.* **2009**, *38*, 606–631. [CrossRef]
63. Honda, T.; Masuda, T.; Yoshida, S.; Arai, M.; Kaneko, S.; Yamashita, M. Synthesis and anti-Influenza virus activity of 7-*O*-alkylated derivatives related to zanamivir. *Bioorg. Med. Chem. Lett.* **2002**, *12*, 1925–1928. [CrossRef]
64. Masuda, T.; Yoshida, S.; Arai, M.; Kaneko, S.; Yamashita, M.; Honda, T. Synthesis and anti-influenza evaluation of polyvalent sialidase inhibitors bearing 4-guanidino-Neu5Ac2en derivatives. *Chem. Pharm. Bull.* **2003**, *51*, 1386–1398. [CrossRef]
65. Masuda, T.; Shibuya, S.; Arai, M.; Yoshida, S.; Tomozawa, T.; Ohno, A.; Yamashita, M.; Honda, T. Synthesis and anti-influenza evaluation of orally active bicyclic ether derivatives related to zanamivir. *Bioorg. Med. Chem. Lett.* **2003**, *13*, 669–673. [CrossRef]
66. Brandstetter, H.H.; Zbiral, E. Strukturelle Abwandlungen an *N*-Acetylneuraminsäure, 2 [Transformations with *N*-acetylneuraminic acid. 2]. *Liebigs Ann. Chem.* **1983**, 2055–2065. [CrossRef]
67. Zbiral, E.; Phadtare, S.; Schmid, W. Strukturelle Abwandlungen an *N*-Acetylneuraminsäure, 6. Synthesen von 7-Oxo- und 8-Oxo-*N*-acetylneuraminsäure-Derivaten. *Liebigs Ann. Chem.* **1987**, 39–43. [CrossRef]
68. Sato, K.; Ikeda, K.; Suzuki, T.; Aoyama, S.; Maki, N.; Suzuki, Y.; Sato, M. Synthesis of 4-*O*-[3-(aryl)prop-2-ynyl]-Neu5Ac2en and its 4-*epi*-analogs modified at C-4 by Sonogashira coupling reaction. *Tetrahedron* **2007**, *63*, 7571–7581. [CrossRef]
69. Isecke, R.; Brossmer, R. Synthesis of 5-*N*- and 9-*N*-thioacylated sialic acids. *Tetrahedron* **1994**, *50*, 7445–7460. [CrossRef]
70. Roy, R.; Pon, R.A. Efficient synthesis of α(2-8)-linked *N*-acetyl and *N*-glycolylneuraminic acid disaccharides from colominic acid. *Glycoconj. J.* **1990**, *7*, 3–12. [CrossRef]
71. Chopra, P.; Madge, P.D.; Thomson, R.J.; Grice, I.D.; von Itzstein, M. Microwave-assisted synthesis of *N*-glycolylneuraminic acid derivatives. *Tetrahedron Lett.* **2013**, *54*, 5558–5561. [CrossRef]
72. Sparks, M.A.; Williams, K.W.; Lukacs, C.; Schrell, A.; Priebe, G.; Spaltenstein, A.; Whitesides, G.M. Synthesis of potential inhibitors of hemagglutination by influenza virus: Chemoenzymic preparation of N-5 analogs of *N*-acetylneuraminic acid. *Tetrahedron* **1993**, *49*, 1–12. [CrossRef]

73. Schreiner, E.; Zbiral, E.; Kleineidam, R.G.; Schauer, R. Structural variations on *N*-acetylneuraminic acid, 20. Synthesis of some 2,3-didehydro-2-deoxysialic acids structurally varied at C-4 and their behavior towards sialidase from *Vibrio cholerae*. *Liebigs Ann. Chem.* **1991**, 129–134. [CrossRef]
74. von Itzstein, M.; Jin, B.; Wu, W.Y.; Chandler, M. A convenient method for the introduction of nitrogen and sulfur at C-4 on a sialic acid analog. *Carbohydr. Res.* **1993**, *244*, 181–185. [CrossRef]
75. Chandler, M.; Bamford, M.J.; Conroy, R.; Lamont, B.; Patel, B.; Patel, V.K.; Steeples, I.P.; Storer, R.; Weir, N.G.; Wright, M.; et al. Synthesis of the potent influenza neuraminidase inhibitor 4-guanidino-Neu5Ac2en. X-Ray molecular structure of 5-acetamido-4-amino-2,6-anhydro-3,4,5-trideoxy-D-*erythro*-L-*gluco*-nononic acid. *J. Chem. Soc. Perkin Trans. 1* **1995**, 1173–1180. [CrossRef]
76. Okamoto, K.; Kondo, T.; Goto, T. Functionalization of 2-deoxy-2,3-dehydro-*N*-acetylneuraminic acid methyl ester. *Bull. Chem. Soc. Jpn.* **1987**, *60*, 631–636. [CrossRef]
77. Ciccotosto, S.; von Itzstein, M. Synthesis of methyl 5-acetamido-3,4,5-trideoxy-4-guanidinyl-D-*glycero*-D-*galacto*-2-nonulopyranosidonic acid (4-deoxy-4-guanidino-Neu5Acα2Me). *Tetrahedron Lett.* **1995**, *36*, 5405–5408. [CrossRef]
78. Warren, L. The thiobarbituric acid assay of sialic acids. *J. Biol. Chem.* **1959**, *234*, 1971–1975. [CrossRef]
79. Blat, Y. Non-competitive inhibition by active site binders. *Chem. Biol. Drug Des.* **2010**, *75*, 535–540. [CrossRef]
80. Pesaresi, A. Mixed and non-competitive enzyme inhibition: Underlying mechanisms and mechanistic irrelevance of the formal two-site model. *J. Enzym. Inhib. Med. Chem.* **2023**, *38*, 2245168. [CrossRef]
81. Mayer, M.; Meyer, B. Characterization of ligand binding by saturation transfer difference NMR spectroscopy. *Angew. Chem. Int. Ed. Engl.* **1999**, *38*, 1784–1788. [CrossRef]
82. Meyer, B.; Peters, T. NMR spectroscopy techniques for screening and identifying ligand binding to protein receptors. *Angew. Chem. Int. Ed. Engl.* **2003**, *42*, 864–890. [CrossRef] [PubMed]
83. Chopra, P. The Development of Carbohydrate-Based Chemical Probes of CMP-*N*-Acetylneuraminate Synthetase. Ph.D. Thesis, Griffith Univeristy, Gold Coast, Australia, 2013. Available online: http://hdl.handle.net/10072/366420 (accessed on 14 July 2025).
84. Alexander, H.E. The haemophilus group. In *Bacterial and Mycotic Infections of Man*, 4th ed.; Dubos, R.J., Hirsch, J.G., Eds.; Pitman Medical Publishing Co.: London, UK, 1965; pp. 724–741.
85. Sambrook, J.; Fritsch, E.F.; Maniatis, T. *Molecular Cloning: A Laboratory Manual*, 2nd ed.; Cold Spring Harbor Laboratory: Cold Spring Harbor, NY, USA, 1989.
86. Power, P.M.; Roddam, L.F.; Dieckelmann, M.; Srikhanta, Y.N.; Tan, Y.C.; Berrington, A.W.; Jennings, M.P. Genetic characterization of pilin glycosylation in *Neisseria meningitidis*. *Microbiology* **2000**, *146*, 967–979. [CrossRef] [PubMed]
87. Steichen, C.T.; Shao, J.Q.; Ketterer, M.R.; Apicella, M.A. Gonococcal Cervicitis: A role for biofilm in pathogenesis. *J. Infect. Dis.* **2008**, *198*, 1856–1861. [CrossRef] [PubMed]

MDPI AG
Grosspeteranlage 5
4052 Basel
Switzerland
Tel.: +41 61 683 77 34

*Molecules* Editorial Office
E-mail: molecules@mdpi.com
www.mdpi.com/journal/molecules

www.ingramcontent.com/pod-product-compliance
Lightning Source LLC
Chambersburg PA
CBHW061805210926
51850CB00006B/614
* 9 7 8 3 7 2 5 8 6 5 1 8 5 *